FIELD EFFECT TRANSISTORS, A COMPREHENSIVE OVERVIEW

FIELD EFFECT TRANSISTORS, A COMPREHENSIVE OVERVIEW

From Basic Concepts to Novel Technologies

POUYA VALIZADEH
Associate Professor of Electrical Engineering
Concordia University, Montreal, Quebec, Canada

WILEY

Published by John Wiley & Sons, Inc., Hoboken, New Jersey
Published simultaneously in Canada

Library of Congress Cataloging-in-Publication Data:

Valizadeh, Pouya, author.
Field effect transistors : a comprehensive overview : from basic concepts to novel technologies / Pouya Valizadeh.
pages cm
Includes bibliographical references and index.
ISBN 978-1-119-15549-2 (cloth)
1. Field-effect transistors. I. Title.
TK7871.95.V35 2016
621.3815′284–dc23

2015032185

Set in 10/12pt Times by SPi Global, Pondicherry, India

10 9 8 7 6 5 4 3 2 1

1 2016

To my parents and to their unequivocal beliefs in social justice.

CONTENTS

INTRODUCTION

CMOS[1] technology, in spite of many gloomy predictions, has still managed to maintain its hegemony over the electronics world. Success in maintaining this status has been made possible through the adoption of a large number of new design techniques in the realization of the structure of modern *MOSFETs*.[2] Among these already adopted solutions, one finds technologies and remedies such as *SOI*[3] technology, *multiple-gate MOSFET*, techniques of incorporation of biaxial (and uniaxial) strain, incorporation of high-*K* dielectrics and *ferroelectric gate* stacks, and ultrathin body *MOSFET* structures. In addition to these, more futuristic approaches as widely ranging as incorporation of mechanical vibration, *spin–orbit coupling*, velocity modulation, and exploration of semiconductors other than *silicon* (such as bulk *III–V*, *II–VI* semiconductors, and *nanostructures* such as *carbon nanotubes*) are gaining a considerable momentum in finding solutions for future semiconductor *chips* of higher speed and functionality.

As a result of all these innovative solutions, *MOSFETs* of today barely resemble their ancestors. Through pure shrinking of the physical dimensions of a *MOSFET* and adoption of the aforementioned innovative solutions, the *drift–diffusion* regime of *carrier transport* is becoming more and more insufficient for expressing the behavior of modern *field effect transistors*. While all this is happening, many of the electrical engineering programs are still presenting the principles of operation of electronic devices, and especially *MOSFETs*, according to the old *bulk planar MOSFET* structure and its applicable *carrier-transport* formalism.

[1] Complementary Metal Oxide Semiconductor.

[2] Metal Oxide Semiconductor Field Effect Transistor.

[3] Silicon on Insulator.

Evolution from the old *FET* concepts to newer ones has been constantly driven by the *scaling* of the *minimum feature size* of these devices. In regard to the issue of size reduction of electronic devices, one of the important outcomes is the predominance of a *scattering*-less *phase-coherent* propagation of *electron waves*. The other major outcome of this *scaling* trend is that *charging* effects, such as *Coulomb blockade*[4] and the *Kondo effect*,[5] are of major significance in the operation of small devices.

After about a decade of teaching courses in the area of electronic devices for electrical engineering students and trying to address the new trends in semiconductor devices through adoption of multiple textbooks and providing supplementary materials to both undergraduate and graduate level students, I have taken it upon myself to prepare a concise manuscript capable of presenting not only the principles but also as many of these new additions as possible.

This volume does not have the ambition of providing an encyclopedia of *field effect transistors*. It only attempts to present what is the most relevant to understanding today's *MOSFETs* and those futuristic trends deemed most interesting (of course in the eyes of the author).

Whereas many of these presented new structures are yet again looking into physics for finding inspiring solutions for the challenges of *MOSFET* technology, the background physics of semiconductors is more than ever deemed necessary for electrical engineering students. This volume, both at its first and at the final chapter, attempts to do justice to this necessity. While the background presented in the first chapter is supposed to serve the discussion in Chapters 3 through 5, the discussion of molecular-scale *field effect transistors*, which is presented in Chapter 6, is supposed to open the eyes of the reader to a futuristic direction. While the material presented in the volume is fit for a one-semester course, this direction should be followed by the reader through taking more courses in the specific subject area.

Although the latest technological solutions are woven within the materials of each chapter, since many of the stand-alone textbooks on *micro-* and *nanofabrication* technology can do a better job in this area, the author has refrained from presenting the technological solutions in a separate chapter.

Each chapter of this volume presents a wealth of diligently selected references (books and research papers alike), which should be taken as a source of further reading. While the author's ambition is to provide a concise volume encompassing this wealth of information, for the use of a senior undergraduate/graduate audience, in his view the education of the graduate students will not be complete without examining the aforementioned references in details.

As part of our discussions in this volume, we will deal with nontraditional materials, structures, and manufacturing tools. In this regard, we look into the electronic properties of different materials, individually and when placed in contact with each other, and will investigate techniques for modification of these electronic properties. In this attempt, old *field effect* device structures (such as the *JFET*[6] and the

[4] Named after Charles-Augustin de Coulomb.

[5] Named after Jun Kondo.

[6] Junction Field Effect Transistor.

long-channel MOSFET) are presented alongside relatively novel (such as *strained MOSFET*, *HFET*,[7] *FinFET*,[8] *CNFET*,[9] *spinFET*,[10] *resonant-gate FET*, *VMT*,[11] and molecular-scale *FET*).

To deliver this material, it is important to not only engage in discussing electronic properties of *bulk* semiconductors but also to pay attention to new dimensions extended through techniques of *strained epitaxy*. Development of *epitaxial* growth techniques such as *MBE*[12] and *MOCVD*,[13] which are the techniques used in realization of planar *heterostructures*[14] and three-dimensional *self-assembled* quantum structures, has been one of the major developments of the past three decades of research on semiconductors. Whereas originally *strained epitaxy* was a topic of interest in more obscure *III–V* technologies, nowadays it is gaining considerable ground in the mainstream *silicon* technology.

It is the author's wish that both the students and practice engineers/research staff with a keen interest in electronic devices will find this volume a step forward in perfecting our knowledge of the operation of these devices. The author is sincerely expecting the feedback of the audience toward alleviating the shortcoming of this first edition.

Last but not least, the author would like to greatly appreciate the assistance of his graduate students Joseph Record, Jean-Lou Gosselin, and Kaveh Rahbardar in editing this text. Assistance of these students has been especially instrumental in the final revision of the text.

Pouya Valizadeh
September 2014
Sainte-Julie, Québec

[7] Heterojunction Field Effect Transistor.
[8] Fin Field Effect Transistor.
[9] Carbon Nanotube Field Effect Transistor.
[10] Spin Field Effect Transistor.
[11] Velocity Modulation Transistor.
[12] Molecular Beam Epitaxy.
[13] Metalorganic Chemical Vapor Deposition.
[14] Electronic structures realized in the form of contact between two or more different semiconductors.

1

ELECTRONIC MATERIALS AND CHARGE TRANSPORT

This chapter presents an overview of the quantum mechanical nature of electrons in a *solid*. Following this discussion, through adoption of a relatively simple language, we will attempt to tackle a number of very complicated phenomena (such as *charge transport*). One of the remarkable achievements of the pioneers of semiconductor electronics has been their success in proposing an intuitive set of simplifying assumptions for reducing the mathematically complex language of quantum mechanics to the closed-form descriptions of the *effective* charged particles. While oftentimes we find it hard to justify these assumptions, they have proven their capabilities for large devices and under low electric fields.

1.1 WAVE/PARTICLE ELECTRONS IN SOLIDS

Early in the twentieth century, a number of important experiments revealed that electrons are not just simply particles in the *Newtonian* sense. Electrons in some of the experiments demonstrated a wavelike nature. These experiments were very similar to the *double-slit* optical experiments. It was demonstrated that electrons coming from the two slits, if not *observed*, produce interference patterns expected from wave propagation only. However, if we try to *observe* the electrons, by means such as the study

Field Effect Transistors, A Comprehensive Overview: From Basic Concepts to Novel Technologies, First Edition. Pouya Valizadeh.

of absorption and emission of light, the *uncertainty*[1] created by these interactions reduces electrons to almost *Newtonian* particles. The same argument can be applied to the movement of electrons in a perfectly crystalline lattice (also known as *monocrystalline* lattice). As a result of these experiments, *wave/particle* duality is attributed to electrons. In this duality, the *de Broglie wavelength*[2] is the *wavelength* assigned to a particle of momentum p (given by $\lambda = h/p$).

Electronic materials, which are suitable for fabrication of high-performance semiconductor devices, come in *monocrystalline solid* forms. The presence of both *long-* and *short-range* orders in the structure of these crystalline *solids* extends many important properties to *charge transport* through these media. However, the assumption of periodicity in crystals is always relative. In real crystals a number of intentional and unintentional mechanisms (such as introduction of impurities, crystal defects, and thermal vibrations) result in *scattering* of electrons after a typical travel distance on the order of 100 Å. Such *scattering* processes can be seen as mechanisms of *observation* of electrons. Of course, the distance paced between successive *scattering* events is dependent on the presence and dominance of different *scattering* mechanisms. The most unavoidable crystal imperfection is rooted in *lattice vibrations*, which are even present in a defectless crystal devoid of impurities at temperatures above 0 K. As a result of these *scattering* events, instead of seeing a three-dimensional large crystal, electrons are only exposed to a small volume on the order of 10^{-18} cm^3 containing only about 3000 atoms, before they are *dephased* by *lattice vibrations*.

In a flawless *monocrystalline* lattice, electrons are predicted by quantum mechanics to travel as *propagating waves*. This results in either a *persistent current* or in *oscillations*.[3] Electron interactions with the vibrations of the lattice (expressed in terms of *observation* by quantum particles known as *phonons*) and the resulting generation of *Joule heat* prevent electrons from acting in such a fashion. The emission of *phonons* with a broad range of energies and *wave vectors* results in *dephasing* of electrons and loss of *coherence*.

Although the *electron–phonon* interactions do not result in total elimination of quantic nature of electrons, they considerably weaken these properties. This reduced degree of quantum nature is often expressed in terms of *perturbation theory*, through employing the so-called *Fermi golden* rule.[4] Throughout this chapter, with the help of analytical models developed on the basis of this theory, familiar characteristics of semiconductors (e.g., their sometimes *Ohmic* behavior[5]) are evaluated. Rivaling the results of this approach, the only way that quantum mechanics can produce an

[1] *Heisenberg*'s *uncertainty principles* (which are named after *Werner Heisenberg*) state that the *uncertainties* of two conjugate variables such as momentum and position (i.e., Δp and Δx, respectively) cannot be both reduced to 0, since $\Delta p \Delta x \geq h/2\pi$. This is also true between uncertainty of energy and time. Where h, named after *Max Planck*, is the *Planck*'s constant, which is equal to 6.63×10^{-34} J s.

[2] Named after *Louis de Broglie*.

[3] These phenomena will be further elaborated in Section 1.7.

[4] Named after *Enrico Fermi*.

[5] Named after *Georg Ohm*.

Ohmic characteristic is when the complicated *many-body Schrödinger equations*[6] for electrons and *phonons* are solved.

Since electrons in *solids* are subject to the laws of quantum mechanics, any discussion of *electron transport* through *solids* requires at least a rudimentary description of the quantum mechanical wave nature of electrons. Although the present text does not seek to present this picture through explicitly invoking the laws of quantum mechanics, a number of important outcomes of such studies are reviewed in this section.

1.1.1 Quantum Description of Electrons

In order to formulate the *wave function* of electrons (which is often represented in form of function $\psi(\vec{r})$ in space) the *time-independent* 3-D *Schrödinger equation* should be solved,

$$\left[-\frac{\hbar^2}{2m}\nabla^2 + U(\vec{r})\right]\psi(\vec{r}) = E\psi(\vec{r}). \tag{1.1}$$

This equation can be presented equivalently through invoking the concept of *Hamiltonian*[7] (i.e., *H*),

$$H\psi(\vec{r}) = E\psi(\vec{r}). \tag{1.2}$$

In (1.1) and (1.2), $\hbar$ is the *modified Planck's constant*,[8] m is the electron mass,[9] E is the energy, and vector $\vec{r}$ represents the spatial coordinates. In these partial differential equations, the potential function $U(\vec{r})$ is assumed to be time independent.

For a constant potential (i.e., the case of an electron in *free space*), the solution is rendered in the form of plane waves,

$$\psi(\vec{r}) = \frac{1}{\sqrt{\Omega}}\exp\left(j\vec{k}\cdot\vec{r}\right) \tag{1.3}$$

where Ω represents the *normalization volume*, which is defined using the square of the amplitude of the *wave function* as the *probability density function*. Evidently, $\vec{k}$ in (1.3) is the *wave vector*. In quantum mechanics, the eigenvalue of momentum is given by $\hbar\vec{k}$.

Oftentimes, semiconductor devices such as *field-effect transistors* (*FETs*) are realized as a 2-D plane through which *charge transport* takes place. Assuming a semiconductor slab in the *x*–*y* plane with an infinitesimal thickness W in the *z*-direction,

[6] Named after *Erwin Schrödinger*.

[7] Named after *William Hamilton*.

[8] $\hbar = h/2\pi$, where $h = 6.62 \times 10^{-34}$ J s.

[9] In here, intentionally the notions of effective mass (m^*, which is to be defined in Section 1.1.2) and the rest mass of electron (i.e., $m_0 = 9.1 \times 10^{-31}$ kg) are not used. m will be later replaced by m^*.

quantum mechanics provides a picture for electrons confined in the z-direction while free to move in the x–y plane. For these quasi 2-D electrons, separation of variables of the 3-D *time-independent Schrödinger equation* results in

$$\psi(\vec{r}) = \phi(z) \cdot \varphi(x,y) = \phi(z) \cdot \frac{1}{\sqrt{A}} \exp\left[j\left(k_x x + k_y y\right)\right] = \phi(z) \cdot \frac{1}{\sqrt{A}} \exp\left(j \vec{k_{\|}} \cdot \vec{\rho}\right) \quad (1.4)$$

in which A is the *normalization area* and $\vec{\rho}$ is a vector in the x–y plane. In the case of a confining potential, which imposes an infinite barrier against the movement of electrons normal to x–y plane, quantization of energy will yield

$$\phi(z) = \sqrt{\frac{2}{W}} \sin(k_n z) = \sqrt{\frac{2}{W}} \sin\left(\frac{n\pi z}{W}\right), \quad n = 1,2\ldots \quad (1.5)$$

This is the case of the so-called infinite *potential well*.

If the confinement is further extended to the y-direction and the confining potential is turned into a simple square well of infinitely high barriers, one will have

$$\psi(\vec{r}) = \phi(y,z) \cdot \varphi(x) = \phi(y,z) \cdot \frac{1}{\sqrt{L}} \exp[j(kx)] \quad (1.6)$$

where

$$\phi(y,z) = \frac{2}{W} \sin\left(\frac{m\pi y}{W}\right) \sin\left(\frac{n\pi z}{W}\right), \quad m,n = 1,2\ldots \quad (1.7)$$

in which L is the *normalization length*. This is the case of the so-called quantum *wire*.

Figure 1.1 provides schematics for a *quantum well* and a *quantum wire*.

What is common between the cases of electron in *free space*, confinement in a *quantum well*, and confinement in a *quantum wire* is that according to (1.3), (1.4), and (1.6) for all three cases, electrons assume a *propagating-wave* nature. What this means is that electrons feel almost free to move, either in 3-D, 2-D, or 1-D space.

As mentioned earlier, confinement of electrons in one and two dimensions results in energy quantization. This is expressed through integer values of n and m.

Assuming a *potential well* in which an electron is only allowed to move between the barriers and not parallel to them, energy quantization is expressed by

$$E_\mathrm{n} = \frac{\hbar^2 \pi^2}{2m^* W^2} n^2, \quad n = 1,2,\ldots \quad (1.8)$$

where, as shown in Figure 1.1, W is the distance between the two barriers.

In this equation, each of the values of energy expressed by E_n represents an allowed *energy level* (or overlooking the *spin* of an electron: *energy state*). Electrons are prohibited to assume all other values of energy. In the case of a finite *potential well*,

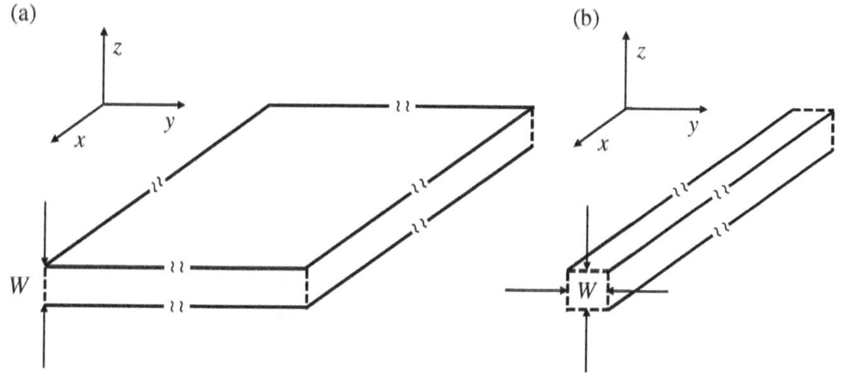

FIGURE 1.1 (a) A 2-D confining quantum well. (b) A 1-D confining quantum wire.

quantization is present to a lesser degree, and a recursive equation should be solved to calculate the allowed energy values.

Extending the degrees of freedom of electrons in the *potential well* results in overlapping of the *wave functions* of the neighboring electrons residing on each of these *energy states*. As a result, due to the restrictions of *Pauli's exclusion principle*,[10] each quantized *energy state* will split into a number of very closely packed *energy states*. As will be seen later on in this section, these split *energy levels*, depending on the degree of confinement, can grow into *bands* or the so-called subbands of energy.

In the form of confinement in a 2-D plane, the *quantum-well* situation especially resembles that of the thin silicon body of a *silicon-on-insulator*[11] *MOSFET*. This device will be seen in Chapter 3. In this context, each of the *energy levels* expressed in (1.8) stands for a *subband* with many allowed *momentum states* (also known as *k-states*) in the *x–y* silicon plane. As shown by (1.8), increasing the confinement (i.e., reducing *W*) increases the separation of the *subbands*. As will be observed later on in this chapter, this is important to the reduction of the chance of electron *scattering* from one *subband* to the next. In this system, through assigning an *effective* value of mass to electrons (i.e., m^*), the total energy written as the sum of kinetic energy and confining energy is given by

$$E\left(\vec{k}\right) = E_n + \frac{\hbar^2 k_{\parallel}^2}{2m^*}, n = 1, 2, 3, \ldots \tag{1.9}$$

where $k_{\parallel}^2 = k_x^2 + k_y^2$. The relationship between the energy and momentum (which is represented by $\hbar k$) is known as the dispersion relationship (also known as *E–k*).

As observed through our brief encounter with the *Schrödinger equation*, knowledge of the *Hamiltonian* and the potential function is pivotal to understanding the

[10] Named after *Wolfgang Pauli*.
[11] Also known as *SOI*.

behavior of electrons in different *solid* structures. The following laws of mechanics govern the time evolution of position (i.e., x) and momentum (i.e., p) in terms of H:

$$\frac{\partial p_i}{\partial t} = -\frac{\partial H(p_i, x_i)}{\partial x_i} \tag{1.10}$$

$$\frac{\partial x_i}{\partial t} = \frac{\partial H(p_i, x_i)}{\partial p_i}. \tag{1.11}$$

The vector of average velocity for electrons is also identified as

$$\vec{v} = \frac{1}{\hbar}\nabla_k E(\vec{k}). \tag{1.12}$$

1.1.2 Band Diagram and Effective-Mass Formalism

The most basic outcome of such a quantum mechanical entity (i.e., electrons in a *solid*) appears in the form of *energy bands* and *forbidden gaps*. The evolution of the *band diagram* composed of these *energy bands* and *forbidden gaps* (and its more detailed version known as the E–k diagram) is deeply rooted in the interaction between the wave-natured electrons and the periodic potential function of the crystalline *solid*. Within each of the allowed *bands*, the variation of energy with momentum is calculated and represented in the form of energy versus momentum (i.e., E–k) diagrams.[12]

A major component of the simplified theory representing electrons in *solids* is the *band theory*, which is the outcome of *effective-mass theory*. *Effective-mass theory* is based on the resemblance of the form of the *electron wave dispersion diagram* at the energy range of interest[13] to that of electrons in *free space*. For an electron in *free space*, since the *effective mass* (i.e., m^*) is clearly a constant (i.e., the rest mass of the electron: m_0), Equation (1.9) reduces to a parabola.

Although in a real semiconductor *charge carriers* are not truly traveling as *propagating waves*, under many important circumstances, they almost behave that way. These almost free *charge carriers*, however, feel a different mass than that of the free electron. Based on the almost parabolic form of the *dispersion diagram*, *effective-mass theory* uses the curvature of this diagram to assign a value of mass to *charge carriers*,

$$m^* = \frac{\hbar^2}{\partial^2 E/\partial k^2}. \tag{1.13}$$

While this simple picture successfully reduces the *wave/particle* electrons to simple particles, it fails to hold when *charge carriers* receive substantial amounts of kinetic energy. Under those circumstances, the resemblance in form to the *dispersion diagram* of a free electron vanishes.

[12] Also referred to as *dispersion diagram*.
[13] This will be clarified shortly.

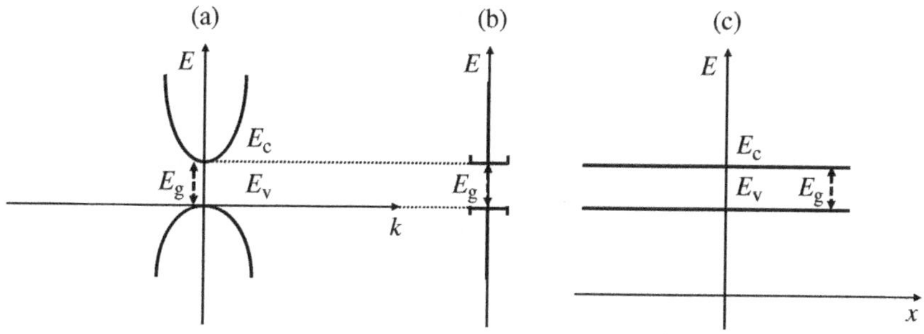

FIGURE 1.2 (a) A simplified typically observed *E–k* diagram among a number of semiconductors. As will be indicated shortly, in this diagram E_c and E_v mark the bottom of the conduction and the top of the valence band and E_g represents the size of the forbidden gap of energy. (b) A 1-D representation of the bandgap. (c) The 2-D band diagram.

In some situations the complete set of information offered in an *E–k* diagrams is needed; however, in most cases of interest to the present volume, we only adopt a selective portion of this information. As will be pointed out shortly, this selection of information in the *E–k* diagram comes from calculating the *E–k* curvature given specific values of momentum matched to the edges of two of the *bands* resulting from splitting of *energy levels*.

Localization of electron *states*, caused by unavoidable imperfections in real semiconductors, can be incorporated into this model through the corrective measure of *scattering* processes. Incorporation of such small perturbations to the perfect crystal is implemented through determining the *scattering* rates using *Fermi golden rule* (or *Born approximation*[14]).

Under these presented conditions, the *E–k* diagram is reduced to merely a one-dimensional energy diagram, which, at least for the sake of presentation, is often expanded into a second dimension (i.e., position). Figure 1.2 illustrates a typical *E–k* diagram and the resulting 1-D and 2-D *band* diagrams. As pointed out in these illustrations, *forbidden gaps* of energy are present among allowed energy *bands*. As shown, the momentum information is not presented in an energy-*band* diagram. Later on in this chapter, through invoking the concept of *momentum relaxation time constant*, we will discuss the significance of this lost bit of information.

The curvatures of the *E–k* diagram and therefore the values of the *effective mass* are different along different directions in a crystal. At this point in our discussion, it is important to point out that in calculating the *subband* energies in (1.8), *effective mass* in the direction of the confining potential should be employed.

1.1.3 Density of States Function

As indicated earlier, the *bands* of energy are composed of very narrowly packed sets of individual *energy states*. However, since at normal temperatures of operation of an

[14] Named after *Max Born*.

electronic device the separation between the individual *states* is much smaller than the average random thermal energy of an electron in a 3-D *solid* (i.e., $3/2kT$[15]), electrons tend to see the *band* as a continuum of *energy states*. Since the occupancy of each individual *energy state* is governed by *Pauli's exclusion principle* (i.e., maximum accommodation for only two electrons that must have opposite spins at each *energy level*), each *band* has a certain maximum capacity for electrons, which is identified by a *density of states* (i.e., *DOS*) *function*. This function is a double-density function, and it essentially expresses the *density of states* per unit energy, per unit volume.

For three-dimensionally moving *charge carriers*, the *DOS* at the bottom of a *band* (i.e., identified by E_c) can be approximated by

$$D_{3D}(E) = \frac{(2m^*)^{\frac{3}{2}}}{2\pi^2\hbar^3}\sqrt{E-E_c}. \tag{1.14}$$

However, in the case of 2-D *charge-carrier* confinement, the two-dimensional *density of states* per unit energy and area is given by

$$D_{2D}(E) = \frac{m^*}{\pi\hbar^2}. \tag{1.15}$$

This is only with the assumption of the first *subband*. Finally, with the same assumption the *DOS* in the one-dimensional case is given by

$$D_{1D}(E) = \frac{\sqrt{2m^*}}{\pi\hbar}\frac{1}{\sqrt{E-E_c}}. \tag{1.16}$$

Figure 1.3 illustrates the dependence of the *DOS* function on energy for the three cases. Due to the direction dependence of *effective mass*, the value of the *effective mass* used in calculation of the *DOS* is very often different than that used in the *transport* problem.

1.1.4 Conduction and Valence Bands

Due to electrons' natural tendency to arrive at minimum enthalpy, electrons occupy *states* of lower energy before filling up *states* of higher energy. In the presence of thermal energy, however, this picture gets slightly distorted. This is caused by thermal excitation of electrons from lower *energy states* to higher *energy states*. In the case of a semiconductor, this renders lower *energy bands* to be full, renders higher *energy bands* to be empty, and also causes two of the adjacent *bands* to be only partially occupied at temperatures above 0 K. At 0 K, however, a semiconductor behaves like an insulator and is composed only of full and empty *bands*. Figure 1.4 illustrates this

[15] This is based on statistical mechanics. According to *equipartition* theorem in 1-D and 2-D cases, this energy is equal to $1/2kT$ and kT, respectively. k is the *Boltzmann* constant, which is equal to 1.38×10^{-23} J/K.

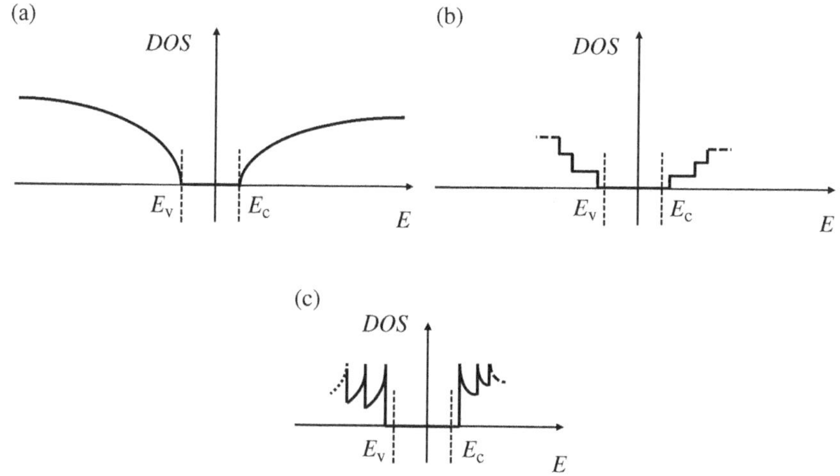

FIGURE 1.3 (a) Typical form of the density of states function in a 3-D semiconductor. (b) Typical form of the *DOS* function in a quantum well. (c) Typical form of the *DOS* function in a quantum wire.

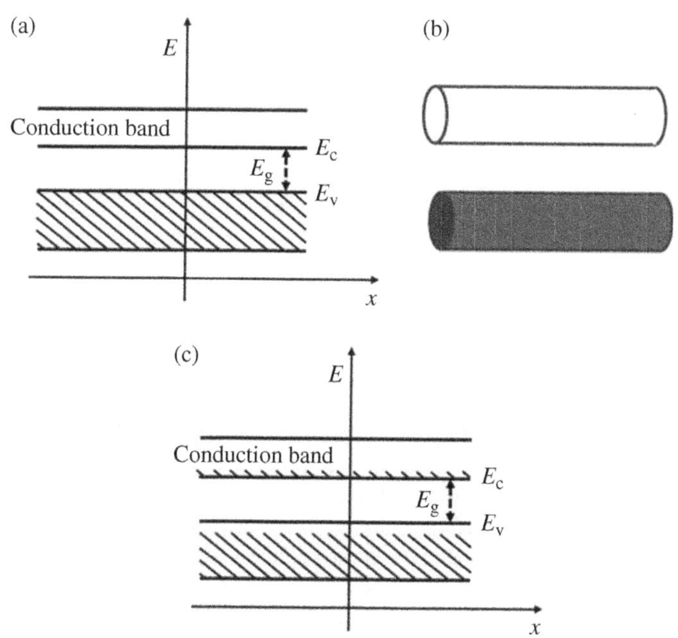

FIGURE 1.4 (a) Occupancy status of the conduction and valence band of a semiconductor at 0 K, where the full portion of a band is hash marked. (b) The water pipe analogous to the band occupancy presented in (a). In this analogy the full portion of the pipe is presented in gray, while the empty portion is white. (c) Occupancy status of the conduction and valence band of a semiconductor at a finite temperature, where the full portion of a band is hash marked. One can see the electron transfer from valence to the conduction band in analogy with water transfer from the lower to the upper pipe.

picture at 0 K and at a temperature T, in analogy with doubly sealed water pipes. According to this picture, since neither a full nor an empty *band* can contribute to net movement of *charge carriers* (i.e., in water pipe analogy: water), among all of the *bands* resulting from the aforementioned quantum mechanical description, only those *bands* that at normal temperature of operation of a device are partially full/partially empty are of prime value.[16]

A partially full *band* is created by means such as thermal excitation of electrons from the top of a lower *band* to the unoccupied *states* at the bottom of a higher *band*. In semiconductors we therefore encounter two partially full/partially empty *bands*. All other *bands* below these remain completely full, and those above remain completely empty. As a result, only two *bands* are worth mentioning: the *valence* and the *conduction band*. The *valence band* is the highest *band* that is full at 0 K, and the *conduction band* is the lowest *band* that is empty at 0 K. The energy difference between the bottom of the *conduction band* (i.e., E_c) and the top of the *valence band* (i.e., E_v) is the *bandgap* (i.e., E_g).

Since the size of the *forbidden gap* is much larger than the average amount of thermal energy acquired by an electron in a three-dimensional *solid*, one can expect such a *charge-carrier* transfer to be far more probable between the top of the *valence band* and the bottom of the *conduction band*. These values of energy are therefore much more important to study than the other *energy levels* in the rest of *conduction* and *valence bands*.

Although approximate equations such as (1.14) present the *DOS* of *conduction* and *valence band*[17] in the form of a square-root law, higher in the *conduction band* (and also lower into the *valence band*) the proportionality of *DOS* to $\sqrt{E-E_c}$ (and $\sqrt{E_v-E}$ in the *valence band*) vanishes. This is due to the more complicated E–k variation at higher values of energy. For these higher energies, numerical methods are needed to evaluate the *DOS*. As will be observed shortly, even for the case of lower energies, in the presence of high concentrations of impurities, there is a need to reevaluate this square-root law.

1.1.5 Band Diagram and Free Charge Carriers

The *band diagram* representation of Figure 1.5, through extending the *bandgap* information in space, substantially helps in implying the relatively free nature of electrons in the *conduction* and *valence band* (i.e., those that are not associated with the nuclei of the atoms in the crystal). This relative freedom is implied since the electrons are now free to be at different positions in the semiconductor without any need for changing their energy value. The energy axis of the *band* diagram imports only the energy difference between the lowest energy value of the *conduction band* (i.e., E_c) and the highest energy value of the *valence band* (i.e., E_v) of the E–k diagram. As a result, as mentioned earlier, momentum and curvature information of the E–k diagram are simply omitted (of course with the exception of the curvature used in calculation of the

[16] Just like the case of a *zero* net movement of liquid in a full, or an empty, doubly sealed water pipe.
[17] In this equation, in case of *valence band* $\sqrt{E-E_c}$ changes to $\sqrt{E_v-E}$.

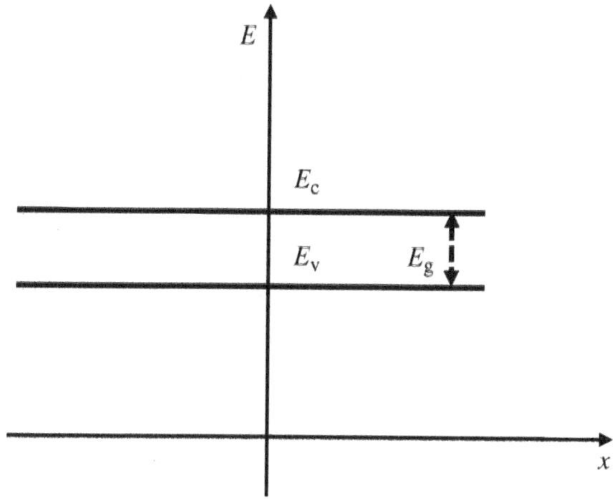

FIGURE 1.5 Band diagram in a semiconductor.

TABLE 1.1 Energy Range of Interest in the Band Structure of Semiconductor for Understanding the Behavior of Various Electronic Devices and under Different Conditions

Electronic Device	Energy Range of Interest within the Bandgap
Low electric-field regions of a transistor	Approximately $2kT$ from the band edges
High electric-field regions of a transistor	Approximately 0.5 eV
Power transistors and devices operating close to breakdown conditions or operating on those basis	The full size of the bandgap

k, is the Boltzmann constant; T, is the temperature in Kelvin.

effective electron mass). While in reality in *charge transport* through electronic devices larger than tens of *nanometers*, the presence of many *scattering* processes results in the loss of *coherence* in *transport* and the randomization of momentum, this simplifying omission of momentum information comes at a very affordable cost. As mentioned earlier, this cost is paid through adoption of the notion of *effective* electron mass.

Knowing that the *band diagram* provides only a selective set of information about the energy and momentum, which is sufficient only for specific applications, Table 1.1 provides a summary of links between the considered range of energy in the *band* structure and various electronic devices. This table can provide a sense of the limitations inherent to the *band diagram.*

1.1.6 Supplementary Notes on Band Diagram

A two-dimensional *band diagram*, such as the one shown in Figure 1.5, is often used while discussing the behavior of electrons in *bulk* semiconductors. However, the

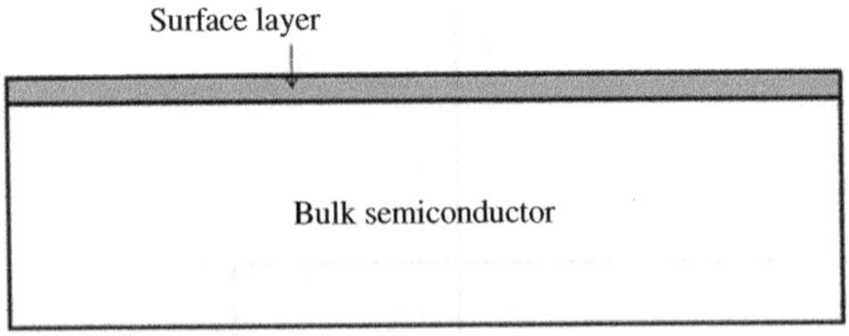

FIGURE 1.6 Surface layer is often treated as a separate layer envisioned at the surface, with a certain concentration of energy levels, known as surface states, within the bandgap of the semiconductor.

one-sided potential imposed on the electrons at the surfaces and interfaces disrupts the periodicity of potential applied on electrons and as a result the formation of the *bandgap*. Nevertheless, with the goal of simplifying the complicated reality of these surfaces and interfaces, oftentimes the aforementioned notion of a *band diagram* is readily extended to the surface. The presence of a surface is then factored into the *band diagram* through improvising an interfacial layer known as *surface layer*. Since most FETs operate through *charge transport* in the vicinity of these surfaces and interfaces, later on in the chapter, we will deal with these nonidealities at greater length. Figure 1.6 shows an example of incorporation of such a layer.

In the operation of semiconductor devices, the presence of *surface states* (which as implied in Fig. 1.6 are present in the *bandgap* of the *surface layer*) often results in hysteresis of the devices' characteristics. In order to avoid this, in high-speed semiconductor devices, bare surfaces should always be avoided. This can be done through implementing the so-called *self-aligned* technologies or even by passivating these surfaces through *deposition* of an insulator layer over them. In later chapters, these technologies will be covered in greater depth.

Additionally, while periodicity of crystal structure is key in developing a *band diagram*, in the description of the behavior of electronic devices, the notion of a *band diagram* is extended to *amorphous* materials (such as dielectrics used in *FET* technologies), which clearly do not possess a periodic structure. In strictest sense such an extension is not allowable. However, in these materials the presence of *short-range* order induces effects that are approximately explainable with the aid of a *band diagram*. One of the most important assets of *silicon* technology has been the possibility of achieving an almost perfect interface between the crystalline *silicon* and *amorphous* SiO_2. Thermal growth of SiO_2 on *silicon* results in a few *dangling bonds* with very little *bond-angle* distortion at the Si/SiO_2 interface. Even the few present *dangling bonds* can be saturated by the presence of *hydrogen* or *fluorine* throughout the *thermal oxidation* process. While SiO_2 is an *amorphous solid*, it exhibits an almost tetrahedral arrangement at short range.

As already pointed out, *bandgap* is an important property of semiconductors. This property is itself dependent on a number of factors including temperature. Table 1.2 summarizes the temperature dependence of the size of the *bandgap* of a number of important semiconductors.

TABLE 1.2 The Temperature Dependence of the Size of the *Bandgap* of a Number of Important Semiconductors

Semiconductor	Temperature Dependence of the Size of the Energy Bandgap
Si	$E_g(\text{eV}) = 1.17 - 4.73\times10^{-4}\cdot T^2/(T+636)$
Ge	$E_g(\text{eV}) = 0.742 - 4.8\times10^{-4}\cdot T^2/(T+235)$
GaAs	$E_g(\text{eV}) = 1.519 - 5.405\times10^{-4}\cdot T^2/(T+204)$ for $0<T<1000$
InAs	$E_g(\text{eV}) = 0.415 - 2.76\times10^{-4}\cdot T^2/(T+83)$ for $0<T<300$
InP	$E_g(\text{eV}) = 1.421 - 4.9\times10^{-4}\cdot T^2/(T+327)$ for $0<T<800$
InSb	$E_g(\text{eV}) = 0.24 - 6\times10^{-4}\cdot T^2/(T+500)$ for $0<T<300$
GaN	$E_g(\text{eV}) = 3.47 - 7.7\times10^{-4}\cdot T^2/(T+600)$
AlN	$E_g(\text{eV}) = 6.21 - 1.799\times10^{-4}\cdot T^2/(T+1462)$ for $0<T<300$

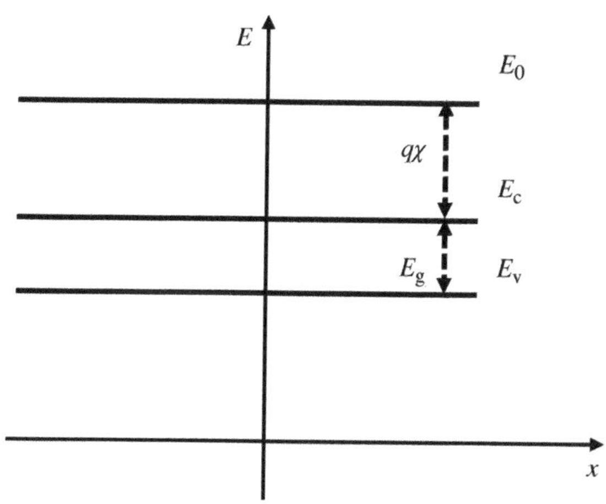

FIGURE 1.7 Identification of electron affinity (i.e., $q\chi$) of a given semiconductor on band diagram.

Another piece of information that is present in a *band diagram* is referred to as *electron affinity* (i.e., $q\chi$). This quantity represents the amount of energy required to set a *conduction-band* electron free from the *bulk* of the semiconductor. This amount of energy is denoted by the difference between the bottom of *conduction band* (i.e., E_c) and *vacuum energy level* (i.e., E_0). Figure 1.7 illustrates this in a *band diagram*. The definition of *electron affinity* is only loosely connected to measurable quantities such as *photothreshold*. This is due to the contribution of other factors including *image forces*[18] and the formation of *surface layer* to these measurable quantities. As a result, *electron affinity* remains as a quantity that can only be used in relative terms with a certain attributed approximation.

[18] These will be dealt with in Chapter 2.

1.1.7 Bond Model

A parallel description often used in explaining the electronic behavior of semiconductors is the *bond model*. The analog of migration of electrons across the *bandgap* in the *bond model* is the breaking away of electrons from covalent bonds. In other words, electrons in the *valence band* represent the electrons in covalent bonds with the atoms in the crystal, while those in the *conduction band* are the ones that have broken away from the covalent bonds and are almost free to roam within the confines of the crystal. While the *bond model* provides a simple picture for the electrons in the *conduction* and *valence bands*, it fails to satisfactorily describe the quantic nature of the *wave/particle* electrons and the role of the *density of states function* in each *band*.

1.2 ELECTRONS, HOLES, AND DOPING IN SEMICONDUCTORS

According to the *energy-band* and *bond* models, in an ideal semiconductor the number of electrons missing from the covalent *bonds* (i.e., missing from the *valence band*) and the number of free electrons (i.e., *conduction-band* electrons) are equal to one another. Such a semiconductor is referred to as *intrinsic* semiconductor. In the operation of electronic devices, however, we deal with semiconductors in which electron concentration in the *conduction band* (i.e., n_0 in cm^{-3})[19] and concentration of energy *states* devoid of electrons in the *valence band* (i.e., p_0 in cm^{-3}) are often not equal to one another. These semiconductors are referred to as *extrinsic* semiconductors.

The process of forming of an *extrinsic* semiconductor is referred to as *doping*. *Doping* occurs when a limited concentration of impurities is introduced into the semiconductor, and impurities are driven to replace some of the compositional atoms of the crystal. There are a few different *doping* processes. In these processes, impurities (also referred to as *dopants*) are either *in situ* incorporated into the crystal structure, as the crystal is being grown, or added in after crystallization. In the latter case, dopants are either *thermally diffused* into the crystal at the gas/solid interface (i.e., due to their higher concentration at the gaseous side) or *implanted* into the crystal in the form of highly energized ions. Each of these techniques has its own advantages in terms of cost/accuracy balance in the creation of the *dopant* profiles. The creation of a certain *dopant* profile (i.e., with precise control on the concentration, depth, and lateral extent of the distribution of impurities) and maintenance of their form throughout the operation of the device (i.e., when the device is exposed to high electric fields and temperatures) are keys to the success of a *doping* process.

1.2.1 Electrons and Holes

In both *doped* and *undoped* semiconductors, empty *energy states* of the *valence band* comprise only a small fraction of the *states* available in that *band*. Likewise, only a

[19] Whereas electron density in metals is in the order of 10^{23} cm^{-3}, in semiconductors this value is at least *two* orders of magnitude smaller.

small fraction of the *states* of the *conduction band* are filled by electrons. Based on the knowledge of *Pauli's exclusion principle* and the limited capacity of each of these *bands* for electrons, *charge transfer* can be envisioned either through movement of electrons or equally well through the transfer of empty *states* (of course in the opposite direction). As previously done in Figure 1.4, if we imagine a doubly sealed partially full water pipe, we can analogously say that the net movement of liquid caused by tilting the pipe is also representable through the movement of bubble in the opposite direction. In dealing with *charge transfer* in the *conduction band*, electrons are fewer in number than the empty *states*. Between the two equivalent presentations of *charge transfer*, it will be much simpler to focus on electrons. The opposite of that happens in the *valence band*. In the *valence band*, instead of electrons, charge transfer is studied through the movement of their complementary profile (i.e., profile of empty *states*). Of course, it will be easier to treat this complementary profile if we use a more addressable name for the empty *states*. Traditionally we call these empty *states* *holes*.[20]

Increasing the temperature of an *undoped* semiconductor causes the generation of electrons and *holes* in pairs (known as *electron–hole pairs* (EHP)). Caused by a number of events including collisions, these *charge carriers* can also go through the inverse of this *generation* process (which is known as *recombination*). During the *recombination* process, electrons fall back into the *valence band* and effectively annihilate an equal number of *holes*. These processes can either take the form of direct *band-to-band* transitions or be *assisted* transitions involving *impurity states* or *excitonic states*.[21] Through the process of *recombination*, the energy difference (i.e., almost equal to E_g) is either emitted in form of *photons* or is nonradiatively passed onto other particles such as *phonons*.

Holes have an equal amount of charge but are opposite in polarity to electrons. *Hole effective mass* is calculated according to the quantum mechanical information expressed in the *E–k* diagram of the *valence band*. *E–k diagrams* or *band diagrams* are developed for a negatively charged particle (i.e., electron). The opposite charge polarity of a *hole* requires us to look at these diagrams upside down while studying *holes*. Whereas the basis of calculation of the *effective mass* is already presented (1.13), it is important to point out that often the *E–k* diagrams are not as simple as illustrated in Figure 1.2. Among the intricacies present in these diagrams are the overlapping profiles of branches known as *degeneracy*. In terms of the presence of *degeneracies*, the structure of the *E–k* diagram of *conduction band*, however, is much simpler than that of the *valence band*.

Among the elements of the *periodic table* of which semiconductors are made of, valence electrons either occupy s-type or p-type orbitals. Even in the crystalline semiconductor made out of these elements, *charge carriers* in the *conduction* and *valence*

[20] These are imaginary charged particles in the *valence band*.

[21] An *exciton* is a *hydrogen* atom-like entity, in which the role of the nucleus is replaced by a *hole*. In the case of an *exciton* in the form of a bond between an electron and a very heavy *hole*, the *exciton* acts not unlike an *impurity level*. However, the picture is more complicated if the *hole* mass is comparable to that of the electron.

band retain much of these s- and p-type characters. This is especially remarkable, when we remind ourselves of the fact that these *charge carriers* are now free *Bloch electrons*[22] and not bound by the nucleus of an atom. This s- and p-type nature plays a very important role in *charge transport* in semiconductors. This nature also extensively simplifies the quantum mechanical description of semiconductors, as now the shape of s- and p-type orbitals can be used as eigenvectors in the matrix representation. This simplification is used in the implementation of *tight-binding* technique.

According to the *tight-binding* technique, the top of the *valence band* is explained in terms of a threefold *degeneracy*, corresponding to that of *p*-orbitals (i.e., p_x, p_y, p_z). With incorporation of *spin degeneracy*, this threefold *degeneracy* grows to a sixfold. The threefold *degeneracy*, or sixfold with reference to *spin degeneracy*, is observed in the form of a twofold *degeneracy* of two bands (one *heavy hole* and one *light hole*) of equal energy at the top of the *valence band* and another *band* with slightly lower electron energy referred to as *split-off band*. The names *heavy* and *light hole* are given in terms of *effective mass* of electrons and *holes* at the top of the *valence band*. In terms of the definition of the *effective mass*, the wider *band* (i.e., of smaller curvature) poses a larger *effective mass* on the *holes*, hence the name *heavy-hole band*. The effects of these *bands* can be incorporated only in the case of incorporation of *relativistic* effects of the problem, which are referred to as *spin–orbit coupling*.

Figure 1.8 provides a schematic representation of the nature of the *valence* and *conduction band* of most semiconductors. On the basis of the typical concave and convex *E–k* diagrams presented in Figure 1.8 (i.e., for *conduction* and *valence bands*, respectively) and considering the opposite polarity of the charge of a *hole*, it can be observed that both electrons and *holes* (i.e., *charge carriers* of the *conduction* and *valence bands*, respectively) possess positive *effective mass*. In spite of commonalities between semiconductors, the curvature at the top of the *E–k* diagram of a *band* is not always negative. In the case of *tellurium*, the *E–k* diagram is of the form shown in Figure 1.9, which prohibits the definition of negative mass for electrons, and as result prohibits the definition of *imaginary* positively charged particles (i.e., *holes*) for missing electrons.

Semiconductors are divided into two groups according to their value of momentum at the *conduction-band* minima: *direct bandgap* and *indirect bandgap*. The *E–k* diagrams for these two groups are illustrated in Figure 1.8. While in the *E–k* diagram of *direct-bandgap* semiconductors the top of the *valence band* and the bottom of *conduction band* coincide at a momentum of 0,[23] *indirect* semiconductors only have the top of their *valence band* defined there. As a result of the difference between the momentum at the top of the *valence band* and the bottom of *conduction band* of *indirect* semiconductors, these semiconductors are not efficient producers of *photons*. This is because another momentum changing collision will be required

[22] Electrons expressed in terms of *propagating waves* belonging to quantized *energy states*. The quantized *energy states* are referred to as *Bloch states*. Named after *Felix Bloch*.

[23] We will identify later in the chapter that this coincidence happens at the Γ-*point* in the so-called *Brillouin zone*.

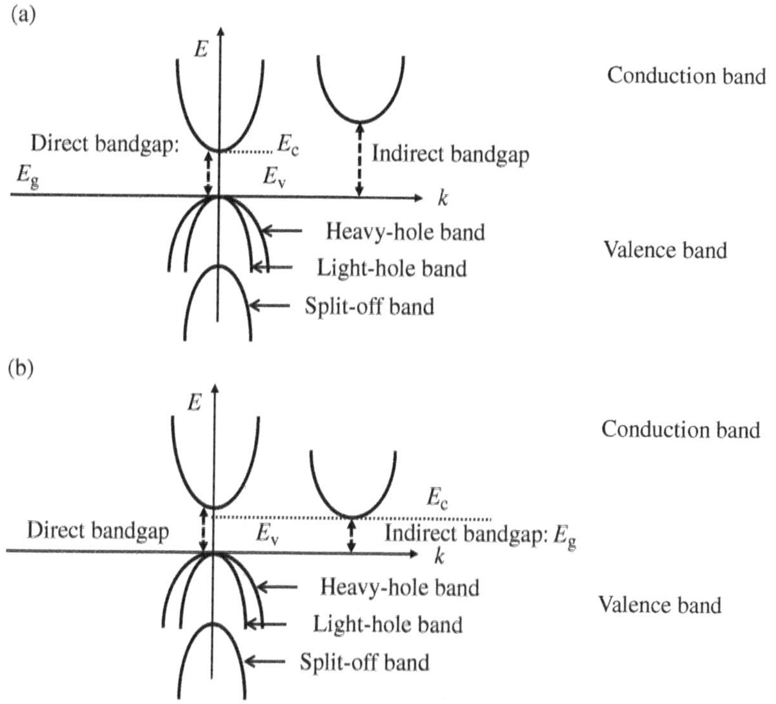

FIGURE 1.8 (a) E–k diagram of a direct-bandgap semiconductor; in this case the direct bandgap is smaller than the indirect bandgaps. (b) E–k diagram of an indirect-bandgap semiconductor; in this case the direct bandgap is larger than one or more of the indirect bandgaps. Top of the valence band is defined at $k = 0$.

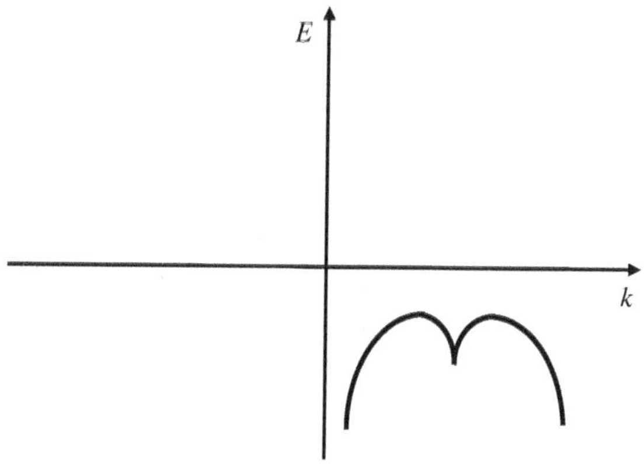

FIGURE 1.9 Top of the valence band in some materials such as tellurium present a positive effective mass for electrons hence at this position prohibits the definition of the positive charge carrier (i.e., hole).

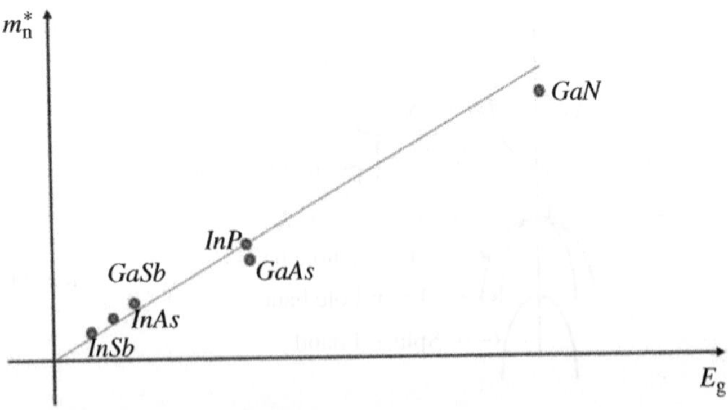

FIGURE 1.10 The prevalent trend between the effective electron mass and the size of the bandgap demonstrated among a few direct-bandgap *III–V* semiconductors. This approximate diagram is presented in linear scales.

besides the *electron–hole* interaction involved in the *recombination process*. In this case, the transitions involve a *photon* (which is responsible for energy change) and a *phonon* (which is responsible for momentum change). This division of responsibilities is caused by the very small momentum of *photons* and very small energies of *phonons*.

Silicon has an *indirect bandgap*,[24] with a sixfold *degenerate conduction-band edge*. *Germanium* is another well-known *indirect* semiconductor.[25] In both of these cases, a strong anisotropy of electron *wave function* persists near the *band edge*, which is caused by the mixing of p-type and s-type orbitals. The *conduction-band* minima of *direct-bandgap* semiconductors, being made of s-type orbitals, have spherically symmetric central cells. However, in the *states* further into the *conduction band*, this spherical symmetry, as the result of increasing contribution of p-type orbitals, fades. This is an important issue to be considered under high kinetic energy conditions.

In contrast to the *conduction-band* edge, the *valence-band* edges of most semiconductors are quite similar. In the case of the *valence band*, the central part of the electron *wave functions* is primarily p-type.[26]

As shown in Figure 1.10, there exists a declining trend between the electron *effective mass* and the size of the *bandgap* of *direct-bandgap* semiconductors.

1.2.2 Doping

Now that we know about *holes*, we can say that in an *intrinsic* semiconductor electron concentration of the *conduction band* (i.e., n_0) is equal to *hole* concentration of the *valence band* (i.e., p_0). In an *extrinsic* semiconductor, this balance is either tilted in

[24] Which, as will be seen later in this chapter, has a *conduction-band* minimum near the *X point* of the *Brillouin zone*.
[25] Whose *band* edge is defined near *L point* of the *Brillouin zone*.
[26] This nature induces a strong *spin–orbit* interaction.

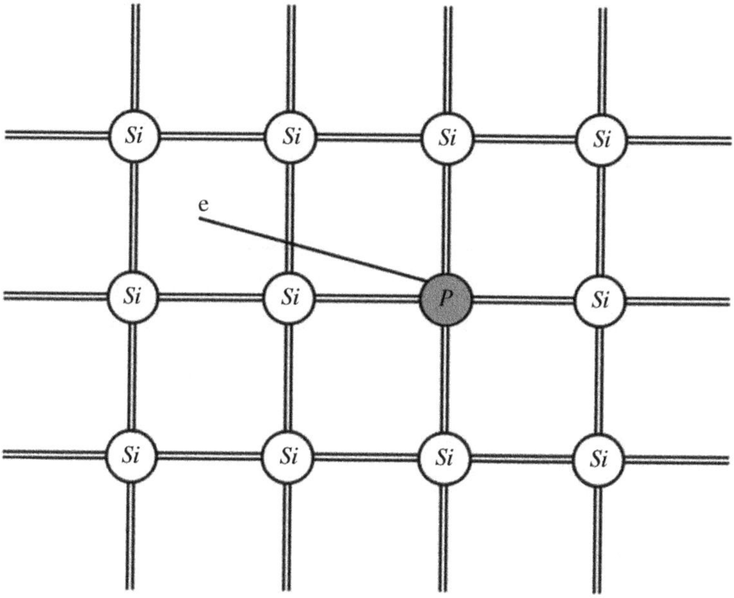

FIGURE 1.11 A 2-D schematic representation of *silicon* crystal upon substitutional doping with phosphorus. Covalent bonds are represented by the double lines, while the circles are representative of the silicon and phosphorus atoms.

favor of electrons (i.e., $n_0 > p_0$ in an n-type semiconductor) or holes (i.e., $p_0 > n_0$ in a p-type semiconductor).

Tilting the intrinsic balance of electron and *hole* concentration upon *doping* can be explained in terms of either the *energy-band* or *bond* models. In this review, we will address the process of *doping* in a specific group *IV* semiconductor (of the *periodic table*): *silicon*.

In an n-type semiconductor: $n_0 > p_0$. This means that not all of the electron population of the *conduction band* has originated from the *valence band*. In this case, the source of the *excess* population of electrons is the loosely bound electrons to the nucleus of *dopants* added to the semiconductor crystal. In *silicon* technology, the *dopants* used for this purpose are group *V* atoms of *phosphorus* and *arsenic*. These group *V* atoms possess one electron in excess of the four that they should share with the nearest neighboring *silicon* atoms while substituting one. This is schematically shown in Figure 1.11. The *ionization energy* of this *excess* electron is much smaller than the *ionization energy* of the *dopant* atom outside the confines of the crystal. This is caused by the overriding effect of the other atoms surrounding the *dopant* in the crystal, which renders an *ionization energy* quite close to that of the *hydrogen* atom. The *effective mass* of this electron and the value of the *permittivity* needed in the calculation of *ionization energy* are, however, modified to those values imposed by the *silicon* crystal. For offering an effective impurity, this *ionization energy* should be small compared to the average thermal energy of electrons at operating temperature

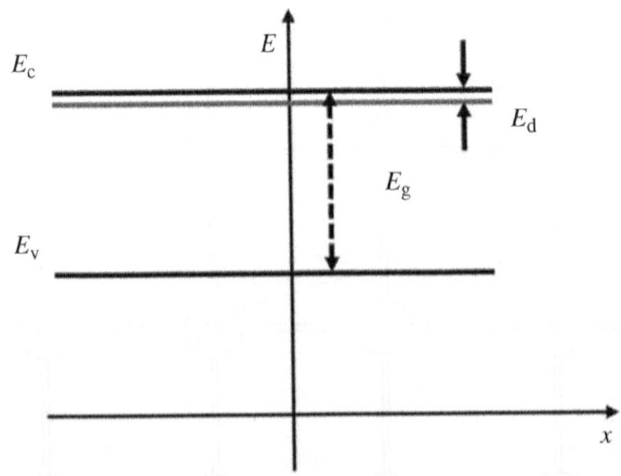

FIGURE 1.12 Band diagram of a semiconductor doped with a shallow donor.

of the device. Under such a condition, *dopants* can be effectively *activated* (i.e., contribute their *excess* electrons to the *conduction band*). This is one of the reasons *P* and *As* are chosen as suitable n-type *dopants* in *silicon* technology as their *ionization energies* are within the range of 20–40 meV.[27] These dopants are referred to as *shallow dopants*. This is in contrast to the *dopants* with higher *ionization energies*, which are known as *deep dopants*.

The other reason for the choice of *As* and *P* in *silicon* technology is the relatively high concentrations at which these atoms can be incorporated into the *silicon* structure under *thermodynamic equilibrium*. Under *thermodynamic equilibrium* there is an upper limit to the concentration of impurities that can be incorporated into a *solid* at a given temperature, known as *solid solubility*. Because the earliest technique of *doping* (i.e., *diffusion doping*) was a *thermal-equilibrium* process, this property played an important role in the choice of these *dopants*. However, in *ion implantation doping*, which is not performed under *thermodynamic equilibrium*, the *solid solubility* is not as important.

While in *silicon* crystal the concentration of *Si* atoms is on the order of 10^{22} cm^{-3} at room temperature, the maximum level of *doping* is only slightly above 10^{20} cm^{-3}. As a result, to a first approximation we can assume that semiconductor properties such as the size of the *bandgap* are not much altered by the *doping* process. In reality, the random nature of *substitutional doping* results in a disrupted periodicity of the crystal (and also a disruption in the formation of *forbidden* energy *gaps*). Due to this disruption, the *doping* process results in an introduction of *energy levels* within the *forbidden* energy gap. If an n-type *dopant* is shallow, the *energy levels* will form closer to the *conduction-band* edge (i.e., E_c). This is the *band diagram* explanation of the n-type *doping* process, which is illustrated in Figure 1.12. With an increasing *doping* level

[27] Since we deal small amounts of energy, in semiconductor electronics instead of using *Joule* as the unit of energy, we use *electron Volt* (i.e., eV). One electron Volt is almost equal to 1.6×10^{-19} J.

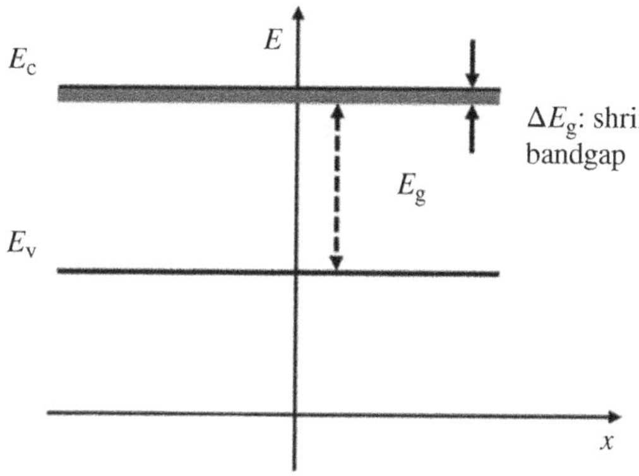

FIGURE 1.13 Schematic depiction of the role of heavy doping in shrinking the bandgap of a semiconductor. The wide gray band of energy stands for the impurity band resulting from the splitting of donor energy levels.

(i.e., number of incorporated *substitutional dopants* per unit volume), these *shallow energy levels* will split into *bands* of energy (often referred to as an *impurity band*). The splitting is caused by the overlapping *wave function* of the electrons residing in the impurity *energy levels*. As shown in Figure 1.13, doping at very high levels, through merging of these *bands* with the *conduction band* (or *valence band* in the case of p-type *doping*), results in *shrinkage of the bandgap*. If the *energy levels* introduced by *dopants* are separated from the *conduction band* by only tens of milli-electron volt, at room temperature electrons residing in those *energy states* (i.e., loosely bound electrons to the nuclei of the n-type impurity) can readily leave these *levels* and jump to the *conduction band* without adding a *hole* to the *valence band*. This is how the balance between the electron and *hole* concentration is broken in an *extrinsic* n-type semiconductor.

The more appropriate name for an n-type *dopant* is *donor* because it donates electron to the *conduction band*. In the *energy-band* model, the *energy levels* provided by these *donors* to the previously *forbidden bandgap* are also known as *donor levels*. Through the process of electron donation to the *conduction band*, the *donor* atom becomes a positively charged ion, and as a result, charge neutrality prevails in the semiconductor. Using the *energy-band* model, we can describe this behavior in terms of the following definition for the *donor level*:

An *energy level* is referred to as *donor*, if it were to be neutral when full and positively charged when empty.

For very highly *doped* semiconductors,[28] when the *impurity band* developed by the *donors* merges with the *conduction band*, the semiconductor adopts a *metallike* behavior and remains highly conductive even at very low temperatures. This formation is also known as an *impurity band tail*. As shown in Figure 1.14, the choice of the name comes from the presence of finite *DOS* at the *band edge* and the gradual, and not sharp, increase of the *DOS* function. This is unlike the predictions of (1.14). Later in

[28] For which a more quantitative definition is to follow in Section 1.3.5.

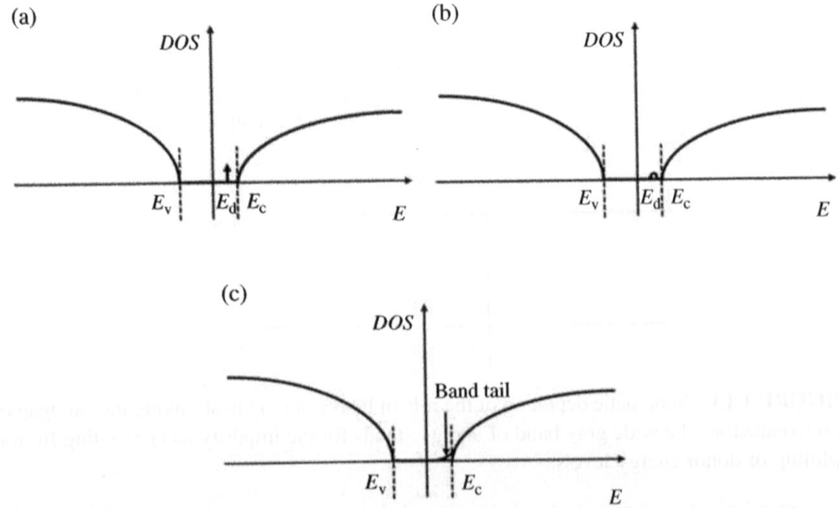

FIGURE 1.14 Schematic depiction of the evolution of the conduction band's *DOS* function upon increase of the donor concentration, from (a) to (c). Scales are linear.

this chapter, we will talk about formation of *band tails* also in the case of the so-called *alloyed* semiconductors.

It is worthwhile mentioning that *arsenic*, due to its larger atomic size than *phosphorus*, has a smaller *thermal diffusion constant* through the *silicon* crystal. Since many processing steps run at very high temperatures, achieving the small n-type *dopant* profiles that are required in modern devices is much easier with the use of *As*. For that reason, in many applications of *Si* technology, *As* has already replaced *P* as the proper *donor*.

So far, with exception of a few points, we have avoided discussing p-type *doping*. A slightly different version of events can explain the intricacies of p-type *doping*. The suitable p-type *dopant* in *Si* technology is the group *III* atom of *boron*. Substitution of *Si* with *B* (which has only three electrons in its valence shell) leaves one out of four broken covalent bonds, originally formed between nearest neighboring *Si* atoms, in need of an electron. The missing electron, however, can be acquired by breaking one electron away from an existing covalent bond in the vicinity of the impurity. In terms of *energy-band* model, this can be explained through the addition of an *energy level* closer to *valence band* in the *forbidden gap*, which is keen to accept electrons. Through receiving electrons from the *valence band*, a *hole* is left behind in that *band*. For this reason, this *energy level* is called an *acceptor level*, and the p-type *dopants* are referred to as *acceptors*. Quite similar to the definition of the *donor energy level*, an *acceptor energy level* is neutral when empty and negatively charged when full. *Boron*, like *arsenic* and *phosphorus*, provides *silicon* with a *shallow impurity level*.

Doping process can take place through substitution of structural atoms of the crystal with *dopants* (also referred to as *direct doping*) or indirectly through incorporation

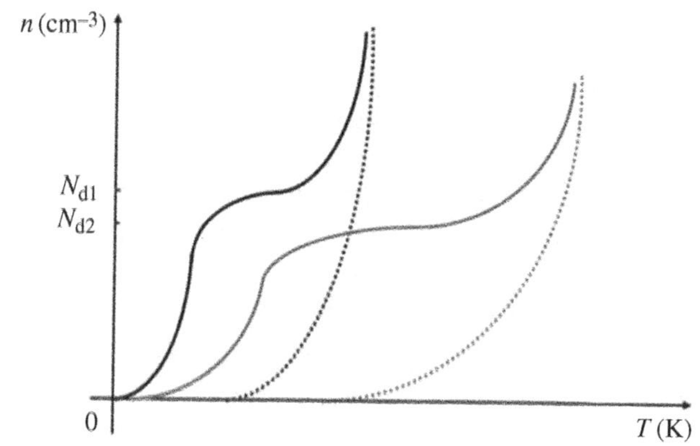

FIGURE 1.15 Variation of the electron concentration versus temperature for two n-doped semiconductors of different bandgaps. The wider bandgap semiconductor is assumed to have been doped more lightly. The levels of doping of the two semiconductors, in increasing order of the size of the bandgap, are indicated by N_{d1} and N_{d2}, respectively. The dashed lines represent the variation of electron concentration in each semiconductor when undoped (i.e., n_i). Diagrams in black represent the smaller bandgap semiconductor, while the gray diagrams represent the wider bandgap semiconductor. Within the temperature range for which the electron concentration reaches a plateau, semiconductor behaves extrinsically. Scales are linear.

of *dopants* into the *interstitial* sites. For *dopants* to contribute to electron and *hole* concentration, however, they need to take *substitutional* positions. As already indicated implicitly, *dopants* also need to be *activated* through *ionizations*. The movement of *dopants* in the crystal and their later *activation* is a process that requires exposure to high temperatures. The process of movement of *dopants* within a crystal can happen through thermal diffusion of *dopant* in the form of *interstitial diffusion*, *substitutional diffusion*, or a mixture of the two (*dopants* sometimes taking *substitutional* positions and sometimes *interstitial* throughout the diffusion process), which is referred to *interstitialcy*.

A *doped* semiconductor under a number of conditions will retain its *intrinsic* characteristics:

1. When the *lattice temperature*[29] is very low and *dopants* are under *carrier freeze-out* (i.e., *not ionized*).
2. When the concentration of *activated donors* and *acceptors* are equal to one another. This case is referred to as *compensative doping*.
3. When *lattice temperature* is high and the number of *thermally induced charge carriers* to *conduction* and *valence band*, in the form of *EHP*, well exceeds the concentration of electrons and *holes* contributed by *donors* and *acceptors*, respectively. As shown in Figure 1.15, the characteristic temperature of this behavior is determined by the size of the *bandgap* of the semiconductor.

[29] It will be identified in Section 1.4.4 why instead of simply speaking of temperature, we are talking about a temperature quantity called *lattice temperature*.

Realization of *extrinsic* semiconductors is a prime requirement in formation of electronic devices. In these devices, the interaction between regions of different *charge-carrier* concentrations provides the chance for enforcing selective control over the movement of *charge carriers*. Such junctions between regions of different *dopant* concentration can be realized through simply *doping* the semiconductor and/or *growing* semiconductors of different *bandgaps* and/or *electron affinities* on top of one another.

1.2.3 Calculation of Ionization Energies in Semiconductors

With regard to the process of *doping*, it is quite important to be able to have means to approximately calculate the energy of *donor* and *acceptor states*. For the loosely bound fifth electron of ordinary *donors* such as *P*, *As*, and *Sb* in *Ge* or *Si* technologies, the *ionization energy*, referenced to the *conduction-band* edge (i.e., E_c), calculated in terms of *hydrogenic* models provides us with

$$E_d = -13.6\left(\frac{m^*}{m_0}\right)\left(\frac{\epsilon_0}{\epsilon}\right)^2 \frac{1}{n^2}\ \text{eV} \tag{1.17}$$

where n is the principal *quantum number*, ϵ is the dielectric constant, and ϵ_0 is the permittivity of the vacuum.

In the above equation, the differences between the *effective mass* of an electron and the free electron mass, and also the *dielectric constant* of a semiconductor and that of the free space, cause the electron orbit for the impurity atom to be much larger than the value in a free atom. As a result, *ionization energy* is much smaller for an impurity in a semiconductor than that of the *hydrogen* atom.

A similar relationship can be developed for *acceptor states*. This approximate framework is, however, not as valid due to the more complicated nature of the *E–k* diagram of the *valence band*. Here the *effective mass* of the electron should be replaced with that of the *hole*. Because of the much higher *effective mass* of *holes* compared to electrons in most semiconductors (with the exception of *Ge*), realizing the full *activation* of *acceptors* requires higher temperatures than those required for *activating donors*. Between *Si* and *Ge*, the *ionization energies* in *Si* for both types of *dopants* are larger. This is due to the smaller dielectric constant and larger *effective mass* in *Si*.

Despite their usefulness, it should be pointed out that the aforementioned formalism of calculation of *donor* and *acceptor energy levels* is only sufficient for evaluation of *ionization energies* within an order of magnitude approximation. However, since more accurate estimations require more intensive quantum mechanical mathematics, this framework is often used for providing the first-order estimate. Although for the considered case of a *hydrogen* atom the potential and, as a result, *effective mass* are spherically symmetric, not all semiconductor *energy bands* have this degree of symmetry. With regard to this case, it is necessary to use an appropriate average of the *effective mass* in different directions. The other simplifying assumption used in the development of the *hydrogenic* model of (1.17) is that the position of the *conduction band* is taken to coincide with E_d when the *quantum number* tends to infinity. While not true in a strict sense, this has been deemed to provide a good estimate. The last limitation that this model suffers from is that it does not distinguish between the *ionization energies* of different *dopants*. Although the impact of the ion core on the *ionization energy* is reduced in a *dopant*-in-semiconductor scenario, this reduced impact still can be felt in terms of the difference in *ionization energy* of different *dopants*.

Table 1.3 provides a detailed list of ionization energies for impurities in *Si*, *Ge*, *GaAs*, *InP*, and *GaN*.

1.3 THERMAL-EQUILIBRIUM STATISTICS

Considering the large number of *energy states* in the *conduction* and *valence band*, studying the state of the electron population of these *bands* (i.e., electron concentration and *hole* concentration in the *conduction* and *valence band*, respectively) can be only addressed statistically. However, before presenting the suitable statistical framework of this problem, it should be mentioned that such a statistical framework loses its validity when statistical fluctuation of the number of *dopants* in an extremely small-size device grows close to the small number of *dopants* incorporated in its volume. As a result, the framework presented in this section faces difficulties in application to *nanoscale* devices.

Evidently, absence of time variation yields a more manageable mathematical framework. The condition set in the study of this time-invariant situation is referred to as *thermal-equilibrium* condition. Under *thermal equilibrium*, the semiconductor is not exposed to any external source of excitation, and at a given temperature (which we will call from now on *lattice temperature*), all thermal processes are counterbalanced. As a result, electron and *hole* concentration among other properties of the semiconductor will remain time independent.

1.3.1 *Fermi–Dirac* Statistics

In Appendix 1.A, with special attention to the restrictions of *Pauli's exclusion principle*, the mathematical derivation of the fundamental outcome of this *thermal-equilibrium* statistics, which is referred to as the *Fermi–Dirac* distribution function,[30] is provided. This temperature-dependent statistical model, irrelevant of the presence of an *energy state* at a given *energy value*, determines the chance of having an electron of that value of energy. As a result of the definition of *Fermi–Dirac* statistics, one can calculate the spatial concentration of electrons within an infinitesimally small range of energy by multiplying the *Fermi–Dirac* distribution function by the known *density of states function* of the *conduction band*.

The process of establishing a state of *thermal equilibrium* is not instantaneous. Instead, for a given crystal at a set temperature, *thermal equilibrium* evolves spontaneously over a certain period of time.

[30] Named after *Enrico Fermi* and *Paul Dirac*.

TABLE 1.3 List of Ionization Energies for a Number of Important Impurities in Si, Ge, GaAs, InP, and GaN

Semiconductor	Ionization Energies of a Number of Important Impurities
Si	Li: $E_c - 0.034$ eV (D)
	Sb: $E_c - 0.043$ eV (D)
	P: $E_c - 0.046$ eV (D)
	As: $E_c - 0.054$ eV (D)
	Be: $E_V + 0.42$ eV (A), $E_v + 0.17$ eV (A)
	Au: $E_c - 0.54$ eV (A), $E_v + 0.35$ eV (D), $E_v + 0.29$ eV (D)
	Cu: $E_v + 0.53$ eV (A), $E_v + 0.4$ eV (A), $E_v + 0.24$ eV (A)
	B: $E_v + 0.044$ eV (A)
	Al: $E_v + 0.069$ eV (A)
	Ga: $E_v + 0.073$ eV (A)
Ge	Li: $E_c - 0.0095$ eV (D)
	Sb: $E_c - 0.0096$ eV (D)
	P: $E_c - 0.012$ eV (D)
	As: $E_c - 0.013$ eV (D)
	Pt: $E_c - 0.23$ eV (A), $E_v + 0.04$ eV (A)
	Au: $E_c - 0.04$ eV (A), $E_c - 0.2$ eV (A), $E_v + 0.15$ eV (A), $E_v + 0.05$ eV (D)
	Cu: $E_c - 0.26$ eV (A), $E_v + 0.32$ eV (A), $E_v + 0.045$ eV (A)
	B: $E_v + 0.010$ eV (A)
	Al: $E_v + 0.010$ eV (A)
	Ga: $E_v + 0.011$ eV (A)
GaAs	Si: $E_c - 0.0058$ eV (D), $E_v + 0.035$ eV (A)
	Ge: $E_c - 0.0061$ eV (D), $E_v + 0.0404$ eV (A)
	O: $E_c - 0.4$ eV (D), $E_v + 0.67$ eV (D)
	C: $E_v + 0.026$ eV (A)
	Be: $E_v + 0.028$ eV (A)
	Mg: $E_v + 0.028$ eV (A)
	Au: $E_v + 0.09$ eV (A)
	Mn: $E_v + 0.095$ eV (A)
	Cu: $E_v + 0.44$ eV (A), $E_v + 0.24$ eV (A), $E_v + 0.19$ eV (A), $E_v + 0.14$ eV (A), $E_v + 0.023$ eV (A)
	Fe: $E_v + 0.52$ eV (A), $E_v + 0.37$ eV (A)
InP	S, Si, Sn, Ge: $E_c - 0.0057$ eV (D)
	C: $E_v + 0.04$ eV (A)
	Hg: $E_v + 0.098$ eV (A)
	Zn: $E_v + 0.035$ eV (A)
	Si: $E_v + 0.03$ eV (A)
	Cu: $E_v + 0.06$ eV (A)
	Be: $E_v + 0.03$ eV (A)
	Mg: $E_v + 0.03$ eV (A)
	Ge: $E_v + 0.021$ eV (A)
	Mn: $E_v + 0.027$ eV (A)
GaN	Si: $E_c - (0.02 - 0.12)$ eV (D) (often instead of a single value, a range has been reported)
	Mg: $E_v + (0.14 - 0.21)$ eV (A) (often instead of a single value, a range has been reported)

(D) and (A) are indicators of energy level of donor and acceptor nature, respectively. The presented information for the first three semiconductors is more complete.

In the description of the *Fermi–Dirac* distribution function, a certain value of energy, known as *Fermi level* (i.e., E_f), serves as the reference value of energy in determination of the probability of occupation of all *energy states*. According to *Fermi–Dirac* statistics, an *energy state* with E_f as its energy value, at any given temperature, has a 50% chance of being occupied by an electron. *Energy states* above E_f have lower than 50% chance of occupation, and those below have higher than 50% chance of occupation. *Fermi–Dirac* statistics is presented by

$$f_D(E) = \frac{1}{1+\exp((E-E_f)/kT)}. \tag{1.18}$$

In this equation, T is the *lattice temperature* in Kelvin and k is the *Boltzmann* constant.[31] At 0 K, $f_D(E)$ reduces to a *Heaviside unitary step function*. Whereas at all temperatures $f_D(E_f) = 1/2$, an increase in temperature causes this step function to spread out further.

1.3.2 *Maxwell–Boltzmann* Statistics

The *Fermi–Dirac* distribution function when $E-E_f \gg kT$ can be reduced to a simpler distribution function, which is referred to as *Maxwell–Boltzmann* distribution function,[32]

$$f_M(E) = \exp\left(-\frac{E-E_f}{kT}\right). \tag{1.19}$$

As shown in Appendix 1.A, the derivation of *Fermi–Dirac* statistics is bound by the constraints of *Pauli's exclusion principle*. Electrons and *holes*, as particles that are required to follow these constraints, are referred to as *Fermions*. Although in the case of *Maxwell–Boltzmann* statistics such a constraint is not present, simplifying *Fermi–Dirac* statistics by the distribution function of (1.19) does not impose any difficulty. The reason is rooted in the condition applied to this approximation (i.e., $E-E_f \gg kT$). Under this condition, the chance of finding an electron at *energy state* E will be very small. As a result the chance of encountering two electrons at the same *state* and clashing with *Pauli's exclusion principle* will be equally small. The semiconductor for which *Maxwell–Boltzmann* statistics is applicable is referred to as a *nondegenerate*[33] semiconductor. Figure 1.16 presents the form of temperature variation of the

[31] $k = 8.62 \times 10^{-5}$ eV/K.

[32] Named after *James Maxwell* and *Ludwig Boltzmann*.

[33] In this case the definition of *degeneracy* is with reference to becoming *metallike*, which as hinted earlier happens when the *dopant* concentration is very high and as a result of overlapping the *impurity-band* and *conduction-band* (likewise *valence band* in case *holes*) *charge-carrier* concentration of *conduction-* or *valence-band* approaches that of the metal. This *degeneracy* is different from the one observed in our discussions of *E–k* diagram.

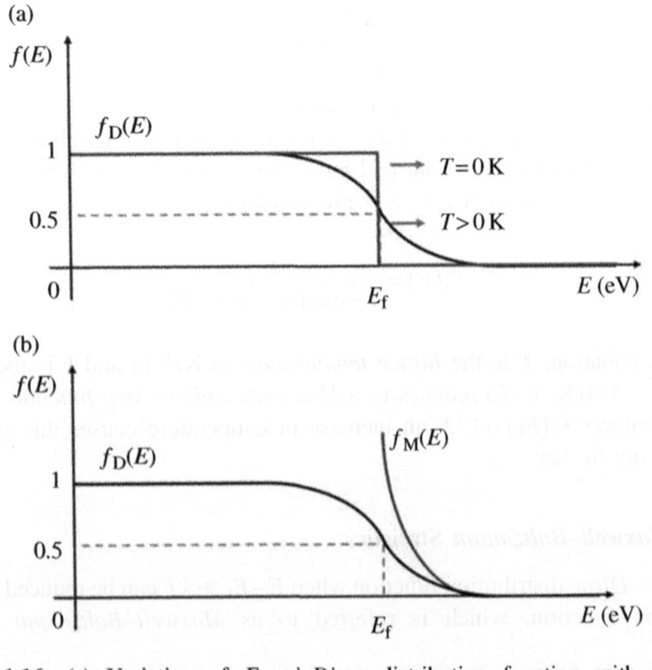

FIGURE 1.16 (a) Variation of *Fermi–Dirac* distribution function with temperature. (b) Comparison between the *Fermi–Dirac* and the *Maxwell–Boltzmann* distribution functions at a temperature above 0 K. Scales are linear.

Fermi–Dirac distribution function and its approximation with the *Maxwell–Boltzmann* distribution function.

For a *nondegenerate* semiconductor the distribution function of (1.18) can be simplified to a *Maxwell–Boltzmann* distribution function. Through adopting a semiclassical approach to the definition of energy such as

$$E = E_c + \frac{1}{2}m^* v^2, \tag{1.20}$$

one can rewrite (1.19) as

$$f_M \cong \exp\left(\frac{E_f - E_c}{kT}\right) \cdot \exp\left(-\frac{m^* v^2}{2kT}\right) = C \cdot \exp\left(-\frac{m^* v^2}{2kT}\right). \tag{1.21}$$

It can be proven that C for a given *doping* level is constant.[34] According to this equation, *charge-carrier* velocities of a *nondegenerate* semiconductor are distributed in the form of a *Gaussian* with an average velocity of 0 and a variance, which is a function of *lattice temperature*. In the three-dimensional setting,

[34] That is, using (1.23) and (1.32) or (1.33).

$$v^2 = v_x^2 + v_y^2 + v_z^2 \tag{1.22}$$

and as a result such a conclusion can also be drawn for *carrier* velocities along each of the axes of the *Cartesian* coordinate system.

1.3.3 Calculating Electron and Hole Concentration in Nondegenerate Semiconductors

As suggested earlier in this section, in calculating *thermal-equilibrium* electron concentration in the *conduction band* (i.e., n_0), one needs to multiply the *Fermi–Dirac* distribution function (or where appropriate the *Maxwell–Boltzmann* distribution function) with the *density of states* function of the *conduction band* and take an integral over the energy width of this *band*. Likewise, in calculating the *thermal-equilibrium hole* concentration in the *valence band* (i.e., p_0), one should multiply the *DOS* function of the *valence band* by the complementary of the *Fermi–Dirac* (or where appropriate the *Maxwell–Boltzmann*) distribution function and take an integral over the energy extent of the *band*. For cases in which *Maxwell–Boltzmann* statistics is applicable, such calculations result in the following closed-form expressions for n_0 and p_0:

$$n_0 = N_c \exp\left(\frac{E_f - E_c}{kT}\right) \tag{1.23}$$

and

$$p_0 = N_v \exp\left(\frac{E_v - E_f}{kT}\right) \tag{1.24}$$

where N_c and N_v are the *effective density of states* at the lower edge of the *conduction band* (i.e., E_c) and the higher edge of the *valence band* (i.e., E_v), respectively. These *effective* values are mathematical tools proposed to yield the above simple expressions.[35] While in reality in a *bulk* semiconductor, the *DOS* at the lower edge of *conduction band* and the higher edge of the *valence band* are either 0[36] or have a very small value (i.e., in case of formation of *band tail*), the values of N_c and N_v are quite large. This is because they *effectively* represent the total number of *states* distributed throughout the *bands* in the form of two *Dirac delta functions* defined at the edge of each *band*. These values, as expressed in (1.25) and (1.26), are dependent on *lattice temperature* and *effective mass* of electrons and *holes* (i.e., for N_c and N_v, respectively):

$$N_c = 2\left(\frac{2\pi m_n^* kT}{h^2}\right)^{3/2} \tag{1.25}$$

[35] In terms of these *effective density of states*, the aforementioned integral is calculated using the *gamma* function.

[36] See (1.14).

TABLE 1.4 A Number of Important Properties of Si, Ge, GaAs, and GaN at Room Temperature

	Semiconductor			
Property	Si	Ge	GaAs	GaN
Lattice type and constant a (Å)	Zinc blende 5.431	Zinc blende 5.658	Zinc blende 5.65325	Wurtzite 3.189 ($c = 5.186$)
Melting point T_m (°C)	1412	937	1237	2500
Bandgap E_g (eV)	1.12	0.67	1.42	3.39
Static relative permittivity ϵ_r	11.7	16	12.9	8.9
Specific heat C_p (J/g/K)	0.7	0.31	0.33	0.49
Thermal conductivity k (W/cm/K)	1.412	0.606	0.455	1.3
Electron mobility μ_n (cm^2/V/s)	1417	3900	8800	1000 (approx.)
Hole mobility μ_p (cm^2/V/s)	471	1900	400	200 (approx.)
Effective conduction-band density of states N_c (cm^{-3})	2.8×10^{19}	1.04×10^{19}	4.7×10^{17}	2.23×10^{18}
Effective valence-band density of states N_v (cm^{-3})	1.04×10^{19}	6×10^{18}	7×10^{18}	4.62×10^{19}
Intrinsic carrier concentration n_i (cm^{-3})	1.45×10^{10}	2.4×10^{13}	9×10^{6}	1.9×10^{-10}
Critical electric-field E_c (V/cm)	3×10^{5}	8×10^{4}	3.5×10^{5}	5×10^{6}
Effective transport electron mass m_n^*	$0.26m_0$	$0.12m_0$	$0.068m_0$	$0.2m_0$
Effective transport hole mass m_p^*	$0.386m_0$	$0.3m_0$	$0.5m_0$	$0.8m_0$
Optical phonon energy E_{op} (meV)	63	37	35	91.2

and

$$N_v = 2\left(\frac{2\pi m_p^* kT}{h^2}\right)^{3/2}. \tag{1.26}$$

Equations (1.23)–(1.26) imply that the *Fermi level* of an *intrinsic* semiconductor is very close the middle of the bandgap but not exactly there since $m_p^* > m_n^*$.

Table 1.4 presents a list of important properties, such as N_c and N_v, for a few well-known semiconductors.

Since in *nondegenerate doping* of a semiconductor, due to low levels of *dopant* concentration, structural properties of the semiconductor remain almost intact, under these conditions N_c and N_v will not change with *doping*. As a result, letting E_i

represent the *Fermi level* of the *intrinsic* semiconductor, (1.23) and (1.24) can be rewritten in terms of E_i and n_i (i.e., *intrinsic* electron and *hole* concentration):

$$n_0 = n_i \exp\left(\frac{E_f - E_i}{kT}\right) \tag{1.27}$$

and

$$p_0 = n_i \exp\left(\frac{E_i - E_f}{kT}\right). \tag{1.28}$$

1.3.4 Mass Action Law

Concentration profiles of electrons and *holes* in a semiconductor are not independent from one another. From (1.27) and (1.28) it is quite obvious that

$$p_0 n_0 = n_i^2. \tag{1.29}$$

This relationship is referred to as the *mass action* law.

Also based on (1.23) and (1.24),

$$n_i = \sqrt{N_c N_v} \exp\left(\frac{-E_g}{2kT}\right). \tag{1.30}$$

On the basis of (1.30), for a semiconductor of smaller *bandgap*, at any given temperature the *intrinsic charge-carrier* concentration (i.e., n_i) is larger. This explains why a wider *bandgap* semiconductor remains *extrinsic* over a wider range of temperatures (Fig. 1.15).

Equation (1.23) shows that as a result of elevation of the *Fermi level* through n-type *doping*, E_f approaches E_c, and E–E_f at the bottom of the *band* will become smaller. When this difference becomes smaller than a few times kT, the approximation of *Fermi–Dirac* statistics by *Maxwell–Boltzmann* statistics is inadmissible. An identical situation will happen when E_f is lowered toward E_v. Under these situations, which are caused by heavy n- and p-type *doping*, approximate equations of (1.23)–(1.28) are no longer valid. A highly doped semiconductor like this is referred to as a *degenerate* semiconductor, since with the increase in *charge-carrier* concentration, the conduction properties of the semiconducting material have essentially *degenerated* into those of a metal.[37] In degenerate semiconductors, electron and *hole* concentrations should be calculated with the use of *Fermi–Dirac integrals*.

[37] We have previously come across this situation while speaking of the merger of the *impurity band* and *conduction* or *valence bands* in a highly *doped* semiconductor.

The *mass action* law is a *thermal-equilibrium* relationship. Whereas simply multiplying (1.27) and (1.28) proved the *mass action* law in a *nondegenerate* semiconductor, for all semiconductors under *thermal equilibrium*, such a relationship is readily extendable. In a general form, this can be easily proven through considering the counterbalancing processes of *generation* and *recombination* of EHP. While the process of *generation* is a thermally induced process (where its rate can be expressed in terms of a function of temperature: $f_1(T)$), the process of *recombination* is not only temperature dependent but also dependent on the population of electrons in the *conduction band* and *holes* in the *valence band*. The temperature-dependent part of the rate of the *recombination* process is expressed in terms of another function of temperature, which we will call $f_2(T)$. By equating the rates of the two counterbalancing processes, under *thermal equilibrium*,

$$G = R \Rightarrow f_1(T) = p_0 n_0 f_2(T) \Rightarrow \frac{f_1(T)}{f_2(T)} = p_0 n_0 = f_3(T). \tag{1.31}$$

Knowing that the function $f_3(T)$ is just a function of temperature and not the *doping* process, it will be evident that $f_3(T) = n_i^2$, which yields the *mass action* law (i.e., (1.29)).

On the basis of the *mass action* law and charge neutrality, one can prove that in a semiconductor *doped* with both *donors* (i.e., N_d in cm^{-3}) and *acceptors* (i.e., N_a in cm^{-3}), if all *dopants* were to be activated,

$$n_0 = \left(\frac{N_d - N_a}{2}\right) + \sqrt{\left(\frac{N_d - N_a}{2}\right)^2 + n_i^2} \quad \text{for } N_d > N_a, \tag{1.32}$$

which yields $n_0 \cong N_d - N_a$ when $N_d - N_a \gg n_i$.

In addition,

$$p_0 = \left(\frac{N_a - N_d}{2}\right) + \sqrt{\left(\frac{N_a - N_d}{2}\right)^2 + n_i^2} \quad \text{for } N_a > N_d, \tag{1.33}$$

which yields $p_0 \cong N_a - N_d$ when $N_a - N_d \gg n_i$.

Example

A *Ge* sample is uniformly *doped* with both *B* and *As* to the levels of 10^{16} and 10^{15} cm^{-3}, respectively. Determine the electron concentration in this sample at both 300 and 500 K. Assume that at both temperatures, all impurities are *activated*.

According to Table 1.4, the *bandgap* of *Ge* at room temperature is 0.67 eV, while $n_i = 2.4 \times 10^{13}$ cm^{-3}.

Based on Table 1.2 the size of the *bandgap* at 500 K is given by

$$E_g(T = 500\,\text{K}) = 0.742 - 4.8 \times 10^{-4} \frac{500^2}{735} \cong 0.579 \text{ eV}.$$

According to this value and the values of N_c and N_v at room temperature (provided in Table 1.4), (1.30) yields the *intrinsic* electron concentration at 500 K:

$$n_i(T = 500\,\text{K}) = \sqrt{1.04\times10^{19}\times6\times10^{18}\left(\frac{500}{300}\right)^3\cdot\exp\left(-\frac{0.579}{2\times0.0258\times\frac{500}{300}}\right)}$$
$$\cong 2.03\times10^{16}\ \text{cm}^{-3}$$

Based on these values, Equation (1.33), and the mass action law, we can calculate the electron concentration given $N_d = 10^{15}\ \text{cm}^{-3}$ and $N_a = 10^{16}\ \text{cm}^{-3}$.

For $T = 300$ K,

$$p_0 = \left(\frac{10^{16}-10^{15}}{2}\right) + \sqrt{\left(\frac{10^{16}-10^{15}}{2}\right)^2 + (2.4\times10^{13})^2} \cong 9\times10^{15}\ \text{cm}^{-3}$$

and

$$n_0 = \frac{n_i^2}{p_0} = \frac{(2.4\times10^{13})^2}{9\times10^{15}} \cong 6.4\times10^{10}\ \text{cm}^{-3}.$$

For $T = 500$ K,

$$p_0 = \left(\frac{10^{16}-10^{15}}{2}\right) + \sqrt{\left(\frac{10^{16}-10^{15}}{2}\right)^2 + (2.03\times10^{16})^2} \cong 2.5\times10^{16}\ \text{cm}^{-3}$$

and

$$n_0 = \frac{n_i^2}{p_0} = \frac{(2.03\times10^{16})^2}{2.5\times10^{16}} \cong 1.6\times10^{16}\ \text{cm}^{-3}.$$

1.3.5 Calculation of Electron and Hole Concentration in a Degenerate Semiconductor

According to the *density of states* functions presented Section 1.1.3 and the *Fermi–Dirac* distribution function of (1.18), the calculation of electron concentration in terms of *Fermi–Dirac integral* yields the following for *bulk* 3-D, 2-D, and 1-D semiconductors. The calculation of *hole* concentration follows the same principles.

In the case of the 3-D semiconductor, using (1.14),

$$n_0 = \int_0^\infty f_D(E)D_{3D}(E)dE = N_{3D}F_{1/2}\left[\frac{E_f - E_c}{kT}\right] = N_{3D}F_{1/2}(\eta_F) \qquad (1.34)$$

where $N_{3D} = 2\left(2\pi m_n^* kT/h^2\right)^{3/2}$ and $F_{1/2}$ is the order ½ *Fermi–Dirac integral*.[38]

As presented in Section 1.3.3, for a *nondegenerate* 3-D semiconductor, this simplifies to the familiar form of

$$n_0 = N_{3D} \exp\left(\frac{E_f - E_c}{kT}\right). \tag{1.35}$$

Similar to this, in the two-dimensional case,

$$n_s = N_{2D} F_0(\eta_F)\ \text{cm}^{-3} \tag{1.36}$$

where

$$F_0(\eta_F) \equiv \frac{1}{\Gamma(1)} \int_0^\infty \frac{d\xi}{1+\exp(\xi-\eta_F)} = \frac{1}{\Gamma(1)} \int_0^\infty \frac{\exp(\eta_F-\xi)d\xi}{1+\exp(\eta_F-\xi)} = -\frac{1}{\Gamma(1)} \text{Ln}(1+\exp(\eta_F-\xi))|_0^\infty. \tag{1.37}$$

As a result, in the case where the second *subband* is much above *Fermi level*,[39]

$$n_s = N_{2D} \text{Ln}\left[1+\exp\left(\frac{E_f - E_1}{kT}\right)\right]. \tag{1.38}$$

E_1 is the bottom of the first *subband* identified in (1.8) and $N_{2D} = \left(m_n^* kT\right)/\pi\hbar^2$. Again, under *nondegenerate* conditions,

$$n_s = N_{2D} \exp\left(\frac{E_f - E_1}{kT}\right). \tag{1.39}$$

For the one-dimensional case,

$$n_L = N_{1D} F_{-1/2}(\eta_F) \tag{1.40}$$

where

$$N_{1D} = \frac{\sqrt{2m^{*kT}/_\pi}}{\hbar}. \tag{1.41}$$

[38] *Fermi–Dirac integral* of order j is defined in terms of the Γ*-function*: $F_j(\eta) \equiv 1/\Gamma(j+1) \int_0^\infty (\xi^j d\xi)/(1+\exp(\xi-\eta))$.

[39] That is, at least 3–4kT above E_f. In this case, only the first *subband* will be having a chance of having electrons.

Under *nondegenerate* conditions, with the assumption of only the first *subband,*

$$n_{\mathrm{L}}(E_{\mathrm{f}}) = N_{1\mathrm{D}} \exp\left(\frac{E_{\mathrm{f}} - E_1}{kT}\right). \tag{1.42}$$

1.3.6 Quasi-Fermi Levels

It should be kept in mind that the definitions of *Fermi–Dirac* statistics and the *Fermi level* are only possible under *thermal equilibrium*. For a semiconductor removed from *thermal equilibrium*, through the application of external excitations (such as electrical, optical, magnetic, or nonuniform thermal excitations), the definition of the *Fermi level* is not possible. However, we tend to preserve the form of the two closed-form Equations (1.27)–(1.28) for calculating the electron and *hole* concentration. This is done through proposing the following equations[40]:

$$n = n_{\mathrm{i}} \exp\left(\frac{E_{\mathrm{fn}} - E_{\mathrm{i}}}{kT}\right) \tag{1.43}$$

and

$$p = n_{\mathrm{i}} \exp\left(\frac{E_{\mathrm{i}} - E_{\mathrm{fp}}}{kT}\right) \tag{1.44}$$

where n and p are the *nonthermal-equilibrium* concentration of electrons and *holes* in the *conduction* and *valence band*, respectively. These values can either be larger than the concentrations under *thermal equilibrium* (i.e., n_0 and p_0) or smaller. E_{fn} and E_{fp} are referred to as *quasi-Fermi levels* of electrons and *holes*, respectively. Unlike the *Fermi level*, there is no physical origin to the *quasi-Fermi* levels. Often *quasi-Fermi levels* are referred to as *imref* (i.e., *Fermi* spelled backward). Evidently, under *thermal equilibrium*, *quasi-Fermi levels* will overlap one another and the *Fermi level*. The degree of deviation from *thermal equilibrium* can be quantified in terms of the difference between the *quasi-Fermi levels* of electrons and *holes*.

1.3.7 Statistics of Dopant Activation Process

Whereas at high enough temperatures all *shallow dopants* are *ionized*, in the low-to-moderate temperature range the degree of *ionization* of *dopants* requires attention. In evaluating this degree of *ionization*, the occupation status of *donor* and *acceptor states* should be studied. In such an investigation *Fermi–Dirac* distribution functions, slightly different from the one presented in (1.18), are employed. The electron distribution functions determining *ionization* status of *donor* and *acceptor states* (i.e., E_{d} and E_{a}) are expressed as

[40] Subscript 0 assigned in the previous equations to electron and *hole* concentrations is reserved for *thermal-equilibrium* situations.

$$f_{d/a} = \frac{1}{1+\beta_{d/a}\exp\left(\frac{E_{d/a}-E_f}{kT}\right)}. \tag{1.45}$$

In here β_d and β_a are often approximated by 1/2 and 2,[41] respectively (Landsberg, 1969, p. 270). Spontaneously arriving at the *thermal-equilibrium* state, which is expressed in (1.45), normally requires a considerably long period of time. This time requirement is due to the fact that in this case the *donor* and *acceptor states* should reach systemic *thermal equilibrium* with *valence* and *conduction band.*

There are important differences in the application of *Fermi–Dirac* statistics to the occupancy of the *conduction* and *valence band* and those of the *donor* and *acceptor energy levels.* In the latter cases, attention should be paid to the fact that while an *energy state* in the *conduction* or *valence band* can accommodate two electrons (i.e., considering *spin degeneracy*[42]), a *donor state* (or an *acceptor state*) can accommodate only one electron (or one *hole*), which is to be donated to the *conduction band* (or *valence band*). This single electron or *hole*, however, can take any of the two *spin* values at its ground state and also excited states. This is the fact that results in the presence of $\beta_{d/a}$ unequal to 1 in (1.45).

In *silicon* technology in the case of a *donor* atom, the *donor state* (i.e., E_d) is occupied if the fifth valence shell electron is occupying it. If *ionized*, this *donor state* will be unoccupied. What makes this process unlike the process of *EHP generation* is that this fifth electron has one of the two *spin-orientation* choices in occupying the *donor state* (i.e., one *spin-up* and one *spin-down*). However, all of the electrons in covalent bonds (i.e., in *valence band*) have such a choice. A *spin state* is associated with the presence of the electron. As a result, in the absence of an electron, the *donor state*, instead of two empty *states*, merely has one. The two *states* of opposite *spin* orientation have the same value of energy and as a result impose a twofold *degeneracy* in the *Fermi–Dirac* distribution function of electrons, which as we saw in (1.45) is represented by β_d^{-1}. In terms of this distribution function, the density of activated *donors* in a 3-D semiconductor is given by

$$n_d = \frac{N_d}{1+2\exp((E_f-E_d)/kT)} \cong N_d \quad \text{for} \quad E_d-E_f \gg kT \tag{1.46}$$

where N_d is the volume density of *donors*. Since for low-to-moderate *dopant* levels the *density of states* function of *dopant states* is representable by a *Dirac delta function*, integration over the product of the *density of states* and the distribution function of (1.45) was not required to be shown explicitly.

This discussion, with an important modification, can also be extended to the *acceptor states*. Since the *valence-band* electrons in many semiconductors are shown to have *wave functions* rooted in p-type orbitals (hence exhibiting a threefold *orbital degeneracy*), β_a is expected to go beyond the pure *spin degeneracy* (addressed by β_d).

[41] As we will see shortly, β_a in many situations is equal to 4.

[42] See the discussions in Section 1.1.2.

This threefold *degeneracy* that impacts the occupation function of *acceptor states* can be partially lifted by the *crystal field* and *spin–orbit* coupling. As a result, either a twofold *degeneracy* between the *heavy-* and *light-hole bands* or a pure *spin-degenerate valence-band edge* will evolve. For these two cases, β_a is equal to 4 or 2, respectively.[43]

As a result, without lifting the *degeneracy* between the *heavy-* and *light-hole bands*,[44] the density of *activated acceptors* is given by

$$p_a = \frac{N_a}{1 + 4\exp((E_a - E_f)/kT)} \tag{1.47}$$

in which β_a is equal to 4.

In the development of this formalism, it has been assumed that the *doping* level has been low enough so that no serious perturbation on the *band* structure of the *host* crystal is induced. As a result, the definition of the *effective charge-carrier* mass is still possible in terms of the simple parabolic *band* approximation near the *band* edges (i.e., calculated in terms of (1.13)). As we took note earlier, with increasing the *doping* level, when the *dopants* are on average separated from one another only by about 10 nm, such a formalism cannot be used. In that case, since *dopants* are not acting independent of one another, electron–electron interactions develop. Caused by *Pauli's exclusion principle* and electron exchange of energy, electrons will spread in their momenta. This spreading is to avoid the overlapping of the *wave function* of individual electrons. As mentioned earlier, this is the cause of the evolution of the *donor* and *acceptor states* into *bands* and eventual formation of *band tails* and shrinkage of the *bandgap*. This issue causes the *optical bandgap* to become larger than the *electronic bandgap*, which is mathematically defined using the *mass action* law (i.e., Equation (1.30)).

1.4 CHARGE-CARRIER TRANSPORT IN SEMICONDUCTORS

On the basis of the *Bloch* theorem, as indicated in Section 1.1, quantum mechanics predicts that in a perfect semiconductor (i.e., devoid of any *lattice vibration* and crystal defects) *charge carriers* travel as *propagating waves* within the boundaries of the 3-D semiconductor. Nevertheless, the presence of nonidealities existing in real semiconductors imposes randomizing *scattering* processes on the trajectory of *charge carriers*, rendering them to be not as free as *Bloch* theorem alludes. Many of these *scattering* processes are *inelastic*.

In electronic devices of relatively large sizes, the randomizing effect of the *scattering* processes paves the way for the description of *charge transport*[45] in terms

[43] In the case of acceptors, the *degeneracy* factor appears as g in the complementary *Fermi–Dirac* function: $\frac{1}{1+\frac{1}{g}\exp((E_f-E)/kT)}$, which is β_a in the *Fermi–Dirac* function of (1.45).

[44] Usually the *split-off band* due to its energy separation from the top of the *valence band* is not considered.

[45] From this point onward instead of speaking of *charge-carrier transport*, we will talk of *charge transport*.

of *averaging*. In order to be able to identify the length scale of the devices to which this language of *charge transport* is applicable, we will identify more quantifiable length scales in Section 1.4.2. The *transport* language used under these circumstances is *drift–diffusion transport*. In this *averaging* formalism, there are *charge carriers* that go through fewer *scattering* processes and obtain higher *carrier* velocities (and therefore energies), which are referred to as *lucky electrons* (or *holes*), and there are those that suffer from *scattering* with a higher frequency than average. As expected, over a shorter travel length, the chance of encountering *lucky electrons* grows, and as a result the *average-based drift–diffusion* formalism fails to account for the full behavior of *charge transport*. Although for modern *deep-submicron* and *nanoscale FETs* such a formalism, which is developed according to the *Drude*[46] model, is insufficient, we will still focus our treatment of *transport* in this chapter on this model. Later on in Chapter 6, we will discuss the *transport* formalism for these smaller-size devices.

As one of the reference scales of length, the concept of *mean free path* is adopted as the length over which an average electron goes through a *scattering* process, resulting in total loss of momentum. Under the condition of *charge transport* in a device of shorter length than the *mean free path*,[47] it is said that *charge carriers* travel *ballistically*. This situation is encountered in *deep-submicron FETs*. Later on in this chapter, other length scales, which are also important to the description of *charge transport*, are reviewed.

While asserting that under *thermal-equilibrium* spatial profiles of electron and *hole* concentration are time independent, it should not be imagined that *charge carriers* at temperatures above 0 K remain stationary. On the contrary, as invoked earlier, according to the laws of statistical mechanics at a temperature T (in Kelvin) per degree of freedom, an average particle possesses an amount of energy equal to $1/2kT$. As a result, in a three-dimensional semiconductor, the average semi-Newtonian charged particle (i.e., with *effective mass* of m^*) has a finite *thermal velocity* (i.e., v_{th}), which can be calculated through the following equation:

$$\frac{1}{2}m^* v_{\text{th}}^2 = \frac{3}{2}kT. \tag{1.48}$$

This random *thermal velocity*, however, results in a *zero net velocity* for the total population of *charge carriers*. It should be mentioned that the above equation, which is based on semiclassical quantum mechanics, is only an approximation and it suffers from improper averaging.

In the presence of an external source of energy (such as an electric field) or a *nonzero* gradient in *charge-carrier* concentration, this random *thermal velocity* can be slightly biased toward a certain trajectory to produce a *nonzero net velocity* of *charge carriers*. This results in electrical current. While (1.48) predicts a random *thermal velocity* in the order of 10^7 cm/s at room temperature, the *net velocity* would be much smaller than this. Therefore, only a semiclassical description of quantum mechanics

[46] Named after *Paul Drude*.

[47] In Section 1.4.2 we will present this condition on a more accurate ground.

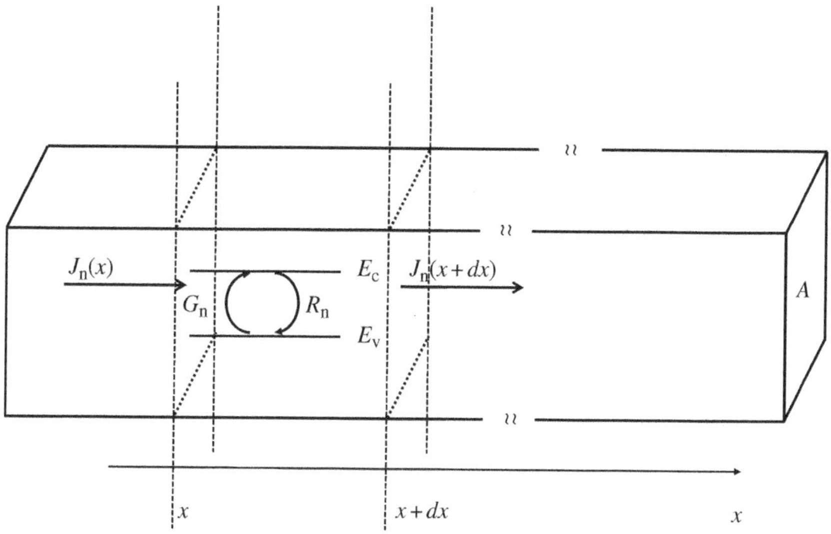

FIGURE 1.17 Schematic depiction of current continuity across a slab of semiconductor.

(certainly without invoking *relativistic mass*) needs to be considered in the formulation of *charge transport*.[48]

1.4.1 Current-Continuity Equation

A capable tool in the description of the *net* movement of *charge carriers* in a relatively large-size semiconductor is the *current-continuity equation*. According to Figure 1.17, the *charge-carrier* concentrations in a semiconductor's *conduction* and *valence bands* evolve with time as functions of *carrier transport* and *generation/recombination* rates of *carriers* (i.e., *G* and *R* in terms of the number of *carriers* per unit time, per unit volume). According to this one-dimensional depiction,

$$\frac{\partial n}{\partial t}A\cdot dx=[F_n(x)-F_n(x+dx)]A+[G_n-R_n]A\cdot dx \tag{1.49}$$

$$\frac{\partial p}{\partial t}A.dx=\left[F_p(x)-F_p(x+dx)\right]A+\left[G_p-R_p\right]A.dx. \tag{1.50}$$

In these equations, $F_{n/p}(x)$ stand for the flow rates of *charge carriers* normal to the cross-sectional area *A*. Since electrons and *holes* are not always created (or annihilated) in pairs,[49] in the above equations (with indication of the appropriate

[48] In semiconductors, due to *Bragg diffraction* and evolution of negative mass, velocities much higher than 10^8 cm/s do not usually transpire.

[49] That is, through *direct band-to-band* processes.

subscripts[50]), the rates of *generation/recombination* of electrons and *holes* are differentiated from one another.

If the flow rates were to change almost linearly (i.e., expressed by the first-order *Taylor* series) between the two ends of the elemental volume $A.dx$, $F_{n/p}(x) - F_{n/p}(x+dx) \cong -\frac{\partial F_{n/p}}{\partial x}dx$. As a result,

$$\frac{\partial n}{\partial t} \cong -\frac{\partial F_n(x)}{\partial x} + [G_n - R_n] \tag{1.51}$$

$$\frac{\partial p}{\partial t} \cong -\frac{\partial F_p(x)}{\partial x} + [G_p - R_p]. \tag{1.52}$$

While in terms of electron and *hole* current densities (i.e., J_n and J_p, in units of A/cm^2), $F_n = J_n/-q$ and $F_p = J_p/q$,[51] we have

$$\frac{\partial n}{\partial t} \cong \frac{1}{q}\frac{\partial J_n(x)}{\partial x} + [G_n - R_n] \tag{1.53}$$

$$\frac{\partial p}{\partial t} \cong -\frac{1}{q}\frac{\partial J_p(x)}{\partial x} + [G_p - R_p]. \tag{1.54}$$

According to the *continuity equation*, in the absence of a considerable *generation/recombination* rate, the time evolution of *charge-carrier* concentration will be purely explainable in terms of the spatial gradient of the electrical current density. The time constant determined in terms of the *charge-transport* processes for removing any amount of *excess charge* concentration in the semiconductor (i.e., beyond charge neutrality) is referred to as the *dielectric relaxation time*. *Dielectric relaxation time* is one of the important metrics in evaluation of *charge-transport* properties of a semiconductor.

1.4.2 Drift–Diffusion Formalism

We shall start our description of the *drift–diffusion* formalism with *drift transport* and also address the limitations of this *carrier-transport* formalism. This mode of *carrier transport*, due to its similarity with *electron transport* through conductors (which is normally expressed in terms of *Ohm*'s law), is relatively easier to understand for a reader not so familiar with semiconductors. According to *Ohm*'s law, for a rectangular slab of semiconductor with homogeneous concentration of *charge carriers*, *conductance* (G in *Siemens*) can be written in terms of its width (W), length (L), thickness (T), and the important material property known as *conductivity* (σ in 1/Ω cm):

$$G = \sigma\frac{W \cdot T}{L}. \tag{1.55}$$

[50] That is, n for *electrons* and p for *holes*.

[51] q is the absolute value of the charge of an electron.

As will be further explored in Chapter 6, with the reduction of the length scale of this slab on the order of a few *mean free paths* at normal temperatures of operation (i.e., into *deep submicrometers*), such an equation will not hold. One recent development in physics of semiconductor devices has taken place regarding determining these minimum dimensions. Physical realization of *deep-submicron* devices (i.e., of dimensions within the range or smaller than the *mean free path*) has led to significant progress in understanding the meaning of *resistance* at *nano-* and *molecular scales* and also the notion of *contact resistance* to these small-scale entities.[52]

With respect to this concept, a relatively new area has evolved in semiconductor electronics, which is referred to as *mesoscopic* electronics. In *mesoscopic* devices, dimensions are intermediate between the atomic scales and scales over which the *Ohmic* behavior of *drift transport* prevails. Although so far we have only talked about this length scale relative to *mean free path*, it is worth elaborating more on the matter. For a conductor to exhibit an *Ohmic* behavior, the dimensions ought to be much larger than each of the following three characteristic length scales (and not just one of them):

- The *de Broglie* wavelength of the electron
- The *mean free path*
- The *phase-relaxation length*

Whereas we are already acquainted with the first two, the last of these length scales indicates the distance that a *wave electron* travels before its initial *phase* information is lost.

These characteristic length scales are dependent on material, processing, temperature, and also external forces. As a result of the shortening of these characteristic length scales with increase in temperature and *bias*, a *mesoscopic* behavior is further encountered at lower temperatures and under lower *biases*. At room temperature, this characteristic is often washed out by the many existing *scattering* processes.

Caused by the tremendous push to realize electronic devices of ever-smaller dimensions, it is worth paying attention to the underlying physics of these length characteristics. This qualitative description paves the way to better understand three important (and often confused or overlooked) time constants in evaluation of the behavior of electronic devices: *momentum relaxation time*, *phase relaxation time*, and *energy relaxation time*.

Not all collisions and interactions that electrons go through are *inelastic*. As a result, *momentum relaxation time* (referred to as τ_m) in terms of collision time constant (i.e., τ_c) can be given by

$$\frac{1}{\tau_m} = \frac{1}{\tau_c}\alpha_m, \tag{1.56}$$

[52] Measuring semiconductor properties requires creating a contact between the *macroscale* world and nano- or micron-scaled semiconductor devices. Starting from Chapter 2, through dealing with the concept of contact, we will address this issue more in-depth.

where α_m is a constant varying between 0 and 1, which indicates the effectiveness of the collision for destruction of momentum. For example, collisions resulting in small-angle *scattering* have very little impact on erasing the momentum information.

In order to get a more hands-on feeling about the length scales over which *mesoscopic transport* prevails, let's consider the case of low-temperature conduction through a 2-D populated channel, where the conductance is entirely determined by the electrons with energies close to the *Fermi level*,[53]

$$v_f = \frac{\hbar k_f}{m^*} = \frac{\hbar}{m^*}\sqrt{2\pi n}. \tag{1.57}$$

In this example, on the basis of (1.56), the *mean free path* is calculated in terms of *Fermi velocity* (i.e., v_f) as[54]

$$L_m = v_f . \tau_m. \tag{1.58}$$

For an electron concentration on the order of 10^{11} cm^{-2}, *Fermi velocity* will be about 10^7 cm/s. As a result, for a *momentum relaxation time* of 100 ps, *mean free path* will amount to about 30 μm. Considering the size of modern semiconductor devices, this is extremely large.[55] As a result, as suggested earlier, application of equations such as (1.55) will not be extendible to devices operating under these regimes, even when the dimensions of the devices are still on micron scale.

In order to magnify the differences in nature of the three aforementioned length scales, it is worth paying attention to the special case of electron–electron *scattering*.[56] Among the many *scattering* events, electron–electron scattering is a *scattering* process that does not impact the *mean free path*. This is due to the fact that in such processes, no *net* loss in momentum of electrons occurs. In this *average* sense, a momentum lost from one electron is a momentum gained for the other. However, unlike the efficiency factor of this process (i.e., α_m), that of *phase relaxation time* is not equal to 0. *Phase relaxation* in general is the byproduct of *scattering* by fluctuating *scatterers*.

Whereas rigid sources of *scattering* such as impurities in general do not contribute to the *phase relaxation* process, magnetic impurities due to their internal degree of freedom (i.e., their time-fluctuating *spin*) serve as *phase randomizers* while *scattering* electrons.

[53] This is due to the fact that under low temperatures, the *Fermi–Dirac* function of (1.18) turns into a *step function*.

[54] Which is the velocity of electrons in *states* matched to *Fermi level*.

[55] In (1.57) electrons are assumed to be in a 2-D channel, which is almost like that of *MOSFETs* to be seen in Chapter 3. On the basis of assumptions of low temperature and occupation of only one *subband* in this channel, Equation (1.15) yields $n = \left(m^*/\pi\hbar^2\right)(E_f - E_1)$. Using the semiclassical definition of energy as $E = \frac{1}{2m}(mv)^2 = \frac{1}{2m}(\hbar k)^2$, this yields the *Fermi wave number* $k_f = \sqrt{2\pi n}$.

[56] In Section 1.7, we will deal with *scattering* mechanisms at a greater length.

In normal temperatures of operation, depending on the semiconductor's crystal structure and electron energy, an electron can propagate without perturbation over an *inelastic mean free path* in the range of 3×10^{-7} to 10^{-5} cm. Because these distances span many *lattice constants*,[57] electrons still see the crystal as a *phase-coherent* quantum mechanical entity. This explains why, between the *scattering* events, *perturbation theory* allows us to see the electrons in light of *Bloch states*, which are defined by the periodic crystal. However, after this unperturbed propagation, electrons, according to *Fermi golden rule*, *scatter* to other *Bloch states* and go through *dephasing*. As a result of this discussion, as indicated in the opening of Section 1.4, on geometric scales much larger than the *inelastic mean free path*, electrons are more or less seen as classical particles.

Evaluation of the aforementioned time constants and modes of *scattering* is extremely important in the formulation of the *Boltzmann transport equation* (*BTE*), of which the *drift–diffusion* formalism dealt with in this chapter is only a special case. For the *BTE* one needs to identify a distribution functional for electrons in terms of time, position, and momentum. Considering the constraint of the *uncertainty principle*, such a definition would be possible only if the geometrical flight length allows the application of the aforementioned semiclassical framework.

According to the earlier discussions and as a result of the *uncertainty principle*, the allowable value of *uncertainty* of the electron *wave vector* will be about 10^5 cm^{-1}, which is matched to the geometrical *uncertainty* in the range of 10^{-5} cm. This degree of accuracy in determining *wave vector* $\vec{k}$ is often deemed sufficient, since the *Brillouin zone*[58] boundary is about 10^8 cm^{-1}. In compliance with these conditions, a distribution functional $f(\vec{k}, \vec{r}, t)$ representing the probability of finding an electron having a *wave vector* between $\vec{k}$ and $\vec{k} + d\vec{k}$ and located in a space between $\vec{r}$ and $\vec{r} + d\vec{r}$ can be identified.

1.4.2.1 Drift Transport So far in our discussions, we have encountered three velocity terms:

- Random *thermal velocity* (which is of *zero net velocity*)
- *Drift velocity* or *net velocity* (which is a slight bias imposed on random *thermal velocity* through application of an external sources of energy)
- *Fermi velocity*

Our aim is now to see why in Equation (1.57) *Fermi velocity*, and not the other velocities, was used in determining the characteristic length scale L_m.

[57] The long-range order in a crystalline material indicates the presence of *unit cells*, which are repeated in the structure. As will be elaborated in Section 1.6, *lattice constant* is an indicatory dimension of these *unit cells*.
[58] Using *Schrödinger equation*, a *real-space* crystal structure can be transformed into *momentum space*. While in the *real space* periodicity of crystal presents itself through the notion of repetition of *unit cells*, in the *momentum space* also the information is repeated in the form of repetition of a building block, which is known as *Brillouin zone*. This is named after *Léon Brillouin*.

In a large homogeneous semiconductor, *carrier transport* is explained in terms of the *drift velocity* (i.e., v_d), which yields the electron *drift* current density as

$$J_n = qnv_{dn}. \tag{1.59}$$

The *drift hole* current density can also be expressed through an equation identical in form to (1.59). However, instead of electron concentration (i.e., n), *hole* concentration (i.e., p), and instead of *drift velocity* of electrons (i.e., v_{dn}), drift velocity of *holes* (i.e., v_{dp}) should be used. In the calculation of current under conditions very close to *thermal equilibrium*, we can still approximately rely on *Fermi–Dirac* statistics. However, for further deviations from *thermal equilibrium*, another distribution functional should be used in the current calculation.

In this equation, through incorporation of electron concentration n, it is implied that all *conduction-band* electrons are contributing to current conduction. However, in the case of conductors, or in *degenerate* semiconductors,[59] such a framework becomes misleading. In these cases, application of an external electric field does not induce a net velocity (i.e., v_d) in all conduction electrons. With the use of *energy-resolved* measurements, it has been shown that only those electrons with energy values within a few kT of the electron *quasi-Fermi level* are carriers of the *net* current. This occurs while the rest of the *conduction-band* electron population maintains at its *zero net velocity*. This finding results in a tremendous simplification in studying the behavior of these *solids* at very low temperatures, since only a very small concentration of conduction electrons are now required to be studied.

This fact highlights the difference between the impact of application of electric field on electrons as *individual* single particles or as an *ensemble* of electrons. From the point of view of an *ensemble* of electrons, the application of an electric field results in moving only a few electrons from *states* with negative *wave number* (i.e., $-k_f$) to *states* with positive *wave number* (i.e., $+k_f$). This is schematically illustrated in Figure 1.18. Accordingly, the *drift transport* equation in this case can be rephrased as

$$J_n = q\left[n\frac{v_{dn}}{v_f}\right]v_f, \tag{1.60}$$

where the quantity in the square brackets is representative of a small fraction of total concentration of electrons with energies within a few kT of *quasi-Fermi level*. These are the electrons that move with *Fermi velocity*. Although from a purely quantitative point of view this equation implies no change in the value of *drift* current density, it conceptually indicates why it is required to talk about *Fermi velocity* in metals as well as in low-temperature/low-dimensional *degenerate* semiconductor systems. As is shown in (1.60), in *degenerate* systems under low temperature, conduction takes place

[59] Which, as indicated earlier in this chapter, refers to the cases where *Fermi level* is positioned above the *conduction-band* edge for a three-dimensional *bulk* semiconductor or above the first *subband level* for the case of a lower-dimensional semiconductor $k_B T \ll E_f - E_1$, which materializes specially at lower temperatures. Many of the *FETs* have *degenerate* channels for current conduction.

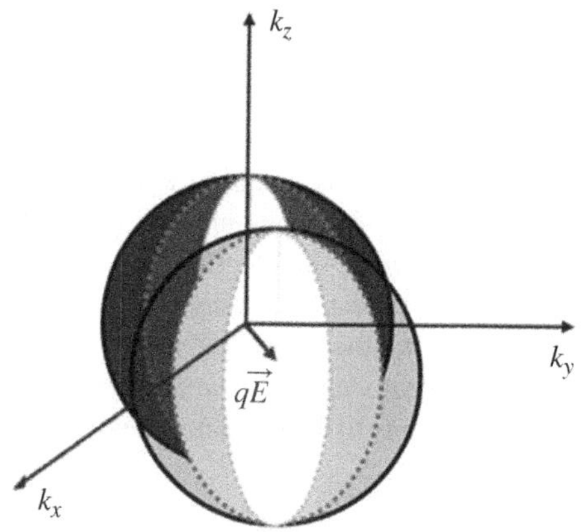

FIGURE 1.18 While prior to the application of an electric field all states within a sphere of radius k_f (i.e., Fermi wave number) are filled at low temperatures, upon application of an electric field $\vec{E}$, this sphere shifts according to the vector $q\vec{E}$. As a result, instead of symmetric filling of the band, states that carry the current along the electric-field vector are filled up to a higher energy. This elimination of symmetry between the states (k_x, k_y, k_z) and ($-k_x$, $-k_y$, $-k_z$) results in current conduction in the range of energies for which the two spheres are nonoverlapping.

through the movement of a small number of electrons at velocities much higher than indicated by the *drift velocity*.

However, aside from this *ballistic transport* in low-dimensional systems and low-temperature transport behavior of *degenerate* systems, transport under low electric fields can be easily described in terms of the *drift* (and to be added in Section 1.4.2.2 *diffusion*) formalism.

As expected, upon application of an electric field, the *net* movement of *charge carriers* is materialized either along the direction of electric field (for *holes*) or opposite of that (for electrons). In *macroscopic* electronic devices, an instantaneous relationship is proposed between the applied electric field and this *net velocity* (i.e., *drift velocity*: v_{dn} and v_{dp}):

$$v_{\mathrm{dn}} = -\mu_{\mathrm{n}} E \tag{1.61}$$

$$v_{\mathrm{dp}} = \mu_{\mathrm{p}} E. \tag{1.62}$$

The proportionality constants of these relationships are referred to as *low-field* electron and *hole mobility* (i.e., μ_{n} and μ_{p} in cm^2/V s, respectively). As will be shown in Section 1.7, these constants are dependent on the frequency of a large number of *scattering* processes that a *carrier* encounters when moving in a semiconductor. As a result, they depend on a long list of parameters including temperature, presence of impurities, presence of internal *polarization* in the semiconductor, etc.

Equations (1.61) and (1.62) certainly have their own limitations. Assumption of such instantaneous relationships even in portions of large-size semiconductor devices (i.e., in regions of rapidly changing electric field) has been proven to be misleading. In these relationships, it is assumed that an average semi-Newtonian electron,[60] while traveling through a piece of semiconductor across which an electric field (i.e., E) has been established, will at any given point experience a force equal to $-qE$. As a result of this force, assuming a constant electric field between any two *inelastic scattering* events (which are separated from one another by an amount of time typically referred to as *mean scattering time*), the electron will receive an amount of energy equal to $qE\lambda$ over the length of one *mean free path* (i.e., λ). If we assume that on average after a time period equal to the so-called mean scattering time (i.e., τ_{cn}, for electrons),[61] an electron undergoes an *inelastic scattering* process (and sees a drop in its field-contributed *drift velocity* to 0), we can write the following Newtonian equation for this particle:

$$qE\tau_{cn} = -m_n^* v_{dn} \tag{1.63}$$

in which m_n^* is the effective mass of electron.

As a result, based on (1.61),

$$\mu_n = \frac{q\tau_{cn}}{m_n^*}. \tag{1.64}$$

Likewise, we can calculate the *low-field hole mobility* as

$$\mu_p = \frac{q\tau_{cp}}{m_p^*}, \tag{1.65}$$

where τ_{cp} and m_p^* are the *mean scattering time* and *effective mass* of *holes*, respectively.

These approximations, in circumstances dealing with large-size electronic devices and when the applied electric field is small, hold well. Nevertheless, at higher electric fields since electrons (and *holes*) move away from the bottom of *conduction band* (and *valence band*), the *effective-mass* values calculated at the *band* boundaries will not be applicable. It is worth noting that in formalizing *drift transport*, if electric field were large, higher powers of E would be required in explaining the *transport*. This is because high electric field appreciably disturbs the equilibrium distribution of velocities of *carriers* (i.e., profile expressed in (1.21)). In addition, increase in energy will not only change the *scattering* rates, and as a result the value of *mean scattering time*, but will also trigger a number of new *scattering* processes. In Section 1.7.3, we will perform a fuller evaluation of a variety of *scattering* processes.

[60] While this discussion is being presented for electrons, in case of *holes*, evidently with the appropriate change of signs, appropriate equations are developed.

[61] It should be emphasized that τ_{cn} (also its counterpart in case of *holes*: τ_{cp}) is given by the *momentum relaxation time* and not the collision time constant in (1.56).

In a simple review of this problem, we see that we have thus far made a good number of assumptions, including:

- The electron, which is a *wave/particle dual*, is a Newtonian particle.
- *Transport* is one-dimensional.
- The electric field is constant between any two *inelastic scattering* events.
- An *inelastic scattering* process happens every τ_{cn} (i.e., *mean scattering time* for electrons) and τ_{cp} (i.e., *mean scattering time* for *holes*).

These assumptions prevent us from developing a thorough understanding of the behavior of the quantum *wave/particle charge carrier.* The processes of receiving (or dissipating) energy of these *wave/particles* are not instantaneous. These processes are instead governed by a *relaxation time constant,* referred to as *energy relaxation time.*

Prompted by the difference in the energy of *phonons* and electrons, the *momentum relaxation time* is often shorter than the *energy relaxation time*. This is because the thermalization of electrons requires a number of inelastic *scattering* events, considering the large electron energy (i.e., electron–*phonon* energy exchange). However, in contrast, even with one *scattering* event, the momentum of the *charge carrier* can be randomized.

As mentioned before, the *low-field mobility* is dependent on the frequency of electron interaction with *scattering* processes. Temperature and presence of *ionized impurities* are two prime contributors to *scattering* processes, on which the *low-field mobility* depends. These contributions result in *ionized-impurity scattering* and *lattice-vibration scattering* (also known as *phonon scattering*). At low temperatures, due to insignificant *lattice vibrations*, *ionized-impurity scattering* is dominant. At higher temperatures, *phonon scattering* plays the dominant role.

As a result, the temperature dependence of the *low-field mobility* takes two completely different characteristics at the temperature extremes. At high temperatures, because of the strengthening of the dominant *scattering* process (i.e., *phonon scattering*), the frequency of *scattering* events increases, and as a result *carrier mobility* deteriorates. At lower temperatures, an increase in temperature results in a *mobility* improvement. This improvement in *mobility* is due to the increased *thermal velocity* of *carriers* and, as a result, reduction in the time spent by them in the vicinity of the dominant *scatterers* (i.e., *ionized impurities*). In semiconductors free of *ionized impurities* (i.e., in the absence of *ionized-impurity scattering*), at lower temperatures *mobility* becomes temperature independent. Whereas depending on the curvature of the *E–k* diagram and the value of mean *scattering time*, the values of electron and *hole mobility* of different semiconductors are different, at higher temperatures the *low-field mobility* becomes almost independent of the semiconductor.

With increase in *dopant* concentration and when *ionized-impurity scattering* is dominant, *carrier mobility* deteriorates. Figure 1.19 schematically illustrates these temperature- and *doping*-dependent variations of *carrier mobility.*

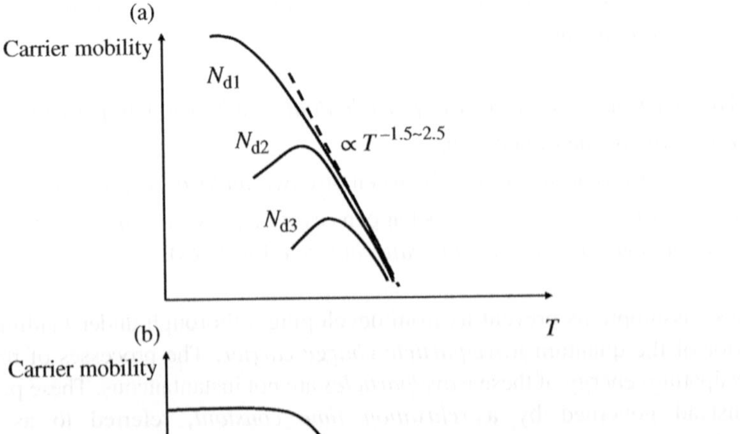

FIGURE 1.19 (a) Typical dependence of carrier mobility on lattice temperature for a semiconductor doped to three different levels of doping: N_{d1}, lightly doped; N_{d2}, moderately doped; N_{d3}, heavily doped. (b) Typical dependence of carrier mobility on the ionized-impurity concentration. Scales are linear.

1.4.2.2 Diffusion Transport Under the aforementioned circumstances, which were identified as suitable for the application of *drift-transport* formalism, the second mode of conduction is *diffusion*. *Diffusion* is a thermal process that, unless there is a gradient in the concentration of *carriers*, produces a *zero net* movement of *charge carriers*. This is because the average *carrier*, although moving with *thermal velocity*, has a random *zero-sum* movement. In the presence of a gradient in *charge-carrier* concentration, however, *carriers* will diffuse from regions of higher concentration to regions of lower concentration.

According to the one-dimensional situation depicted in Figure 1.20, with a few assumptions a simple equation can be produced for this *net* movement of charge in the presence of a *carrier* concentration gradient (while electric field is 0). In this one-dimensional situation, *carriers* at any given point have a 50% chance of traveling toward the right or the left side. Taking two points to the right and to the left of a given reference point, distanced from the reference point by the *mean free path*, one can assume that an average *carrier* starting at each of these two points and coming to the reference point will not go through any *inelastic scattering* event before reaching $x=0$. As a result, in the case of electrons, the net flux of *carriers* at $x=0$ will be equal to

$$F=\frac{1}{2}v_{\text{th}}n(x-\lambda)-\frac{1}{2}v_{\text{th}}n(x+\lambda). \tag{1.66}$$

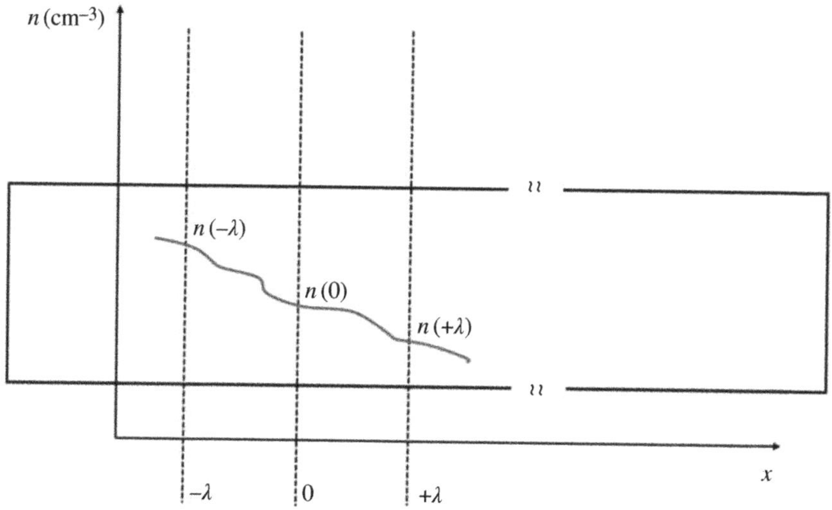

FIGURE 1.20 Schematic depiction of an arbitrary 1-D dependent carrier concentration profile (i.e., along the x-axis) on the cross section of a slab of semiconductor.

Assuming that the electron concentration gradient is small enough so that $n(x)$ can be approximated by the first-order *Taylor* series, we can write

$$F \cong -\lambda v_{\text{th}} \frac{dn}{dx}. \tag{1.67}$$

Since for electrons *diffusion current* and flow rate of electrons can be related as $J_{\text{Diff-n}} = -qF$, the *diffusion*-current density can be written as

$$J_{\text{Diff-n}} \cong q\lambda v_{\text{th}} \frac{dn}{dx}. \tag{1.68}$$

The factor λv_{th} is referred to as the *diffusion constant* of electrons (D_{n} in units of cm^2/s).

For the case of holes, with appropriate modifications, the following equation is developed:

$$J_{\text{Diff-p}} \cong -q\lambda v_{\text{th}} \frac{dp}{dx}. \tag{1.69}$$

1.4.2.3 Einstein Relationship It is important to understand that the constants of *drift* and *diffusion* in terms of current are not independent of one another. At conditions close to *thermal equilibrium* (i.e., low E) and for *nondegenerate* semiconductors, a simple derivation will result in the famous *Einstein relationship*[62] between these constants.[63]

According to our earlier discussions,

$$D_{\mathrm{n}} = \lambda v_{\mathrm{th}} = v_{\mathrm{th}}^2 \tau_{\mathrm{cn}} = v_{\mathrm{th}}^2 \frac{\mu_{\mathrm{n}} m_{\mathrm{n}}^*}{q}. \tag{1.70}$$

For a one-dimensional semiconductor, with no electric field, according to the *equipartition* theorem,

$$\frac{1}{2} m_{\mathrm{n}}^* v_{\mathrm{th}}^2 = \frac{1}{2} kT. \tag{1.71}$$

As a result, based on (1.70) and (1.71),

$$D_{\mathrm{n}} = \mu_{\mathrm{n}} \frac{kT}{q}. \tag{1.72}$$

An identical relationship can also be established between the *diffusion constant* of *holes* and their *low-field mobility*. As a result, between the four transport factors of *drift–diffusion* formalism, we have

$$\frac{D_{\mathrm{n}}}{\mu_{\mathrm{n}}} = \frac{D_{\mathrm{p}}}{\mu_{\mathrm{p}}} = \frac{kT}{q}, \tag{1.73}$$

which, as indicated already, is referred to as *Einstein relationship*.

This version of *Einstein relationship* is only valid for a *nondegenerate* semiconductor (i.e., the *Maxwellian* semiconductor expressed in (1.21)). However, as proven in Appendix 1.B, for a *degenerate* semiconductor at low electric fields, in terms of *Dirac integrals*, we will have

$$\frac{D}{\mu} = \frac{kT}{q}\left[2F_{3/2}(\eta)/3F_{1/2}(\eta)\right], \tag{1.74}$$

where

$$F_j(\eta) \equiv \int_0^{\infty} \frac{\xi^j d\xi}{1+\exp(\xi-\eta)} \quad \text{and} \quad \eta = \frac{E_{\mathrm{f}} - E_{\mathrm{c}}}{kT}. \tag{1.75}$$

Notice that $\mathrm{F}_j(\eta)$, which we saw in Section 1.3.5, is the scaled version of $F_j(\eta)$ by $\Gamma(j+1)$.

[62] Named after *Albert Einstein.*

[63] In Appendix 1.B, we will show how a more sophisticated version of this relationship is also valid without these conditions.

According to the *drift–diffusion* formalism, the total electron and *hole* current density terms can be written as

$$J_n = qn\mu_n E + qD_n \frac{dn}{dx} \tag{1.76}$$

$$J_p = qp\mu_p E - qD_p \frac{dp}{dx}. \tag{1.77}$$

Through applying the definitions of *quasi-Fermi levels* to a *nondegenerate* semiconductor, we can combine the *drift-* and *diffusion*-current terms in one equation:

$$J_n = n\mu_n \frac{dE_{fn}}{dx}. \tag{1.78}$$

Likewise we can write

$$J_p = p\mu_p \frac{dE_{fp}}{dx}. \tag{1.79}$$

The *diffusion* process, in addition to being instigated by the presence of a finite gradient in *dopant* and *charge-carrier* concentration, can also be provoked by the existence of a temperature gradient. This can be clearly seen through the temperature dependence of the *quasi-Fermi levels* presented in Equations (1.43) and (1.44).

With explicit indication of transport due to a temperature gradient, *drift–diffusion* transport of electrons can be formulated as

$$J = nq\mu E + qD\nabla n + qS\nabla T, \tag{1.80}$$

where S is referred to as *Soret coefficient.*[64]

Although we have already provided the *drift–diffusion* formalism in a unified form, it is worth emphasizing again that, especially due to the restrictions of *Pauli's exclusion principle*, for *degenerate* semiconductors, one cannot assume that *drift* and *diffusion* are independent processes.

Example

Along the length of a slab of *n-doped germanium* with a peak *dopant* concentration of 10^{15} cm^{-3}, a constant electric field of 100 V/cm is applied. The length of the slab is 1 μm. Assuming that at room temperature along this sample a current of 10 A/cm^2 is flowing in the direction of the electric field, determine the spatial distribution profile of the *dopants*. Room-temperature values of electron and *hole mobility* everywhere along the length of this sample are taken equal to 1000 and 800 cm^2/V s, respectively.

[64] *Soret* coefficient is defined as $S_n = \mu_n (k/q) n$, and it is named after *Charles Soret*.

Since the sample is *n-doped*, we start with assuming that the current is being carried by electrons only. Taking the electric field to be in the $+x$ direction, in terms of (1.76),

$$J_n = qn\mu_n E + qD_n \frac{\partial n}{\partial x} = +10 \text{ A/cm}^2.$$

Since $\mu_n = 1000 \text{ cm}^2/\text{V s}$, according to *mass action law*,

$$\frac{D_n}{\mu_n} = \frac{kT}{q} \Rightarrow D_n \cong 25.8 \text{ cm}^2/\text{s}.$$

Based on these values,

$$1.6 \times 10^{-19} \left[n(x) \times 1000 \times 100 + 25.8 \frac{\partial n}{\partial x} \right] = 10,$$

which yields

$$n(x) = A.\exp(-3.88 \times 10^3 x) + 6.25 \times 10^{14} \text{ cm}^{-3}.$$

We assume first that $N_d(x) \cong n(x)$ and then evaluate the validity of this assumption.

Since the peak of the *donor* concentration is 10^{15} cm^{-3}, $n(x = 0)$ is set to this value. Accordingly,

$$N_d(x) = 3.75 \times 10^{14}.\exp(-3.88 \times 10^3 x) + 6.25 \times 10^{14} \text{ cm}^{-3}.$$

Based on this distribution profile, along the length of the sample stretching from $x = 0$ to $x = 10^{-4}$ cm, *dopant* concentration changes from 10^{15} to 8.79×10^{14} cm^{-3}.

Within this range everywhere N_d is at least an order of magnitude greater than $n_i(T = 300 \text{ K}) = 2.4 \times 10^{13}$ cm^{-3}, which validates our earlier assumption of $N_d(x) \cong n(x)$.

We still have to validate another assumption (i.e., if electrons are the predominant charge carriers). Based on the electron concentration profile, the *hole* concentration profile is evaluated as

$$p(x) = \frac{(2.4 \times 10^{13})^2}{3.75 \times 10^{14}.\exp(-3.88 \times 10^3 x) + 6.25 \times 10^{14}}.$$

According to which,

$$J_p = qp\mu_p E - qD_p \frac{\partial p}{\partial x},$$

where $\mu_p = 800 \text{ cm}^2/\text{V s}$ and $D_p = 800 \times 25.8 \times 10^{-3} \cong 20.64 \text{ cm}^2/\text{s}$.

Based on these, since both $n \ll p$ and $(\partial p/\partial x) \ll (\partial n/\partial x)$, J_{p} would be much smaller than J_{n}, which sets electrons as the predominant charge carriers.

1.4.3 Characterization of Low Electric-Field Transport Parameters

Hall measurement is one of the most capable tools in characterizing semiconductors. This tool can be used to not only measure the *carrier mobility* in a semiconductor but also to reveal the *type* of its majority *carriers* and their concentration. Figure 1.21 illustrates a *Hall* measurement setup. In this setup in presence of both an electric field (which is induced due to the application of *bias V*) and the permanent magnetic field, while *charge carriers* are traversing along, or opposite to, the direction of the electric field (i.e., for the *holes* and electrons, respectively), they are deflected to the same transverse side of the semiconducting slab. In this deflection, charge neutrality will be imbalanced, which results in induction of a second electric field normal to the first. The balancing act between the *Lorentz* force imposed on the moving *charge carriers* by the magnetic field and the force exerted by this secondary electric field (which is referred to as *Hall* field) renders an equilibrium state. Through measuring the polarity and the magnitude of V_{H}, one can gain insight into the majority *carrier type*, *mobility*, and concentration. For example, for an n-type sample the balance of forces yields

$$E_{\mathrm{H}} = -\frac{J_{\mathrm{n}}}{qn}B. \tag{1.81}$$

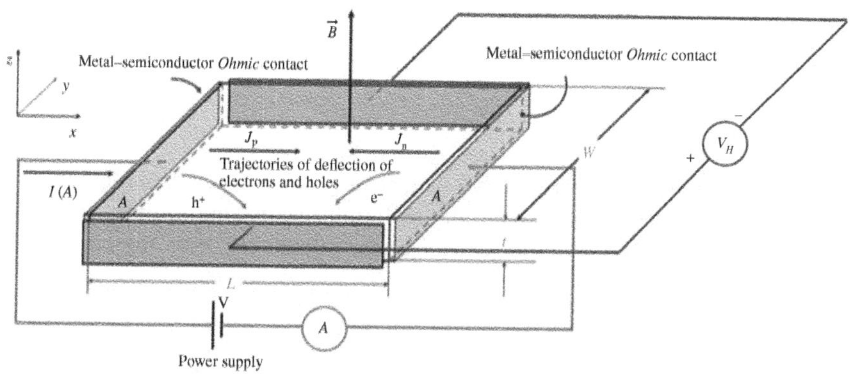

FIGURE 1.21 Schematic depiction of a simple Hall measurement setup. A uniform magnetic field with flux density $\vec{B}$ is applied perpendicular to the x–y plane. According to the polarity of the connections to the power supply, a current in the positive x-direction is fed normal to the y–z plane of the slab with contact area indicated as $W \times t$. Based on the indicated directions of deflection for electrons and holes, in the case of p- and n-doped semiconductor, a positive or a negative Hall potential (i.e., V_{H}) will be observed on the voltmeter, respectively. Accordingly, the direction of the resulting Hall electric field E_{H} defined along the y-axis will depend on whether the semiconductor is n- or p-type doped.

Likewise, in a p-type sample we have

$$E_H = \frac{J_p}{qp} B. \tag{1.82}$$

Clearly V_H, which is equal to $E_H \cdot W$, is positive in a p-type sample and negative in an n-type sample. E_H is often expressed as $R_H \cdot J \cdot B$.

In the analysis of *Hall* measurement data, we have to pay attention to the inherent assumptions of *drift-current transport*. According to this first-order approximation expressed in (1.81) and (1.82), we have assumed that all *carriers* have the same *mean scattering time*. In reality, however, a proper averaging of *mean scattering time* must be adopted. Formal analysis by the *Boltzmann transport equation*, in the case of electrons, results in

$$\mu = \frac{q}{m^*} \cdot \frac{\langle v^2 \tau \rangle}{\langle v^2 \rangle} \tag{1.83}$$

$$R_H = -\frac{1}{qn} \cdot \frac{\langle v^2 \tau^2 \rangle \langle v^2 \rangle}{\langle v^2 \tau \rangle^2}. \tag{1.84}$$

The symbol <> represents the averaging process over the *Boltzmann* distribution of *carriers*. As a result, the product of R_H and *conductivity* (i.e., σ in units of $\Omega^{-1}\ \mathrm{cm}^{-1}$) is different from the low-field *mobility* (i.e., *drift mobility*) that we have talked about in this section. This product is referred to as *Hall mobility*.

As a result of these relationships, the ratio of *Hall mobility* (i.e., μ_H) versus *drift mobility* (i.e., μ) can be written as

$$\frac{\mu_H}{\mu} = \frac{R_H . \sigma}{\mu} = \frac{\langle v^2 \tau^2 \rangle \langle v^2 \rangle}{\langle v^2 \tau \rangle^2}. \tag{1.85}$$

This ratio, although not equal to 1, is generally close to 1. Nonuniformity in the distribution of current, temperature nonuniformity across the sample, and lack of an ideal *Ohmic* contact[65] to the semiconductor sample are often the other sources of ambiguity in *Hall* measurement data.

1.4.4 High Electric-Field Drift Transport

As identified earlier, under high electric fields the semilinear relationship between the drift velocity and the electric field vanishes (i.e., (1.61) and (1.62)). Throughout

[65] As shown in Figure 1.21 and more clearly explored in Chapter 2, a metal–semiconductor contact between the outside world (in this case power supply) and the semiconductor is required. The main characters of such a contact are that it presents very little potential drop at the contact-site and behaviorally it follows *Ohm*'s law (hence the name Ohmic contact).

this section we have already named a few culprits, which we now intend to summarize.

The increase in the amount of *excess* kinetic energy gained by the average electron (or *hole*) from the electric field results in further drift of the *charge carrier* into the *conduction band* (or *valence band*). Consequently, the *effective mass* of the *charge carrier* and its *mobility* change. In addition, in some semiconductors, this can cause the migration of electrons from *lower valleys* to *higher valleys*[66] of *conduction band*. At the same time, the increase in energy of the *charge carriers* plays an important role in modifying the *scattering* rates and also in activating *scattering* processes that were not present in the case of lower *carrier* energies. This *excess* of energy under high electric-field conditions is often expressed in terms of a quantity referred to as *electron temperature* (i.e., T_e). This definition is in analogy with the statistical mechanics, notion of kT (where T is the *lattice temperature*) as a representation of average thermal energy of an electron. In this discussion, one should be careful to distinguish between the notions of *lattice temperature* and that of *electron temperature*.

1.4.4.1 Electron Temperature versus Lattice Temperature In order to understand the development of *electron temperature* beyond *lattice temperature* at high electric fields, we shall start with a review of our understanding of *thermal equilibrium*. Under *thermal equilibrium*, the equality of *charge carriers'* emission and absorption of energy results in a *zero-sum* gain of energy. This equality is maintained through equality of emission and absorption rates of *phonons*, which are quantum particles representative of *lattice vibrations*. Under this condition, the energy distribution is *Maxwellian*. However, applying an electric field across a semiconductor removes the system from the state of *thermal equilibrium* (as now *charge carriers* begin to gain additional energy from the electric field while still emitting some of it through the *phonon emission* process). As a result, not only for high electric fields but also for low-to-moderate values of electric field, electrons will become more energized than they were under *thermal equilibrium*. This *excess* of energy beyond *thermal equilibrium* is what is explained through the notion of *electron temperature*. In connection to this *excess* energy, it must be indicated that even under moderate electric-field conditions, the increase in energy is less than the amount predicted by the laws of electrostatics. This is because the emission rate of *phonons* is now also on the rise, which will eventually reach steady state for a given electric field by matching the absorption rate. This description partially explains the *bowing* in the *drift velocity* versus electric-field characteristics (also known as v_d–E characteristics) at medium values of electric field (Fig. 1.22).

Figure 1.22, in addition to demonstrating the bowing of v_d–E characteristics, shows two other important phenomena: (i) saturation of *drift velocity* at high electric fields and (ii) peaking of the v_d–E and evolution of a *negative differential mobility*[67] region

[66] Also referred to as *satellite valley*.

[67] *Differential mobility* is defined unlike the *linear mobility* and is calculated around a given value of electric field as $\Delta v/\Delta E$.

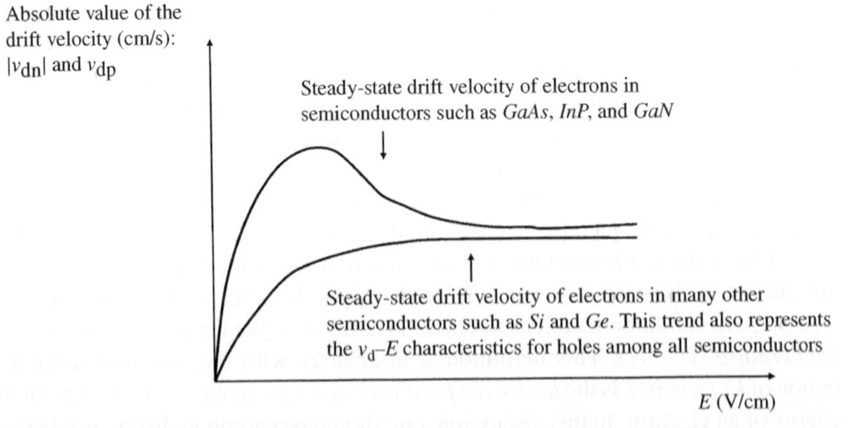

FIGURE 1.22 A behavioral depiction of v_d–E characteristic for electrons and holes in different semiconductors. Scales are linear.

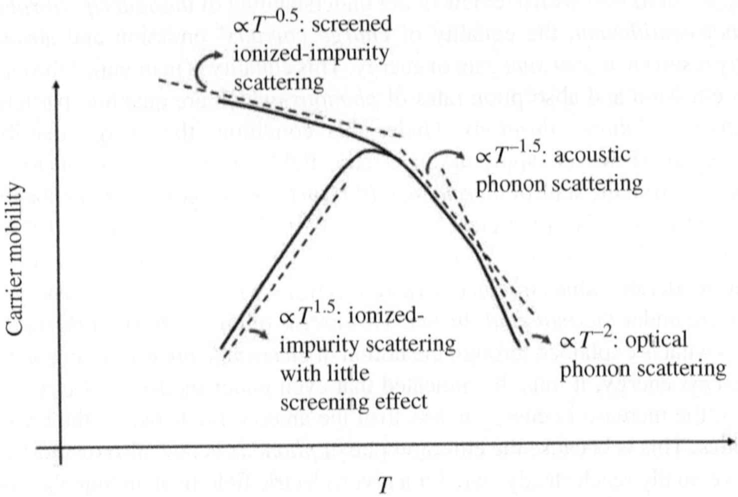

FIGURE 1.23 Correlation between the typically observed temperature dependences of low-field carrier mobility in semiconductors and the dominance of scattering processes. Scales are linear.

in the intermediate range of electric field only for electrons (and not *holes*) in some semiconductors.

While for small electric fields the *low-field carrier mobility* remains descriptive of the achievable *carrier drift velocity*, Figure 1.23 behaviorally presents the dependence of this parameter on temperature under the dominance of a number of *scattering* processes (which will be dealt with especially in Section 1.7).

For moderate values of electric field, which do not result in transfer of electrons from one *valley* to the next, we can equate the rates of emission and absorption of energy at a given electric field in order to quantify the *electron temperature*. For the relatively simpler cases of *Ge* and *Si*, which are semiconductors with no *intervalley* electron transfer, balancing the rate equations results in the following definition for the ratio of *electron temperature* to *lattice temperature*:

$$\frac{T_e}{T} = \frac{1}{2}\left[1 + \sqrt{1 + \frac{3\pi}{8}\left(\frac{\mu_n E}{c_s}\right)^2}\right]. \tag{1.86}$$

In terms of which,

$$v_{dn} = -\mu_n E\sqrt{\frac{T}{T_e}}. \tag{1.87}$$

In these equations, μ_n is the low-field electron *mobility* and c_s represents the velocity of sound. As shown in (1.86) and (1.87), for moderate fields (i.e., when the linearly calculated *drift velocity* is comparable to the sound velocity), the *electron temperature* increases beyond the *lattice temperature*. As a result, *drift velocity* will become smaller than the linearly predicted value of $\mu_n E$.

Other more general frameworks have also been developed for determining the *electron temperature*. With certain approximations, it can still be proven that the distribution function of electrons under conditions far from *thermal equilibrium* follows the form of *Maxwell–Boltzmann* statistics. The major difference is that the *lattice temperature* (i.e., T) should be replaced by the electron temperature (i.e., T_e), where approximately

$$T_e \approx T\left(1 + \frac{E^2}{E_C^2}\right). \tag{1.88}$$

In (1.88) E is the electric field and E_C is called the *critical* electric field (which for the case of electrons in *silicon* is about 10^4 V/cm). At this *critical* electric field, T_e is approximately equal to $2T$. $3/2kT_e$ is a measure of average energy of electron. This can result from the *Boltzmann transport equation*, assumption of *E–k* diagram, and *scattering* in terms of the *Fermi golden rule*.

In the aforementioned derivation of the *Einstein relationship* in Section 1.4.2.3, it has been assumed that the first-order *Taylor* series representation of electron concentration is sufficient. However, in certain situations, such as under high electric fields, this picture is not necessarily acceptable. Obviously, the aforementioned derivation was performed from a *thermal-equilibrium* perspective in which it was assumed that kinetic energy is equal to thermal energy. Such an assumption, which ignores *drift energy*, is not permissible under high electric fields. However, under high electric fields, one can establish a relationship close to the *Einstein relationship* by simply replacing the *lattice temperature* with the *electron temperature* (i.e., $(D/\mu) = (kT_e/q)$). This is valid only for *nondegenerate* (i.e., *Maxwellian*) semiconductors.

The electric fields encountered in a transistor can be very well in excess of E_C. As a result, *electron temperature* can be much higher than *lattice temperature*. However, we should avoid confusing the *electron temperature* with the actual device temperature. A transistor with thousands of degrees of *electron temperature* is still cool to the touch. For electrons to leave the crystal and have their temperature felt, they have to overcome a *work function*. This *work function* itself is typically several *electron volts*, which interestingly enough corresponds to thousands of degrees *Kelvin* (i.e., just like the T_e). It should also be mentioned that the earlier formalism loses validity at extremely high fields (i.e., $E \sim 10^6$ V/cm). At such high electric fields (and as a result high energies), the *density of states* function becomes extremely nonlinear. However, in the derivation of the above, a *DOS* function such as (1.14) has been employed. For those high electric-field cases, numerical techniques such as the *Monte Carlo* method should be used to calculate the *electron temperature*.

As a result of the discussions on *electron temperature*, it should be indicated that in the definition of *quasi-Fermi levels*, these *energy levels* are functions of T_e and not the *lattice temperature*.

1.4.4.2 Steady-State Velocity Overshoot and Saturation At higher electric fields, increase in the *carrier* energy (and *carrier temperature*) triggers the interaction of *carriers* with other sources of *scattering* known as *optical phonons*.[68] This is an interaction that is, however, not incorporated into the *balance* equations employed in the derivation of (1.86) and (1.87). The *optical-phonon emission* process, being a very efficient process of removing *excess* energy (beyond that of *optical phonons*) from *carriers*, results in a *drift velocity* quite independent of electric field (i.e., saturation of v_d–E). For all semiconductors, such a field-independent saturation of *drift velocity* is observed for both electrons and *holes* at high electric fields (Fig. 1.22). The onset of this saturating behavior is determined by the *optical-phonon emission* energy of the given semiconductor.

Saturation velocity, being determined by *optical-phonon* emission *scattering*, is essentially independent of *doping* concentration. It is also quite independent of the semiconductor itself. The reason for the similarity of saturation velocity of different semiconductors is that at high energies, corresponding to the high electric fields, the *DOS* functions of all semiconductors are quite similar. In spite of this, the saturation velocity decreases with increase in temperature, since under these conditions at lower electric fields, electrons acquire a sufficient amount of energy for triggering the *optical-phonon* emission process.

As pointed out on Figure 1.22, for the case of electrons in only *direct-bandgap* semiconductors, a more interesting characteristic emerges before *velocity saturation* takes over. In these semiconductors, due to the migration of the energetic[69] electron

[68] We will pay a more in-depth attention to these *scatterers*, among others, in Section 1.7.3.7.
[69] Called also *hot* electrons, in connection to the notion of *electron temperature*.

from lower *valleys* of lower *effective mass*[70] to higher *valleys* of higher *effective mass*, a region of *negative differential mobility* develops. This is because while such a transfer is taking place, electron energy is still insufficient for interaction with *optical phonons*. To maintain the value of kinetic energy as a consequence of this migration, since electrons are feeling heavier, the *drift velocity* will become less as the electric field is increasing. This region of *negative differential mobility* provides fascinating opportunities for designing microwave sources and oscillators in forms such as *Gunn diodes*. Due to the absence of multiple *valleys* in the *valence band*, *negative differential mobility* is not observed on *drift-transport* characteristics of *holes*.

The saturation velocity for the case of *Si* and *Ge*, which have a simple saturating v_d–E characteristic, is

$$v_s = \sqrt{\frac{8\ E_p}{3\pi m_0}} \cong 10^7\ \text{cm/s}, \tag{1.89}$$

where E_p is the *optical-phonon* energy (which as indicated in Table 1.4 is 63 meV in the case of *Si*).

Oftentimes, approximate closed-form analytical models are used to explain the variation of *drift velocity* versus electric field in the regions of small, medium, and high electric fields. One such an empirical relationship used for *Si* is

$$v_d = \frac{\mu E}{\left[1 + (\mu E/v_s)^{C_2}\right]^{1/C_2}}. \tag{1.90}$$

The constant C_2 is a temperature-dependent fitting parameter, which is in the range of two for electrons and one in the case of *holes*. Evidently, (1.90) represents the *drift velocity* as an absolute value.

For the case of the *direct-bandgap* semiconductor of *GaAs*, which is the most investigated *III–V* semiconductor,[71] due to the presence of *negative differential mobility*, the v_d–E characteristic is more complicated (i.e., as suggested in Fig. 1.22). As indicated already, in order to explain the hump in this characteristic, careful knowledge of the *E–k* diagram and also *optical-phonon* energy is required. While in the lower *valley*, which is located at the center of *Brillouin zone* (i.e., Γ-*valley*), electron *mobilities* as high of 8000 cm²/V s can be achieved at room temperature, the satellite *valley* (located along the <*111*> axes) offers an electron *mobility* in the range of 100 cm²/V s. This satellite *valley* is located 0.3 eV above the *lower valley*. The electron *effective mass* in the lower and the satellite *valleys* are equal to $0.068m_0$ and $0.55m_0$, respectively.

[70] That is, the central valley of the *Brillouin zone*: Γ valley.

[71] Unlike *silicon*, which is a monoatomic semiconductor, *GaAs* is a *compoundsemiconductor* composed of atoms from groups *III* and *V* of the *periodic table* (hence the name *III–V*).

1.4.4.3 Nonsteady-State Velocity Over- and Undershoot With only one exception mentioned in the opening of Section 1.4, all of our discussions are thus far concerned with steady-state *drift velocity*. In order to arrive at steady state, *charge carriers* need to go through a sufficient number of *scattering* events. For devices of shorter length scale than the *mean free path*, however, such a steady-state condition will not be achieved. Under such circumstances, *drift velocity* is observed to reach values larger than the steady-state values. This is referred to as *velocity overshoot*. This condition is instigated by what was called earlier in this section *ballistic transport*. Under this condition, in analogy to our discussion on *low-field mobility*, we can say that velocity increases with time and as a result with distance according to $\approx(qEt/m^*)$. However, it should be emphasized that these values, which are higher than the steady-state values, are only attained momentarily, within a limited span of space and time. Besides the phenomenon of *velocity overshoot*, an *undershoot* in the nonsteady-state *drift velocity* is also possible.

Generally speaking, these nonsteady-state phenomena are not just encountered in *nanoscale* devices. Instead they are encountered when *carriers* are suddenly exposed to a large variation of electric field. For a Γ-*valley* electron suddenly exposed to a high electric field, electron energy can very well exceed the *intervalley* separation, while due to the short time of travel, an electron does not get a chance to *scatter* to a higher *valley*. As a result, *drift velocity* exceeds the value it could obtain under steady state and in the higher *valley* (i.e., of higher *effective mass*). This is the case of the velocity *overshoot*. However, if an electron originally occupying the higher *valley* (of higher *effective mass*) is suddenly introduced into a region in which electric field is suddenly reduced, electron velocity remains much lower than the steady-state value that it could obtain from that electric-field strength. This happens because the electron has not received a chance to *scatter* to the lower *valley*. This is the case of the velocity *undershoot*.

1.4.4.4 Summary of Observations on the Choice of Carrier-Transport Formalism As observed through the so far presented discussions in Section 1.4, *charge-carrier transport* is a sophisticated problem. In *carrier-transport* problem, interactions of *charge carriers* with *scattering* potential of a variety of sources (e.g., *ionized impurities*, *acoustic phonons*, *optical phonons*, *polar optical phonons*, *piezoelectric polarization*, etc.),[72] internally induced and externally applied electric fields are investigated. While in large-size electronic devices, oftentimes phenomenological description of *carrier transport* in terms of the *drift–diffusion* formalism is capable of providing sufficient insight into the problem, in smaller-size devices (and also where the electric-field variation is strong), effects such as *nonsteady-state velocity overshoot* (or *undershoot*) and *ballistic transport* are required to be considered.[73]

[72] Only some of which are named prior to this in the text. Section 1.7 will deal with a larger number of these processes at a deeper level.

[73] Later on, where appropriate, we will add more items to this list, such as *real-space transfer*.

The ongoing trend in the design of *transistors* of scaled dimensions has rendered the *drift–diffusion* formalism more and more insufficient. As an intermediate step in incorporation of some of these *carrier-transport* effects, steady-state models of *drift* characteristics have been used to incorporate effects such as *transferred electron* and *velocity saturation* (which as noticed earlier are due to take over at intermediate to high electric fields). In these models of *carrier transport*, *mobility* and *diffusion constant* are defined as functions of local electric field. Accordingly, even beyond low electric fields, the *drift velocity* in terms of the instantaneous relationship to electric field is identified using (1.61) and (1.62). On the basis of this *high-field* linear *mobility*, *Einstein relationship* is then employed to produce the high electric-field *diffusion constant*.

However, as it was mentioned earlier, reduction of the size of the *channel* along which *carriers* are to fly (compared to *mean free path*) renders even these intermediate models insufficient. This is because under these circumstances, the steady-state assumption vanishes. At the same token *carrier transit time* becomes comparable to *energy relaxation* and *momentum relaxation time constants*. Under these conditions, *carrier* distribution functions determined by the local value of electric field are no longer valid, and *carrier transport* both in time and in space is *nonlocal*. As a result, the device's current characteristics cannot be explained by the continuity equations (i.e., Equations (1.49) and (1.50)).

Now that we have presented some of the limitations of different formalisms of *carrier transport* in semiconductor devices, it is time to point out that in design and analysis of electronic devices, the first step is the selection of the appropriate *transport* formalism. This process of selection is done in terms of comparison of device sizes to those characteristic length scales identified in Section 1.4.2 and also the *carrier* flight time with the relevant timescales. In the case of *silicon* as the semiconductor medium of *carrier transport*, this is summarized in Figure 1.24. In this regard, for a rapidly changing potential on scale of the *lattice constant*, only a full quantum mechanical treatment, and not a semiclassical *effective-mass* approach, will be sufficient. In this quantum mechanical treatment, the *wave equation* with incorporation of all potentials (i.e., including *scattering*, externally applied and internally induced) should be solved.

So far in our discussion, we have invoked a few properties of *Si* and *GaAs*. Table 1.5 provides an essential summary of a few of their important *carrier-transport* properties.

1.4.5 Thermionic and Field Emission

In macroscale devices, not all components of *carrier transport* are explained through the *drift–diffusion* formalism. There are two other frequently encountered modes of current conduction in semiconductor devices: *thermionic* and *field emission*. *Thermionic* and *field emission* modes evolve in association with the presence of a potential barrier.

In the case of the *thermionic emission*, existence of a potential barrier restricts the movement of majority *charge carriers* normal to the barrier, especially if the *carrier*

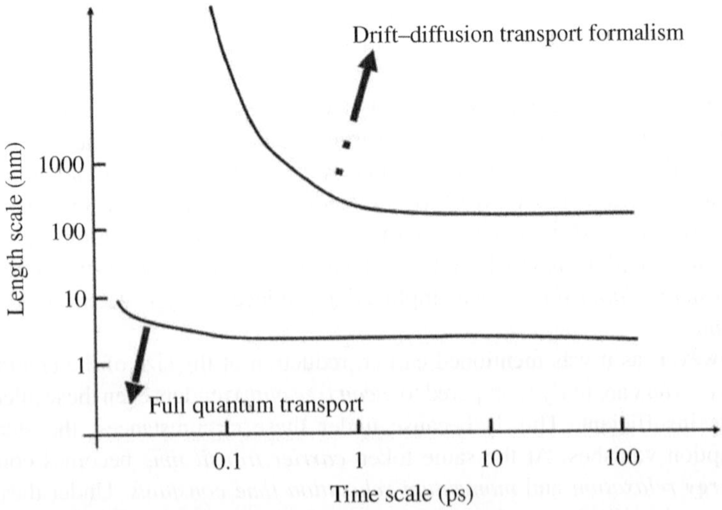

FIGURE 1.24 Approximate boundaries for the applicability of drift–diffusion carrier transport and the necessity for dealing with the full quantum mechanical description of transport in a silicon medium, as a function of time and length scale. Although, more or less such trends are applicable to other semiconductor media, the indicated values are dependent on the medium. Adapted from Lundstrom (2000, p. 347). Copyright 2000, Cambridge University Press. Reprinted with the permission of the Cambridge University Press.

TABLE 1.5 A Summary of Important Transport Properties of Si and GaAs

	Semiconductor	
Property	Si	GaAs
Longitudinal acoustic velocity (cm/s)	9.04×10^5	5.24×10^5
Transverse acoustic velocity (cm/s)	5.34×10^5	3.0×10^5
Electron acoustic deformation potential (eV)	9.5	7.01 (Γ-*valley*), 9.2 (*L-valley*), and 9.0 (*X-valley*)
Electron optical deformation potential (eV/cm)	—	3.0×10^8 (*L-valley*)
Hole acoustic deformation potential (eV)	5.0	3.5
Hole optical deformation potential (eV/cm)	6.0×10^8	6.48×10^8

temperature is low (and as a result the kinetic energy of the *charge carrier* is less than the barrier height). Nevertheless, energized (i.e., *hot*) *charge carriers* can rise over the barrier and move across it. It is due to this temperature dependence that the mechanism of *carrier transport* is referred to as *thermionic emission*. In this mode of *transport*, the critical parameter is the height of the potential barrier and not its shape. An example is provided in Figure 1.25.

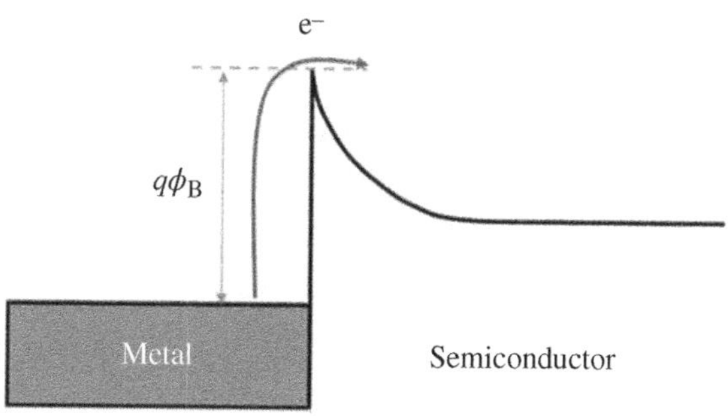

FIGURE 1.25 An example of thermionic emission across a potential barrier of height $q\phi_B$ formed between a metal and a semiconductor. We will talk about these contacts in details in Section 2.2.

Thermionic emission is observed in a variety of semiconductor devices including *Schottky barrier diodes.*[74] Ideally, for this mechanism to be the controlling mode of *transport*, the *carrier transport* in the barrier is supposed to be without *scattering*. In other words, *carriers* injected into the barrier should go through this material *ballistically* (i.e., the barrier width is to be smaller than the *mean free path*), instead of following the *drift–diffusion transport*. In addition to this, injected *carriers* should be moved out of the barrier on the opposite interface (which is often an interface with a metal), without *scattering*.

In order to quantify the *thermionic* current, we should remind ourselves of the fact that according to *Fermi–Dirac* statistics, while the number of *carriers* with energy values deeper into the *conduction* and *valence band* decreases with the separation from E_c and E_v, these *carrier* concentrations are not equal to 0. In addition, the *carrier* concentration in the *states* deeper above E_c, or below E_v, increases as the *carrier temperature* increases. As a result, the integrated number of *carriers* with energy values above the barrier height increases with this *temperature*. These *carriers*, which are not confined by the barrier, can now participate in current conduction through the *thermionic emission* process. The electron current density over a barrier of height $q\phi_B$ can be quantified as

$$J_n = A^* \cdot T^2 \cdot \exp\left(\frac{-q\phi_B}{kT}\right), \tag{1.91}$$

where A^*, which is often referred to as *effective Richardson constant*,[75] is defined as

$$A^* \equiv \frac{4\pi q m_n^* k^2}{h^3}. \tag{1.92}$$

[74] Which will be explored in Chapter 2. These *diodes* are named after *Walter Schottky*.
[75] Named after *Owen Richardson*.

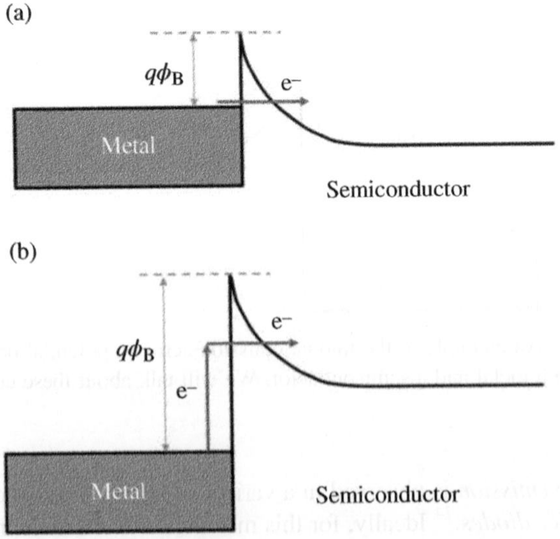

FIGURE 1.26 (a) An example of field emission across a potential barrier formed between a metal and a semiconductor. (b) An example of field-assisted thermionic emission across a potential barrier of height $q\phi_B$.

As suggested earlier, the *thermionic emission* process is not the only mode of *transport* through a potential barrier. Transport through such a barrier can be assisted, or even overwhelmed, by another mode of *carrier transport,* which is rooted in the quantum mechanical *tunneling*. These modes, which are known as *field emission* and *field-assisted thermionic emission*, are illustrated in Figure 1.26. In those cases, A^* will be modified by processes such as *reflection* and *tunneling* of the *electron wave*.

Quantum *tunneling* is rooted in the *wave* nature of electrons. According to quantum mechanics, even *charge carriers* with less energy than the barrier height exhibit a finite probability of *tunneling* through the barrier. According to the shape of the barrier, at each energy value, the effective thickness of the barrier that *carriers* need to *tunnel* through is different. Hence, unlike *thermionic emission* this mode of *transport* is quite dependent on the shape of the barrier. The smaller the effective width of the barrier at a certain energy value, the larger is the *tunneling* probability for *charge carriers* with that much energy. This is an exponential dependence.

Although the potential barriers often encountered in semiconductor devices do not have a rectangular form, for the sake of simplicity, it is worth considering the simple case of a rectangular barrier of height V_0 and of thickness W to describe the quantum *tunneling* process. This situation is illustrated in Figure 1.27. According to quantum mechanics, in this one-dimensional scenario, electrons on either side of the potential barrier are expressed in terms of *propagating waves*. However, at positions within the thickness of the barrier, the time-independent portion of the *wave function* follows a decaying exponential represented by $\exp(-|kx|)$, where *wave number* $k=\sqrt{(2m^*(V_0-E))/\hbar^2}$. In this case, the energy of the electron (i.e., E) is less than

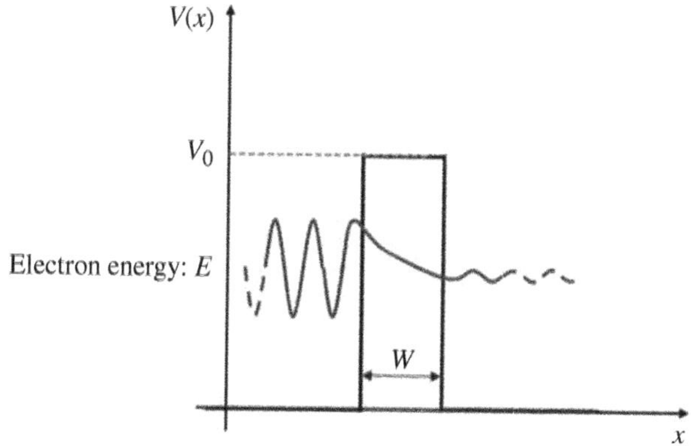

FIGURE 1.27 Quantum tunneling across a rectangular potential barrier of height V_0.

the barrier height V_0. Establishing the boundary conditions (i.e., in terms of continuity, single valuedness, and finiteness of the *wave function* and its first derivative) results in *tunneling* probability of

$$T_t = \left[1 + \frac{V_0^2 \cdot \sinh(|k|W)}{4E(V_0 - E)}\right]^{-1} \approx \frac{16E(V_0 - E)}{V_0^2} \cdot \exp\left(-2\sqrt{\frac{2m^*(V_0 - E)}{\hbar^2}}W\right). \quad (1.93)$$

In the case of more complicated barrier shapes, *WKB*[76] approximation is employed as a simplification over solution of the *Schrödinger* equation. This approximation is applicable where potential $V(x)$ is not varying rapidly with position. In terms of this approximation, according to the schematics of Figure 1.28, the *tunneling* probability can be calculated by

$$T_t \approx \exp\left\{-2\int_{x_1}^{x_2} |k(x)|dx\right\} = \exp\left\{-2\int_{x_1}^{x_2} \sqrt{\frac{2m^*}{\hbar^2}[V(x) - E]dx}\right\}. \quad (1.94)$$

Knowing the *tunneling* probability one can calculate the *tunneling* current density J_t by integration versus energy over the product of three components: the *tunneling* probability, the number of *carriers* at each value of energy at the originating side of the barrier, and the number of empty *states* at that value of energy on the receiving side of the barrier:

$$J_t = \frac{qm^*}{2\pi^2\hbar^3}\int F_E \cdot N_E \cdot T_t \cdot (1 - F_R) \cdot N_R dE. \quad (1.95)$$

[76] Named after *Gregor Wentzel, Hendrik Anthony Kramers*, and *Léon Brillouin*.

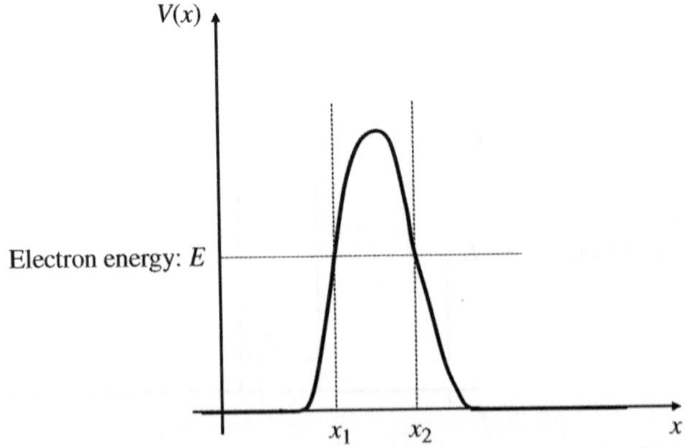

FIGURE 1.28 Quantum tunneling of an electron of energy E across an arbitrary potential barrier.

F_E, F_R, N_E, and N_R stand for the *Fermi–Dirac* distributions and *densities of states* in the emitting and receiving sides of the barrier, respectively. This framework describes the mode of *transport*, which is purely explicable in terms of quantum *tunneling* (i.e., *field emission*).

Transport through a potential barrier can also take place through a mixture of *thermionic-* and *field emission* processes. In this fashion, whereas *carriers* are *thermally* raised behind the barrier, depending on the shape of the barrier, they see an effective reduction in the barrier width and as a result an improvement in *tunneling* probability. As suggested earlier, this mode of *carrier transport* is known as *thermionic field emission* or *field-assisted thermionic emission*.

1.5 BREAKDOWN IN SEMICONDUCTORS

In discussing the high electric-field effects on electrons, the other mechanism that is of interest is *impact ionization*. *Intervalley scattering* is not the only outcome of electrons becoming *hot* (i.e., rise in T_e). As electrons become *hot*, they also become capable of breaking covalent bonds and producing *EHP*. This is shown in Figure 1.29. This happens in the form of an *avalanche*. One *hot charge carrier* produces two more *charge carriers*, and if the high electric-field region is large enough, each of these *charge carriers* will develop a chance for producing an extra *EHP*. As a result, *carrier* concentration will multiply itself. Due to this, the other name for this process of *EHP generation* is *avalanche breakdown*. The minimum amount of energy needed for a *hot carrier* to instigate this process should be larger than the *bandgap* of the semiconductor. Assuming the same *effective mass* for electrons and *holes*, this energy can be easily proven to be 50% in excess of the *bandgap* (i.e., $1.5 \times E_g$).

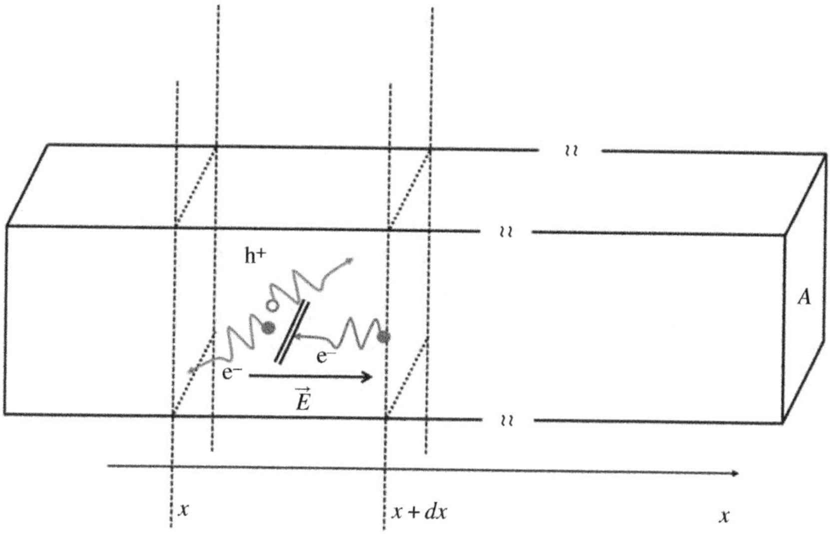

FIGURE 1.29 Generation of an electron–hole pair upon collision of a hot electron generated within the high electric-field section of a semiconductor slab with a covalent bond.

The *avalanche* multiplication process is characterized by an *ionization rate*, which is defined by the number of *EHP* generated while the electron is traversing a unit of distance. Assuming the electron velocity to be v_n, this rate can be written as

$$\alpha_n = \frac{1}{n}\frac{dn}{d(tv_n)} = \frac{1}{nv_n}\frac{dn}{dt}. \tag{1.96}$$

In terms of this *ionization rate* and a similarly defined rate for *holes*, the time variation of electron and *hole* concentrations can be written as

$$\frac{dn}{dt} = \frac{dp}{dt} = \alpha_n n v_n + \alpha_p p v_p = \frac{\alpha_n J_n}{q} + \frac{\alpha_p J_p}{q}. \tag{1.97}$$

According to the *continuity equation*, we can also have

$$\frac{dJ_n}{dx} = \alpha_n J_n + \alpha_p J_p \tag{1.98}$$

and

$$\frac{dJ_p}{dx} = -\alpha_n J_n - \alpha_p J_p. \tag{1.99}$$

Based on these, the spatial derivative of $J_n + J_p$ remains 0, resulting in

$$\frac{dJ_n}{dx} = -\frac{dJ_p}{dx}. \tag{1.100}$$

As expected, the *ionization rates* have a strong dependence on electric field. The following physical expression represents this dependence:

$$\alpha(E) = \frac{qE}{\varepsilon_I}\exp\left\{-\frac{E_I}{E[1+(E/E_P)]+E_T}\right\}. \tag{1.101}$$

In this equation ε_I is the *high-field effective ionization threshold energy*. E_T, E_P, and E_I stand for the electric-field strength needed by *carriers* to overcome the decelerating effects of *thermal*, *optical phonon*, and *ionized-impurity scattering*, respectively. In the case of *silicon*, ε_I for electrons and *holes* is equal to 3.6 and 5 eV, respectively. For a limited range of electric field, the above equation can be replaced by the following simpler equations:

$$\alpha(E) = \frac{qE}{\varepsilon_I}\exp\left(-\frac{E_I}{E}\right), \quad \text{if } E_T < E < E_P \tag{1.102}$$

or

$$\alpha(E) = \frac{qE}{\varepsilon_I}\exp\left(-\frac{E_I \cdot E_P}{E^2}\right), \quad \text{if } E > E_P \text{ and } E > \sqrt{E_P \cdot E_T}. \tag{1.103}$$

Figure 1.30 depicts the experimentally observed *ionization rates* for *Ge*, *Si*, *SiC*, and *GaN*.

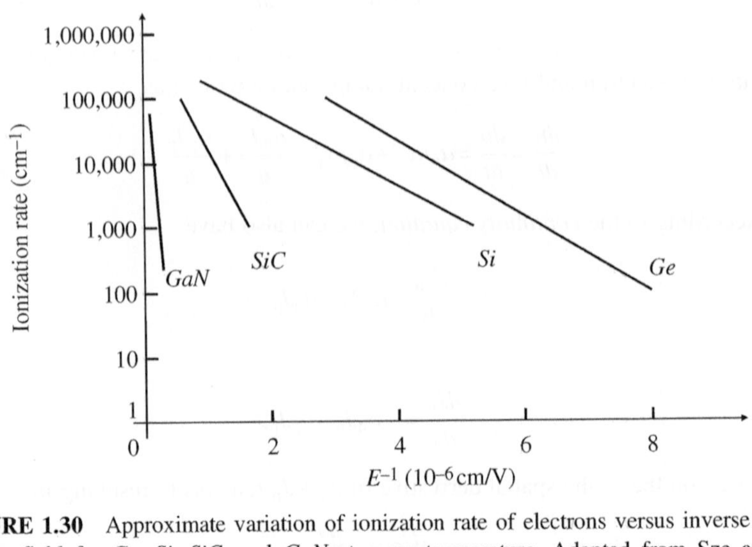

FIGURE 1.30 Approximate variation of ionization rate of electrons versus inverse of the electric field for *Ge*, *Si*, *SiC*, and *GaN* at room temperature. Adapted from Sze and Ng (2006, p. 41). Copyright 2006, John Wiley and Sons. Reprinted with the permission of the John Wiley and Sons.

In electronic devices, in connection to the process of *avalanching*, we often speak of *avalanche breakdown voltage*. This refers to the voltage difference applied to the terminals of a device that causes the induction of electric fields beyond the onset of an *avalanche* process. Production of large concentrations of *charge carriers* induces heating and either permanent structural damage to the electronic device or temporary malfunction.

As expected from (1.101), for larger *bandgap* materials the *ionization rates* are smaller. As a result, wide *bandgap* semiconductors such as *SiC*, *GaN*, *AlN*, and *ZnO* are deemed suitable for high electric-field/high-voltage applications. In addition to *bandgap*, *optical-phonon* energy is also an important factor in determining the *breakdown voltage*, which as suggested earlier is the minimum voltage needed to create a rush of current. The higher the *optical-phonon energy*, the higher the chance of *charge carriers* becoming *hot*, and as a result, the smaller would be the applied voltage causing *breakdown*. *GaN*, which is a wide *bandgap* material, possesses a large value of *optical-phonon energy*. These two factors promote adversary agenda with regard to the *breakdown voltage*.

With respect to *ionization rates*, it is also important to mention that at a constant electric-field *ionization energy* is a declining function of temperature. This is due to the fact that at high temperatures, *phonon scattering* deteriorates the chance of *carriers* becoming *hot*.

The other mechanism instigating *breakdown* at high electric fields is *Zener*[77] *breakdown*. In this mechanism, the high electric field directly breaks covalent bonds and produces a large number of *EHPs* needed to create a rush in current.

The *impact ionization* process, unlike the *Zener breakdown* mechanism, is an avalanching process, which is why it requires *carriers* to travel a distance multiple times longer than the *mean free path*. As a result of this requirement, *avalanche breakdown* can only happen in devices that have a relatively long high electric-field region.

1.6 CRYSTALLINITY AND SEMICONDUCTOR MATERIALS

At the beginning of this chapter, it was pointed out that while from the point of view of crystallinity semiconductors can be found in *amorphous*, *polycrystalline*, or *monocrystalline* forms, *monocrystalline* semiconductors are often preferred in electronic applications. An interesting question to address in dealing with semiconductor devices, which is at the root of semiconductor electronics, is: why should we prefer *monocrystalline* semiconductors over *polycrystalline* or *amorphous*?

The reason for this choice is the short wavelength of electrons. As a result of this short wavelength, *charge-carrier transport* is critically dependent on the atomic arrangement in the *solid*. The higher the crystalline order, the better the *transport* properties. As invoked earlier, according to *Bloch* theorem, for a perfect *monocrystalline*

[77] Named after *Clarence Zener*.

solid electrons travel as *propagating waves*, and the atoms in the crystal essentially pose no cause for *scattering* of the *electron wave*. However, in contrast to electrons, *photons* have a longer wavelength and are not impacted by the mere *short-range* order of the *host* material. As a result, noncrystalline materials such as glasses are suitable for optical applications while not so for electronic applications.

An important line of research in the area of semiconductor electronics focuses on crystalline growth of semiconductor materials. Interaction of these crystals with *X-rays* in the form of *diffraction* is one of the capable tools for studying the degree of crystallinity of a *solid*. *X-rays* are suitable for this purpose because their wavelengths are very close to the *lattice constants* of the semiconductor crystals.

The structure of an ideal crystal can be explained in terms of *copying* and *pasting* of a building block or a so-called unit cell according to a certain *translation vector*,

$$\vec{r'} = \vec{r} + u_1\,\vec{a_1} + u_2\,\vec{a_2} + u_3\,\vec{a_3}, \tag{1.104}$$

where u_1, u_2, and u_3 are the three arbitrary integers, $\vec{r}$ is the position of an arbitrary point in the building block, and $\vec{a_1}$, $\vec{a_2}$, and $\vec{a_3}$ are referred to as the *translation vectors*. This is schematically presented in Figure 1.31.

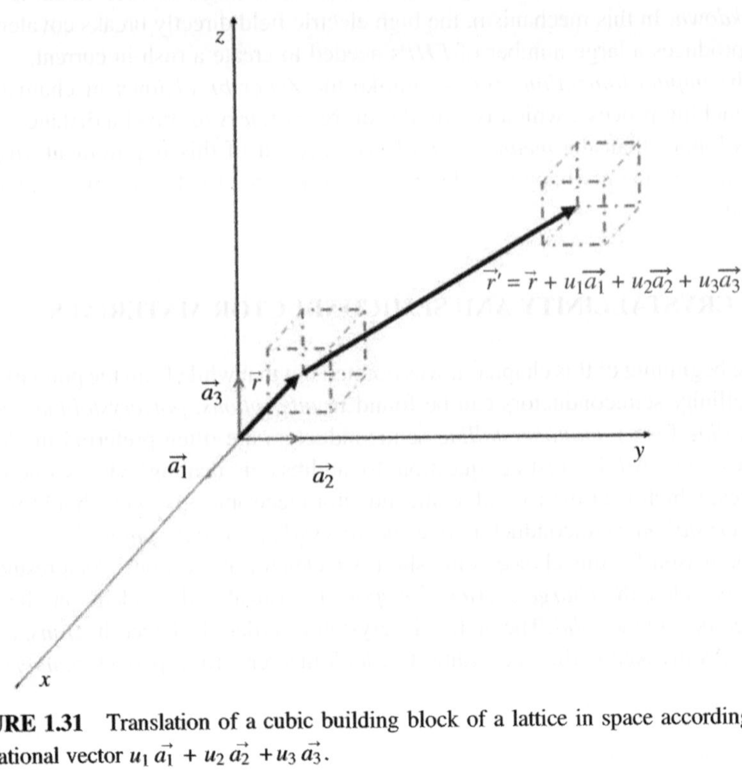

FIGURE 1.31 Translation of a cubic building block of a lattice in space according to the translational vector $u_1\,\vec{a_1} + u_2\,\vec{a_2} + u_3\,\vec{a_3}$.

1.6.1 Bravais Lattices

In the representation of (1.104), considering all possible integer values taken by u_1, u_2, and u_3, an infinite three-dimensional space has to be completely filled (i.e., with no voids). This requirement imposes certain restrictions on the *degree of symmetry* of the *unit cells* of a crystal. For example, while one, two, three, four, or sixfold rotational symmetry is allowed, five and seven are inadmissible (see Fig. 1.32). The degrees of symmetry for a given semiconducting crystal have considerable importance in its *carrier-transport* properties.

These building blocks are often explained in terms of a *lattice* and a *basis*. While *lattice* is the abstract arrangement of points in space representing a block, *basis* refers to the atoms assigned to these points. An infinite number of building blocks can be proposed for constructing a crystal. This is shown in Figure 1.33. The smallest of all these building blocks (or *unit cells*) is referred to as a *primitive cell*. While the number of atoms assigned to a *primitive cell* and its volume (which is identified by the translation vectors as $\vec{a_1} \cdot \vec{a_2} \times \vec{a_3}$) are unique, the vectors representing the *primitive cell* are by no means unique. An easy way to envision a *primitive cell* is through following the *Wigner–Seitz*[78] procedure.

In the *Wigner–Seitz* procedure, the boundaries of the *primitive cell* are determined by the intersection of imagined planes normal to the lines connecting a given *lattice* point to all its neighboring points. In the case of a two-dimensional *lattice*, this is illustrated in Figure 1.34.

Generally speaking, different *lattice types*, which are explained in terms of *lattice* and *basis*, are referred to as *Bravais*[79] *lattices*. In the case of three-dimensional

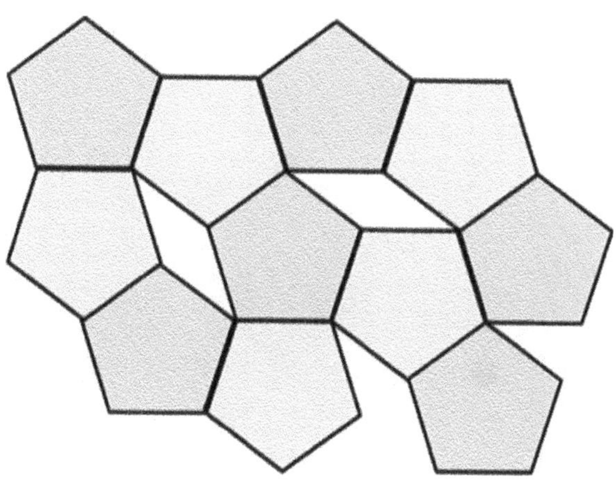

FIGURE 1.32 Schematic demonstration of inability of fivefold symmetry in filling of an infinite space.

[78] The *primitive cells* produced according to this procedure, after *Eugene Wigner* and *Frederick Seitz*, are referred to as *Wigner–Seitz cells*.
[79] Named after *Auguste Bravais*.

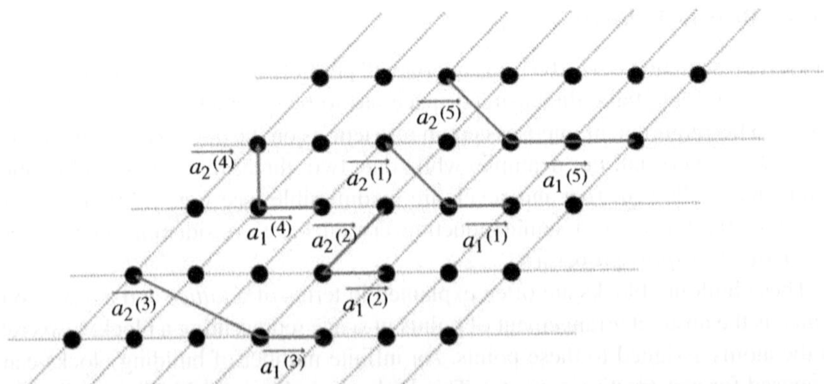

FIGURE 1.33 A schematic depiction of lattice and basis on a two-dimensional *Bravais lattice*. Atoms are represented by the dark circles. While as examples the parallelograms represented by a pair of vector $\vec{a_1}^n$ and $\vec{a_2}^n$ for $n = 1\text{–}4$ represent various primitive cells for this 2-D crystal, the parallelogram represented by a_1^5 and a_2^5 owing to its twice as large the area does not represent a primitive cell.

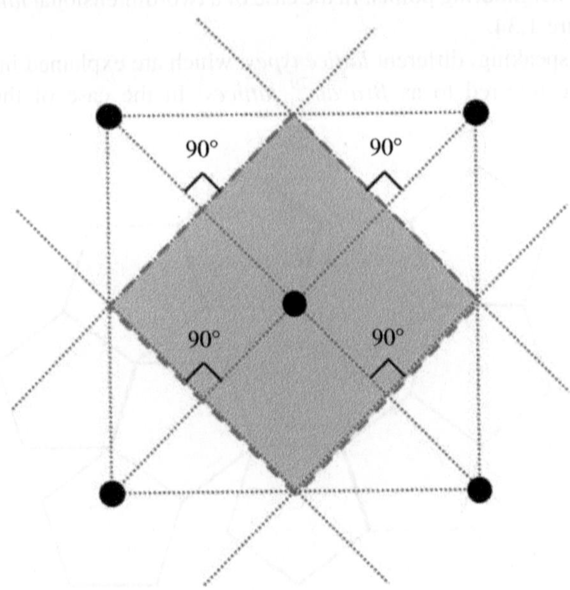

FIGURE 1.34 Drawing of the Wigner–Seitz primitive cell for a two-dimensional lattice. In this case, instead of envisioning a volume defined by the intersection of normal planes passing through the middle of the lines connecting neighboring atoms of the crystal, a primitive cell is formed by the intersection of normal lines passing through these points on the plane of the 2-D crystal. This plane is marked in gray, where atoms are presented by the dark circles.

TABLE 1.6 The Fourteen 3-D Lattice Types

Crystalline System	Number of Admissible Lattices	Relationships between the Sizes of the Axes of the Cell (a_1, a_2, a_3) and the Angles Defined Sequentially between These Axes (α, β, and γ)
Triclinic	1	$a_1 \neq a_2 \neq a_3$ $\alpha \neq \beta \neq \gamma$
Monoclinic	2	$a_1 \neq a_2 \neq a_3$ $\alpha = \gamma = 90° \neq \beta$
Orthorhombic	4	$a_1 \neq a_2 \neq a_3$ $\alpha = \beta = \gamma = 90°$
Tetragonal	2	$a_1 = a_2 \neq a_3$ $\alpha = \beta = \gamma = 90°$
Cubic	3	$a_1 = a_2 = a_3$ $\alpha = \beta = \gamma = 90°$
Trigonal	1	$a_1 = a_2 = a_3$ $\alpha = \beta = \gamma < 120°, \neq 90°$
Hexagonal	1	$a_1 = a_2 \neq a_3$ $\alpha = \beta = 90°$ $\gamma = 120°$

lattices, the number of *lattice types* is restricted to 14. This is due to the limitations posed by the *point symmetry groups*. These *lattice types* are indicated in Table 1.6. Out of these 14 *lattice types*, only *two* (which belong to *cubic* and *hexagonal* systems) are of interest in solid-state electronics and optoelectronics.

1.6.1.1 Hexagonal Crystals In the *hexagonal* system, the *unit cell* is in the form of a right prism. Figure 1.35a illustrates this right hexagonal prism through indication of 14 atoms, 7 arranged on each of the two basal hexagons. The separation between the two basal hexagons is referred to as *c* (which clearly denotes the height of the hexagonal prism). In order to clearly imagine this structure, envision *seven* identical atoms on a plane, *six* of which are centered at the *six* corners of a hexagon and *one* sitting right in the middle. If we imagine the atoms as hard spheres, in the so-called *hexagonal close-packed* (or *hcp*) structure, these spheres are supposed to tangentially touch each other. Now looking from above normal to this plane, we see cavities among the spheres. Obviously in the right prism, from this angle the *seven* atoms of the top basal hexagon will also have their centers coincide with those of the bottom basal hexagon.

The unit cells of many important semiconductors are formed through interpenetration of two of these hexagonal prisms according to a certain *translation vector*. As a result of this interpenetration, a second layer of atoms will also appear between the two basal hexagons of the first prism. As suggested in Figure 1.35b, the centers of these atoms coincide with the identified set of cavities. In terms of the vectors identified in Figure 1.35a, the second hexagonal prism is displaced by $2\vec{a_1}/3 + \vec{a_2}/3 + \vec{a_3}/2$. As shown in this figure, $\vec{a_3}$ has a magnitude equal to c and is identified normal to the basal hexagon. $\vec{a_1}$ and $\vec{a_2}$ defined in the basal plane have an

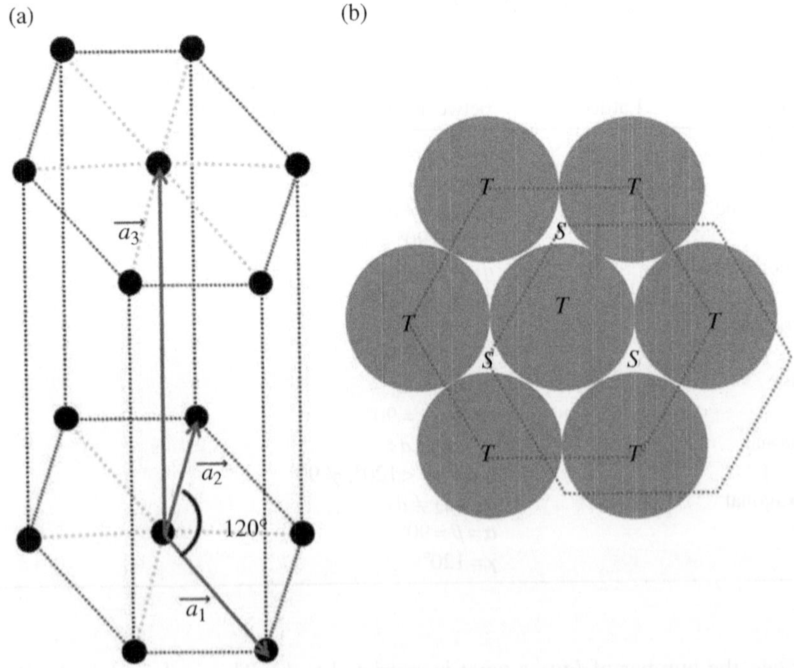

FIGURE 1.35 (a) A right hexagonal prism with the indication of placement of atoms as dark circles. (b) Position of the center of the atoms on the first, second, and third layer of a wurtzite crystal. While the gray circles represent the atoms of the first layer, letters *S* and *T* stand for the position of the center of the atoms on the second and third layer from a top perspective, respectively.

equal magnitude, which we denote as *a*, defined at 120° from one another. This crystal structure is known as *wurtzite*.[80] In order to guarantee *maximum sphere packing*, *c*/*a* should be equal to $\sqrt{8/3}$. This condition identifies an ideal *hcp* crystal.

1.6.1.2 Cubic Crystals The *cubic* system can be in the forms of *simple cubic*, *body-centered cubic* (or *BCC*), and *face-centered cubic* (or *FCC*). Shared between these three *lattice types* is a cube with edge length *a* (which is known as the *lattice constant*). While in a *unit cell* of *simple cubic* crystals only eight atoms are present (each centered at one of the eight corners of the cube), in *BCC* an extra atom is added to the middle of the cube. In *FCC*, six extra atoms are centered in the middle of the six faces of the cube. The atomic arrangements of these *cubic lattice types* are illustrated in Figure 1.36. A number of important characteristics of *cubic lattices* are summarized in Table 1.7.

[80] Named after *Charles-Adolphe Wurtz*.

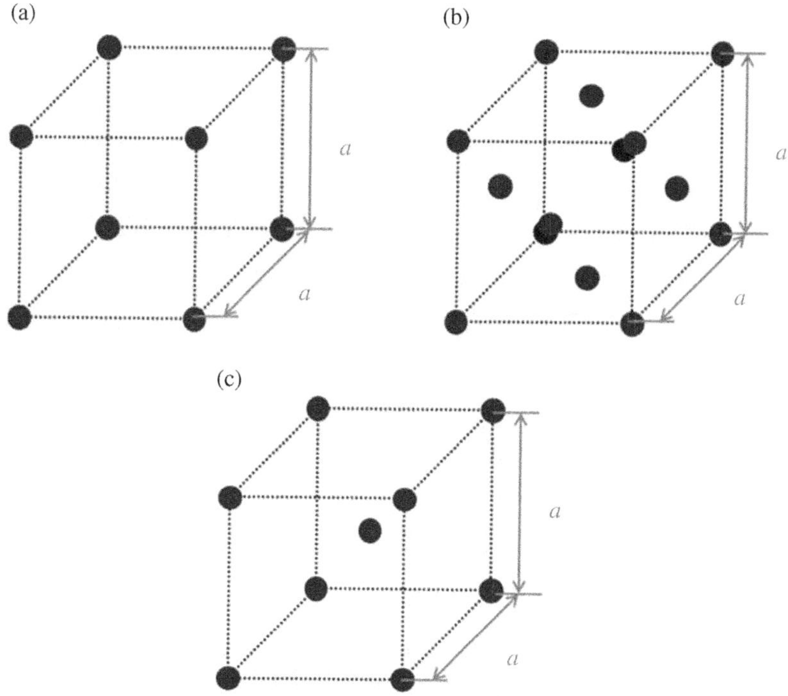

FIGURE 1.36 Atomic arrangements in the unit cells of (a) a simple cubic, (b) a face-centered cubic, and (c) a body-centered cubic crystal. Atoms are marked by full circles.

TABLE 1.7 Summary of the Characteristics of Cubic Lattice Types

	Lattice Type		
Characteristic	Simple Cubic	BCC	FCC
Volume of the conventional cell in terms of the lattice constant *a*	a^3	a^3	a^3
Volume of the primitive cell	a^3	$0.5a^3$	$0.25a^3$
Number of nearest neighbors	6	8	12
Nearest neighbor distance	*a*	0.866*a*	0.707*a*
Maximum packing ratio	0.524	0.680	0.740

Among the *cubic* crystals, *FCC Bravais lattice* is the *lattice type* that explains the largest number of semiconductors of interest in optoelectronics and electronics (e.g., *Si*, *Ge*, *GaAs*, *AlAs*, *InP*, etc.). However, as suggested already, not all semiconductors of interest are explained by the *FCC Bravais lattices*. Many semiconductors, and also metals, crystallize in form of *hcp* structures (e.g., *BN*, *AlN*, *GaN*, *SiC*, etc.).[81]

[81] Some semiconductors such as *GaN* can be grown in both *cubic* and *hexagonal* forms.

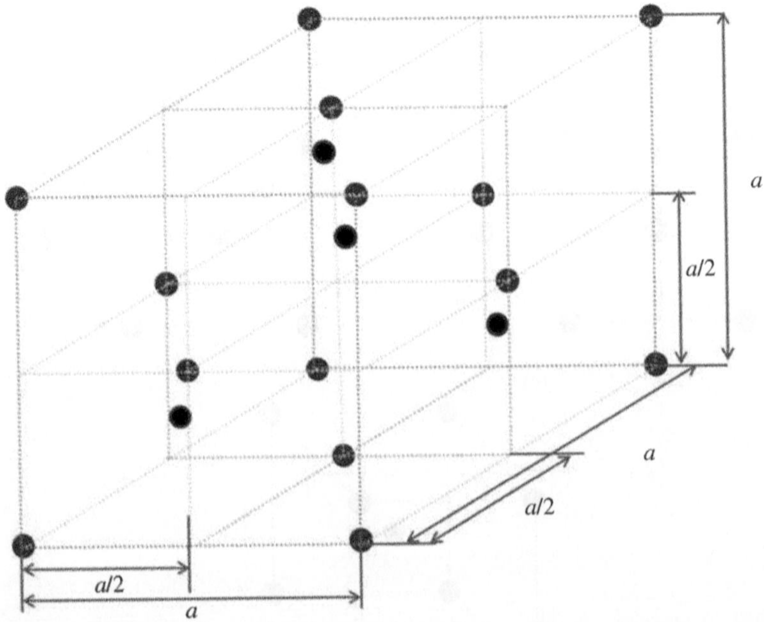

FIGURE 1.37 Placement of atoms on a diamond crystal unit cell. The gray full circles are sitting at the *FCC* positions. The other four atoms indicated by the darker circles are placed in the middle of the four inner cubes of edge length *a*/2.

Silicon, which is the most popular semiconductor, has a *lattice type* that results from interpenetration of *two* identical *FCC lattices*. Each of these *FCC lattices* is referred to as a *sublattice*. The *translation vector* between the two *FCC sublattices*, in terms of the *lattice constant a*, is (*a*/4, *a*/4, *a*/4). The crystal structure of *silicon*, which is constructed in this fashion, is referred to as *diamond* (since this is also the crystal structure of diamond). This is illustrated in Figure 1.37. *Germanium* also crystallizes in this form, although with a different *lattice constant*.

Whereas other important semiconductors such as *GaAs* and *InP* are also formed through interpenetration of *two FCC sublattices*, in their cases the *two sublattices* are not identical. In these cases, one *sublattice* has metallic atoms from group *III* of the *periodic table* (e.g., *Ga* and *In*) as its *bases*, while the other one has group *V* atoms (hence the name *III–V* semiconductors). These crystals are referred to as *zinc blende*.[82]

In contrast to *silicon* and *germanium*, which are known as *elemental* semiconductors, those semiconductors in which more than one atom is present in their *undoped* structures are referred to *compound* semiconductors. There are *compound* semiconductors that are *cubic* and those that are *hexagonal*. There are those such as *GaAs* that are *III–V* and those such as *ZnO* that are *II–VI*. There are those such as *InP* that are

[82] Named after the mineral *zinc blende* (*sphalerite*).

binaries and those such as $Ga_xIn_{1-x}P$ (i.e., x part *Ga*, $1-x$ part *In*) that are *ternaries*.[83]

Example

Calculate the volume density of the atoms in a diamond crystal of *lattice constant a*.

Since according to Figure 1.37 in a *diamond* crystal out of the 18 atoms associated with the *cubic unit cell* of volume a^3 only *four* completely reside within this volume while eight atoms share only 1/8th of their volume with the cube and the other six share half of their volume, the volume density is expressed as $\frac{(4+(8/8)+(6/2))=8}{a^3}$.

In this evaluation we have taken the atoms as hard spheres.

1.6.1.3 Miller Indexing System Due to differences in atomic arrangement along different directions of a *lattice*, *carrier transport* will be dependent on the alignment of the designed *channel* of the device with the crystalline axes. For that reason, it is important to use an indexing system to distinguish between different directions in a crystal. This system is known as the *Miller index*.[84] This indexing system is used to denote different planes and directions in a crystal.

To identify the *Miller indices* for a plane in a *cubic* crystal, the following procedure is used:

- Choose a *Cartesian* coordinate.
- Find the intersection of the given plane with the three axes of the coordinate.
- Construct a vector composed of the inverses of the intersections with x-, y-, and z-axis.
- Multiply this vector by the smallest common denominator.
- The resulting vector, which is often referred to as (*hkl*), refers to the aforementioned plane.

In a *cubic* crystal direction normal to the plane (*hkl*) is denoted by [*hkl*]. This is not generally extendable to all crystalline types. This procedure is shown schematically in Figure 1.38a.

In this procedure, the choice of the *Cartesian* coordinate was an arbitrary one. Due to this and the existing degrees of *symmetry* among different crystals, there are a number of equivalent planes and directions in a crystal. The existence of these equivalencies is a source for *degeneracy* in crystals, with which we have already made an acquaintance in Section 1.2.1. As a result of this *degeneracy*, {*hkl*} refers to a group of identical planes, and <*hkl*> denotes a group of identical directions.

Since many semiconductors crystallize in the hexagonal form, it is also of interest to indicate how the planes and directions in the *hexagonal unit cells* are indexed. The adopted indexing system in these cases is represented in terms of four digit indices known as *Miller–Bravais* indices. In dealing with *hexagonal* crystals, instead of orthogonal

[83] We can equally well have more than three atoms in a *compound*.
[84] Named after *William Miller*.

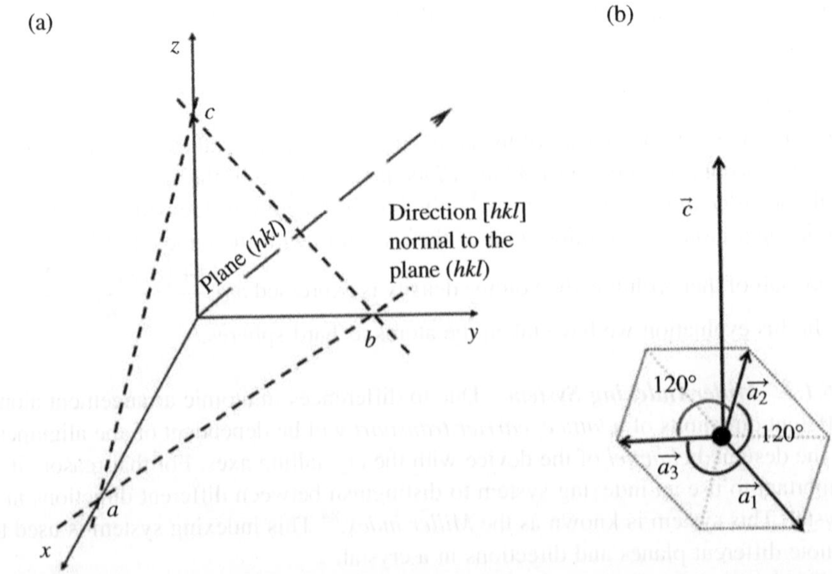

FIGURE 1.38 (a) Definition of *Miller indices* in a cubic system. Vector (*h,k,l*) is created by multiplying the vector (1/*a*, 1/*b*, 1/*c*) by the smallest common denominator between the three elements of the vector. (b) Coordinate system used in identifying *Miller–Bravais* indices in a hexagonal system.

coordinates, three nonorthogonal *basis vectors* (i.e., $\vec{a_1}$, $\vec{a_2}$, and $\vec{a_3}$) in the basal plane of the *hexagonal unit cell* and one *height vector* (i.e., $\vec{c}$) along the height of this hexagon are chosen. Directions of these vectors are identified in Figure 1.38b. In terms of the previously mentioned procedure identified for evaluation of *Miller indices* in cubic crystals, *Miller–Bravais* indices of *hexagonal* crystals are identified by finding out the intersection of a plane with these four vectors. Following that procedure, designations for a plane and the direction perpendicular to that plane are identified by (*hkil*) and [*hkil*], respectively. In evaluating these, care should be taken that the amplitude of unit vector $\vec{c}$ is different from that of the three other vectors. As an example, the plane parallel to the base of the unit cell has *Miller–Bravais* indices of (0001), while a face plane on the side that intersects $\vec{a_1}$, $\vec{a_2}$, $\vec{a_3}$, and $\vec{c}$, at 1, ∞, −1, and ∞ (which are scaled with the amplitudes of these vectors), is called $(10\bar{1}0)$. Orientations identified in Figure 1.38b for $\vec{a_1}$, $\vec{a_2}$, and $\vec{a_3}$ indicate that $h + k + i$ is always equal to 0.

1.6.2 Strain and Techniques of Epitaxy

Originally, *epitaxial* crystal growth techniques such as *MOCVD*[85] and *MBE*[86] were developed for growing planar crystals on a substrate. More recently, there has been

[85] Which, as identified earlier, stands for *metal organic chemical vapor deposition*.
[86] Also indicated earlier to stand for *molecular beam epitaxy*.

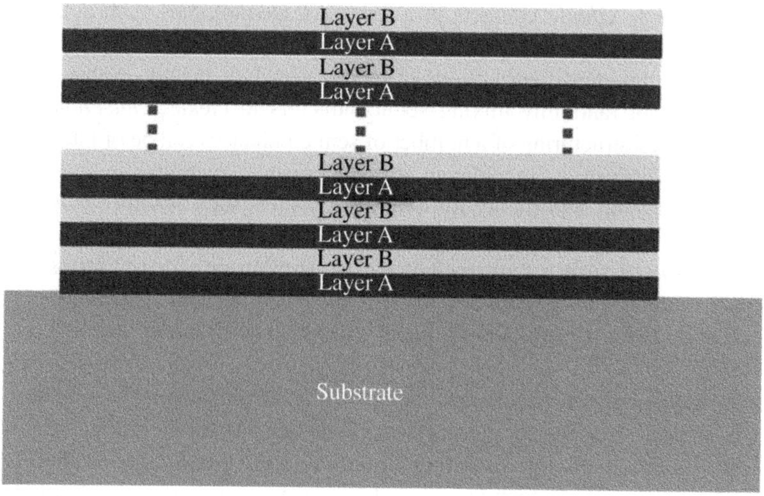

FIGURE 1.39 A superlattice is formed by growing a stack of a large number of alternating thin layers of at least two different semiconductors on a substrate.

an increasing interest in selective *overgrowth* (i.e., growth over the windows opened through a *masking* material). For local *overgrowth* applications, an advantage of *epitaxial* crystal growth techniques relying on chemical reactions (e.g., *vapor deposition epitaxial* process of *MOCVD*) over *MBE* is that they can take advantage of lateral temperature control for realizing local area growth. Based on this principle, for some materials laser-assisted local area growth has been proven possible.

Over the past three decades, techniques of *epitaxy*, either through realizing local area or planar crystal growth, have contributed significantly to realization of new opportunities in the *FET* technologies.

1.6.2.1 Ordered and Random Mixing of Semiconductors Whereas fundamentals such as free-energy minimization dictate the natural choice of the *growth mode* and *lattice constant* of crystals, the advances made with crystal growth techniques have made it possible to grow artificial crystals with structures other than those that can be found based on natural laws. Such a possibility is the outcome of the availability of techniques that enable atomic placement with exact precision.

With this degree of precision, the so-called *superlattice* structures have been grown. Realization of *superlattices* takes place in the form of *epitaxial* growth of alternating layers of different crystalline materials. While in each of these layers the crystalline structure is defined according to the laws of nature, control over the *pitch* and the *period* of these alternating layers helps in realizing an arbitrary degree of periodicity along the direction normal to the interface. This is schematically presented in Figure 1.39.

As the periodicity of the crystal impacts the *carrier-transport* properties, realization of *superlattices* provides a new dimension in engineering the *carrier transport*. These structures have been used in a variety of applications such as realization of *drain* and *source Ohmic* contacts in a few *FET* technologies such as *AlGaN/GaN HFET* technology[87] or production of high-frequency *signals*.

One of the important contributions of crystal growth techniques is that they provide the possibility of randomly mixing semiconductors to create *semiconductor alloys*. Whereas ordered structuring of a number of semiconductors can be of interest in forms such as *superlattices* (i.e., in realizing artificial crystals), random alloying of semiconductors provides the opportunity for tuning parameters such as *lattice constant*, *bandgap*, and *effective carrier mass* between the elements of the *alloyed* semiconductor. The *lattice constant* of such a *semiconductor alloy* is often provided in terms of a linear weighted average between those of the parent semiconductors. This approximation is referred to as *Vegard's law*. We will deal with this law in further detail shortly.

1.6.2.2 Coherent and Incoherent Growth of Heterojunctions In *epitaxy*, crystal growth is performed over a crystalline template. An *epitaxial* junction formed between two different semiconductors is referred to as a *heterojunction*. This is in contrast to *homojunctions*, which are formed between two pieces of one semiconductor that are, for example, merely *doped* differently. The terms *heterostructure*, *heterointerface*, and *heteroepitaxy* are also used in this context. Depending on the matching of the *lateral lattice constant* of the template (which is often the substrate) and the *freestanding*[88] *lateral lattice constant* of the *overgrowing* crystal, one can either have a *lattice matched* or a *lattice mismatched* mode of crystal growth. Figure 1.40 offers a schematic representation of these two modes of *epitaxial* growth. *Lattice-matched* (also known as *coherent*) growth, which is the result of the similarity of the two *lattice constants*, produces the highest-quality *heterojunction* between the two crystals. On the other hand, the presence of *lattice mismatch* results in formation of a large number of *dangling bonds* at the *heterointerface* and also crystal faults in the *overgrown* layer.

While only a few of the known semiconductors are *lattice matched* to one another, there is a strong demand for realization of *heterojunctions* between a variety of *lattice-mismatched* semiconductors. In order to avoid the unwanted effects of *lattice-mismatched* growth, techniques of *strained epitaxy* have been developed in a large number of semiconductor families. Through engineering the *lateral lattice constant* of the overgrowing crystals to match that of the template (i.e., often substrate), these techniques realize the so-called *pseudomorphic* mode of *epitaxy*.

1.6.2.2.1 Strain Calculation Depending on the sign of the algebraic difference between the *lateral lattice constant* of the substrate and that of the *freestanding* overgrown material, a built-in strain in either *tensile* or *compressive* form is induced at the *pseudomorphically* grown *heterointerface*. The presence of this strain, if *tensile*, limits the maximum thickness of the overgrown material. This is because the increasing of the thickness of the overlayer builds up strain energy, eventually surpassing the bonding energy of this film. Consequently, if not instantaneously, overtime cracks will appear in the film. In contrast, *compressive* strain poses no limit of this kind

[87] To be dealt with in detail in Chapter 5.

[88] That is, when grown as bulk and not in contact with another crystal.

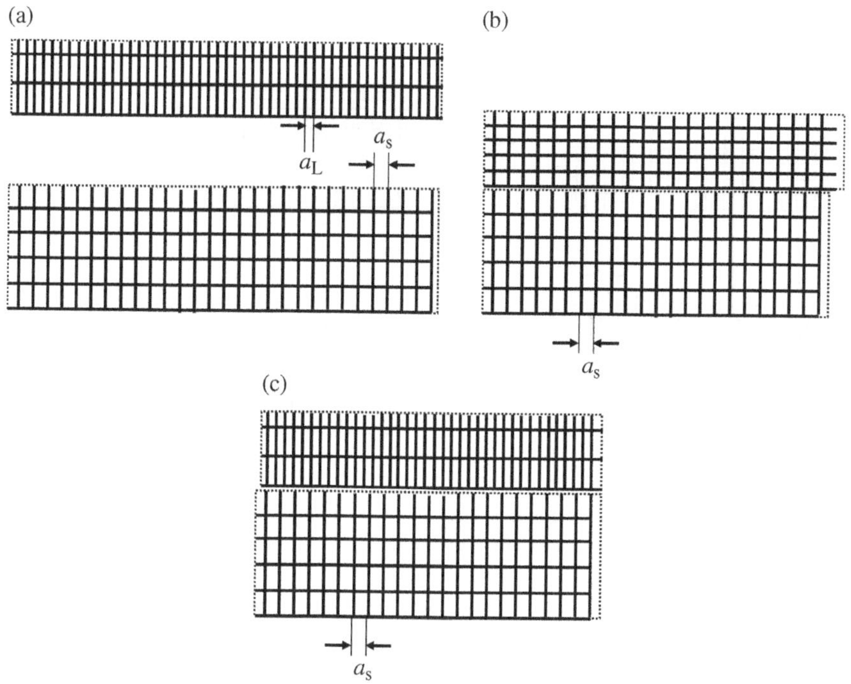

FIGURE 1.40 (a) Two lattice-mismatched 2-D lattices. (b) Strained epitaxial growth. Under this mode of tensile-strained epitaxy, the lateral lattice constant of the overlayer is expanded to match that of the substrate (i.e., a_S). (c) Lattice-mismatched epitaxial growth.

on film thickness. A film grown under *compressive* strain, however, eventually relaxes to its own *freestanding lattice constant* as its thickness grows beyond a critical value. Figure 1.41 provides schematics for these growth modes.

One of the recent developments in the area of semiconductor devices has taken shape in the form of engineering this *heteroepitaxial* strain for realizing different levels of charge concentration in the vicinity of the *pseudomorphically* grown *heterojunctions*.[89] This property has been used especially in polar *III-Nitride* technology, which enjoys large *piezoelectric coefficients* in all its *binaries* (i.e., *AlN*, *GaN*, and *InN*) and their *alloys*.

Another recent development in this area is the realization of *compliant* substrates. In these substrates, it is not just the overgrown material whose *lattice constant* is being modified. Researches on development of *compliant* substrates, which contribute to decoupling the strain induced at the *wafer* substrate interface from the *heterojunctions*, and also development of substrates for *heterostructures* such as *AlGaN/GaN*, which traditionally lack a viable *freestanding lattice-matched* substrate, are some of the major activities in this area. Such modes of *epitaxy* are also seen as major hopes

[89] We will provide an in-depth analysis of this situation in Chapter 5.

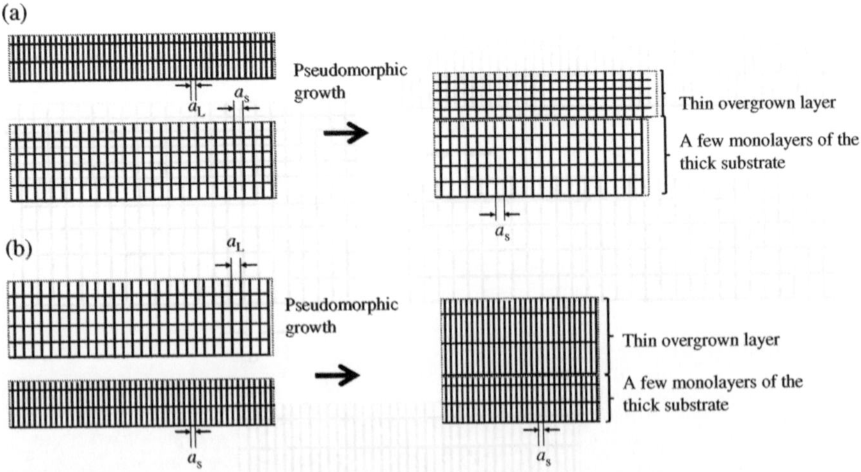

FIGURE 1.41 Two-dimensional schematic depictions of pseudomorphic growth under (a) tensile and (b) compressive strained epitaxy.

for integration of devices and circuits of different semiconductor technologies on the same *chip*.

In order to explain the outcomes of *lattice-mismatched epitaxy*, it is worth acquainting ourselves with its language of the *strain tensor*.[90] In a planar *epitaxial* growth, the overgrown layer is *biaxially strained* in the plane of substrate (indicated by $\epsilon_{\parallel}$) while *uniaxially strained* in a direction normal to the substrate (indicated by $\epsilon_{\perp}$). For a thick *noncompliant* substrate, the amount of strain in terms of *lateral lattice constant* of the substrate (i.e., a_S) and that of the *freestanding* overgrowing crystal (i.e., a_L) is calculated as

$$\epsilon_{\parallel} = \frac{a_S}{a_L} - 1 = \epsilon. \tag{1.105}$$

Since there is only *in-plane* stress, the amount of perpendicular strain $\epsilon_{\perp}$ will be calculated in terms of *in-plane* strain and *Poisson's ratio*[91] (i.e., σ),

$$\epsilon_{\perp} = \frac{-\epsilon_{\parallel}}{\sigma}. \tag{1.106}$$

According to this situation of *zero* stress along the growth direction, for a *strained growth* over a (001) substrate of *FCC* type, it can be shown that

[90] Within a more general context, the notion of *strain tensor* is presented in Appendix 1.C.

[91] Named after *Siméon Poisson*.

$$
\begin{aligned}
\sigma &= \frac{c_{11}}{2c_{12}} \\
\epsilon_{xx} &= \epsilon_{\|} \\
\epsilon_{yy} &= \epsilon_{\|} \\
\epsilon_{zz} &= -\frac{2c_{12}}{c_{11}}\epsilon_{\|}\,. \\
\epsilon_{xy} &= 0 \\
\epsilon_{yz} &= 0 \\
\epsilon_{zx} &= 0
\end{aligned}
\tag{1.107}
$$

c_{ij}'s present in (1.107) and the rest of this discussion are the *elastic constants*.

Such relationships are obviously developed on the basis of the knowledge of the plane over which growth is taking place. As another example in case of growth over a *(111) FCC* substrate, we have

$$
\begin{aligned}
\sigma &= \frac{c_{11}+2c_{12}+4c_{44}}{2c_{11}+4c_{12}-4c_{44}} \\
\epsilon_{xx} &= \left[\frac{2}{3}-\frac{1}{3}\left(\frac{2c_{11}+4c_{12}-4c_{44}}{c_{11}+2c_{12}+4c_{44}}\right)\right]\epsilon_{\|} \\
\epsilon_{yy} &= \epsilon_{xx} \\
\epsilon_{zz} &= \epsilon_{xx} \\
\epsilon_{xy} &= \left[-\frac{1}{3}-\frac{1}{3}\left(\frac{2c_{11}+4c_{12}-4c_{44}}{c_{11}+2c_{12}+4c_{44}}\right)\right]\epsilon_{\|} \\
\epsilon_{yz} &= \epsilon_{xy} \\
\epsilon_{zx} &= \epsilon_{yz}\,.
\end{aligned}
\tag{1.108}
$$

It can be observed from (1.107) and (1.108) that while in *(001)* growth the *strain tensor* is diagonal, in *(111)* growth (among a few other directions) the *strain tensor* has nondiagonal terms. The distortion caused by *strained epitaxy* to the *cubic* lattice, depending on the *growth orientation*, can produce a reduced degree of *crystal symmetry*. The nondiagonal terms introduced into the *strain tensor* by the reduced *crystal symmetry* are often used in production of built-in *polarization* fields in the *heterostructure*. Since we have already made a connection between the *degree of symmetry* and *transport*-related properties such as the presence of *degeneracies* at the *band* edges, we can expect the *strained epitaxy* to be capable of extending opportunities for engineering these properties. We will deal with these aspects in further details later in Chapter 4.

As we have already mentioned, due to the possession of large *piezoelectric coefficients*, *strained growth* has even more important impacts on electronic devices realized in *hcp pseudomorphic heterostructures* of *III-Nitride* semiconductors. These *heterostructures* are often realized in the form of $Al_xGa_{1-x}N$ or $In_xGa_{1-x}N$ *alloyed ternaries*[92] on a thick *relaxed GaN* layer, grown along *c-axis* of the *hexagonal* prism. In this case, the *strain tensor* is given by

[92] As an example, *alloyed ternary* of $Al_xGa_{1-x}N$ is created by random mixing of *AlN* and *GaN binaries* in proportions identified by $x(AlN) + (1-x)GaN$.

$$\begin{aligned}\epsilon_{xx} = \epsilon_{yy} &= \frac{a_S}{a_L} - 1 \\ \epsilon_{zz} &= -\frac{2c_{13}}{c_{33}}\epsilon_{xx}\end{aligned} \tag{1.109}$$

In Chapter 5, we will establish a connection between this component of strain and problem of charge induction into the *heterostructure's quantum well* formed at the *heterointerface*. Recently, a growing degree of interest is also observed on *a-axis* growth in this family. This endeavor has been developing with the goal of reducing the amount of *polarization*-induced charge.

1.6.2.2.2 Band Diagram Engineering The *band* structure of a semiconductor, which is naturally defined through its chemical composition and crystalline structure, can be engineered through a number of different ways including:

- Alloying two or more semiconductors
- Implementing quantum confinement in *heterostructures*
- Implementing built-in strain in *pseudomorphically* grown *heterojunctions*

BAND ENGINEERING THROUGH ALLOYING In the discussion of *alloyed epitaxy*, we have already indicated that a weighted averaging law can produce a first-order approximation for the properties of the semiconductor *alloy*. In regard to the application of this law (i.e., *Vegard's law*), it should be emphasized that this law of weighted averaging is applicable only when the *alloy* is random and the components in the *alloy* have the same crystalline structures. This law is not extendable to *phase-separated alloys*, which are *alloys* in which components of the *alloy* are separated into regions. Obviously, this need not be a *superlattice* for which case we have already hinted that *Vegard's law* is not applicable.

Most of the *alloys* used in semiconductor electronics are *random alloys*. In these *alloys*, in the absence of periodicity in the background crystal potential, the description of the *electron wave* in terms of a *traveling wave* (i.e., defined by *Bloch* theorem) is not possible. Instead, the *wave function* and the *probability density function* are position dependent. As will be discussed later in this chapter, this is a cause for *scattering* and degradation of *carrier mobility* in *alloyed* crystals.

In the application of *Vegard's law*, which is motivated by weighted averaging of the *virtual crystal approximation*, it should be appreciated that in most *alloys* bowing effects (i.e., in excess of linear averaging) arise from the increasing disorder due to *alloying*. This bowing is usually modeled through adding a parabolic term to *Vegard*'s linear weighted average. In the *virtual crystal approximation*, the *effective carrier mass* is defined in terms of the *effective carrier masses* in the parent semiconductors (i.e., A and B) of the *alloy* A_xB_{1-x}[93]:

[93] That is, x part A and $1-x$ part B.

$$\frac{1}{m^*_{\text{alloy}}} = \frac{x}{m^*_{\text{A}}} + \frac{1-x}{m^*_{\text{B}}}. \tag{1.110}$$

This is because in terms of the *Vegard's law*,

$$E_{\text{alloy}}(k) = \frac{\hbar^2 k^2}{2m^*_{\text{alloy}}} = x\frac{\hbar^2 k^2}{2m^*_{\text{A}}} + (1-x)\frac{\hbar^2 k^2}{2m^*_{\text{B}}}. \tag{1.111}$$

Based on (1.111), the law of weighted averaging used to determine the *electron affinity* and *bandgap* of a *direct-bandgap* alloy is formulated as

$$q\chi_{\text{alloy}} = xq\chi_{\text{A}} + (1-x)q\chi_{\text{B}} \tag{1.112}$$

$$E_{\text{g-alloy}} = xE_{\text{g-A}} + (1-x)E_{\text{g-B}}. \tag{1.113}$$

The *lattice constant* also approximately follows the same relationship.

As mentioned, Equation (1.113) can only be used when parent semiconductors are both *direct*. Otherwise, (1.111) should be used to evaluate the bottom of the *conduction band* for all values of k and then to identify the *bandgap* as the smallest energy gap between the *conduction* and *valence band*.

BAND ENGINEERING THROUGH QUANTUM CONFINEMENT AND STRAIN As indicated earlier in this section, the implementation of *carrier* confinement is another way for altering the *density of states* functions and the semiconductor *band* structure. Nowadays, with an increased intensity, a combination of crystal growth (i.e., *MOCVD* and *MBE*) and processing techniques are being explored for realizing 2-D, 1-D, and 0-D confined *carriers* in semiconductor *heterostructures*. Implementation of *heterostructures* in different material systems has created a variety of new possibilities for the design of electronic and optoelectronic devices. Differences between the size of the *bandgap* and the *electron affinity* of different semiconductors provide three different forms of *band lineup* between the crystals from which a *heterostructure* can be created (i.e., nested *bandgap*: *Type-I*, staggered *bandgap*: *Type-II*, and offset *bandgap*: *Type-III*). The three types of *band lineup* are depicted in Figure 1.42. The various possibilities for *band lineup* are among the determining factors for the choice of materials in designs involving *carrier* confinement.

Among the types of *band lineup*, *Type-I* is the one that is most studied for electronic applications. Especially in Chapter 5, an in-depth analysis of *Type-I band lineup* with respect to *FETs* is provided.

Type-II band lineup, because of its small *effective bandgap*,[94] is of considerable interest in the design of long wavelength optoelectronic devices. An important observation of *Type-II band lineup* is the small spatial separation of electrons and *holes*.

[94] Which is essentially formed between the *conduction-band* edge of one semiconductor and the *valence-band* edge of the other.

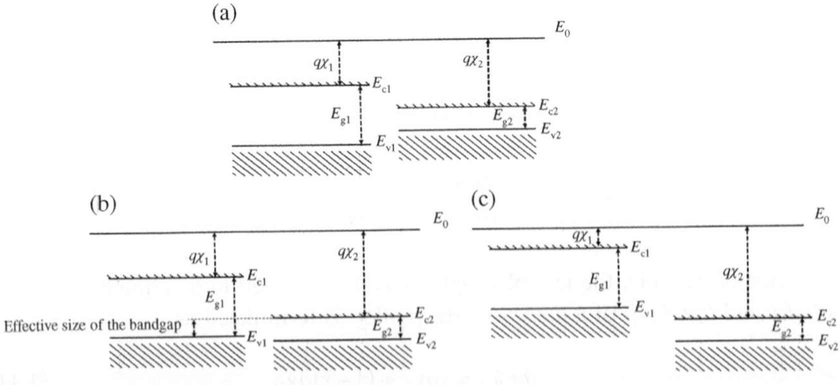

FIGURE 1.42 Types of band lineup: (a) *Type-I*, (b) *Type-II*, and (c) *Type-III*. Occupation of the bands by electrons is indicated by the hash marks.

In this case as demonstrated in Figure 1.42b, electrons and *holes* are accumulated on the opposite sides of the *heterojunction*. This can be very important in the design of *infrared detectors* when it comes to the consideration of instantaneous *EHP recombination*.

One should also be aware that, while oftentimes the definition of *electron affinity* is employed in envisioning the type of *band lineup* in a *heterostructure*, this definition is often not so accurate when two different semiconductors form a *heterostructure*. This is due to the presence of *charge sharing* across the *heterointerface* atoms. Although a few theoretical techniques have been developed to decide the type of *band lineup*, due to the complexity of these techniques, experiments are often used to shed light on the type of *band lineup*.

According to our discussions in Section 1.1.3 in terms of *subband* energy *levels*, the *density of states* functions in a *quantum well* is formulated as

$$N(E) = \sum_i \frac{m^*}{\pi\hbar^2} U(E - E_i) \text{ for conduction band} \tag{1.114}$$

$$N(E) = \sum_i \sum_{j=1}^{2} \frac{m_j^*}{\pi\hbar^2} U\left(E_{ij} - E\right) \text{ for valence band,} \tag{1.115}$$

where U is the *Heaviside* step function and E_i is the *subband* energy *level*. The presence of double subscripts in Equation (1.115) points to the lift in *degeneracy* of *heavy-* and *light-hole subbands* even in the absence of strain.

Often in simple descriptions of *quantum wells*, the *conduction-band states* are seen as pure s-type *states* and a simple *effective-mass* theory like the one suggested in (1.13) is employed. For a more accurate description, a full *band* structure, which is referred to as an *eight-band model*, is employed. As a result of this more sophisticated calculation, it is observed that while the *conduction-band states* in *unstrained heterostructures* are not affected by the incorporation of this fuller mode, in *highly strained*

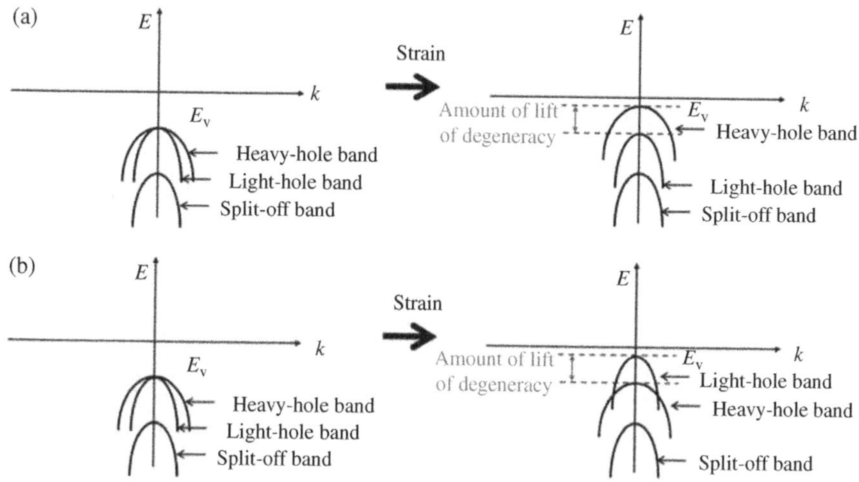

FIGURE 1.43 Lifting the valence-band degeneracy of heavy- and light-hole bands through the application of (a) compressive and (b) tensile biaxial strain, depicted on the *E*–*k* diagram.

heterostructures the more complicated calculation is definitely required. An example of this would be the case of so-called self-assembled quantum dots. These are 3-D *quantum wells* that are created as the result of large *lattice mismatch* between the substrate and the *freestanding lattice constant* of the overgrown material. Excessive *tensile* strain induces a 3-D mode of growth instead of a planar growth. The density of these islands of 3-D growth is determined by the material properties and *lattice mismatch*. These *self-assembled quantum dots* have been realized between *AlAs* and a *GaAs* substrate.

The description provided in (1.115) of *subbands* for *valence-band states* is only approximately valid. While *heavy-hole* and *light-hole states* are pure *states* at $k = 0$, they strongly mix away from $k = 0$.

Built-in strain has been successfully used to lift *degeneracies* in the *band* edges, change the character of *band*-edge *wave functions*, and engineer the *density of states* at the *band* edges.

In *direct-bandgap* semiconductors, with regard to the *conduction band*, strain only moves the position of the *band edge* and has a very limited impact on the *carrier mass*. Because the bottom of *conduction band* is not *degenerate*, this shifting of the *band edge* does not result in lifting any *degeneracy* or change in the *DOS* and *effective mass*. However, while the *valence-band* edge of semiconductors are *degenerate*, strain causes lifting of that *degeneracy*. The amount of lift in *degeneracy* caused by quantum confinement is usually about 10–15 meV and is less than the amount of shift due to strain. *Compressive biaxial* strain raises the *band edge* and lifts the *degeneracy* on the order of 100 meV. In this case, the *HH band* is lifted above *LH band*. Under *tensile biaxial* strain this order is reversed. This is depicted in Figure 1.43.

In *indirect-bandgap* semiconductors, due to the presence of *degeneracy* in the *conduction-band* edge, the role of strain in modifying the *conduction-band* characteristics is much different. Whereas the bottom of *conduction band* of a *direct* semiconductor is in the Γ-*valley* (which has no *degeneracy*), for the case of the *indirect* semiconductor *Si*, this is located close to the *X-valley*. According to the form of the *unit cell* of the *diamond* crystal of *Si* and *Ge*, there are six equivalent faces (or, consistently, six equivalent *X-valleys* in their *Brillouin zone* or *E–k* diagram).

As a result, in contrast to the *direct-bandgap* semiconductors, the *conduction band* of *indirect alloy* of Si_xGe_{1-x} is significantly altered by the strain imposed through *pseudomorphic* growth on *Si* substrate. In the case of *(001)* growth, strain lifts in the *degeneracy* between the six equivalent *valleys* of the *conduction band*. This *biaxial compressive* strain causes breaking of the sixfold *degeneracy* into a fourfold (*in-plane*) and a twofold (*out of plane*) *degeneracy*, where the fourfold *degeneracy* is lower in energy. The resulting reduction of *DOS* at the bottom of the *conduction band* helps with reducing the *effective electron mass*. An additional result of this lift in *degeneracy* is a rapid reduction of the size of the *bandgap* of *SiGe alloy* with the *Ge mole fraction* (i.e., $1-x$). In the context of improving the *carrier-transport* properties of the *channel* of the modern *FET* technologies, this concept will be heavily dealt with in Chapter 4.

Despite the differences in the *conduction-band* structures, due to the similarity of the *valence bands* of *direct* and *indirect* semiconductors, the aforementioned discussion on the impact of strain on the *valence band* of *direct* semiconductors is readily extendible to *indirect* semiconductors.

In the *valence band*, lift in *degeneracy* is also accompanied by large changes in the *band* curvature. In the edge *states* of this *band*, strain can cause the *DOS effective mass* to be scaled down by as large as a factor of *three*.

We have already talked about the consequences of the formation of the bottom of the *conduction band* of *Si* near the *X-valley*. In this case, constant-energy surfaces for electrons form *six* ellipsoids along the *<100>* directions. The ellipsoidal form is expressed in terms of two values of effective mass. For an *x–y* surface, whereas the *two* ellipsoids in the *z*-direction have an *effective mass* (referred to as *longitudinal*), which is equal to $m_l = 0.98m_0$, the other *four* ellipsoids' electron *effective mass* is the *transverse effective mass* (given by $m_t = 0.19m_0$). Among these, in determining the *subband* energies, the *effective mass* in the direction of the confining potential is employed.

In calculating the *DOS* function, it is easy to incorporate ellipsoidal shape of the constant-energy surfaces. In the case of *silicon*, m^* must be replaced by $\left(m_l^* m_t^{*2}\right)^{1/3}$. Multiplying this by 6 takes care of the sixfold *degeneracy* inside the *Brillouin zone* of *silicon*.[95] In the case of *germanium*, because the *conduction band* is defined at the *L-valley*, an eightfold *degeneracy* is present. Hence, rather than *six degenerate* ellipsoids, *eight* are present. However, since only *one-half* of each ellipsoid falls inside the *Brillouin zone*, the overall *degeneracy* will be only fourfold. Instead of multiplying the mass by 6 (as is the case with *silicon*), it should only be multiplied by 4.

[95] In the calculation of *conductivity* in *silicon*, however, $\frac{1}{m_n^*} = \frac{1}{3}\left(\frac{1}{m_l^*} + \frac{2}{m_t^*}\right)$.

1.7 QUANTUM TRANSPORT PHENOMENA AND SCATTERING MECHANISMS IN SEMICONDUCTORS

Whereas *scattering* processes are of interest to physicists in studying the electron–electron and electron–matter interactions at a fundamental level, electrical engineers need to know how these processes contribute to the problem of *charge-carrier transport* in a semiconductor. In evaluation of the *conductivity* of semiconductors, accurate understanding of the motion of electrons in the *solid* (i.e., including the *scattering* events) is required. It is only in the presence of *scattering* events (i.e., imperfections) that the known notion of *conductivity* in terms of the *Ohm*'s law is extendible to semiconductors. Without these imperfections, as mentioned earlier, *ballistic transport* and a number of other important modes of *persistent current* or *oscillations* will prevail.

In order to establish the connection between *conductivity* and *scattering*, the *velocity–field* relationships of different semiconductors are studied. These relationships, as encountered in Section 1.4, are often illustrated in terms of the *drift* process of *charge carriers* under steady-state conditions. According to those discussions, in the description of the steady-state *drift-transport* characteristics, three regions are identified:

1. When the electric field is low and *drift velocity* changes as a linear function of the electric field: $\vec{v_d} = \mu \vec{E}$. Under this condition, the *Boltzmann transport equation* yields an analytical form.
2. When the electric field takes on moderate values (usually larger than 1 kV/cm). Under this regime, $v_d - E$ does not follow a linear trend. Numerical methods are often used in evaluation of *carrier transport* under this regime. Interpolation of an analytical relationship between the *drift velocity* and electric field is often used to produce an analytical basis for evaluation of *carrier transport*.
3. When the electric field exceeds the breakdown field of the semiconductor (usually larger than 10^5 kV/cm). Under this regime, either due to *impact ionization* or electron *tunneling* from *band* to *band* (i.e., *Zener* breakdown), the semiconductor breaks down.

As mentioned in Section 1.4, in order to reach steady state, an electron needs to undergo several tens of collisions. Considering the collision times on the order of *picoseconds* and assuming electrons to travel at velocities as high as 10^7 cm/s, the traveling distance to reach steady state will amount to *micron*-size distances. As a result, in small submicron devices the steady-state *transport* formalism becomes less and less applicable.

It has been already indicated in this chapter that electrons can be *scattered* through a variety of *elastic* and *inelastic scattering* processes. Two major categories of these processes are *ionized-impurity scattering* and lattice-vibration (or *phonon*) *scattering*. It should also be mentioned that *lattice vibrations* themselves (i.e., *phonons*) are also *scattered* by *elastic scatterers* such as *ionized* and *neutral impurities*, the presence of different atomic isotopes, and surfaces. This is the cause for the slow propagation of heat in a semiconductor, which is identified in terms of *heat resistivity*. While ideally in a *solid* heat should propagate with the *velocity of sound*, these *scattering* mechanisms cause the *heat transfer* to become much slower.

1.7.1 Quantum Phenomena in Carrier Transport: A Snapshot

We already know that in the presence of *scattering* events, such as *lattice vibrations* (which result in *inelastic scattering*), a few of the important predictions of quantum mechanics are not realizable. Among these is the *Bloch oscillation*. This mode of oscillation originates from reflection of *electron waves* at the boundaries of the *Brillouin zone*. According to *Bloch theorem*, electrons in a perfectly periodic crystal follow the *bands* to the edges of the *Brillouin zone* and then return. This situation for an arbitrary *E–k* diagram is depicted in Figure 1.44. Considering the small size of the *Brillouin zone*, these oscillators are expected to operate at very high frequencies.

However, in the presence of *phonon scattering*, due to the *inelastic* nature of the *scattering* process, electrons do not get a chance to reach the boundaries of the *Brillouin zone*. As a result all attempts to realize a *Bloch oscillator* have remained unsuccessful.

Another mode of oscillation envisioned for *perfect* semiconductors is referred to as *Esaki–Tsu oscillation*. Although these oscillations are also never realized in practice, the theoretical drive behind them prompted major advances in *epitaxial* growth of *superlattices*. As shown later, quantum mechanics predicts that the application of a constant *DC* electric field to a periodic structure (i.e., with periodic *E–k* diagram) results in very high-frequency oscillations. The frequency of these oscillations is determined proportional to the *lattice constant* of the periodic structure. These oscillators in theory are capable of generating very high-frequency tunable *signals*.

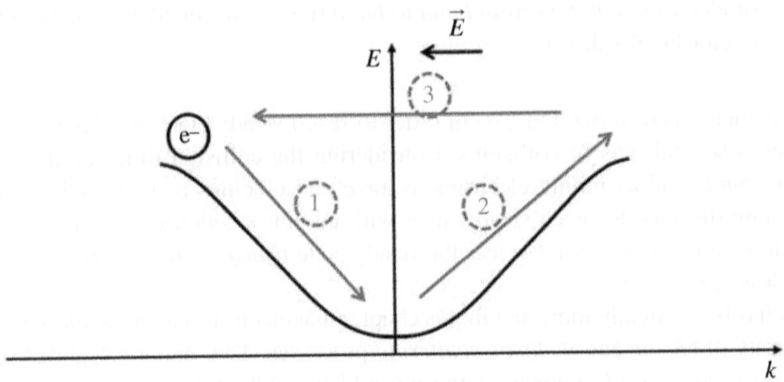

FIGURE 1.44 Trajectory of movement of an electron in an arbitrary band. In a perfect crystal, electron after climbing up the band tail, reflects upon reaching the zone edge. This reflection looks like scattering by a reciprocal lattice vector. The numbers are supposed to provide a snapshot of the position of electron in sequential instances of time.

In order to gain a first-order mathematical insight into the *Esaki–Tsu oscillators*, we will consider the *band* structure of a one-dimensional crystal. This *E–k* diagram similarly applies to a simple cubic three-dimensional crystal,

$$E(k) = E_0(1 - \cos ka). \tag{1.116}$$

As suggested in (1.12), for electrons in the *conduction band*, we can calculate their *group velocity* as

$$v = \frac{1}{\hbar}\nabla_k E(k) = \frac{1}{\hbar}\frac{dE}{dk} = \frac{E_0 a}{\hbar}\sin(ka). \tag{1.117}$$

In addition, in terms of the definition of *momentum* in quantum mechanics, we know that $\hbar(dk/dt) = -q\mathrm{E}$. This equality results in

$$k = -q\mathrm{E}\frac{t}{\hbar} + k_0. \tag{1.118}$$

Assuming that the electron starts at *zero* momentum at $t = 0$, we will have

$$v = -\frac{E_0 a}{\hbar}\sin\left(q\mathrm{E}a\frac{t}{\hbar}\right). \tag{1.119}$$

According to (1.119), the current density created by the constant electric field E is a sinusoidal *signal* of angular frequency: $(q\mathrm{E}a)/\hbar$. Clearly this angular frequency can be tuned by changing the period of the structure of the crystal and its *lattice constant*: *a*. As a result, realization of a *superlattice* with a very large period is expected to result in materialization of a very high-frequency oscillation.

1.7.2 Drude's Model: A Close-UP

Due to the presence of a large number of interfering factors (including dependence of *scattering* on the momentum of electrons and *phonons*, presence of many electrons and as a result importance of consideration of *many-body interactions*, requirement for statistical evaluation of electron propagation, and the role played by the *band* structure in *transport*), the theory of *carrier transport* is quite complicated. However, as long as the electric field is weak to moderate, instead of a detailed description of all these intricacies, simpler models are usually adopted.

1.7.2.1 Boltzmann Transport Theory The *Boltzmann transport equation* (i.e., *BTE*) is the foundation of these simpler formalisms, through which the *mean scattering time constant* needed in calculation of low-field *mobility* is evaluated (see (1.64) and (1.65)). According to this equation, upon application of an external perturbation to a system under *thermal equilibrium*, the distribution functions of *charge carriers* can still be represented in terms of the *Fermi–Dirac* distribution function (i.e., *thermal-equilibrium* distribution). As we took note in Section 1.3.1, such a distribution function shows the spread of electrons in *energy space* (and as result *momentum space*). According to *BTE*, this distribution function is then used to evaluate all of the *transport* properties of *charge carriers*.

Assuming $f_{\vec{k}}(\vec{r})$ as the local occupation function of electron in *momentum state* $\vec{k}$ and in position $\vec{r}$, the first step in *Boltzmann transport theory* is to study the time evolution of this function. This time evolution is prompted by the *thermal motion* of the electrons, the *drift* of electrons due to an external force, or due to *scattering* between different *momentum states*. For these three different causes, we can explain the time evolution of $f_{\vec{k}}(\vec{r})$ in terms of the following equations:

1. *Thermal motion* of the electrons

$$\left.\frac{\partial f_{\mathrm{k}}}{\partial t}\right|_{\mathrm{diff}} = -\frac{\partial f_{\mathrm{k}}}{\partial \vec{r}} \cdot \vec{v}_{\mathrm{k}}, \tag{1.120}$$

where $\vec{v}_{\mathrm{k}}$ is the velocity of a *carrier* in the *state* $\vec{k}$.

2. *Drift* of electrons due to an external electromagnetic force

$$\left.\frac{\partial f_{\mathrm{k}}}{\partial t}\right|_{\text{external force}} = -\frac{q}{\hbar}\left[\vec{E} + \vec{v} \times \vec{B}\right] \cdot \frac{\partial f_{\mathrm{k}}}{\partial \vec{k}}. \tag{1.121}$$

3. *Scattering* between different *momentum states*

$$\left.\frac{\partial f_{\mathrm{k}}}{\partial t}\right|_{\text{scattering}} = \int \left[f_{\mathrm{k}'}(1-f_{\mathrm{k}})W(k',k) - f_{\mathrm{k}}(1-f_{\mathrm{k}'})W(k,k')\right]\frac{d^3k'}{(2\pi)^3}, \tag{1.122}$$

where $W(k, k')$ represents the rate of *scattering* from *momentum state* k to k'. According to the process of *microscopic reversibility*, the *scattering* rates between the before and after *momentum states*, as long as *scattering* is elastic, are equal:

$$W(k,k') = W(k',k). \tag{1.123}$$

The inclusion of the *scattering* rate in *BTE* is rooted in the wave nature of electrons.

The role of time constant τ in *Boltzmann transport theory* is to model the time constant for relaxation of the aforementioned perturbations. This is based on the so-called relaxation time approximation. As suggested earlier, this time constant can be calculated according to the *scattering* rate of different *scattering* processes. Under steady-state conditions, the time evolutions created by the aforementioned sources cancel one another out.

In formulation of the *scattering* rate, attention should be given to the nature of the collision processes, which cause the *scattering*. For example, *alloy scattering* (which is present in semiconductor *alloys*, due to their deficient periodicity) and *impurity scattering* are results of collisions in which electron's energy remains unchanged (i.e., *elastic* collisions). However, many *scattering* processes are *inelastic*. For example, *phonon scattering* due to the change caused in the electron energy is an *inelastic scattering* process. Instigated by the domination of different *scattering* processes in *carrier transport* and a number of other complications listed in Section 1.4.3, *Hall mobility* can be quite different from *drift mobility* (see (1.85)).

In the case of *elastic scattering* processes in parabolic *bands*, the calculation of *relaxation time* is quite trivial. However, many of the *scattering* processes are not so lenient.

As long as the energy gained from the electric field is smaller than the thermal energy, *carrier transport* stays under the linear regime expressed in terms of the low-field *mobility*. However, with further increase of energy, the simple approximations used in solving the *Boltzmann transport equation* will become insufficient. As mentioned earlier in this section, under this regime complex numerical techniques such as the *balance equation* and *Monte Carlo* method are often used. The continuity equations, (1.49) and (1.50), are two of these so-called balance equations. *Monte Carlo* method treats electrons as particles whose *scattering* events between *Bloch states* are described probabilistically through the so-called Fermi golden rule. In this numerical technique, *carrier transport* is seen as periods of free flight and instantaneous *scattering* events, which are accurately described in terms of the probability of the involved *scattering* processes.

Monte Carlo has been proven to be a versatile technique in addressing a variety of *transport*-related problems, such as evaluation of steady-state *drift-transport* characteristic, *electron temperature*, *valley* occupation, distribution in *k-space*, noise, *ballistic transport*, transit time, *carrier* injection/*thermalization* process,[96] and *impact ionization*.

1.7.2.2 Drude's Model *Drude*'s model of conduction, which is devised on the basis of the *Boltzmann transport equation*, is the model that we have so far adopted in this volume (i.e., in the definition of low-field *diffusion constant* and *mobility*). Now that we are further acquainted with the foundations of this model, before getting too involved in a rigorous discussion of *scattering* rates of different *scattering* processes, it seems quite instructive to pay a closer visit to this model. As implicitly suggested in Section 1.4, *Drude*'s model is built on the following assumptions:

1. All electrons move with the same velocity v, which in terms of applied force F_0 is identified as $\hbar k = mv = F_0 t$.
2. Instead of taking the whole *band* structure into account, the *effective-mass* notion (i.e., m^*) has been introduced to take the position of the mass of an electron.

[96] *Thermalization* of electrons refers to the process of electrons losing energy and coming back to the bottom of *conduction band* through emitting *phonons*.

3. *Scattering* processes are envisioned as a friction force in the above equation of motion (i.e., F_f), which is defined in terms of the relaxation time constant of the friction force: $F_f \cong (mv)/\tau$. In terms of the *Boltzmann transport equation*, the presence of this friction force is presented through the modified electron distribution function: $f_{\vec{k}}(\vec{r}) = f^{0}_{\vec{k}+\frac{q\tau E}{\hbar}}(\vec{r})$.

As a result of these assumptions, the total equation of motion is given by[97]

$$m^*\frac{dv}{dt} = F_0 - F_f = F_0 - \frac{m^* v}{\tau}, \tag{1.124}$$

where in the absence of an applied force (i.e., F_0),

$$m^*\frac{dv}{dt} = -\frac{m^* v}{\tau} \tag{1.125}$$

and

$$\vec{v} \propto \exp\left(-\frac{t}{\tau}\right). \tag{1.126}$$

As a result, τ will also be the time constant in decaying $\vec{v}$. The current density is then evaluated by multiplying this velocity by the charge of an electron and the electron concentration.

In this framework, in the presence of a number of independent *scattering* processes, the total *scattering* rate can be calculated in terms of *mean scattering* times of individual *scattering* processes

$$\frac{1}{\tau_{\text{tot}}} = \sum_i \frac{1}{\tau_i}. \tag{1.127}$$

Only when the various *scattering* rates follow the same energy dependence (i.e., the same *effective mass*), the above equation results in an overall low-field *mobility* that follows *Matthiessen's rule*[98]:

$$\frac{1}{\mu_{\text{tot}}} = \sum_i \frac{1}{\mu_i} \tag{1.128}$$

However, the above condition is usually not satisfied. In spite of this, *Matthiessen*'s rule has been widely used and found to be reasonably accurate.

[97] As suggested in (1.121), magnetic force can be easily incorporated into this equation (i.e., $m^*\frac{dv}{dt} = -q\left(E + \vec{V}\times\vec{B}\right) - \frac{m^* v}{\tau}$).

[98] Named after *Augustus Matthiessen*.

As examples for the application of *Drude's theory*, in the presence of only an electric field, it is instructive to apply (1.124) to the following two special cases:

1. When the electric field is finite but time independent.
   Due to lack of time dependence in the external source of energy, this case refers to a steady-state situation where $\frac{dv}{dt}=0$ and $m^* \vec{v} = -q\vec{E}\tau$. This relationship results in a *DC* current of

$$J = -qnv = \frac{q^2 \tau n}{m^*} E \tag{1.129}$$

   with no chance of *persistent current* and *oscillation* (since in *Drude*'s model the full E–k diagram has been overlooked by the notion of the *effective mass*). This special case refers to the case that we have investigated earlier in our discussions in Section 1.4. In this case, the *conductivity* σ is observed to be a scalar value given by $\sigma = qn\mu$, where $\mu = q\tau/m^*$.
2. When the electric field is a low-frequency field.
   The definition of the "low frequency" is given in relation to the *energy relaxation time*. In this case, $E \equiv E_{ac}\exp(j\omega t) + E_{ac}\exp(-j\omega t)$. In solving for $\vec{v}$, the velocity vector is written as $\vec{v} = v_0\exp(j\omega t) + v_0\exp(-j\omega t)$, which results in

$$m^* j\omega v_0 e^{j\omega t} = -qE_{ac}\exp(j\omega t) - m^* v_0 \frac{\exp(j\omega t)}{\tau}. \tag{1.130}$$

   Accordingly,

$$-v_0 = \frac{qE_{ac}}{jm^*\omega + (m^*/\tau)} = \frac{\mu E_{ac}}{1 + j\tau\omega}. \tag{1.131}$$

   This result indicates that the *drift velocity* is equal to the *DC* velocity divided by $1 + j\tau\omega$. As a result, at frequencies comparable to $1/\tau$ (usually greater than 10 GHz), the semiconductor instead of a purely resistive behavior expresses a resistive–inductive behavior. The imaginary part can be seen as a contribution of free electrons to the *dielectric constant* of the semiconductor. The other interesting point to mention is that the real part is inversely proportional to the relaxation time at very high frequencies, meaning that for this frequency range, the *conductivity* is 0 even in the absence of *scattering*.

1.7.3 Major Scattering Processes

According to *Boltzmann transport theory*, *scattering* rates between different *states* of *momentum space* are to be evaluated for a range of important *scattering* processes. Before embarking any deeper into this discussion, it is worth pointing out that *momentum space* (which is also referred to as *k-space*) is really a *Fourier space*.[99] In this analogy, it should be emphasized that in treating a *scattering* problem, a *scattering* matrix should be established between each initial *state* and a range of final *states* of an electron. These initial and final *states* are *plane wave states* that can be calculated by taking the *Fourier transform* of the potential (i.e., solving the *Schrödinger equation* for *carrier wave function*). Analogous to the role of *Fourier series*, the transformation from almost periodic *real space* to *k-space* sometimes greatly simplifies the equations. As a simple example, the *Poisson equation* according to this transformation will be simplified as

$$\nabla^2\phi = -\frac{\rho}{\epsilon} \rightarrow \phi = \frac{\rho}{\epsilon k^2}. \tag{1.132}$$

At this point in our discussion, without getting into details, we present a few important facts such as *scattering* rates to explain different *scattering* processes that are encountered by electrons in semiconductors. A few of these processes, such as *ionized-impurity scattering* and *phonon scattering*, have been introduced in Section 1.4. In addressing *ionized-impurity scattering* and the *Coulombic* potential imposed by these impurities on *charge carriers*, one needs to pay attention to the issue of *charge screening*. As an example, we can consider the case of a positive charge placed in an *electron gas*. Movement of electrons around this charge can essentially *screen off* its potential. As will be seen in this section, in the case of *ionized-impurity scattering* and also *carrier–carrier scattering*, *screening* can be dealt with by replacing the *Coulombic* potential of a *point charge* by a *screened Coulombic* potential.

1.7.3.1 Ionized-Impurity Scattering In this case, the *scattering* rate given by the *Fermi golden rule* in terms of *Dirac's delta function* is represented by

$$W(k,k') = \frac{2\pi}{\hbar}\left(\frac{Zq^2}{V\epsilon}\right)\frac{\delta(E_k - E_{k'})}{\left(4k^2\sin^2(\theta/2) + \lambda^2\right)^2}, \tag{1.133}$$

where Zq, V, and θ represent the charge of the impurity, volume, and polar *scattering* angle, respectively. ϵ is the semiconductor's permittivity. In this equation, λ is defined as

$$\lambda = \sqrt{\frac{n_0 q^2}{\epsilon kT}}, \tag{1.134}$$

where n_0 is the mean background *carrier* concentration.

In the presence of a large *free carrier* concentration, a *screened Coulombic potential* has been used in this framework:

$$\phi(r) = \frac{q}{4\pi\epsilon r}\exp(-\lambda r). \tag{1.135}$$

[99] Named after *Joseph Fourier*.

For the situations without *screening* (i.e., low *free carrier* concentration), the rate equation is evaluated when λ tends toward 0, for which case

$$W(k,k') \propto \frac{1}{16k^4 \sin^4(\theta/2)}. \quad (1.136)$$

However, under strong *screening* $\lambda \rightarrow \infty$ and

$$W(k,k') \propto \frac{1}{\lambda^4}. \quad (1.137)$$

On this basis, one can see that while forward *scattering* is dominant in the case of weak *screening*, for strong *screening* angular dependence is not present. As a result, because forward *scattering* does not imply *carrier mobility* reduction, the *ionized-impurity scattering* in presence of weak *screening* is less important.

In the formalism presented in (1.133), *ionized impurities* are treated independent of one another (i.e., the average distance >10 nm). This assumption is not extendable to heavily *doped* semiconductors (i.e., $>10^{18}\ \mathrm{cm}^{-3}$) for which case the *ionized-impurity scattering* is more complex. For a *degenerate* semiconductor, the *screening parameter* λ should be changed to

$$\lambda = \sqrt{\frac{3n_0 q^2}{2\epsilon E_F}}, \quad (1.138)$$

where E_F refers to the *Fermi energy level* measured from the *band* edge.

The Equation (1.138) results in a faster drop in *mobility* with the concentration of *ionized impurities*. This is due to the effect of multi-impurity *scattering*.

In *ionized-impurity scattering*,

$$\mu \propto \left(Z^2 N_i\right)^{-1} \quad (1.139)$$

where N_i is the concentration of *ionized impurities* and Z, as indicated earlier, is the charge of the *donor*. The presence of Z in this relationship explains one of the reasons why impurities that can offer more than *one* electron from each atom are not used as proper *dopants*. In order to explain this further, we can imagine *dopant* atoms that offer *two* electrons instead of *one*. In this case, to achieve the same electron concentration, Z^2 will become *four* times larger, whereas N_i will be only divided by a factor of 2. As a result, according to this change of *dopants*, *mobility* will be reduced by 50%.

As indicated in Section 1.4.2.1, *carrier mobility* as a function of temperature follows an improving characteristic (which is of the form $\mu \propto T^{3/2}$), when *ionized-impurity scattering* is dominant. This is a distinguishing feature of this process of *scattering*.

1.7.3.2 Alloy Scattering We have already suggested that the root of this *scattering* mechanism is the presence of disorder in the crystal's potential. According to the *hard sphere model*, the *scattering* potential is represented by

$$\Delta V(r) = \begin{vmatrix} V_0 & \text{for } |r| \le r_0 \\ 0 & \text{for } |r| > r_0 \end{vmatrix} \tag{1.140}$$

in which r_0 and V_0 refer to interatomic distance and the maximum potential difference between any two points in the *solid*, respectively. Using the *Fermi golden rule*, the *scattering* rate of an *alloy scattering* process is represented by

$$W(k) = \frac{2\pi}{\hbar} \sum_{k'} |M_{kk'}|^2 \delta(E_k - E_{k'}) \tag{1.141}$$

where matrix elements are given by

$$M_{kk'} = \int \exp\left[j\left(\vec{k} - \vec{k'} \right) \cdot \vec{r} \right] \Delta V(r) d^3 r. \tag{1.142}$$

Using the fact that *scattering* potential extends only to a *unit cell*, the exponential term in the above equation tends toward one. As a result, in *alloy scattering* there is no angular dependence because there is no k and k' dependence on the matrix elements. Hence, for an isotropic *density of states* function, there will be no angular dependence of *scattering* rate. After performing a proper ensemble averaging, *relaxation time* of *alloy scattering* results in

$$\frac{1}{\langle\langle \tau \rangle\rangle} = \frac{3\pi^3}{8\hbar} V_{\text{unit}} V_0^2 x(1-x) \frac{m^{*3/2} (kT)^{1/2}}{\sqrt{2}\pi^2 \hbar^3} \frac{1}{0.75} \tag{1.143}$$

where V_{unit} is the volume of the unit cell.

This equation shows that the *carrier mobility* in terms of *alloy scattering* degrades with temperature as

$$\mu \propto T^{-1/2}. \tag{1.144}$$

In Equation (1.143), V_0 is on the order of 0.5 eV. The assumptions behind these equations are that there are no clusters formed in the *alloy* and that the smallest region over which a disorder is present is a *unit cell*.

1.7.3.3 Neutral Impurity Scattering The presence of neutral impurities and defects is another reason for *carrier scattering* in semiconductors. This presence can be caused by *substitutional* impurities and *dopants* that are not *activated*. *Scattering* by neutral impurities can be seen in the same light as *alloy scattering* (i.e., in the form of a disturbance to the periodicity of the crystal potential seen by *Bloch states*). The resulting *scattering* rate, not unlike the case of *alloy scattering*, is given by

$$W(k) = \frac{2\pi}{\hbar} \left(\frac{4\pi}{3} r_0^3 V_0 \right)^2 N(E_k) \tag{1.145}$$

in which V_0 denotes the *scattering* potential and r_0 stands for the radius of the hard sphere representing the defect. Accordingly, the *scattering* time constant is given by

$$\frac{1}{\langle\langle\tau\rangle\rangle} = N_{\text{imp}} \frac{2\pi}{\hbar} \left(\frac{4\pi}{3} r_0^3 V_0 \right)^2 \frac{m^{*3/2} (kT)^{1/2}}{\sqrt{2}\pi^2 \hbar^3} \frac{1}{0.75} \tag{1.146}$$

where N_{imp} stands for the concentration of neutral impurities.

Only in the presence of a very large concentration of these types of impurities ($>10^{18}\ \text{cm}^{-3}$) is the *neutral impurity scattering* mechanism worthy of consideration.

1.7.3.4 Interface Roughness Scattering The *channel* of the majority of *FETs* is formed in the vicinity of an interface (i.e., of either two semiconductors, the interface of a semiconductor and a metal or the interface of a semiconductor and an insulator). As a result, a *scattering* process often encountered in the *channel* of these *transistors* is the *interface-roughness scattering*. The degree of roughness at these interfaces is dependent on fabrication technology. Roughness at these interfaces imposes an element of disturbance on the potential felt by *Bloch states*. This is in addition to the role played by the *surface states* (as either neutral or charged *states*), which were discussed in Section 1.1.6. The *scattering* rate caused by these interface potential bumps in an *inversion MOSFET channel* to be explored in Chapter 3 is formulated as

$$W(k) = \frac{1}{A} \frac{2\pi}{\hbar} \frac{1}{4\pi^2} \int_0^{2\pi} d\theta \int_0^{\infty} q dq |M(k,k')|^2 \delta(E_k - E_{k'}) \tag{1.147}$$

where A is the area and $q = 2k\sin(\theta/2)$.

In this equation, it is assumed that half of the electric field drops across the insulator.

1.7.3.5 Carrier–Carrier Scattering Whereas the *scattering* sources so far discussed in this section are fixed in time and in space, not all *scattering* sources operate in this way. *Carrier–carrier scattering* is one of the *scattering* processes that belongs to this latter group. *Scattering* processes that are fixed in time and in space result in *elastic* scattering (i.e., *zero* change in *carrier* energy), which is due to the large mass of *scatterers* in those processes. This is not the case in *carrier–carrier scattering*. *Carrier–carrier scattering* can be in the form of *electron–hole scattering* or *electron–electron scattering* (which is analogous to *hole–hole scattering* in the *valence band*). Due to the *Fermionic* nature of electrons, *scattering* of two identical *carriers* is a more complex process. *Carrier–carrier scattering* gains significant importance only when *carrier* concentration exceeds $10^{18}\ \text{cm}^{-3}$.

As an approximation, assuming that the *effective mass* of a *hole* is much larger than that of an electron, we can envision the role of *electron–hole scattering* on electrons in a p-type semiconductor (*minority-carrier scattering*) by multiplying the *scattering* rate due to *ionized acceptor impurities* by a factor of 2.

1.7.3.6 Auger Processes and Impact-Ionization Scattering An important characteristic of these two *scattering* processes is that they result in a change in *carrier* concentration in the *conduction* and *valence band.* While in *Auger*[100] process, *scattering* results in *recombination* of an electron and a *hole* (and as a result reduction of *carrier* concentration in both *conduction* and *valence band*), *impact ionization* results in an increase in *carrier* concentration in both *bands* (i.e., through *EHP generation*). We have already visited the process of *impact ionization* while addressing the *breakdown* processes in Section 1.5.

Auger is the inverse process of *impact ionization*. This is a *nonradiative recombination* process and is detested in optoelectronic *photon* generators. According to this process, even in *direct-bandgap* semiconductors, an electron and a *hole* can recombine without *generation* of a *photon*. This is possible since in this interaction the energy is being transferred to another *carrier* or to a *phonon*. These processes can be assisted by *Coulombic* interactions (i.e., *electron–electron scattering*), by *phonons*, or by *trap states*. Only in high-purity *direct-bandgap* semiconductors is the *photon generation* dominant.

In *impact-ionization* process, as noted in Section 1.5, *carriers* gain energies in excess of the size of the *bandgap* (from sources such as the electric field). This can cause a high-energy electron in the *conduction band* to *scatter* an electron from the *valence band* (i.e., to break a covalent *bond* and produce an *EHP*). Each of these three *carriers* can then go through the same process and as a result instigate an avalanche in the number of *carriers* in both *bands*.

As a result of the requirement for the conservation of momentum and energy, rates of *Auger* and *impact-ionization* process are obviously dependent both on the size of the *bandgap* and temperature. Reduction of the size of the *bandgap* among different semiconductors results in an exponential increase in the rate of both processes. The rate of *Auger* process increases with the number of *carriers* and likewise with temperature. The *impact ionization* rate, however, reduces with increasing temperature. This is because of the increase in the rates of other *inelastic scattering* processes and the subsequent increased difficulty for *carriers* to gain energy from the electric field. In semiconductors with *bandgaps* larger than 1.5 eV, *Auger* process has a limited presence.

Complete calculation of the rate of *impact-ionization* process requires knowledge from the entire *band* structure. In the case of *Auger* process, the threshold value of energy can be approximated by

$$E_{\mathrm{I}} = \frac{1+2\mu}{1+\mu} E_{\mathrm{g}}, \text{ where } \mu = \frac{m_{\mathrm{n}}^*}{m_{\mathrm{p}}^*}. \tag{1.148}$$

However, due to the anisotropic nature of the *bands*, at higher energies this threshold has a strong angular dependence.

[100] Named after *Pierre-Victor Auger*.

In the case of parabolic *bands*, *Ridley* has successfully formulated the *impact-ionization* rate in terms of a simple closed-form expression:

$$W_{\text{impurity}} = 4.139 \times 10^{16} \times \left\{ \frac{\sqrt[4]{m_n^* m_p^*}}{m_0} \left(\frac{m_n^*}{m_0} + \mu \right) \left(\frac{\epsilon_0}{\epsilon} \right)^2 \left[\frac{E_1}{E_g} - (1+\mu) \right] \right\}. \qquad (1.149)$$

1.7.3.7 Lattice Vibration or Phonon Scattering Vibration of the crystal *lattice* can be seen as a hindrance on the periodicity of the *lattice* and, as a result, a source of *scattering* between the *Bloch states*.

As indicated before, crystallization in a particular crystalline form is the result of the tendency of a system to achieve minimum energy. Due to this tendency, as the atoms start to move about their *lattice* points, for example, due to exertion of thermal energy, a restoring force will be developed. This combination results in *lattice* vibrations. Depending on the similarity, or lack thereof, of atoms in *sublattices* of the crystal, their charge status, and the direction of the movement of these *sublattices*, *lattice* vibrations can have a few different modes (e.g., *acoustic*, *optical*, *polar*, *piezoelectric*). As indicated earlier in this chapter, lattice vibrations are expressed in terms of quantum particles known as *phonons*.

1.7.3.7.1 Optical and Acoustic Vibrations As a result of *lattice* vibrations, the position of atoms within the crystal can then be expressed as *traveling waves*. Solving the force equations for these *traveling waves* yields certain *dispersion* characteristics (i.e., ω or E vs. k dependence, where k is the *wave number*). An example is depicted in Figure 1.45. In this figure, a is the equilibrium distance between identical atoms along the direction of vibration.

In Figure 1.45, *two* branches of vibration can be identified. In the lower-frequency branch, known as the *acoustic phonon branch*, the frequency of vibration tends toward 0 as the *wave number* reduces to 0. This is the result of the movement of the *two sublattices* of the crystal along the same direction with equal velocity at any given time. However, if keeping the condition on velocity we reverse the direction of the movement of one *sublattice* versus the other, a high-frequency oscillation results (i.e., at k equal to 0). The branch identifying with this high-frequency mode is known as the *optical-phonon branch* (due to its higher frequency). This *branch* even for k equal to 0 is not producing 0 in its *dispersion* characteristic.

The *acoustical branch* represents the propagation of *sound waves* in a crystal. Using the *dispersion* characteristic of this branch, *sound velocity* is given by

$$v_s = \frac{d\omega}{dk} = \sqrt{\frac{C}{M_{av}}} a \qquad (1.150)$$

where M_{av}, a, and C stand for the average mass of the two atoms, minimum separation between identical planes in the crystal, and *spring constant* of the vibratory system, respectively.

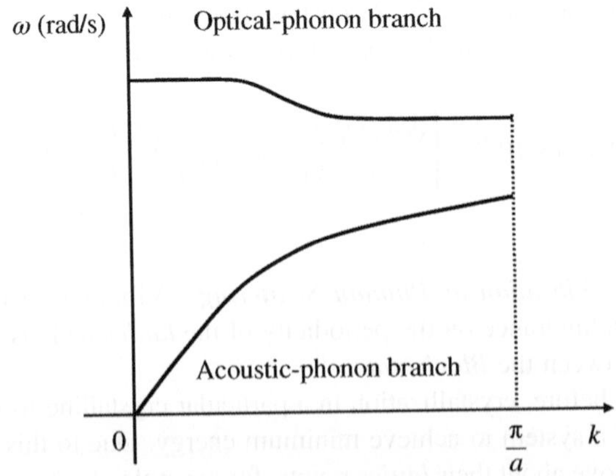

FIGURE 1.45 Schematic depiction of the dispersion diagram of a 1-D diatomic lattice, expressing the optical- and acoustic-phonon branches.

Whereas in Figure 1.45 we envisioned the vibrations in a one-dimensional *lattice*, in the case of three-dimensional *lattices*, a few other branches also appear on the *dispersion* characteristic. In reality, for a three-dimensional crystal, for each *wave vector* *one longitudinal* and *two transverse modes* of vibration will be present. This is true in the case of both *optical* and *acoustical* branches. Due to the difference in arrangement of atoms along different directions of most crystals, frequencies of vibration differ between the *longitudinal* and *transverse* branches.

DEFORMATION POTENTIAL AND BANDGAP VARIATION The distorting effect of *phonons* on the crystal, among a number of other appearances, takes the form of inducing a *deformation* potential. This *deformation* potential can be envisioned through the variation of the semiconductor *bandgap* (as the *lattice deformation* is causing the *interatomic* distance to change). According to this description, the displacement of the *lattice* by an amount u results in changing the energy of the *conduction* or *valence bands* in the form

$$\Delta E_{\mathrm{c,v}} = E_{\mathrm{c,v}}(a) - E_{\mathrm{c,v}}\left(a + \frac{du}{dx}a\right). \tag{1.151}$$

Here a is the *lattice constant*, which in the description of (1.151) has been assumed to be smaller than the wavelength of the *phonon* (i.e., *phonon* wavelength spans over many *lattice constants*, so that the displacement of the *lattice* can be taken for the expansion and contraction of the whole crystal). Through *Taylor* series expansion, (1.151) results in

$$|\Delta E_{\mathrm{c,v}}| = \frac{dE_{\mathrm{c,v}}}{da}.\frac{du}{dx}a. \tag{1.152}$$

WAVE-FUNCTION SYMMETRY AND CRYSTAL VIBRATIONS In a 3-D *lattice* the volumetric changes induced by *acoustic* and *optical phonons* are fundamentally different. Therefore, *optical-phonon scattering* has been shown to be very sensitive to the symmetry of the *band* structure (i.e., in the range relevant to *carrier scattering*). This has very important implications on *carrier mobility*. As a result of this sensitivity, if an electron is *scattered* close to the Γ-*valley* minimum (e.g., in *GaAs*) or near the *X-valley* minima and has a spherically symmetric *wave function*, *optical deformation potential scattering* is forbidden. The existence of this symmetry at the *conduction band* minimum of *GaAs* and lack thereof in *Si* result in superior electron *mobility* in *GaAs*.

However, while for both *GaAs* and *Si* the top of the *valence band* happens at the Γ-*point*, for these cases the *hole wave function* does not have spherical symmetry (i.e., they possess *sp* rather than *s*-orbital nature). Therefore, the *optical deformation potential* is present for *holes* in both semiconductors, which results in low *hole mobility* for both cases.

POLAR OPTICAL AND PIEZOELECTRIC VIBRATIONS In *ionic crystals*, in which the atoms of the *two sublattices* are different (e.g., in the *zinc blende* crystal of *GaAs*), *optical* vibrations also induce *vibrating polarization fields*. This is due to the vibration of *cation* and *anion sublattices* in opposite directions to one another. Presence of these vibrating fields is important in *longitudinal* vibrations but not in *transverse* vibrations. Hence, in the case of *longitudinal* vibration, there is an additional restoring force due to the *long-range polarization*. As an example, Figure 1.46 shows the dispersion relationship of *GaAs*, in which the frequency of the *longitudinal optical* mode is higher than the *transverse optical* mode.

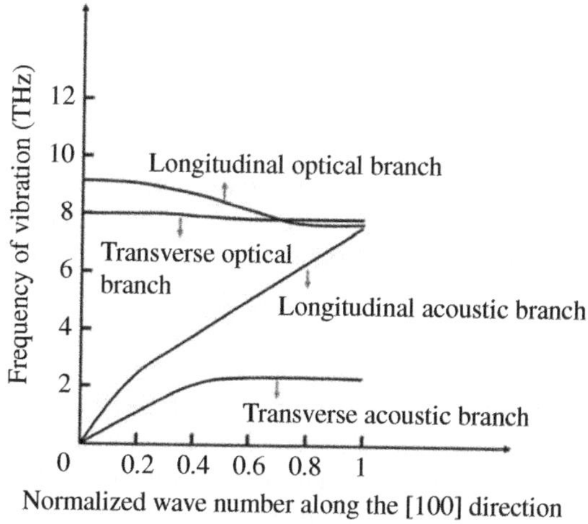

FIGURE 1.46 Approximate representation of phonon dispersion characteristics in *GaAs*. Adapted from Singh (2003, p. 228). Copyright 2003, Cambridge University Press. Reprinted with the permission of the Cambridge University Press.

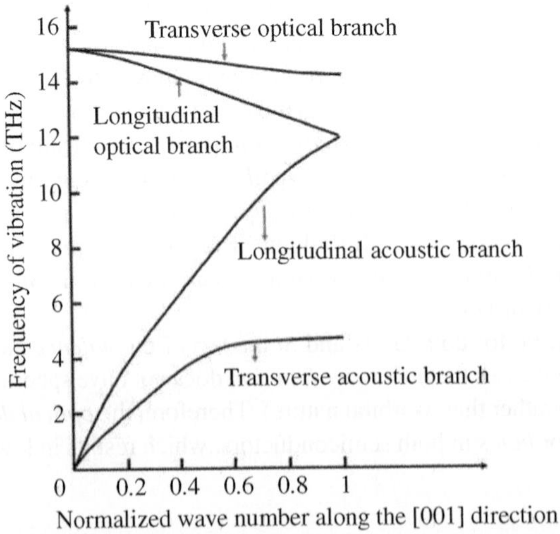

FIGURE 1.47 Approximate representation of phonon dispersion characteristics in *Si*. Adapted from Singh (2003, p. 228). Copyright 2003, Cambridge University Press. Reprinted with the permission of the Cambridge University Press.

These *vibrating polarization fields* are indicated by the presence of *polar optical phonons*. While *polar optical phonons* are not present in group *IV* semiconductors, in *III–V* semiconductors they play a very important role. This is why in group *IV* semiconductors there is no split between the *transverse* and *longitudinal optical* branches of the *dispersion* characteristic. Figure 1.47 illustrates this situation.

Optical phonons (especially *transverse optical phonons*) are much less dispersive than *acoustic phonons*. This is especially true at low k values. As shown in Figure 1.47, *longitudinal optical* and *longitudinal acoustic dispersions* tend toward the same value of angular frequency as k increases. However, this is not the case with transverse modes.

So far, we have only spoken of the role of *optical phonons* in inducing *polar* vibrations. Acoustic vibrations also cause *polar* vibrations, which are referred to as *piezoelectric effects*. *Piezoelectric scattering*, while much weaker than *polar optical-phonon scattering*, only becomes important in very high-purity samples at low temperatures.

1.7.3.7.2 Phonon Distribution Function Although we have already mentioned that *phonons* are quantum particles, at this point in our discussion, it is worthwhile to explain the quantum nature of these particles a little further. According to quantum mechanics, oscillation energy of crystal vibrations is not continuous. The *quantum oscillator* has a minimum energy of $\hbar\omega/2$, while the energy changes only in $\hbar\omega$ quanta, referred to as a *phonon*. With regard to this *quanta*, quantum number n refers to the occupation number of *phonons* in the system. Just like electrons, in order to evaluate the number of *phonons* in a given mode ω_k and at a given temperature T, a distribution function is required. However, because *phonons* do not obey *Pauli's exclusion principle*, they do not follow *Fermi–Dirac* statistics. Instead, the proper distribution function for them is defined in terms of *Bose–Einstein* statistics. Hence, *phonons*, rather than being *Fermions*, are referred to as *Bosons*. Because *phonons* are *Bosons*, their *occupation number* at *thermal equilibrium*, which is denoted as

$$\langle n_\omega \rangle = \frac{1}{\exp\frac{\hbar\omega}{kT} - 1}, \tag{1.153}$$

is not a probability of occupation unlike the case of *Fermions*.

As expressed by this distribution function, the number of *phonons* increases with the temperature, since *vibration* also becomes stronger at higher temperatures.

Obviously, at low temperatures the occupancy of the *optical phonons* will be very small. This is due to the fact that for any value of *k*, the energy of an *optical phonon* is large unlike that of the *acoustic phonon*. As a result, *acoustic-phonon scattering* is present even at low temperatures. For *optical-phonon scattering*, temperature or energy of the *carriers* must be beyond a certain value. This threshold of energy is determined by the *band* structure of the crystal.[101]

1.7.3.7.3 Quantum Mechanical Foundations of Phonon Scattering Electrons and *phonons* are treated similarly by solving the *Schrödinger equation* in a periodic potential. However, there is a qualitative difference between the two cases, which results from the difference in their *de Broglie wavelength*. In the case of electrons, only when the dimensions of the *quantum well* approach the *de Broglie wavelength* of the electron (i.e., ~10 nm) the *band offsets* at semiconductor *heterostructures* render the *heterostructure* effects important. However, the equivalent length scale in the case of *phonons* is about a few monolayers. This difference results in the development of *phonon* modes, associated with *interfaces* and *superlattices* of small periods.

As an example of the similarity between electrons and *phonons*, we can take a look at interface between *AlAs* and *GaAs*. This structure has a *Type-I band lineup*, which restricts the movement of *carriers* normal to the *heterointerface*. It also restricts the movement of the *optical phonons* from one material to the next. This is because the *optical* branches of the *dispersion* characteristics of *GaAs* and *AlAs* do not overlap. This is a fact that we will appreciate more after covering the materials in Chapter 2, where we get ourselves further acquainted with *junctions* and *interfaces*.

Phonons, *lattice* vibrations, and *band* structure are, to a first-order approximation, separated from one another. This treatment overlooks the obvious variations of the *band* diagram depicted in Figure 1.48, which are expressed in equations such as (1.151) and (1.152). As a result of this approximation, *scattering* by *phonons* is treated merely as a *perturbation*. Frequencies of lattice vibrations (i.e., *phonons*) are in the range of *terahertz*. Yet, in dealing with this electronic system, even in presence of *phonons*, it has been proven possible to imagine a *band* structure defined by a

[101] Previously in Section 1.4.4.2, we talked about *optical phonons* in relation to saturation of *drift velocity*.

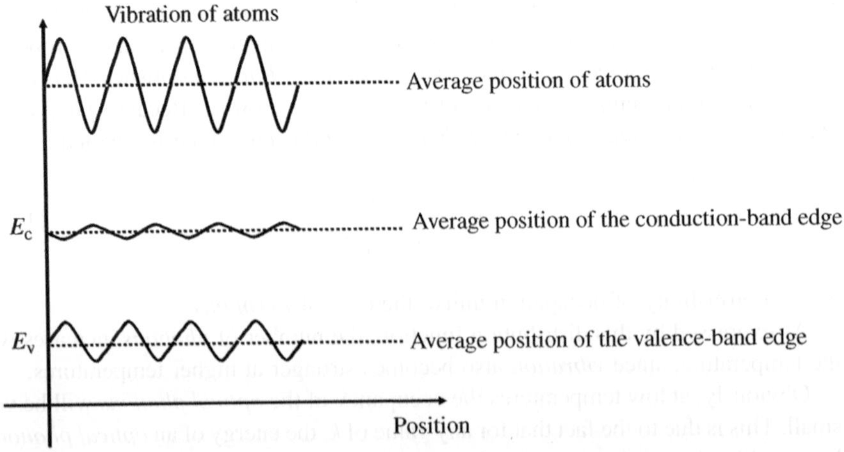

FIGURE 1.48 Schematic depiction of position dependence of band edges due to lattice vibrations. Adapted from Singh (2003, p. 228). Copyright 2003, Cambridge University Press. Reprinted with the permission of the Cambridge University Press.

time-independent periodic potential. Due to the much smaller mass of electrons than atoms, which gives them a *de Broglie wavelength* on the order of 10 nm, this is a valid approximation.

As a result of the small *wavelength*, the electron frequency will be in the order of 10^{16} Hz (i.e., $\frac{c}{\lambda} = \frac{3 \times 10^8}{10^{-8}} = 3 \times 10^{16}$ Hz). This is about four orders of magnitude higher than the *phonon* frequency, which renders the *Hamiltonian* describing the system almost stationary. This approximation is referred to as *adiabatic approximation*.

In the *adiabatic approximation*, electrons see the impact of *phonon*-caused energy fluctuations in terms of *scattering* between existing *states*. These interactions are seen as processes of *emission* and *absorption* of *phonons*, depending on whether a loss or a gain in electron energy results. The *scattering* rate is deduced based on the *Fermi golden rule*.

Through *phonon scattering*, both the energy and momentum of *charge carriers* are changed. While the *wave vectors* of electrons and *phonons* are similar, the energy of an electron is much larger (i.e., due to larger frequency). These are very consequential points in establishing the energy and momentum conservation.

As a result of the small energy of *phonons*, under low electric fields, *interband scattering* of *holes* (and not electrons) is probable. This will be in the form of *scattering* between the *two degenerate heavy-* and *light-hole bands*. However, due to the small amount of energy of *phonons*, even *interband scattering* to the split-off band, which is only separated by several hundred meV, is not possible. For *interband scattering* to occur to the *split-off band*, the electric field should be much stronger. However, because of the much smaller difference between the *degenerate bands* and *split-off band* at the top of the *valence band* of *silicon* (i.e., 44 meV), this semiconductor is an exception. This is one of the contributing factors to the deterioration of *hole transport* in *silicon*.

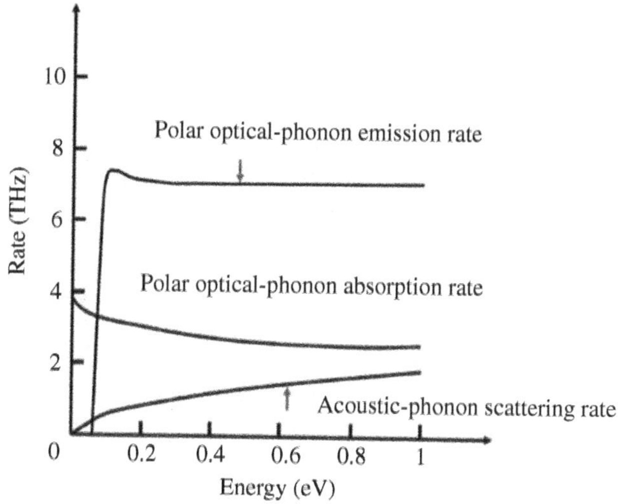

FIGURE 1.49 Comparison of absorption and emission rates of polar optical phonons with acoustic-phonon scattering rate of *GaAs* at room temperature. Adapted from Singh (2003, p. 228). Copyright 2003, Cambridge University Press. Reprinted with the permission of the Cambridge University Press.

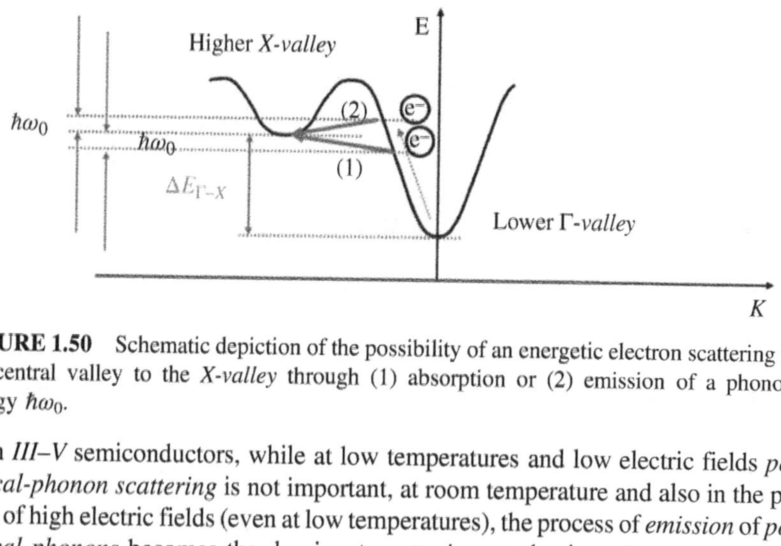

FIGURE 1.50 Schematic depiction of the possibility of an energetic electron scattering from the central valley to the *X-valley* through (1) absorption or (2) emission of a phonon of energy $\hbar\omega_0$.

In *III–V* semiconductors, while at low temperatures and low electric fields *polar optical-phonon scattering* is not important, at room temperature and also in the presence of high electric fields (even at low temperatures), the process of *emission* of *polar optical phonons* becomes the dominant *scattering* mechanism. As an example see Figure 1.49 for an approximate depiction of the rates of *acoustic* and *polar optical-phonon scattering* processes in *GaAs* at room temperature.

The processes of *phonon emission* and *absorption* in *intervalley scattering* are illustrated in Figure 1.50. As mentioned in Section 1.4.2.2, the *intervalley scattering* process plays a very important role in manifesting *negative differential mobility* in some semiconductors.

Now that we have visited a number of important *scattering* mechanisms, it is time to point out that there are important differences between *ionized-impurity scattering* and *phonon scattering*. The first important difference is that in regard to *phonon scattering*, we have both *absorption* and *emission* processes, which result in variation of the *wave functions* of both the *phonon* and the electron. The second important difference between the *phonon scattering* and *ionized-impurity scattering* is rooted in the formation of a *scattering* potential. As we have seen already in this section, the distorting effect of *phonons* on the crystal *lattice* can take a number of different forms (i.e., *deformation* potential, *piezoelectric* potential, and *polar optical* potential). Due to these differences, dealing with *ionized-impurity scattering* is much less complicated.

1.7.3.8 Carrier Scattering in Lower-Dimensional Systems So far in our discussions, we have maintained a focus on *carrier transport* in three-dimensional systems. In lower-dimensional systems, in addition to the aforementioned processes of *scattering*, due to the differences in *DOS* functions, a few additional *scattering* mechanisms become important.

As a special case, we have already identified *interface-roughness scattering* as a *scattering* process important only to two-dimensional *carrier transport*. Such a *scattering* mechanism is also sometimes present when *carrier transport* is confined to a one-dimensional *channel* (i.e., a *quantum wire*). In such lower-dimensional systems, due to the development of energy *subbands* (i.e., which are expressed by the modification in the 2-D *DOS* presented in (1.114) and (1.115) and also the so-called minibands in smaller lower-dimensional systems), both *intra*- and *intersubband scattering* are present. In a 2-D system, the larger the energy difference between the *subbands*, the less probable would be the *intersubband scattering*.

The *DOS* in 1-D systems (such as *quantum wires*) is qualitatively different from 3-D and 2-D systems, which results in very important implications on *carrier transport*. As illustrated in Figure 1.51, in the case of a 1-D system, *intrasubband scattering* is very restricted, while *elastic scattering* is considered. As depicted in this figure, in *elastic scattering* electron can *scatter* only to a *state* with the same or opposite momentum (i.e., k and $-k$). While the former results in no change on *transport*, the latter event requires a very short-ranged potential to conserve momentum. This requirement severely limits the *scattering*. Consequently, *mobilities* as high as 10^7 cm^2/V s are predicted for *quantum wires*. However, due to fabrication and crystal growth difficulties, this has not been experimentally observed yet.

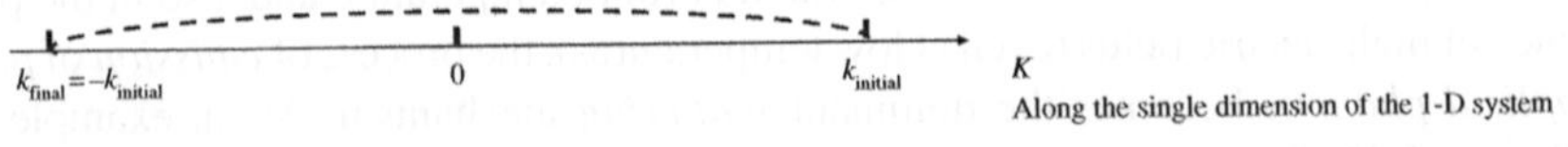

FIGURE 1.51 In the event of elastic scattering in a 1-D semiconductor, the equal energy surface has only two viable k-states. In the case of 2-D or a 3-D system, this equal energy surface evolves into the surface of a circle or a sphere, respectively, of radius $k_{initial}$ containing many more k-states available to scattering. Adapted from Singh (2003, p. 228). Copyright 2003, Cambridge University Press. Reprinted with the permission of the Cambridge University Press.

FURTHER READING

Following references have been used in preparation of the materials presented in this chapter. They are also suggested as sources for further reading.

SOLID-STATE THEORY

S. Datta, *Quantum Phenomena*, Addison-Wesley, Reading, MA, 1989.

S. Datta, *Electronic Transport in Mesoscopic Systems*, Cambridge University Press, Cambridge, UK, 1995.

K. Hess, *Advanced Theory of Solid State Devices*, IEEE Press, Piscataway, NJ, 2000.

C. Kittle, *Introduction to Solid State Physics*, John Wiley & Sons, Inc., Hoboken, NJ, 2005.

P. T. Landsberg (Ed), *Solid State Theory: Methods and Applications*, John Wiley & Sons, Inc., Hoboken, NJ, 1969.

M. Lundstrom, *Fundamentals of Carrier Transport*, Cambridge University Press, Cambridge, UK, 2000.

J. Singh, *Electronic and Optoelectronic Properties of Semiconductor Structures*, Cambridge University Press, Cambridge, UK, 2003.

PHYSICS OF SEMICONDUCTOR DEVICES

R. S. Muller, T. I. Kamins, and M. Chan, *Device Electronics for Integrated Circuits*, John Wiley & Sons, Inc., Hoboken, NJ, 2003.

M. Lundstrom and J. Guo, *Nanoscale Transistors: Device Physics, Modeling and Simulation*, Springer, New York, 2006.

W. Shockley, *Electrons and Holes in Semiconductors with Applications to Transistor Electronics*, Van Nostrand, New York, 1950.

S. M. Sze and K. K. Ng, *Physics of Semiconductor Devices*, John Wiley & Sons, Inc., Hoboken, NJ, 2007.

S. Wang, *Fundamentals of Semiconductor Theory and Device Physics*, Prentice Hall, Englewood Cliffs, NJ, 1989.

SEMICONDUCTOR MATERIALS AND HETEROSTRUCTURES

S. Adachi, *Properties of Semiconductor Alloys: Group IV, III-V, and II-VI Semiconductors*, John Wiley & Sons, Inc., Hoboken, NJ, 2009.

Ioffe Physico-Technical Institute, Electronic Archive of New Semiconductor Materials, Characteristics and Properties: http://www.ioffe.rssi.ru/SVA/NSM/. Accessed September 18, 2015.

M. Levinshtein, S. Rumyantsev, and M. Shur, *Properties of Advanced Semiconductor Materials: GaN, AlN, InN, BN, SiC, SiGe*, John Wiley & Sons, Inc., Hoboken, NJ, 2001.

Y. Sun, S. Thompson, and T. Nishida, *Strain Effect in Semiconductors: Theory and Device Applications*, Springer, New York, 2009.

PROBLEMS

1.1 Using the time-independent 3-D Schrödinger equation, prove that for an electron in a 3-D infinite potential well described by $U(x,y,z)=\begin{cases} 0 & \text{for } 0\le x\le W,\ 0\le y\le W, 0\le z\le W \\ \infty & \text{elsewhere} \end{cases}$, the allowed energy levels are expressed by $E=\frac{\hbar^2\pi^2}{2m^* W^2}(n^2+m^2+p^2)$ where n, m, and p are integers.

Wave function and its first spatial derivative are continuous, single-valued functions.

1.2 Demonstrate that the Fermi level of an intrinsic semiconductor instead of residing in the middle of the bandgap has a slight offset from this position, which is determined by the following expression: $\frac{3}{4}kTLn\left(\frac{m_p^*}{m_n^*}\right)$.

1.3 According to (1.21) calculate the average thermal energy of Maxwellian particles having a single degree of freedom. What is the thermal energy if the particles were allowed to move in the 2-D or the 3-D space?

1.4 Calculate the effective density of states at the lower edge of the conduction-band N_c, for silicon at room temperature. Notice the difference in the definitions of transport and *DOS* effective masses and take the transverse and longitudinal effective mass to be equal to $0.19m_0$ and $0.98m_0$, respectively.

1.5 Prove Equation (1.32).

1.6 For a silicon sample doped with arsenic, only a quarter of the impurities are ionized at room temperature. Elaborate on the cause(s) of the observation and describe how the electron concentration is calculated in this case.

1.7 Prove Equations (1.78) and (1.79).

1.8 Formulate the Hall effect when the semiconductor is behaving semi-intrinsically. Presentation of the Section 1.4.3 was provided with the assumption of a significant extrinsic characteristic.

1.9 Perform a literature survey on the values of the magnetic flux density often used in Hall effect measurements and indicate why Hall measurements are not performed under high magnetic fields.

1.10 Employing *WKB* approximation, calculate the tunneling probability of an electron of energy 1 eV across a rectangular barrier of height 5 eV and width 1 nm. Repeat this problem for a triangular barrier of the same height and also the same width at the energy level of the electron.

1.11 Prove that assuming the same effective mass for electrons and holes, an electron requires 50% more kinetic energy than the size of the bandgap to instigate impact ionization.

1.12 Prove that to guarantee maximum sphere packing, the *c/a* ratio of the hexagonal prism of a wurtzite crystal should be equal to $\sqrt{8/3}$.

1.13 Calculate the entries of Table 1.7.

1.14 Indicate the plane $(\bar{1}100)$ on a hexagonal unit cell.

1.15 Perform a literature survey to determine the type of the band lineup between the following pairs:

a. *InP/InGaAs*.

b. *AlGaN/GaN*.

c. *GaSb/InAs*.

1.16 Perform a literature survey to determine which material system is more suitable for solar cell applications:

a. AlInGaAs

b. AlInGaN

Determine the energy coverage of the bandgap of each compound and compare to solar spectrum.

1.17 In strained epitaxy we often simplistically define the critical thickness of the overlayer causing the generation of dislocations within this film as $\frac{a_S}{2|\epsilon|}$. Considering the pseudomorphic growth of $In_{0.2}Ga_{0.8}As$ over a thick relaxed layer of *GaAs*, determine the critical thickness of *InGaAs*. Perform the calculations at room temperature for which $a_{InAs} = 6.058$ Å and $a_{GaAs} = 5.653$ Å.

1.18 Through producing schematic depictions of a 1-D lattice, demonstrate why the dispersion diagram of the acoustic-phonon branch tends toward 0 at small values of wave number while for the optical-phonon branch the energy remains sizable (see Fig. 1.45).

APPENDIX 1.A DERIVATION OF *FERMI–DIRAC* STATISTICS

The key point in the derivation of *Fermi–Dirac* statistics is to acknowledge that *Fermions* (electrons included) are indistinguishable particles obeying *Pauli's exclusion principle*. According to this definition, the total number of ways that N_i *Fermions* can be arranged among g_i *states* corresponding to *i*th energy *level* (which are determined in terms of *density of states* function) is given by

$$P_i = \frac{g_i!}{N_i!(g_i - N_i)!}. \tag{1A.1}$$

According to *Pauli's exclusion principle*, g_i is greater than or equal to N_i.

Since the *density of states* function is a double-density function (i.e., representing the number of *states* per unit energy per unit volume), in studying the *thermal-equilibrium* distribution of *Fermions* among these *states*, we should not only consider the *states* corresponding to the *i*th energy *level* but also all those other *states* assigned to other values of energy. Accordingly, considering a total of N electrons, where N_i of which take the *i*th *energy level*, we will have

$$N=\sum_{i=1}^{n} N_i \tag{1A.2}$$

where n is the number of energy-wise distinguishable *levels*.

In terms of this distribution, the total energy is given by

$$E=\sum_{i=1}^{n} N_i E_i \tag{1A.3}$$

where E_i is the energy of the *i*th *level*.

As a result of (1A.2), the total number of ways of arranging $N_1, N_2, \ldots N_n$ indistinguishable particles among n energy *levels* would be

$$P=\prod_{i=1}^{n} P_i=\prod_{i=1}^{n} \frac{g_i!}{N_i!(g_i-N_i)!}. \tag{1A.4}$$

On the basis of (1A.2) and (1A.3), in order to identify the distribution function presenting the least amount of systemic energy (i.e., representative of *thermal equilibrium*), the most likely distribution function should be selected. Toward that end N_i's should be arranged so that P is maximized (i.e., representing the most plausible distribution) for a fixed N.

According to the discussions of Section 1.3.5, N_i can be replaced by $g_i f_i$. This substitution results in

$$P=\prod_{i=1}^{n} \frac{g_i!}{(g_i f_i)!(g_i-g_i f_i)!}. \tag{1A.5}$$

Using *Stirling*'s approximation of *factorials* and assuming that n is greater than 20,

$$n! \cong \sqrt{2\pi n}\, n^n \exp\left[-n+\frac{1}{12n}\right], \tag{1A.6}$$

which gives

$$Ln(n!) \cong nLn(n)-n+\frac{1}{2}Ln(2\pi n). \tag{1A.7}$$

Neglecting the third term on the right side of (1A.7), (1A.5) can be rewritten as

$$Ln(P) \cong \sum_{i=1}^{n} Ln(P_i)=\sum_{i=1}^{n} [g_i Ln(g_i)-g_i f_i Ln(g_i f_i)-g_i(1-f_i)Ln(g_i-g_i f_i)]. \tag{1A.8}$$

This has been proven to be a valid approximation.

To maximize P, according to the *Lagrange* method of *undetermined multipliers* and Equations (1A.2) and (1A.3), the following function should be maximized by the appropriate choice of α and β to get P or $Ln(P)$ maximized

$$f(g_i f_i) = Ln(P) - \alpha \sum_i g_i f_i - \beta \sum_i E_i g_i f_i. \tag{1A.9}$$

As a result, the derivative of function $f(g_i f_i)$ versus $g_i f_i$ is set to 0 in order to calculate α and β:

$$\frac{\partial}{\partial(f_i g_i)} \left[Ln(P) - \alpha \sum_i g_i f_i - \beta \sum_i E_i g_i f_i \right] = 0. \tag{1A.10}$$

Assuming fixed g_i (which is determined by the *density of states* function of the semiconductor) and through the following mathematical manipulation, f_i results in

$$\begin{aligned} &-Ln(g_i f_i) + Ln(g_i - f_i g_i) - \alpha - E_i \beta = 0 \\ &\Rightarrow Ln\left(\frac{g_i - g_i f_i}{g_i f_i}\right) = \alpha + \beta E_i \Rightarrow f_i = \frac{1}{1 + \exp(\alpha + \beta E_i)}. \end{aligned} \tag{1A.11}$$

As expressed below, the function f_i is a *Fermi–Dirac* distribution function with $E_f = -\alpha/\beta$:

$$f_D(E_i) = \frac{1}{1 + \exp\left(\dfrac{E_i - E_f}{1/\beta}\right)}. \tag{1A.12}$$

Knowing that $E = \sum_{i=1}^{n} E_i f_i g_i$,

$$dE = \sum_{i=1}^{n} E_i d(f_i g_i) + \sum_{i=1}^{n} f_i g_i d(E_i). \tag{1A.13}$$

Besides from (1A.10),

$$\frac{1}{\beta}\partial(LnP) = \sum_{i=1}^{n} E_i d(f_i g_i) + \frac{\alpha}{\beta} \sum_{i=1}^{n} d(f_i g_i) \tag{1A.14}$$

where $E_f = -\alpha/\beta$ and $\sum_{i=1}^{n} d(f_i g_i) = dN$.

As a result, (1A.14) can be rewritten as

$$dE = \frac{1}{\beta}\partial Ln(P) + E_f dN + \sum_{i=1}^{n} f_i g_i d(E_i). \tag{1A.15}$$

Through incorporating dV as a volumetric variation,

$$dE = \frac{1}{\beta}\partial Ln(P) + E_f dN + \sum_{i=1}^{n} f_i g_i \frac{dE_i}{dV} dV. \tag{1A.16}$$

This relationship is similar to thermodynamic identity of $dE = TdS - PdV + \mu dN$. According to this analogy, $\beta = 1/kT$, $S = kLnP$, and μ (the energy of the *Fermions*) is represented by E_f.

As a result (1A.12) evolves into the familiar form of

$$f_D(E) = \frac{1}{1 + \exp\left(\frac{E - E_f}{kT}\right)} \tag{1A.17}$$

FURTHER READING

G. Fournet, English edition edited by S. Chomet, *Solid State Electronics (first published in French: Physique Électronique des Solids)*, Iliffe Books, London, 1968.

APPENDIX 1.B DERIVATION OF EINSTEIN RELATIONSHIP IN DEGENERATE SEMICONDUCTORS

In the case of one-dimensional *carrier transport* and according to the definition of *thermal equilibrium*,

$$J_n = qn\mu_n E + qD_n \frac{dn}{dx} = 0 \tag{1B.1}$$

where $E = \frac{1}{q}\frac{dE_i}{dx}$ or equivalently $E = \frac{1}{q}\frac{dE_c}{dx}$.

In addition, based on the discussions in Section 1.3.5, $n = N_c F_{1/2}(\eta_F)$ where $\eta_F = \frac{E_f - E_c}{kT}$.

Resulting from the definition of electric field E and n,

$$\frac{dn}{dx} = -\frac{1}{kT}\frac{dn}{d\eta_F}\frac{dE_i}{dx} = -\frac{q}{kT}\frac{dn}{d\eta_F}E. \tag{1B.2}$$

Based on (1B.1) and (1B.2), the following general form of the *Einstein* relationship results:

$$\frac{D_n}{\mu_n} = \frac{kT}{q}\frac{n}{(dn/d\eta_F)} \tag{1B.3}$$

in which $n = N_c F_{1/2}(\eta_F)$ and $F_{1/2}(\eta) = (2/\sqrt{\pi})F_{1/2}(\eta)$ where $F_{1/2}(\eta) = \int_0^\infty \frac{\xi^{1/2} d\xi}{(1 + e^{\xi - \eta})}$.

A relationship identical in form to (1B.3) can also be developed for *holes*. In general,

$$\mathrm{F}_j(\eta) \equiv \frac{1}{\Gamma(j+1)}\int_0^{\infty}\frac{\xi^j d\xi}{(1+e^{\xi-\eta})} \tag{1B.4}$$

$$\mathrm{F}_j(\eta) \rightarrow e^{\eta} \text{ as } \eta \rightarrow -\infty \tag{1B.5}$$

$$\frac{d}{d\eta}\mathrm{F}_j = \mathrm{F}_{j-1}(\eta) \tag{1B.6}$$

$$\mathrm{F}_{1/2}(\eta) \cong [e^{-\eta}+\xi(\eta)]^{-1} \tag{1B.7}$$

where

$$\xi(\eta) = 3\sqrt{\frac{\pi}{2}}\left[(\eta+2.13)+\left(|\eta-2.13|^{2.4}+9.6\right)^{5/12}\right]^{-3/2} \tag{1B.8}$$

with a maximum error of $\sim \pm 0.5\%$,

$$\eta \cong \frac{Ln(U)}{1-U^2} + \frac{(3\sqrt{\pi}U/4)^{2/3}}{1+\left[0.24+1.08(3\sqrt{\pi}U/4)^{2/3}\right]^{-2}} \tag{1B.9}$$

where $U \equiv \mathrm{F}_{1/2}(\eta)$ with a maximum error of $\sim \pm 0.5\%$.

A convenient approximation for $(n)/(dn/d\eta_{\mathrm{F}})$ is presented in Nilsson (1978), in the form of $\mathrm{F}_{1/2}(\eta_{\mathrm{F}})/\mathrm{F}_{-1/2}(\eta_{\mathrm{F}})$.

In its *nondegenerate* limit, n becomes equal to $N_c \exp(\eta_{\mathrm{F}})$ and $(n)/(dn/d\eta_{\mathrm{F}})$ tends toward *one*. As a result (1B.3) takes on to its familiar form,

$$\frac{D_n}{\mu_n} = \frac{kT}{q}. \tag{1B.10}$$

While this derivation relies on *thermal-equilibrium* conditions, it can be shown that the *Einstein* relationship is also valid under *nonthermal-equilibrium* conditions.

FURTHER READING

N. G. Nilsson, "Empirical approximation applied to generalized Einstein relation for degenerate semiconductors," Phy Stat Sol (a), vol. **50**, 1978b, K43–K45.

R. F. Pierret, *Advanced Semiconductor Fundamentals, vol. VI Modular Series in Solid State Devices*, Prentice Hall, Upper Saddle River, NJ, 2003.

APPENDIX 1.C STRAIN TENSOR

Deformations in the crystal structure result in induction of strain. Strain is defined in terms of the relative change in the *lattice constants* with regard to the freestanding *lattice constants* of a crystal. Figure 1.C.1 illustrates a simple 2-D *lattice* under strain. The outcomes of this pictorial insight can be mathematically represented in the 3-D form with the use of the *strain tensor*.

As illustrated in Figure 1.C.1a, in the 2-D case we can use two *unit vectors* $\hat{x}$ and $\hat{y}$ of the *Cartesian* coordinate system to represent the *unstrained lattice*. In a simple *lattice*, these vectors correspond to the *basis vectors* of the *lattice*. A small uniform deformation of the *lattice* results in distortion of the *unit vectors* both in magnitude and orientation (see Fig. 1.C.1b). These distortions result in a new set of vectors identified by

$$\hat{x'} = (1+\epsilon_{xx})\hat{x} + \epsilon_{xy}\hat{y} + \epsilon_{xz}\hat{z} \tag{1C.1}$$

$$\hat{y'} = \epsilon_{yx}\hat{x} + (1+\epsilon_{yy})\hat{y} + \epsilon_{yz}\hat{z}. \tag{1C.2}$$

In the 3-D case, we also have a distortion with regard to the z-axis of the coordinate system,

$$\hat{z'} = \epsilon_{zx}\hat{x} + \epsilon_{zy}\hat{y} + (1+\epsilon_{zz})\hat{z}. \tag{1C.3}$$

Elements ϵ_{ij} are referred to as *strain coefficients*. These coefficients are dimensionless. The 3×3 matrix composed of these elements is referred to as *strain tensor*:

$$\bar{\bar{\varepsilon}} = \begin{bmatrix} \epsilon_{xx} & \epsilon_{xy} & \epsilon_{xz} \\ \epsilon_{yx} & \epsilon_{yy} & \epsilon_{yz} \\ \epsilon_{zx} & \epsilon_{zy} & \epsilon_{zz} \end{bmatrix}. \tag{1C.4}$$

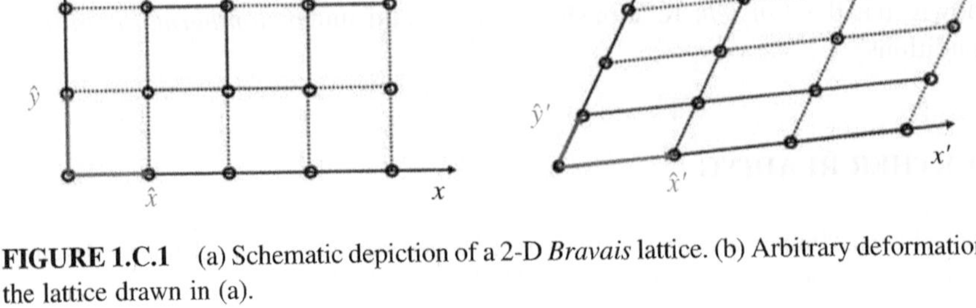

FIGURE 1.C.1 (a) Schematic depiction of a 2-D *Bravais* lattice. (b) Arbitrary deformation of the lattice drawn in (a).

Generally speaking, *tensor* is a mathematical notion used to describe a linear relationship between two physical quantities. Depending on the degrees of freedom, a *tensor* can be a *scalar* quantity (i.e., *zero rank*), a vector (i.e., *first rank*), or as in this case a matrix (i.e., *second rank*). As an example, we can look into the strain-caused deformation with regard to a *lattice point* represented by the vector $\vec{r} = x\hat{x} + y\hat{y} + z\hat{z}$. Considering a uniform deformation to evolve this point to $\vec{r'} = x\hat{x'} + y\hat{y'} + z\hat{z'}$ for a general varying strain, the *strain tensor* is formulated as

$$\epsilon_{ij} = \frac{\partial u_i}{\partial x_j} \tag{1C.5}$$

where $u_i = u_x, u_y, u_z$, $x_j = x, y, z$.

In this definition, u_i refers to the displacement of the *lattice* point (e.g., $\vec{r}$) along x_i.[102] According to this definition, it is obvious that without rotation, the *strain tensor* is symmetric and

$$\epsilon_{ij} = \epsilon_{ji} = \frac{1}{2}\left(\frac{\partial u_i}{\partial x_j} + \frac{\partial u_j}{\partial x_i}\right). \tag{1C.6}$$

Oftentimes, instead of the aforementioned set of *strain tensor* components, the following are used:

$$\begin{cases} e_{xx} = \epsilon_{xx}; e_{yy} = \epsilon_{yy}; e_{zz} = \epsilon_{zz} \\ e_{xy} = \hat{x'} \cdot \hat{y'} = \epsilon_{xy} + \epsilon_{yx} \\ e_{yz} = \hat{y'} \cdot \hat{z'} = \epsilon_{yz} + \epsilon_{zy} \\ e_{zx} = \hat{z'} \cdot \hat{x'} = \epsilon_{zx} + \epsilon_{xz} \end{cases}. \tag{1C.7}$$

As presented in (1C.7), the *strain components* e_{xy}, e_{yz}, and e_{zx} are defined with respect to changes of angle between the *basis* vectors. This definition is provided with neglecting the terms of order ϵ_{ij}^2 in the small-strain approximation.

The above six *strain coefficients*, shown by an array $e = \{e_{xx}, e_{yy}, e_{zz}, e_{yz}, e_{zx}, e_{xy}\}$, offer a complete definition of strain. This set provides a more convenient way for describing the relationship between the strain and the strain-related physical quantities. The form of presentation expressed in (1C.4), however, gets complicated very quickly. This is because the relationship between two *second-rank tensors* (i.e., one for the *strain tensor* and one for the strain-related physical quantity) is representable through a *fourth-rank tensor*. However, describing each of the two *second-rank tensors* by a vector only requires dealing with a *second-rank tensor* for evaluation of the interactions.

[102] According to (1C.1)–(1C.3), $\vec{r'} = [x(1+\epsilon_{xx}) + y\epsilon_{yx} + z\epsilon_{zx}]\hat{x} + [x\epsilon_{xy} + y(1+\epsilon_{yy}) + z\epsilon_{zy}]\hat{y} + [x\epsilon_{xz} + y\epsilon_{yz} + z(1+\epsilon_{zz})]\hat{z}$. Hence, as an example, $u_x = x(1+\epsilon_{xx}) + y\epsilon_{yx} + z\epsilon_{zx} - x$, and as a result $\partial u_x / \partial x$ would be equal to ϵ_{xx}.

In terms of the *first-rank tensor* description of (1C.7), the *dilation* of the *unit cell* of the crystal can be simply evaluated through calculating the volume of the *unit cell* by

$$V' = \hat{x}'.\hat{y}' \times \hat{z}' = 1 + e_{xx} + e_{yy} + e_{zz}. \tag{1C.8}$$

Equation (1C.8) shows the *dilation* δ as

$$\delta = \frac{\delta V}{V} = e_{xx} + e_{yy} + e_{zz}. \tag{1C.9}$$

Interestingly enough, this is the *trace* of the *strain tensor*. This *dilation* is the same as the negative of the *hydrostatic pressure*.

It is worthwhile indicating that under a *hydrostatic pressure P*, the *shear* stress is 0 and the stress along any principal direction is equal to $-P$,

$$\tau = \begin{bmatrix} -P & 0 & 0 \\ 0 & -P & 0 \\ 0 & 0 & -P \end{bmatrix}. \tag{1C.10}$$

According to this sign convention, the *tensile* stress is indicated with a positive sign, while the *compressive* stress is indicated in terms of negative values.

In the case of the uniaxial stress T along the [*001*] direction, all stress components but τ_{zz} are 0, and $\tau_{zz} = T$. In other words,

$$\tau = \begin{bmatrix} 0 & 0 & 0 \\ 0 & 0 & 0 \\ 0 & 0 & T \end{bmatrix}. \tag{1C.11}$$

2

JUNCTIONS

In realizing semiconductor devices, creating contacts in a variety of forms (e.g., semiconductor–semiconductor *homo-* and *hetero*junctions, metal–semiconductor junctions, and metal–insulator–semiconductor junctions) is pivotal. In this chapter, we will focus on these forms of contacts and will study the important characteristics of each of them.

2.1 CONTACTS UNDER THERMAL EQUILIBRIUM

Developing the *thermal-equilibrium band diagram* is the first step in analyzing a junction. Excluding *deep-submicron* contacts, the procedure for producing such a *band diagram* is applicable to all semiconductors and their contacts under *thermal equilibrium*.

The first-order analysis of this *band diagram*, which is very often used in analyzing junctions, is based on what is known as *Schottky's depletion approximation* (i.e., *SDA*). This theory, despite its simplicity, is very descriptive of the first-order physics of the junction. However, in many cases it will be necessary to incorporate other components into this description to make it adequate.

For any two separate electronic systems under *thermal equilibrium* that originally have no interaction with one another, upon forming a contact individual states of

Field Effect Transistors, A Comprehensive Overview: From Basic Concepts to Novel Technologies, First Edition. Pouya Valizadeh.

thermal equilibrium will be disturbed. The cause for this disturbance is the energy difference between *charge carriers* in the two originally separate systems. Upon forming the contact, *charge carriers* tend to move to the side in which the energy is lower. Such a net movement of *carriers* breaks the two originally separate states of *thermal equilibrium* and generates an imbalance into charge neutrality. The resulting imbalance induces a built-in electric field, which opposes further movement of *carriers* across the interface. These two opposite tendencies cause the establishment of a new state of *thermal equilibrium* between the two sides of the junction.

Under this new *thermal-equilibrium* state, we can express the electron concentration $n(E)$ and concentration of vacancies $v(E)$ at a given energy level E in terms of the *density of states* function (i.e., $D(E)$) and the *Fermi level* of the system (i.e., E_f):

$$n_{1,2}(E) = D_{1,2}(E) f_{D_{1,2}}(E) \tag{2.1}$$

$$v_{1,2}(E) = D_{1,2}(E)\left[1 - f_{D_{1,2}}(E)\right] \tag{2.2}$$

where according to our discussions in Chapter 1,

$$f_{D_{1,2}}(E) = \frac{1}{1 + \exp\left[\left(E - E_{f_{1,2}}\right)/kT\right]}. \tag{2.3}$$

Corresponding to the schematics of the junction depicted in Figure 2.1, in (2.1)–(2.3), subscripts 1 and 2 refer to system-1 and system-2, respectively.

Because electrons are *Fermions*, the rate of electron transfer from system-1 to system-2 is expressed in proportion to $n_1(E) \cdot v_2(E)$. Likewise, the transfer rate of

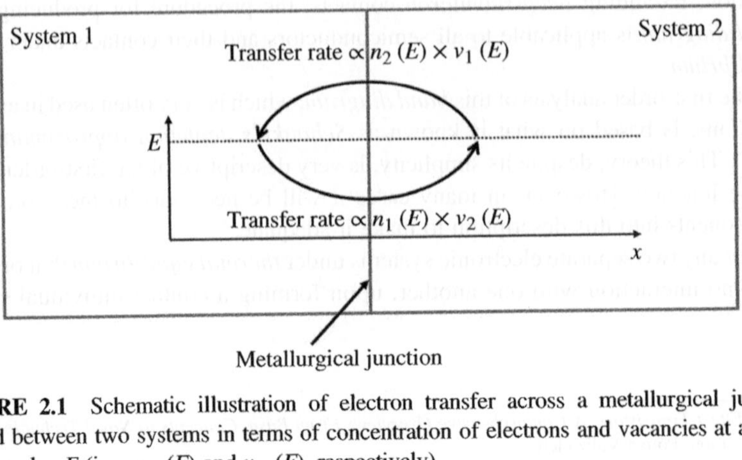

FIGURE 2.1 Schematic illustration of electron transfer across a metallurgical junction formed between two systems in terms of concentration of electrons and vacancies at a given energy value E (i.e., $n_{1,2}(E)$ and $v_{1,2}(E)$, respectively).

electrons, at energy *level E*, from system-2 to system-1 is proportional to $n_2(E) \cdot v_1(E)$. As a result, at *thermal equilibrium*, for any energy *level* and position,

$$n_1 v_2 = n_2 v_1 \tag{2.4}$$

or equivalently[1]

$$f_{D_1} D_1 [1 - f_{D_2}(E)] D_2 = f_{D_2} D_2 [1 - f_{D_1}(E)] D_1. \tag{2.5}$$

According to (2.5), at all *energy levels*,

$$f_{D_1}(E) = f_{D_2}(E), \tag{2.6}$$

which according to (2.3) produces

$$E_{f_1} = E_{f_2}. \tag{2.7}$$

The result expressed in (2.7) conveys that when *thermal equilibrium* is established between the two sides of a junction, the spatial gradient of *Fermi level* will become 0. Since in the derivation of (2.7) from (2.4) the two systems were approached as abstract entities (i.e., without invoking any of the special properties of metals, insulators, or semiconductors), this result is applicable to any junction at *thermal equilibrium* (i.e., formed between any number of layers with a different nature of *density of states* functions). As a result, having a constant *Fermi level* can be treated as an equivalent definition for *thermal equilibrium*.

2.2 METAL–SEMICONDUCTOR CONTACTS

Metal–semiconductor (*MS*) interconnects are often encountered in interfaces between different semiconductor devices of an *integrated circuit* and also in the *IC*'s interfaces with the external board-level components of the system. Late in the 1980s, understanding the MS contact, while being one of the earliest forms of contact explored in semiconductor technology, became an obstacle against the ongoing *scaling* trend of *field effect transistor* technology. This was due to a change of paradigm in establishing contacts between metals and semiconductors where the semiconductor size was reduced to *deep submicron*. The model, which is presented in this section, deals with the description of the contacts that are not in *deep submicron* or *nano*scales. Later on, in Chapter 6, we will talk about the roots of this paradigm change as dimensions get smaller and briefly describe the needed formalisms and results.

[1] For simplicity, energy arguments are not explicitly indicated in this equation.

2.2.1 Band Diagram of an *MS* Junction

Whereas in a semiconductor system the energy of electrons in the *conduction band* is represented by E_c, in the case of a metal, the *Fermi level* serves this purpose. As shown in Figure 2.2, by placing the two systems in intimate contact, due to the difference in energy of the electrons on the two sides of the *metallurgical junction*, these *charge carriers* will spontaneously start to move from where they are at higher energies to where their energy is lower. In the case presented in Figure 2.2, this transfer is from the n-type semiconductor to the metal. As expressed in Section 2.1, this charge transfer between the two systems will break the original states of *thermal equilibrium* and will drive the formation of a new state of *thermal equilibrium*.

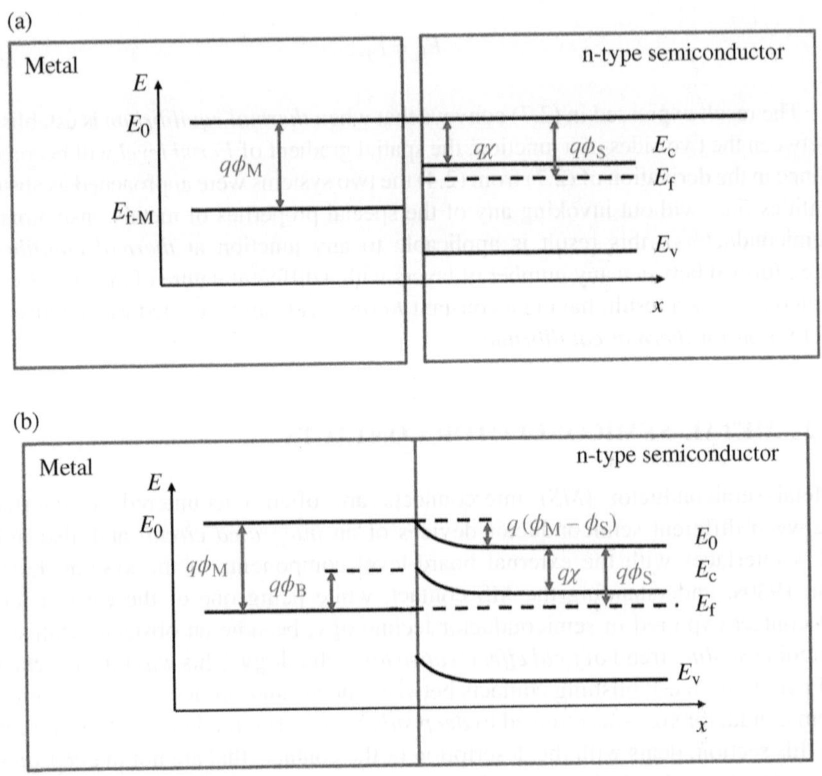

FIGURE 2.2 (a) An example of band lineup between a metal and an n-type semiconductor prior to contact formation. Difference between the metal and the semiconductor work function is highlighted in this figure. (b) Resulting band diagram upon formation of the contact. Upon forming this contact, electrons have been partially transferred across the contact from the semiconductor into the metal.

In the case of MS contacts, *band diagrams* for two different types of contacts are encountered:

1. Those in which majority *carriers* of the semiconductor need to overcome a potential barrier before they can arrive at the metal side (also known as *blocking*)
2. Those in which such a barrier is not present (also known as *nonblocking* or *Ohmic*)

According to *Schottky*'s approximation, the description of an *MS* junction starts by dealing with the process of junction formation. This is the process of combining two originally independent systems (i.e., one metallic and one semiconducting) at *thermal equilibrium* by placing the two in very close contact so that *charge carriers* can move across the interface. In this contact, it is assumed that the two materials do not mix with each other (either physically or chemically) and that they simply adhere to one another on a contact plane, one side of which is metal and the other side is a semiconductor. Also, in this first-order approximation, the rearrangement of the atoms at this *metallurgical junction* and the existence of unsatisfied *dangling bonds* at the interface are neglected. Upon forming such a junction, the two originally independent systems under *thermal equilibrium* will see their equilibrium status revoked. However, caused by *charge-carrier* transfer between metal and semiconductor, spontaneously, but not instantaneously, the unified MS system will establish a new state of *thermal equilibrium*. Forming this new state of *thermal equilibrium* is at the heart of drawing the *band diagram*.

In order to proceed toward drawing the *band diagram*, we have to be reminded of yet another important property of this material system known as the metal *work function*. For any metal, there is a certain minimum value of energy required to release an electron from the metal. This amount of energy is referred to as the metal *work function* and is defined by the difference between the *vacuum energy level* (i.e., E_0) and the *Fermi level* of the metal. Not unlike metals, a *work function* can also be defined in the case of a semiconductor. However, in contrast to metals, since the position of the *Fermi level* in semiconductors depends on the *type* and concentration of *dopants*, the semiconductor *work function* is not a constant of material.

Based on these definitions, the following steps outline the procedure for drawing the *thermal-equilibrium band diagram* of an *MS* junction:

1. Draw a horizontal line representing the constant *Fermi level.*
2. Identify the *metallurgical junction* between the metal and the semiconductor.
3. Using the definitions of metal *work function* and *electron affinity* as properties of material, on both sides of the junction, establish the position of *vacuum energy level* with respect to the constant *Fermi level* drawn in step 1, at points far from the *metallurgical junction.*

   While on the metal side of the *metallurgical junction*, due to the high *conductivity* of the metal and definition of *metal work function* as a property of the metal, a constant *vacuum level* runs parallel to E_f; on the semiconductor side the situation is a little more delicate. On the semiconductor side, first, far from the *metallurgical junction*, the position of *vacuum energy level* is identified based on the information on *electron affinity* (i.e., $q\chi = E_0 - E_c$) and the position of the *Fermi level* within the *bandgap*. As presented in Chapter 1, the latter quantity can be deduced based on *Maxwell–Boltzmann* statistics, or where needed *Fermi–Dirac* statistics, and the activated *dopant* concentration.

4. After identifying the *vacuum energy levels* on the metal and semiconductor sides (and as a result *band* edges of the semiconductor far from the *metallurgical junction*), we have to smoothly connect the *vacuum energy level* of the semiconductor to that of the metal (i.e., at the *metallurgical junction*).

   This is because *vacuum energy level* at any given point should be single valued. If not, we face a contradiction with the conservation of energy (i.e., a free *vacuum-level* electron moves over an extremely short distance and gains energy).

5. Having found the *vacuum energy level* across the junction and knowing that on the semiconductor side the *bandgap* and *electron affinity* at any given point are constants, we can draw the variation of the energy at the bottom of *conduction* and top of the *valence band* (i.e., E_c and E_v) of the semiconductor in the vicinity of the junction.

As illustrated in Figure 2.2b, the resulting *band bending* in this *MS* junction imposes a potential barrier against the movement of majority *carriers* of the semiconductor (i.e., in this case: electrons) to the opposite side of the junction. As mentioned earlier, this *band diagram* is representative of a *blocking* contact to an n-type semiconductor.

According to the aforementioned procedure, throughout the length of the region over which the *vacuum energy levels* of the two sides were smoothly connected to each other, we can only have approximate values of energy for the *conduction-* and *valence-band* edges. In order to evaluate the validity of this approximation, this length span needs to be compared to a metric known as *Debye length*.[2]

As pointed out earlier, this derivation of the *band diagram* is made possible through the assumption of a sharp interface between the two materials and by neglecting the presence of any *dangling bonds* at the interface. To this list we should also add the absence of *surface states* and *surface layers*. As will be shown later, in the presence of *surface states*, independent of the earlier procedure, the *Fermi level* might be pinned to certain values of energy identified by these *states*.

Figure 2.3 presents an extended list of *work functions* for a variety of metals. Figure 2.4 presents a list of *electron affinities* and *bandgaps* for a number of popular semiconductors. As illustrated through the numbers presented in these figures, depending on the metal *work function*, semiconductor *electron affinity*, *doping* type, and concentration, one can expect two different directions of *band bending* on the semiconductor side of an *MS* junction.

[2] Named after *Peter Debye*. More discussion on the topic of *Debye length* is presented in Appendix 2.A.

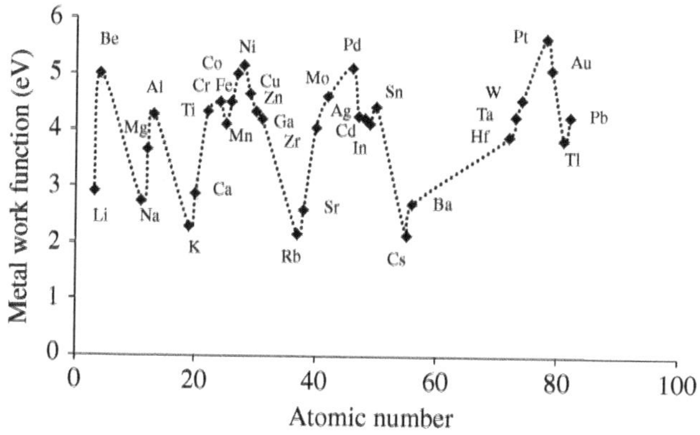

FIGURE 2.3 Work functions of a few metals according to their atomic number.

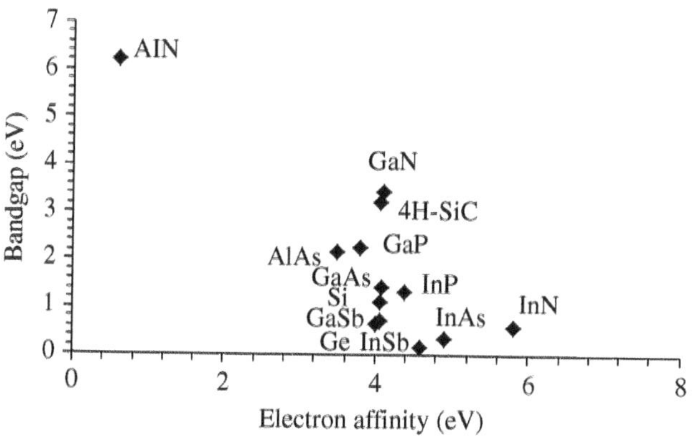

FIGURE 2.4 Variation of room-temperature bandgap and electron affinity among a few famous semiconductors.

According to *Maxwell–Boltzmann* statistics, bending of the *conduction band* away from the *Fermi level* represents both a decrease in electron concentration and an increase in *hole* concentration in the vicinity of the interface with the metal. This has different meanings in the cases of contact to an n-type or a p-type semiconductor. The first case presents a contact in which majority *carriers* are driven away from the vicinity of the interface with metal. In the second case majority *carriers* are accumulated toward this interface. The opposite of this scenario happens when it is the *valence band* that is bending away from the *Fermi level*. The *band diagrams* of the previously four identified situations are presented in Figure 2.5.

The *blocking* versus *nonblocking* character of an *MS* junction is dictated by the reduction or accumulation of majority *carriers* in the vicinity of the interface. In *blocking* contacts majority *carriers* are driven away from the surface of the semiconductor. Accumulation of majority *carriers* in the vicinity of the interface with metal

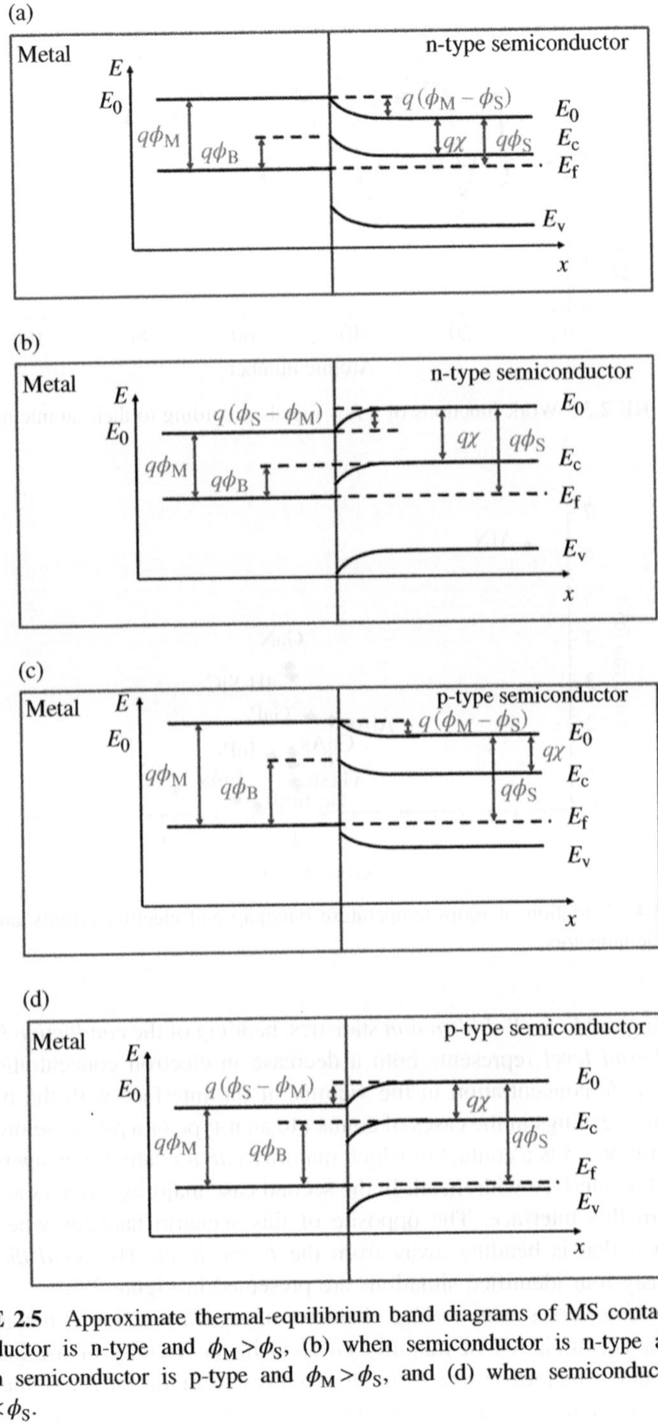

FIGURE 2.5 Approximate thermal-equilibrium band diagrams of MS contacts, (a) when semiconductor is n-type and $\phi_M > \phi_S$, (b) when semiconductor is n-type and $\phi_M < \phi_S$, (c) when semiconductor is p-type and $\phi_M > \phi_S$, and (d) when semiconductor is p-type and $\phi_M < \phi_S$.

happens in a *nonblocking* contact. Reduction in majority *carrier* concentration toward the interface imposes a potential barrier on *bulk* majority *carriers* to arrive at the opposite side of the junction.

As illustrated in Figure 2.5, in both *blocking* and *nonblocking Schottky* contacts, metal's electrons face a potential barrier against moving to the semiconductor side. This potential barrier is defined in terms of the semiconductor *electron affinity* and metal *work function* (i.e., $q\phi_M - q\chi$).

Figure 2.5 clearly shows that in order to establish *blocking* contacts to n-type semiconductors, metals with *work functions* larger than the semiconductor's are needed (i.e., $\phi_M > \phi_s$). However, in order to establish *blocking* contacts to p-type semiconductors, metals with small *work functions* are required (i.e., $\phi_M < \phi_s$). Otherwise, *nonblocking* contacts will result. Nevertheless, not all of the metals indicated in Figure 2.3 are compatible with mainstream semiconductor processing. For that reason, oftentimes in realization of *Ohmic* contacts to n-type semiconductors, other techniques such as formation of *tunneling*, *alloyed*, or *superlattice* contacts are employed to realize the required *nonblocking* character.[3]

Since *ionized impurities* are not mobile, the development of regions of either degraded or enhanced concentration of majority *carriers* toward the interface causes the elimination of charge neutrality in these regions. The disturbance on charge neutrality evidently causes the induction of a built-in electric field in the vicinity of the junction. Caused by the high *conductivity* of the metal, these charge profiles are balanced by an ideally zero-thickness profile of opposite polarity on the metal side of the interface. Such a profile results from the transfer of *carriers* across the junction in the course of the process of reestablishing *thermal equilibrium*.

2.2.2 *SDA*

According to *SDA*, on the semiconductor's side of the *metallurgical junction* of a *blocking MS* contact, the reduction in majority *carrier* concentration materializes in the form of a complete *carrier* depletion up to a certain distance from the junction. This is referred to as *depletion-region width*: x_d. Since *Maxwell–Boltzmann* statistics establishes an exponential dependence between the *band banding* and the *carrier* concentration, such a simple framework extends an appropriate degree of approximation.

According to this theory, whereas across the *depletion-region width* all *charge carriers* are eliminated, beyond the boundary of this region, *carrier* concentration is undisturbed. This description results in the division of the semiconductor into two regions:

1. A region extended to x_d from the *metallurgical junction*, across which there exists a charge concentration profile matched to the profile of ionized *dopants*
2. A region immediately past the *depletion-region width*, across which charge neutrality has remained completely undisturbed

[3] These alternatives are explored later in this volume.

The first region is referred to as the *space charge* or *depletion* region, and the second region is called the *neutral* region. Everywhere across the junction, minority *carrier* concentration is assumed to be much smaller than either the *ionized-impurity* concentration of the *dopants* or the majority *carrier* concentration.

2.2.2.1 *Blocking Schottky Barrier* In this section, we will apply *SDA* to a *blocking MS* junction formed on a planar interface to a uniformly *doped* n-type semiconductor (with *doping* level N_d) and deal with the profiling of charge concentration, potential, and electric field. The *blocking* contacts formed to homogenously *doped* semiconductors are referred to as *Schottky barriers*.

For this junction, according to *SDA*, Figure 2.6a presents the charge concentration profile on the semiconductor side of the junction as

$$\rho(x)=\begin{cases} qN_d & 0<x\leq x_d \\ 0 & x_d<x \end{cases}. \tag{2.8}$$

Based on this charge concentration profile, *Gauss*'s law is used to calculate the profile of the electric field. The boundary condition used in this integration process is set by the assumption of *zero* electric field across the *neutral* region. Assuming the *dielectric constant* of the semiconductor is ϵ_s, the resulting electric field shows a linear spatial variation

$$\nabla\cdot E=\frac{\rho(x)}{\epsilon_s}\Rightarrow E(x)=\frac{qN_d}{\epsilon_s}(x-x_d). \tag{2.9}$$

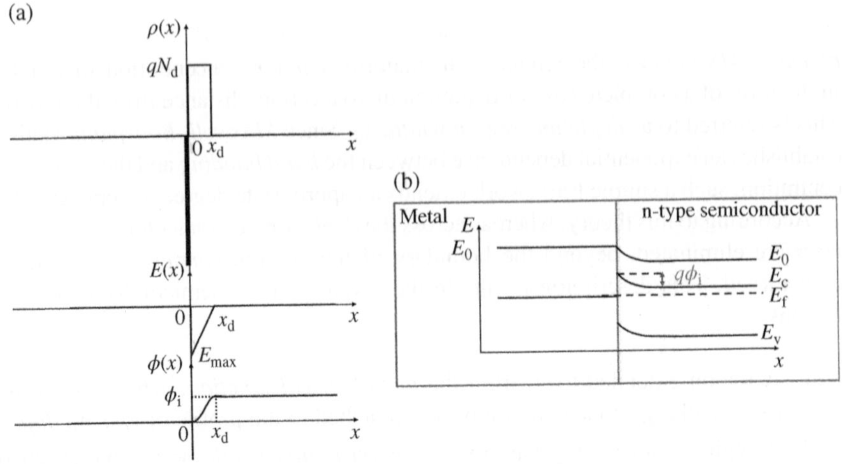

FIGURE 2.6 (a) Variation of the charge concentration, electric field, and potential across a blocking *Schottky barrier* made to an n-type semiconductor under thermal equilibrium, according to *SDA*. (b) Demonstration of the built-in potential $q\phi_i$ according to the band diagram of this junction.

As expressed in (2.9), in this type of barrier, the maximum built-in electric field, which is defined as $(-qN_d x_d)/\epsilon_s$, is induced at the *metallurgical junction*. According to this electric-field profile, assuming *zero* potential on the metal side and using the definition of the potential in the neutral region of the semiconductor as $\phi_n = (E_f - E_i)/q$, the potential profile of the *depletion* region can be readily calculated as

$$E = -\nabla\phi \Rightarrow \phi(x) = -\frac{qN_d}{2\epsilon_s}(x - x_d)^2 + \phi_n. \tag{2.10}$$

Based on (2.10), the potential difference between the *bulk* of the semiconductor (i.e., *neutral* region) and the metal will be equal to

$$\phi_i = \phi(x_d) - \phi(0) = \frac{1}{2}\frac{qN_d x_d^2}{\epsilon_s}. \tag{2.11}$$

In compliance with (2.9) and (2.10), Figure 2.6a demonstrates the variation of electric field and potential across this *Schottky* barrier. As shown in this figure, the potential difference ϕ_i, which is referred to as built-in potential, is equal to $-1/2(E_{max} x_d)$. This quantity represents the total potential drop across the *MS* junction. According to (2.11), given by the *Maxwell–Boltzmann* statistics, the built-in potential of the *Schottky* barrier made to a *nondegenerate* semiconductor is also equal to

$$\phi_i = \phi_M - \phi_S = \phi_M - \left(\chi + \frac{E_g}{2q} - \phi_n\right), \tag{2.12}$$

while $\phi_n = \dfrac{E_f - E_i}{q} = \dfrac{kT}{q} Ln\left(\dfrac{N_d}{n_i}\right)$.

As a result of (2.11), the *depletion-region* width as a function of the built-in potential can be written as

$$x_d = \sqrt{\frac{2\epsilon_s \phi_i}{qN_d}}. \tag{2.13}$$

In addition, the *space charge* per unit area of the junction is given by

$$Q_s = \sqrt{2\epsilon_s \phi_i q N_d}. \tag{2.14}$$

As shown in Figure 2.6b, the amount of *band banding* between the *depletion* region and the *neutral* region is equal to $q\phi_i$. Ideally speaking, as shown in (2.12), this is a potential barrier that is tunable via varying the *dopant* concentration of the semiconductor. However, from the same ideal perspective, the size of the potential barrier observed by the metal's electrons is a fixed quantity (i.e., $q(\phi_M - \chi)$).

The two aforementioned potential barriers are clearly created by the absence of available energy *levels*, in a region between the metal and the semiconductor, for a certain range of energy corresponding to the *conduction-band* edge. As a result, in the absence of quantum *tunneling*, *conduction-band* electrons have to acquire enough energy to rise over this energy *gap* in order to transfer from one side to the other. The presence of this energy *gap* causes the *blocking* nature of the contact. However, for larger values of *dopant* concentration (i.e., N_d), x_d as shown in (2.13) reduces. As a result, the probability of quantum *tunneling* exponentially increases. In the case where quantum *tunneling* dominates, electrons will not be confined, and the contact exhibits an *Ohmic* characteristic. As alluded to earlier, this is one of the popular forms of *Ohmic* contacts made to n-type semiconductors.

Example

Consider an *MS* junction between *aluminum* and *silicon*, in which *silicon* is *doped* with *boron* according to the following distribution profile:

$$N_a(x) = \begin{cases} 10^{12}\ \text{cm}^{-3} & 0 \le x \le 5\ \text{nm} \\ 10^{16}\ \text{cm}^{-3} & 5\ \text{nm} \le x \end{cases}$$

where $x = 0$ represents the *metallurgical junction*.

For this junction, calculate the *depletion-region width*. Take the *Al work function* as 4.1 eV and *silicon's electron affinity* equal to 4.05 eV. Perform the calculations at room temperature.

Considering the size of the *bandgap* (i.e., 1.12 eV), the *electron affinity*, and the p-type character of *dopants* in *silicon*, ϕ_M is smaller than ϕ_S. Hence, according to Figure 2.5, the junction represents a *blocking* contact.

Since within the first 5 nm from the *MS* contact the *dopant* concentration is small, we expect the *depletion region* to extend beyond this region's boundary and the variation of the electric field across the first 5 nm to be very small. We also consider the potential drop across this thin region to be negligible. All these are assumptions that require validation.

According to the aforementioned assumptions, we first evaluate the built-in potential according to the *bulk* semiconductor's *work function*:

$$q\phi_i = q(\phi_S - \phi_M) = q\chi + \frac{E_g}{2} + (E_i - E_f) - q\phi_M$$
$$= 4.05 + \frac{1.12}{2} + 0.0258 Ln\left(\frac{10^{16}}{1.45 \times 10^{10}}\right) - 4.1 \cong 0.86\ \text{eV}.$$

Since we have assumed all this built-in potential to drop across the *bulk*, according to (2.13) we can write

$$x_d = \sqrt{\frac{2\epsilon_s\phi_i}{qN_a}} = \sqrt{\frac{2 \times 11.7 \times 8.85 \times 10^{-14} \times 0.86}{1.6 \times 10^{-19} \times 10^{16}}} \cong 3.33 \times 10^{-5}\text{cm, or } 0.333\ \mu\text{m}$$

where, according to Table 1.4, the *relative permittivity* of *silicon* is equal to 11.7.

The calculated value of the *depletion-region width* is much larger than the 5 nm thickness of the lightly *doped* interface, which validates one of our assumptions.

Based on this value of x_d and our earlier assumptions,

$$E_{max} = \frac{2\phi_i}{x_d} = \frac{2 \times 0.86}{3.33 \times 10^{-5}} \cong 5.16 \times 10^4 \text{ V cm}^{-1},$$

which for the negligible thickness of the surface layer provides the following potential drop across this region:

$$\Delta\phi = E_{max} \times 5 \times 10^{-7} \cong 25.8 \text{ mV}.$$

Since $\Delta\phi \ll \phi_i$ our other assumption is also validated.

2.2.2.1.1 Application of Bias across a Schottky Barrier *SDA* also provides us with a simple framework for dealing with *biased* junctions. While *biasing* a junction shakes off the constraints of *thermal equilibrium* (i.e., constant *Fermi level* throughout the system), *SDA* assumes that certain aspects of that state can persist as long as the *bias* is small. In this fashion, *biased* junctions are expressed by assuming that *thermal equilibrium* is revoked only in the vicinity of the junction. This is while on the metal side and away from the *metallurgical junction* on the semiconductor side, two separate states of *thermal equilibrium* are formed.

Resulting from this assumption, *SDA* asserts that while the *Fermi level* of the metal can maintain the same *level* as defined under *unbiased thermal equilibrium*, the *Fermi level* of the *bulk* of the semiconductor moves up, or down, with respect to this *unbiased level*. The amount of the shift is equal to qV_a, V_a being the applied potential to the semiconductor with respect to the grounded metal. For a positive V_a, the *Fermi level* moves down by the aforementioned amount, and for a negative V_a, it will move up by that amount. These two *bias* scenarios, in the case of a *Schottky* barrier made to an n-type semiconductor, are expressed in Figure 2.7. As shown in this figure, a positive V_a will result in enhancing the potential barrier size and reducing the probability of electron transfer from semiconductor to metal, while *hole* transfer is promoted. This case is referred to as *reverse bias*. However, majority *carriers* (i.e., electrons) see a smaller potential barrier when V_a is negative (i.e., *forward bias*). The size of this potential barrier can never be reduced to 0. For the presented framework to apply to this case, V_a should be smaller than the built-in potential. The indicated asymmetric dependence on *bias* voltage is the root of the rectifying property of blocking *MS* junctions.

This approximation still assumes that the semiconductor is divided between a *depletion* region and a *neutral* region. However, caused by the presence of two, and not just one, states of *thermal equilibrium*, establishment of a net flow of *carriers* (i.e., majority *carriers* in case of *forward bias* and minority *carriers* in case of *reverse bias*) creates a charge imbalance even deep within the semiconductor. Nevertheless, as long as the *bias* is limited, the charge concentration profile in this region will be negligible compared to that of the *depletion* region. In this case, while we still apply the condition of charge neutrality, oftentimes instead of the *neutral* region, this region

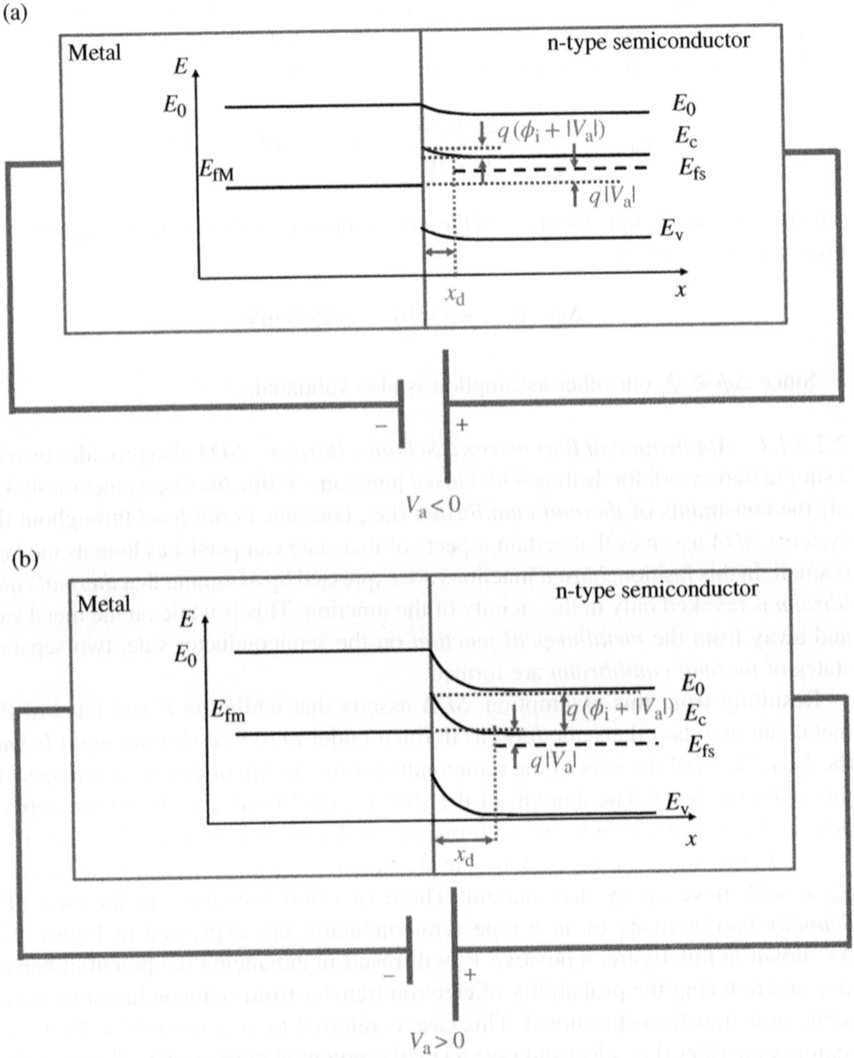

FIGURE 2.7 Band diagrams of a blocking Schottky contact made to an n-type semiconductor under (a) forward bias and (b) under reverse bias. *MS* bias V_{MS} is equal to $-V_a$.

is referred to as *quasi-neutral*. This is precisely in compliance with the previously indicated assumption of *thermal equilibrium* in this region. Although, strictly speaking it would be more appropriate to speak of *quasi-thermal equilibrium* in this region rather than *thermal equilibrium*.

As we have already taken note in Section 2.2.1, such a behavior is not observed in *biased nonblocking* (or *Ohmic*) contacts. In those contacts, in the absence of the *depletion* region and a potential barrier for majority *carriers*, imposing an electric field across the *MS* junction results predominantly in a potential drop across the *bulk* semiconductor (i.e., away from the junction). This renders the definition of the state of *thermal equilibrium* and *Fermi level* inadmissible across the semiconductor. These junctions, granted by their name, demonstrate a potential drop of *Ohmic* behavior across the semiconductor, especially for small electric fields. We have already seen in Section 1.4.4 that increasing of the electric field results in variation of the semiconductor resistivity with electric field and hence deviation from a linear (i.e., *Ohmic*) dependence of current on voltage.

Whereas in a *blocking Schottky* contact *bias* modifies the potential barrier on the semiconductor side, to the first order, the height of the potential barrier seen by metal electrons remains *bias* independent. According to the *band diagrams* shown in Figure 2.7, the *depletion-region width* and Q_s increase with *reverse bias*, while they reduce under *forward bias*. The *SDA* formalism expresses these changes simply by replacing ϕ_i by $\phi_i + V_a$ in (2.13) and (2.14). In other words, *SDA* assumes all of the *bias* potential to drop across the *depletion* region. Under *biased* conditions, because of the removal from the state of *thermal equilibrium*, across this region only *quasi-Fermi levels* are defined. Here it has been implicitly assumed that a perfect *nonblocking MS* junction has been established between the voltage source and the semiconductor. Otherwise, part of V_a would have been dropped across that junction. The implicit presence of this second *nonblocking MS* junction also indicates why in connecting a *resistor* between the two sides of *MS* junction, the presence of the built-in potential is preventing a current to flow through the *resistor*. This is because the effective potential seen across the terminals of the *resistor* is *zero*. In this situation, the built-in potential simply drops across the *nonblocking MS* contact, while *thermal equilibrium* is established throughout the system composed of two junctions and the *resistor*.

As alluded to earlier, the presented perception of a *biased MS* junction is more accurate in the case of a *blocking Schottky* contact under *reverse bias* and also to a certain extent in a *forward-biased* junction (*blocking* or *nonblocking*) under *low-level injection*.[4]

2.2.2.1.2 Capacitance–Voltage Characteristic of Biased Schottky Barriers According to what is shown in Figure 2.5, upon formation of an *MS* junction, a *capacitor* is formed between the semiconductor's *depletion* (or *accumulation*) regions and a thin layer of charge of opposite polarity formed at the metal side of the interface. While in *blocking* junctions, the component of the charge on the semiconductor side is stored in the depletion region, in *nonblocking MS* contacts this role is played by the *accumulation* region. The presence of this *capacitor* gives a *resistive–capacitive* (*RC*) nature to *MS* junctions. This *capacitive* nature is especially important in *blocking* contacts in which the resistive part is very inhibitive.

[4] *Low-level injection* refers to cases in which concentration of excess majority *carriers* (i.e., injected or induced beyond *thermal equilibrium*) is much smaller than the *thermal-equilibrium* value of majority *carrier* concentration.

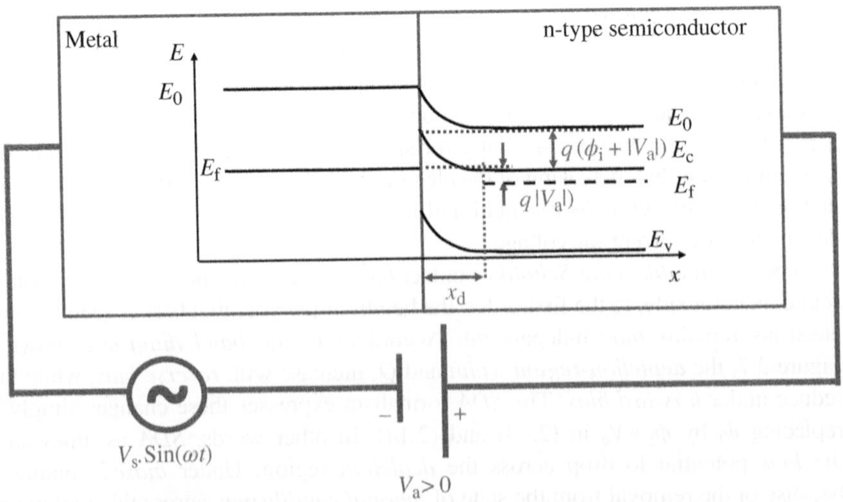

FIGURE 2.8 Bias configuration for *C–V* measurement.

In a *blocking MS* junction, the *capacitive* behavior established between the metal and the *depletion* region of the semiconductor is presented in terms of a *depletion capacitance*. One of the capable techniques for characterization of semiconductor junctions is referred to as *capacitance–voltage* (*C–V*) characterization. In application of this technique to *blocking MS* junctions, variation of the *small-signal depletion capacitance* formed across these junctions with *reverse bias* is studied. These studies are used in evaluation of parameters such as built-in potential and *dopant* concentration. As shown in Figure 2.8, in these measurements, while the *blocking* contact is *reverse biased*, a low-frequency sinusoidal *signal* of small amplitude is overriding the *bias* voltage. In terms of this *bias* configuration, an *RLC meter* is used to measure the value of the *small-signal depletion capacitance* at different values of *reverse bias*. The value of *reverse bias* voltage is either swept very slowly or modified in steps.

As presented in Section 2.2.2.1.1, as a result of the application of *bias* voltage (i.e., V_{MS}) between metal and semiconductor, the per unit area charge stored in the *depletion* region is given by

$$Q_s = \sqrt{2\epsilon_s q N_d(\phi_i - V_{MS})}. \tag{2.15}$$

As a result, in the *bias* configuration presented in Figure 2.8, the *small-signal depletion capacitance* per unit area can be calculated as

$$C = \left|\frac{\partial Q_s}{\partial V_{MS}}\right| = \sqrt{\frac{\epsilon_s q N_d}{2(\phi_i - V_{MS})}} = \frac{\epsilon_s}{x_d}. \tag{2.16}$$

The last form of this presentation is the generic form of the *small-signal capacitance*.[5] As a result of (2.16), we have

$$\phi_i - V_{MS} = \frac{\epsilon_s q N_d}{2C^2}. \tag{2.17}$$

Whereas this relationship applies to the case of *Schottky barriers*, where the semiconductor is uniformly *doped*, a similar but more generic framework can also be established for the case of a *blocking MS* junction formed to a semiconductor with an arbitrary *doping* profile. In this case,

$$Q_s = q\int_0^{x_d} N(x)dx \Rightarrow dQ_s = qN(x_d)dx = -CdV_{MS} \tag{2.18}$$

where applying to the definition of *small-signal capacitance* as ϵ_s/x_d,

$$\frac{dx}{dV_{MS}} = \frac{dx}{dC} \times \frac{dC}{dV_{MS}} = -\left(\frac{\epsilon_s}{C^2}\right)\frac{dC}{dV_{MS}} \tag{2.19}$$

results in

$$N(x_d) = \frac{C^3}{q\epsilon_s(dC/dV_{MS})}. \tag{2.20}$$

However, since $\dfrac{d(1/C^2)}{dV_{MS}} = -\dfrac{2}{C^3}\dfrac{dC}{dV_{MS}}$, (2.20) can be rewritten as

$$N(x_d) = \frac{-2}{q\epsilon_s(d(1/C^2)/dV_{MS})}. \tag{2.21}$$

What is common between (2.17) and (2.21) is that both of these equations imply that the slope of the plot of $1/C^2$ versus *reverse bias* contains information on the *dopant* concentration. Since the *depletion-region width* x_d is an expanding function of reverse *bias*, (2.21) shows that such a plot can reveal the *dopant* concentration at various depths of the semiconductor. Since the expansion of x_d is also linked to the increase in the electric field, junction *breakdown* sets a limit to the maximum depth

[5] To be more complete in the description of *C–V* characteristics, it should be indicated that the *permittivity* of the semiconductor is itself dependent on the frequency of the *small-signal* stimulus. Between limiting cases of $\omega \to 0$, $\omega \to \infty$ (i.e., *static* and *high-frequency dielectric constants*) *polar* semiconductors such as *GaAs* have different *dielectric constants*. This is due to contributions of the *polar* nature of the crystal *lattice* of these materials. *Si* and *Ge*, being *nonpolar*, do not have this change of *relative dielectric constant* with frequency.

of characterization. This general dependence is used in several commercial semiconductor profilers to determine the *dopant* concentration as a function of depth.

Obviously, such relationships between reverse *bias* and *small-signal depletion capacitance* are established within the confines of *SDA*. However, it is pleasantly surprising that (2.17) and (2.21) offer a very reliable framework for characterizing the *dopant* concentration. In contrast to the information taken from the slope of $1/C^2$ versus *reverse-bias* plot, the intersection of this plot with the voltage axis, contrary to (2.17), does not offer an accurate read on built-in potential (i.e., ϕ_i). This is because *SDA* completely overlooks the *surface states* and the effect of the image-charge force, which is caused by the presence of the metal (known as *Schottky barrier lowering*).[6] Both of these are very important in determining the built-in potential.

It should be reemphasized that this discussion has been based on the application of reverse *bias* across *blocking MS* junctions. As noted earlier in Section 2.2.2.1.1 *forward-biasing MS* junctions result in a potential drop across the *bulk* of the semiconductor, which is essentially modeled as a resistor (at least under low electric fields) and an *excess* injected charge profile into the bulk semiconductor. The latter is expressed in terms of a *diffusion capacitance* across this region.[7] This *capacitance* is formed in series with the *junction capacitance* (in this case: *accumulation capacitance*). As a result of this series combination, the *junction capacitance* loses its dominance.

2.2.2.1.3 Current–Voltage Characteristics of Schottky Barriers Understanding the current–voltage characteristic of *blocking MS* junctions is very important in the design and evaluation of electronic devices. As seen in Section 2.2.2.1.1, application of a positive or negative *bias* across a *Schottky barrier* results in establishment of a net current across the junction. This current is either carried by majority *carriers* (in case of *forward bias*) or by minority *carriers* (in case of *reverse bias*). However, based on *current continuity* both of these situations are contrary to *SDA*'s assumption of depletion of *charge carriers* from the vicinity of the junction. As a result, in the evaluation of current across *biased* junctions, this aspect of *SDA* should be modified. According to this modified framework, whereas the semiconductor within the vicinity of the junction is not devoid of *charge carriers*, their concentration, and also their spatial gradient, is very limited. Therefore, via establishing *current continuity*, current can be evaluated at this junction in terms of *drift transport*.

Since this *drift* current is carried by the *charge carriers* present at the junction, at a qualitative level, the current–voltage characteristic of a *Schottky barrier* made to an n-type semiconductor can be evaluated in terms of electron concentration in the vicinity of the junction and its *bias* dependence. According to Figure 2.9, under the *thermal equilibrium*, the electron concentration on the semiconductor side of this junction is given by

$$n_s = N_c \exp\left(-\frac{q\phi_B}{kT}\right) \cong N_d \exp\left(-\frac{q\phi_i}{kT}\right). \tag{2.22}$$

[6] Both of these topics are presented in Section 2.2.2.1.4.

[7] As will be observed later in this chapter, *diffusion* is the mode of *transport* of these excess *carriers*.

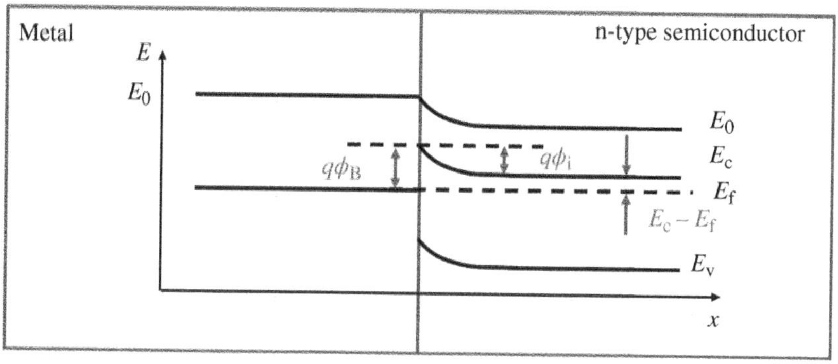

FIGURE 2.9 Band diagram of a blocking *MS* contact made to an n-doped semiconductor under thermal equilibrium.

In terms of this electron concentration, without *biasing* the junction (according to the definition of *zero* net current across the junction under *thermal equilibrium*), we will have

$$|J_{MS}| = |J_{SM}| = KN_d \exp\left(-\frac{q\phi_i}{kT}\right) \tag{2.23}$$

in which K is a proportionality constant and J_{MS} and J_{SM} are metal-to-semiconductor and semiconductor-to-metal current densities, respectively.

The application of *bias* across the junction, however, creates an imbalance between these two current terms. Whereas the potential barrier seen by metal's electrons according to *SDA* is *bias* independent, the size of the barrier on the semiconductor side is *bias* dependent. This dependence results in modification of (2.22) to

$$n_s = N_d \exp\left(-\frac{q(\phi_i - V_{MS})}{kT}\right). \tag{2.24}$$

According to the single-sided dependence of the potential barrier height of the *Schottky barrier* on the *bias*, (2.23) predicts a net current density of

$$\begin{aligned} J = J_{MS} - J_{SM} = NK_d \exp\left(-\frac{q(\phi_i - V_{MS})}{kT}\right) - KN_d \exp\left(-\frac{q\phi_i}{kT}\right) \\ \Rightarrow J = J_0\left[\exp\left(\frac{qV_{MS}}{kT}\right) - 1\right]. \end{aligned} \tag{2.25}$$

This equation is referred to as the *ideal diode equation*. Whereas (2.25) predicts an exponential improvement in current density with *forward bias*, under *reverse bias* it predicts a *bias*-independent current density.

Deficiency in predicting the experimentally observed *bias* dependence of the *reverse* current demands a more rigorous evaluation of the current across these junctions. In order to do justice to the current–voltage characteristic of a *Schottky barrier*, integration over both the *drift* and the *diffusion* current components over the width of the *depletion* region is carried out. For the *drift–diffusion* formalism to apply, we assume that the length of this integration region is longer than a few times the *mean free path*. In addition, it is assumed that the electric field is weak enough so that the values of low-field *carrier mobility* and *diffusion constant* can be used.[8]

Based on (1.76) and under the assumption of no *generation–recombination* (*G–R*), current as a function of position across the depletion region can be expressed by a constant current density

$$J_x = qD_n\left[-\frac{qn}{kT}\frac{d\phi}{dx} + \frac{dn}{dx}\right]. \tag{2.26}$$

Through integrating over the width of the *depletion region*, we will have

$$J_x\int_0^{x_d} \exp\left(-\frac{q\phi}{kT}\right)dx = qD_n\left[n\cdot\exp\left(-\frac{q\phi}{kT}\right)\right]_0^{x_d} \tag{2.27}$$

where the boundary conditions as expressed earlier are

$$\phi(0) = 0 \ \text{ and } \ \phi(x_d) = \phi_i - V_{MS}. \tag{2.28}$$

Equivalently, in terms of height of the potential barrier on the metal side (i.e., ϕ_B),

$$\phi(x_d) = \phi_B - \phi_n - V_{MS} \tag{2.29}$$

where, as shown in Figure 2.7, $q\phi_n = E_c - E_f$.

According to *Maxwell–Boltzmann* statistics, we also know that

$$n(0) = N_c\exp\left(-\frac{q\phi_B}{kT}\right) \ \text{ and } \ n(x_d) = N_c\exp\left(-\frac{q\phi_n}{kT}\right). \tag{2.30}$$

On the basis of (2.28)–(2.30), (2.27) simplifies to

$$J_x = qD_nN_c\exp\left(-\frac{q\phi_B}{kT}\right)\left[\exp\left(\frac{qV_{MS}}{kT}\right)-1\right]\Big/\int_0^{x_d}\exp\left(-\frac{\phi(x)}{kT}\right)dx. \tag{2.31}$$

[8] In Section 2.5 we will explain an alternative physical approach based on *thermionic emission* of *carriers* from the metal. This approach, which was originally adopted by *Hans Bethe*, proves that even without the simplifying assumption of *drift–diffusion* (which are identified previously), we can obtain an identical dependence.

Whereas (2.31) is a generic equation applicable to all *blocking MS* junctions, for the specific case of a *Schottky barrier* where

$$\phi(x) = \frac{qN_d}{\epsilon_s} x\left(x_d - \frac{x}{2}\right) \quad 0 < x < x_d, \tag{2.32}$$

Equation (2.31), reduces to

$$J_x = J_s\left[\exp\left(\frac{qV_{MS}}{kT}\right) - 1\right] \tag{2.33}$$

where

$$J_s = \frac{q^2 D_n N_c}{kT}\left[\frac{2q(\phi_i - V_{MS})N_a}{\epsilon_s}\right]^{1/2} \exp\left(\frac{-q\phi_B}{kT}\right). \tag{2.34}$$

Unlike the *reverse current* of the *ideal diode equation* (i.e., J_0), (2.34) correctly predicts a *bias*-dependent reverse current density (i.e., J_s). An implicit assumption of this derivation is the lack of observation of the *bias* dependence of x_d.

Quite often, instead of employing the laborious framework of (2.31), the current–voltage characteristic of a *Schottky diode* is expressed through interpolating the following heuristic equation among the experimental data:

$$J_x = J_s'\left[\exp\left(\frac{qV_{MS}}{nkT}\right) - 1\right] \tag{2.35}$$

in which J_s' is independent of voltage and n is a parameter that takes values usually found to be between 1.02 and 1.15.

It is quite obvious that the equations explored in this section are also restricted by the confines of *SDA* and the other identified assumptions in this section. These equations remain descriptive as long as the current is low and the system can be explained in the form of a state of *quasi-thermal equilibrium* in the *quasi-neutral* region. Increasing the *current density*, however, due to potential drop across the semiconductor *bulk*, results in invalidating of this assumption. As a result, for large *forward biases*, the *bulk* semiconductor effectively acts as a *resistor*. This *resistive* behavior eventually becomes dominant at high current densities. For that reason, the effective potential barrier height seen by the semiconductor's majority *carriers* can never be reduced to zero.

2.2.2.1.4 Corrections to SDA So far in this chapter, we have assumed that the potential barrier faced by metal electrons (i.e., $q\phi_B$) is *bias* independent. However, this is not exactly true. Application of *bias* and inducing change in charge concentration of the *depletion* region effectively result in changing the *image-charge* profile of opposite polarity effectively formed on the metal side of the interface. The effects of this *image-charge* profile have so far been absent from our discussions. The potential barrier height, especially in the case of a *reverse-biased MS* junction, manifests a small dependence on the *bias*. This effect was originally explained by *Walter Schottky* in his studies on electron emission into a vacuum.

The other phenomena, which we have so far overlooked in the calculation of $q\phi_B$ and also the built-in potential (i.e., ϕ_i), are the presence of *surface states* and dangling bonds at the interface with the metal. These effects are often capable of rendering a built-in potential quite independent of the metal *work function*, unlike the predictions of Figure 2.5.

Schottky Barrier Lowering According to electrostatics, in an *MS* junction, free electrons in the semiconductor (with considering the *effective mass* and *relative permittivity* of semiconductor) effectively feel the presence of the conducting metal plate in the form of an opposite *image charge* placed at an equivalent distance behind the interface. In evaluation of the variation of energy of these free electrons versus position, the *Coulombic* interaction with this *image charge*, in addition to the energy felt by electrons due to the presence of the externally applied or internally induced electric field, should be taken into account. Assuming that the electric field induced by the *reverse bias* (which is identified by *E*), in the vicinity of the *metallurgical junction* is independent of depth, we can evaluate the energy of electrons as

$$E(x) = -\frac{q^2}{4\pi\epsilon_s(2x)^2}x - qEx. \tag{2.36}$$

This situation is depicted in Figure 2.10.

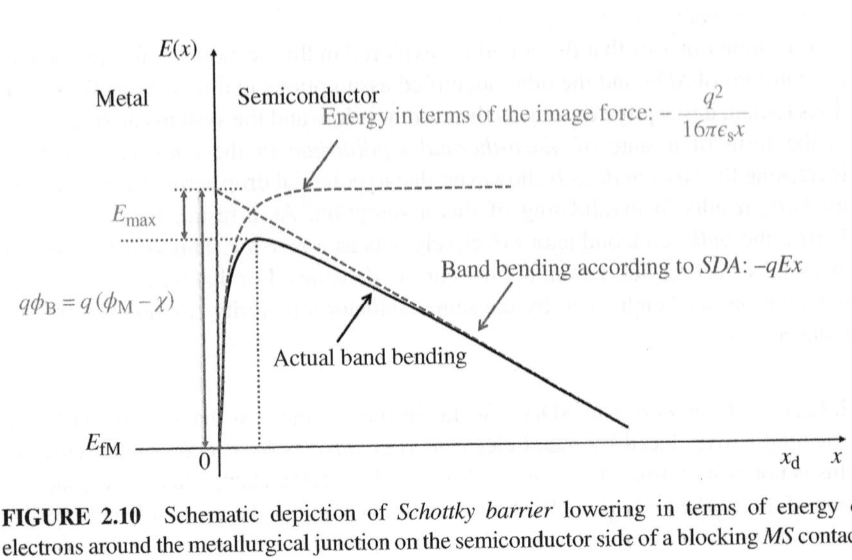

FIGURE 2.10 Schematic depiction of *Schottky barrier* lowering in terms of energy of electrons around the metallurgical junction on the semiconductor side of a blocking *MS* contact.

As shown in Figure 2.10, Equation (2.36) maximizes the energy at a point close to, but different from, the metallurgical contact. This maximum energy is given by

$$E_{\mathrm{max}} = -\sqrt{\frac{q^3 E}{4\pi\epsilon_s}}. \tag{2.37}$$

According to Figure 2.10, this is the amount of reduction in $q\phi_B$ rather than the previously indicated value of $q(\phi_M - \chi)$ on Figure 2.5. This reduction of potential barrier height with increasing E (*reverse bias*) is referred to as *Schottky barrier lowering*. According to *Maxwell–Boltzmann* statistics, larger values of E_{max} result in exponentially larger amounts of *reverse* current passing through the junction. The reduction in potential barrier height causes an exponential increase in the number of electrons with energies higher than the potential barrier, which are therefore capable of surmounting it.

Having known that according to *SDA* electric field is proportional to the square root of $\phi_i - V_{\mathrm{MS}}$, we can expect this current to depend exponentially on the *fourth* root of the *reverse voltage* at large *biases*. However, such a characteristic is only encountered when the *reverse* current caused by nonidealities in the form of G–*R* processes of free *carriers* in the *depletion* region (which we have not factored into our calculations of current) is negligible. Otherwise, the dependence of the *reverse* current on voltage is much more gradual.

SURFACE STATES AND SURFACE LAYER As indicated earlier, our first-order model has neglected issues such as the existence of dangling *bonds*, the rearrangement of atoms at the crystal's surfaces, and surface *states* at the interface between the metal and the semiconductor. It has been experimentally observed that the presence of these nonidealities can essentially dissociate the height of the *Schottky* potential barrier from the metal *work function*.

The surface of a semiconductor possesses properties dissimilar from the *bulk*. An important contributor to these properties is the presence of what is known as *Shockley–Tamm states*[9] within the forbidden *bandgap* of the semiconductor surface. These *states* have a density on the order of the surface density of atoms (i.e., $N_0^{2/3}$ cm^{-2}, where N_0 is the volumetric atom density). Formation of these *states* at the surface is due to the presence of a one-sided crystal potential at the surface and therefore disruption in the periodic background potential imposed on the electrons. Due to this, the surface arrangement of atoms also becomes different from the *bulk*. This rearrangement imposes an additional perturbation on the periodic potential. In *diamond lattices*, the *Shockley–Tamm states* have a profile that peaks at about *one-third* of the *bandgap* above the *valence-band* edge. The presence of these *states* at the bare surface of the semiconductor or when it is placed in contact with another layer imposes important changes to its characteristics. Figure 2.11 schematically depicts the distribution of these *states* within the *bandgap* of a *diamond lattice*-like *silicon*.

[9] Named after *William Shockley* and *Igor Tamm*.

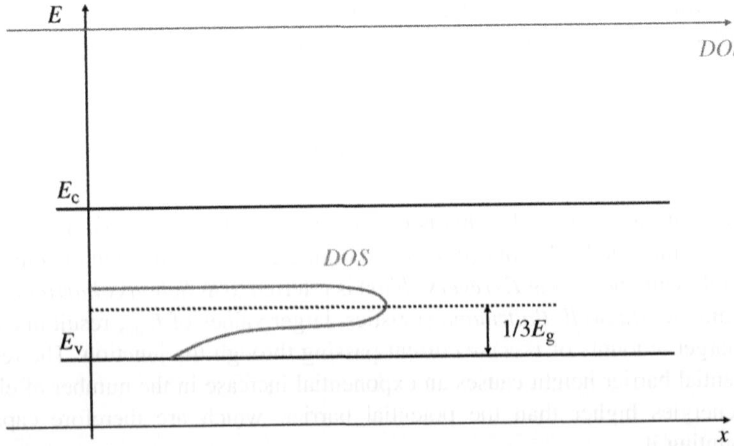

FIGURE 2.11 Approximate depiction of the density of states function of the *Shockley–Tamm* states at the surface of *silicon*.

While *Shockley–Tamm states* are present even in the purest of semiconductor surfaces, in the real world, the presence of unsatisfied dangling *bonds* and the bonding of impurity atoms to them can increase the profile of *surface states* even further. This situation is more convoluted in case of interfaces. So far we have inspected the interface of a metal and a semiconductor as though we have metal on one side and semiconductor on the other. In reality there is an interface layer within which the two materials mix. The formation of zones of intermediate materials with thicknesses several times the atomic scale is another cause for additional *states* within the *bandgap* of the semiconductor. Those *interface states*, which are physically close to the *bulk*, reach *thermal equilibrium* with the *bulk*. These are referred to as *fast interface states* because as the potential is changing rapidly, they tend to preserve this *thermal equilibrium*. Those *states*, which are more remote from the *bulk*, are much slower in reaction to a change in potential and as a result are referred to as *slow states*. The generally accepted demarcation frequency between the *slow* and *fast states* is 1 kHz.

These nonideal *states* can be of different characteristics. Broadly speaking, they can be *trap levels* or *recombination centers*. A *surface state* has a *trap level* character when it simply takes part in taking *charge carriers* out of the *conduction* or *valence band* and not releasing them back to the *energy bands* (unless energized, e.g., through *UV illumination* of the sample). However, some *surface states*, rather than keeping *charge carriers*, serve the purpose of stepping stones between the *conduction* and *valence band*. This can be envisioned as a process of entrapping both an electron and a *hole* and *recombining* the two *charge carriers*. Such *surface states* are said to have a *G–R center* (or for short *recombination center*) character. In agreement with the definitions offered in Chapter 1, from the charge status point of view, a *surface state* can be a *donor* or an *acceptor*. An *acceptor surface state* is a *state* that is neutral when empty and negatively charged when full. Likewise, a *donor state* is neutral when full and positively charged when empty. The charge status of *surface states* and their temporal variations demand modifications to our earlier treatment of *MS* junctions in terms of *SDA*. In the presence of a large concentration of these *states*, they can mask other factors such as the metal *work function* in evaluation of the potential barrier height. This describes why it was mentioned earlier that in *Schottky barriers*, it is often observed that the barrier height is quite independent of the metal *work function*.

To incorporate *surface states* in the evaluation of junctions, or even at bare *surfaces* of semiconductors, often a layer with distinguishing properties (i.e., finite *density of states* within the *bandgap*) is assumed to be sitting on top of the semiconductor of ideal *bandgap*. This layer is *several* to *ten atomic layers* in thickness. The small thickness of this so-called surface layer paves the way for quantum *tunneling* of *carriers* through it. Its presence also affects the evaluation of potential barrier height in a semiconductor junction in terms of a potential drop.

Under *thermal equilibrium*, this layer and the *states* present in it need to establish a unified *Fermi level* with the *bulk* semiconductor. In terms of this state, the charge status of the *surface states* incorporated into the *surface layer* can form *depletion*, *accumulation*, or *inversion*[10] regions at the surface of the ideal semiconductor. Evolution of these regions is not unlike what was earlier observed in Section 2.2.1 in terms of the *band* bending between the semiconductor *bandgap* and metal in an *MS* junction. For example, the presence of a *surface layer* with large density of *acceptor-type states* at the surface of an n-type semiconductor causes migration of the *electrons* from the ideal semiconductor surface to this layer. As a result, *depletion* occurs at the ideal semiconductor surface. The presence of a large density of *surface states* can *pin* the *Fermi level* to these *states*. This is depicted in Figure 2.12. This is due to the limitation in the number of *charge carriers* capable of satisfying the *surface states*, therefore repositioning the *Fermi level* within the process of establishing *thermal equilibrium* between metal, *surface layer*, and the *bulk* semiconductor (i.e., ideal semiconductor). One can see this more obviously if paying attention to the exponential dependence between the *carrier* concentration and a shift in position of the *Fermi level*. It is precisely due to *Fermi-level pinning* that, in practice, most *Schottky* barrier heights only slightly depend on metal *work function*.

2.2.2.2 A Few Other Forms of Metal Semiconductor Junctions Whereas so far in Section 2.2, we have been dealing with *Schottky* barriers, the approximate framework presented in Section 2.2.2.1 can also be extended to *MS* junctions in which the semiconductor is not uniformly *doped*. In this section we will deal with another well-known *MS* barrier referred to as the *Mott* barrier. In addition, we will pay *nonblocking MS* junctions a closer visit.

[10] We have not so far talked about *inversion*. While we will deal with this concept at length in Chapter 3, it would be beneficial at this point to mention that this is a condition at which due to *band bending* in part of a semiconductor injection of *carriers* from elsewhere results in a mismatch between the type of the present *dopants* and majority *carriers* (e.g., majority *carriers* are electrons in a place where semiconductor is *doped* with *acceptors*).

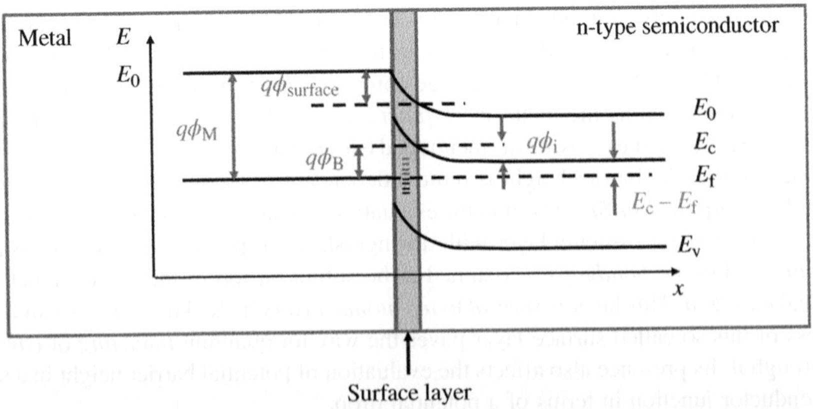

FIGURE 2.12 Schematic demonstration of *Fermi-level* pinning in case of a blocking *MS* contact due to the existence of surface states. The distribution of these states within the bandgap of the semiconductor is often considered in terms of a *DOS* function within a very thin surface layer positioned at the ideal surface of a semiconductor. In this figure, the *DOS* of the acceptor surface states is presented in terms of the horizontal lines drawn within the bandgap of this layer. Dark lines represent full states, while gray lines stand for the empty states. Since the surface layer is very thin, electrons can tunnel across this layer to migrate from metal to semiconductor. This explains the definition of the barrier height $q\phi_B$ in the above figure. As shown, across the surface layer a potential drop of $\phi_{surface}$ occurs. For large *DOS* within the surface layer, $\phi_{surface}$ will absorb most of expected change from using a metal with larger work function, hence rendering an almost invariable ϕ_B.

2.2.2.2.1 Mott Barrier Another form of *MS* barriers, which is often encountered in semiconductor devices, is referred to as a *Mott* barrier. This junction is in the form of a contact between a metal and a semiconductor in which there is an abrupt change in *doping* from a low concentration at the contact to a very high concentration shortly away from the contact. Because of the small thickness of the *lightly doped* region of a *Mott* barrier, essentially *no* electric-field line terminates in this part. As a result, electric field in the *lightly doped* region is assumed *constant*. The resulting linearly varying potential across this layer, due to the small thickness of the layer, usually results in *no* consequential drop in potential. As a result, in a *Mott* barrier, most of the potential drop happens in the *highly doped* semiconductor. However, caused by the very large *dopant* concentration in this region, the *depletion-region width* is only slightly in excess of the thickness of the *lightly doped* layer. Figure 2.13 presents a schematic representation of the *band bending* and electric-field profile in such a junction.

This type of contact is encountered when metal is contacting an *epitaxially* grown *lightly doped* layer on top of a *highly doped* substrate. As we will see later on in Chapters 4 and 5, *epitaxial* growth is becoming quite important in modern *CMOS* technology.

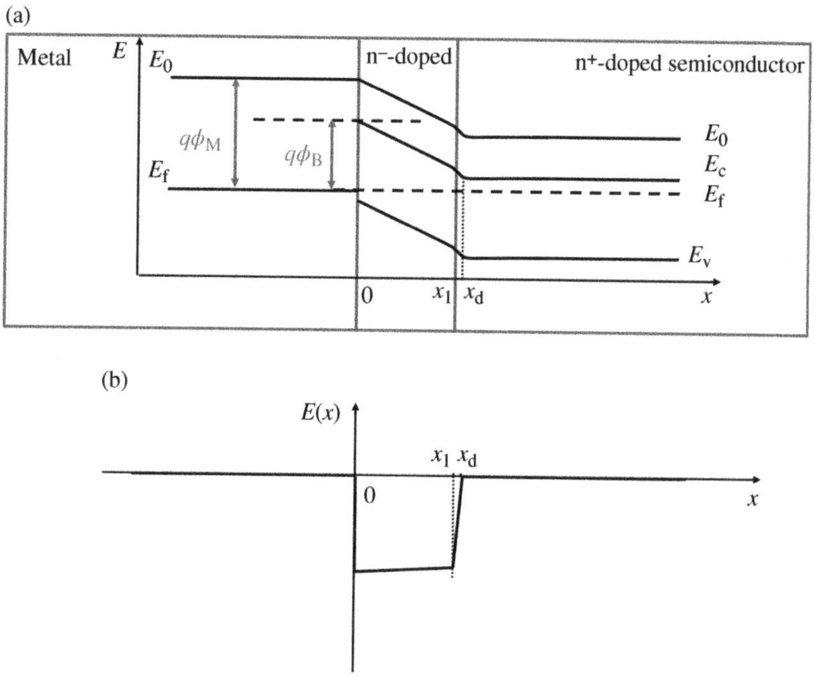

FIGURE 2.13 (a) Band bending at a *Mott* barrier under thermal equilibrium. (b) Variation of electric field across a *Mott* barrier. First region's width is exaggerated.

For the *Mott* barrier presented in Figure 2.13, while under *bias*, according to earlier approximations and *SDA*, we will have

$$\phi(x)=(\phi_i-V_{\mathrm{MS}})\frac{x}{x_{\mathrm{d}}}\quad 0<x<x_{\mathrm{d}}. \tag{2.38}$$

On the basis of (2.27) and the *current-continuity* equation, the current–voltage characteristic of such a junction can be represented as

$$J_{\mathrm{x}}=J_{\mathrm{M}}\left[\exp\left(\frac{qV_{\mathrm{MS}}}{kT}\right)-1\right] \tag{2.39}$$

where

$$J_{\mathrm{M}}=\frac{q^2D_{\mathrm{n}}N_{\mathrm{c}}(\phi_{\mathrm{i}}-V_{\mathrm{MS}})\exp(-\phi_{\mathrm{B}}/kT)}{x_{\mathrm{d}}kT\{1-\exp[(-q(\phi_{\mathrm{i}}-V_{\mathrm{MS}}))/kT]\}}. \tag{2.40}$$

Comparing (2.34) and (2.40), J_{M} shows a stronger dependence on V_{MS} than J_{s}. Equation (2.39) is obviously limited by all the limitations of *SDA* that we spoke of earlier in Section 2.2.2.1.4. This equation, not unlike the equation developed for the current density of a *Schottky barrier*, offers a good approximation as long as V_{MS} is much smaller than the built-in potential (i.e., $\phi_M - \phi_s$, where $q\phi_s$ is the *work function* of the *highly doped* semiconductor).

2.2.2.2.2 *Nonblocking* M–S *Contacts* As it was alluded earlier, *Ohmic* contacts can be created mainly in three different forms:

1. *Schottky nonblocking* contacts. Considering the range of available metal *work functions*, there is a certain level of difficulty in achieving this type of a contact to n-type *silicon*.
2. *Tunneling Ohmic* contacts. Most *Ohmic* contacts to n-type *silicon* are of this form. As indicated earlier, this technique takes advantage of quantum *tunneling* by *doping* the semiconductor very highly at *degenerate* levels to reduce the *depletion-region width* of the otherwise *blocking* contact.
3. *Alloyed Ohmic* contacts. This is primarily used to create *Ohmic* contacts in *III–V* semiconductor technologies. Rapidly increasing the temperature at the surface of the semiconductor using a processing technique known as *rapid thermal annealing* (*RTA*),[11] causes mixing of the semiconductor and metal forming an *alloyed* conducting film. This *alloy* formation alleviates the potential barrier of the *MS* junction, yielding an *Ohmic* behavior.

As expressed earlier, current flow in *nonblocking MS* junctions is not contact limited. Consequently, unlike *blocking MS* junctions, the result of the application of a voltage cannot be studied through the variation of potential drop across the charge-imbalanced region formed in the vicinity of the junction (i.e., *accumulation* region). The charge imbalance in this case is caused by the excess concentration of mobile charges and not fixed ionized impurities. Whereas application of *Gauss*'s law to the charge profile of this region can be used in profiling the electric field and potential variation across the junction, these results do not constitute the primary factors in studying the behavior of these junctions.

Since the voltage drop across an *Ohmic* contact is negligible, without excessive power dissipation the *Ohmic* contact can be described by a state of *quasi-thermal equilibrium*, even when current is flowing across the contact.

To be more accurate in our description of *Ohmic* contacts, we must assert that the *Ohmic* behavior is limited within a range of current densities. In order to quantify this range, consider the case of the *nonblocking Schottky* contact depicted in Figure 2.14. In this special case, the semiconductor's *band* edges lie parallel to *Fermi level* throughout the semiconductor (i.e., $\phi_M - \chi = \phi_n$). This type of a contact, with total

[11] In *RTA* techniques since the semiconductor *wafer* is being heated up from the surface, instead of heating the whole *wafer* from the substrate, the exposure of the semiconductor to thermal energy (usually expressed in terms of *thermal budget*) is very limited. This causes minimum unwanted diffusion/redistribution of *dopants* within the semiconductor. In *RTA* techniques, both ramping-up and ramping-down times are very short.

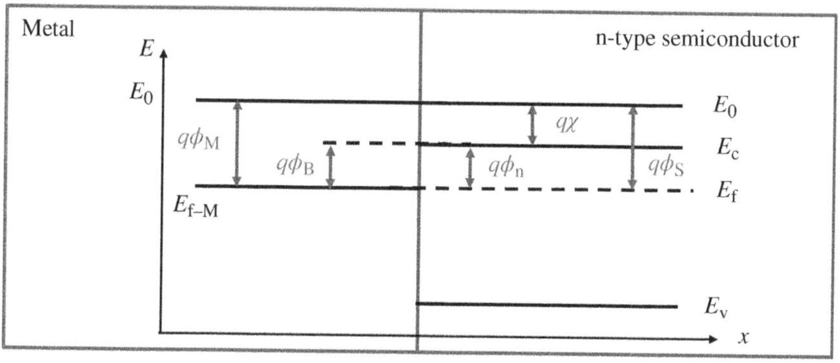

FIGURE 2.14 Thermal-equilibrium band diagram of a neutral contact.

elimination of the potential barrier on the semiconductor side, is referred to as a *neutral contact.* In this case, as shown in the *band* diagram of Figure 2.14, the electron concentration is constant throughout the semiconductor.

Under *bias*, for current densities less than $qn_0v_{th}/4$ (which is the *thermally induced* current density in a solid along any random direction), in which n_0 is the electron concentration and v_{th} is average electron *thermal velocity*, the *neutral contact* behaves as an *Ohmic* contact. However, for higher electric fields, this thermal *diffusion* current will not be sufficient, and the *MS* contact loses its *Ohmic thermal-equilibrium* characteristic. In other forms of *nonblocking Schottky* contacts, the current limit for *Ohmic* behavior is different from $qn_0v_{th}/4$ by a factor of $\exp\{q[(\phi_M-\chi)-\phi_n]/kT\}$. This is due to the *Maxwell–Boltzmann* nature of electron distribution. As we can see from the previous discussion, when $(\phi_M-\chi)<\phi_n$ we have an exponentially higher chance for achieving an *Ohmic* contact. The opposite is true when $(\phi_M-\chi)>\phi_n$.

In the earlier discussion, it was indicated that *thermally induced* current density in a solid along any random direction is equal to $qn_0v_{th}/4$. While in *thermal equilibrium* the net flux of *charge carriers* is 0, the flux along any given direction can be calculated to be equal to this value. In order to establish this relationship, we consider a spherical coordinate shown in Figure 2.15. The volume element of the spherical coordinate system is $r^2\sin(\theta)drd\theta d\phi$. In this context, for each velocity and direction increment, we consider dn_v electrons where, in terms of the surface area of the sphere (and volume of a shell), $dn_v=4\pi r^2dr$. According to this, the three-dimensional concentration of electrons can be expressed in terms of $d^3n_{\theta,\phi,v}=r^2\sin(\theta)drd\theta d\phi=(1/4\pi)dn_v\sin(\theta)d\theta d\phi$. Where integration over $0<\theta<\pi$ and $0<\phi<2\pi$ results in cancellation of 4π from the denominator and gives $n_0=\iiint d^3n_{\theta,\phi,v}=\int_0^\infty dn_v$. According to this distribution, the average velocity, independent of direction, can be calculated as

$$\langle v\rangle=\frac{1}{n_0}\iint\int_{0\,0\,0}^{\infty\;\pi\;2\pi}\frac{v}{4\pi}dn_v\sin\theta\cdot d\theta d\phi=\frac{1}{n_0}\int_0^\infty vdn_v. \tag{2.41}$$

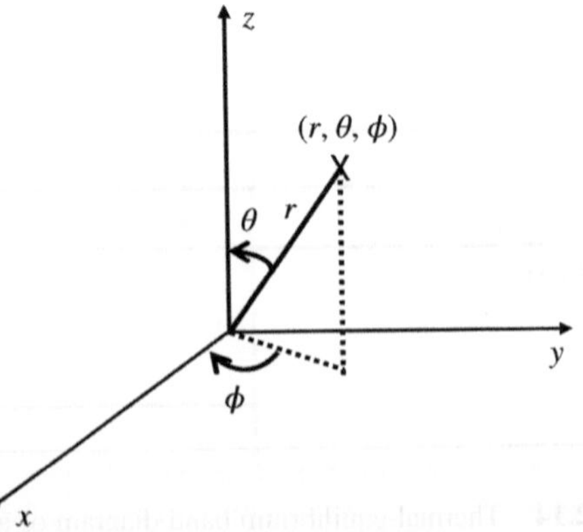

FIGURE 2.15 Spherical coordinate system.

Here, v represents only the amplitude of the velocity vector, however, according to the definition of *thermal velocity* $\langle v \rangle = v_{th}$.

The average velocity along any given direction, for example, the positive *x-direction*, can then be written as[12]

$$\langle v \cdot \cos\theta \rangle = \frac{1}{n_0} \int\!\!\int\!\!\int_{0\,0\,0}^{\infty\,\frac{\pi}{2}\,2\pi} \frac{v \cdot \cos\theta}{4\pi} dn_v \sin\theta \cdot d\theta d\phi = \frac{1}{4n_0} \int_0^{\infty} v dn_v. \tag{2.42}$$

According to (2.41), (2.42) reduces to $v_{th}/4$. As a result, thermal electron flux along a single direction (such as the positive *x-direction*) results in a current density equal to $qn_0(v_{th}/4)$.

In our discussions on *SDA*, we have already mentioned that our description of the *depletion* region is valid within a few length scales known as *Debye length* (L_D). This is an appropriate length scale that describes the extent of the electrical effects at the boundary. Within a length span of $\sqrt{2}L_D$ from the interface, 50% of the *space charge* of the semiconductor lies. According to this definition, for an *MS Ohmic* contact,

$$L_D = \left(\frac{\epsilon_S kT}{q^2 n_S} \right)^{1/2} \tag{2.43}$$

where n_s is the excess electron concentration in the semiconductor at the interface. For any other contact such a length scale can be developed.

[12] Because we are only looking at the positive *x-direction*: $0 < \theta < \pi/2$.

2.3 *P–N* JUNCTIONS

Generally speaking, *p–n* junctions are the junctions formed between two semiconductors among which the majority *carrier type* is opposite. The difference between the *p*- and *n*-side of the junction does not need to be restricted to the *doping type*. These junctions can also be realized between two different semiconductors. This latter group of *p–n* junctions is referred to as *p–n heterojunctions*.

Since *p–n* junctions, not unlike *MS* junctions, possess current-rectifying and sometimes asymmetric *carrier-transport* properties, both of these junctions to various degrees have been explored in the realization of gated control of the *channel's carrier* population in *field effect transistor* technologies. Due to this, following the discussion of *MS* junctions in Section 2.2, in this section we will go through the essentials of the operation of *p–n* junctions.

2.3.1 Thermal-Equilibrium Band Diagram of *P–N* Junctions

The framework for drawing the *band diagram*, which was presented earlier in Section 2.2 for MS systems, can be easily extended to all other types of contacts under *thermal equilibrium*. As a result, we start our discussion of *p–n homo-* and *heterojunctions* from this point. However, before embarking on this task, we should remind ourselves that in developing such diagrams, we have made a few important assumptions in terms of the definition of the *electron affinity* and *bandgap* (for the latter, especially at the surface of the semiconductor). In addition, it is worth mentioning that especially for *compound alloyed* semiconductors, the edges of the *conduction* and *valence band* are not defined as sharply as indicated by the *density of states* functions such as (1.14). As pointed out in Chapter 1, in these cases instead of a sharp definition of *band* edges, semiconductors present a decaying *density of states* spread within a range of energy in the *bandgap* referred to as a *band tail*. This challenges the validity of the approximations involved in the definition of the *bandgap* and their application to drawing the *band diagram*. In the first-order investigations, these effects are merely treated as secondary phenomena and are ignored.

In terms of what was proven in Section 2.1, assumption of a constant *Fermi level* throughout the system is the first step in drawing the *thermal-equilibrium band diagram*. After this step, *Maxwell–Boltzmann* statistics, or if required *Fermi–Dirac* statistics, allows us to estimate the position of the *band edges* away from the junction (i.e., with regards to this *Fermi level*). This is done on the basis of the knowledge on the *activated dopant* concentration profile. Analogous to the discussions in Section 2.2.1, junctions are assumed to be sharp and planar. This assumption conveys that the two sides of the interface, without any chemical or physical mixing, meet one another at a plane with a clean interface devoid of *defects* and dangling *bonds*. According to this definition, knowledge of the *electron affinity* and *bandgap* on the two sides of the junction, in addition to the requirement for establishing a continuous *vacuum level* (i.e., E_0), provide the essentials for estimating the energy of the *band edges* throughout the semiconductor–semiconductor system.

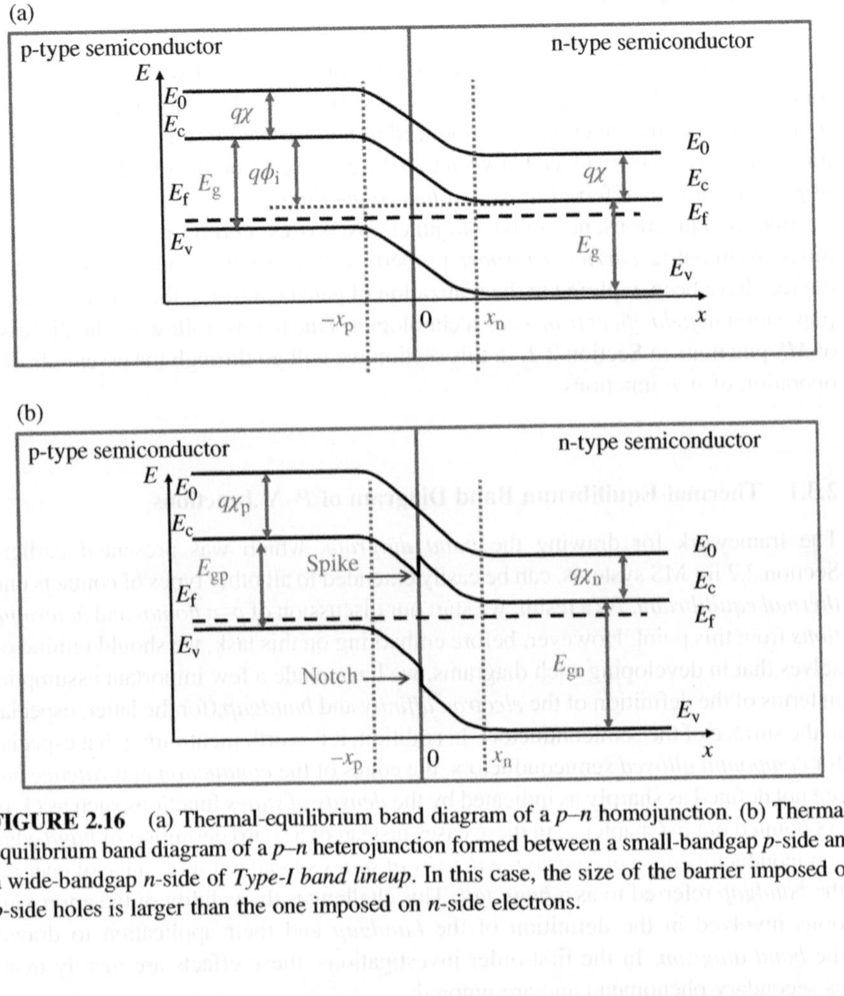

FIGURE 2.16 (a) Thermal-equilibrium band diagram of a *p–n* homojunction. (b) Thermal-equilibrium band diagram of a *p–n* heterojunction formed between a small-bandgap *p*-side and a wide-bandgap *n*-side of *Type-I band lineup*. In this case, the size of the barrier imposed on *p*-side holes is larger than the one imposed on *n*-side electrons.

Figure 2.16 presents two examples of *band diagrams* drawn for *p–n* junctions: one for a *p–n homojunction* and one for a *p–n heterojunction*. As observed in Figure 2.16a, in a *p–n homojunction* conformity of the *bandgap* and *electron affinity* at the two sides of the junction enforces the continuity of *conduction-* and *valence-band* edges. However, since one or both of these values at the two sides of a *heterojunction* can be different, the required continuity of *vacuum level* (i.e., E_0) does not guarantee the continuity of either E_c or E_v at the *heterointerface*. The result of this observation at the *heterointerface* can appear in the form of *notches* and *spikes* in one or both of these *band* edges. This is shown in Figure 2.16b.

In the *band diagrams* presented in Figure 2.16, we have simply assumed that both sides of the junctions are uniformly *doped*. As illustrated in Figure 2.16, we observe that majority *carriers* from both sides of a *p–n* junction face a potential barrier for moving across the *metallurgical junction*. However, minority *carriers*, as illustrated by the drop in their energy, are encouraged to move past the barrier.

Figure 2.16b presents only one of the varieties of the *band diagrams* expected at *p–n heterojunctions*. As pointed out in Section 1.6.2.2.2, in terms of the three *types* of *band lineup*, which were depicted in Figure 1.42, three *types* of *p–n heterojunctions* can be expected:

1. *Type-I*: the *bandgap* of the semiconductor with smaller *bandgap* is nested within the *bandgap* of the wider-*bandgap* semiconductor.
2. *Type-II*: the *bandgaps* of the two semiconductors are placed staggered with respect to one another.
3. *Type-III*: the two bandgaps are completely offset.

As shown in Figure 2.16, while *homojunctions* present an equivalent potential barrier for both types of majority *carriers*, *heterojunctions* offer an asymmetry between the sizes of these two barriers. This possibility is of great relevance in realization of electronic devices such as *heterostructure bipolar transistors* (i.e., *HBTs*).

2.3.2 Calculation of Potential across *P–N* Junctions and *SDA*

In the treatment of *p–n* junctions, not unlike *MS* junctions, we start from *SDA*. In this approximation, while most of the potential variation is assumed to take place in the vicinity of the junction, *quasi-neutrality* across the bulk of the *p-* and *n-*sides is assumed. The approximation of *quasi-neutrality* states that in regions of slowly varying *dopant* concentration, *quasi-neutrality* is preserved. In other words, the internally induced electric field created by the mismatch between the spatial profile of *carriers* and *ionized impurities* is assumed negligible. In order to validate this assumption, we can evaluate the electric-field strength in semiconductors with a gradual variation of *dopant* concentration and compare the results to the externally induced electric field.

As a generic example, an arbitrary profile of *dopant* concentration is shown in Figure 2.17. In this treatment, a *nondegenerate* semiconductor is assumed. The presence of a gradient in *dopant* concentration serves as a driving force for *diffusion* of *charge carriers* and, as a result, forms a mismatch between the profiles of mobile charges and fixed ionized impurities. This imbalance in charge neutrality in turn serves as the driving force for establishing *thermal equilibrium* and reducing the electron and *hole* currents to 0. This happens through inducing an electric field and an opposing *drift-current* term. As a result of this balance, under *thermal equilibrium*,

$$\begin{cases} J_{\mathrm{n}}=0 \\ J_{\mathrm{p}}=0 \end{cases} \Rightarrow \begin{cases} qn\mu_{\mathrm{n}}E_{\mathrm{x}}+qD_{\mathrm{n}}\dfrac{dn}{dx}=0 \\ \\ qp\mu_{\mathrm{p}}E_{\mathrm{x}}-qD_{\mathrm{p}}\dfrac{dp}{dx}=0 \end{cases} \Rightarrow E_{\mathrm{x}}=-\frac{kT}{qn}\frac{dn}{dx}=\frac{kT}{qp}\frac{dp}{dx}. \tag{2.44}$$

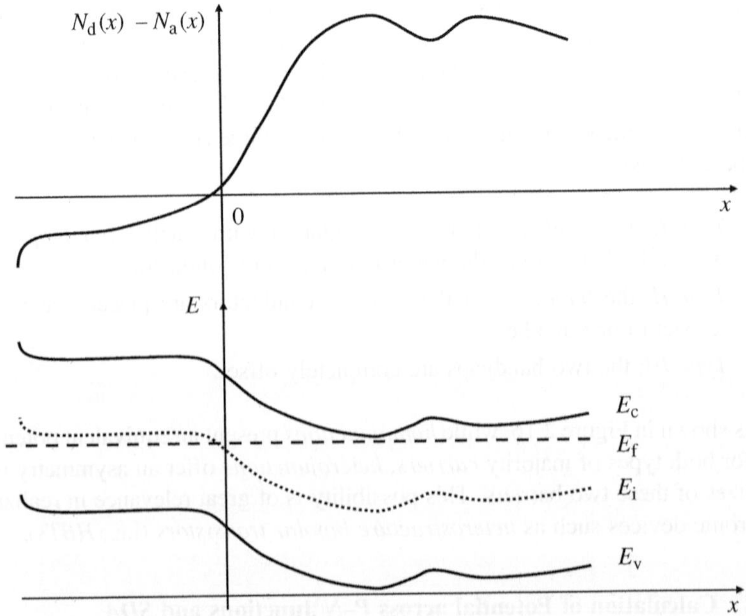

FIGURE 2.17 Arbitrary distribution of impurities in a semiconductor and demonstration of the resulting induction of the built-in electric field on the basis of the thermal-equilibrium band diagram. The built-in electric field is defined proportional to the spatial gradient of the intrinsic Fermi level, E_i.

In terms of this internally induced electric field (i.e., E_x), the potential difference between any two points in the semiconductor can be calculated on the basis of

$$E_x = -\frac{d\phi}{dx} \Rightarrow \begin{cases} d\phi = \dfrac{kT}{q}\dfrac{dn}{n} \\ d\phi = -\dfrac{kT}{q}\dfrac{dp}{p} \end{cases}. \tag{2.45}$$

For a small spatial gradient in the *dopant* concentration, despite the separation among the mobile and fixed charge profiles, we can still approximately replace n by N_d (assuming a highly n-doped semiconductor and $N_d \gg n_i$) and p by N_a (assuming a highly *p-doped* semiconductor and $N_a \gg n_i$), in (2.44) and (2.45). This is simply the assumption of *quasi-neutrality*, which results in the following equations for calculating the internally induced electric field:

$$E_x = -\frac{kT}{qN_d}\frac{dN_d}{dx} \tag{2.46}$$

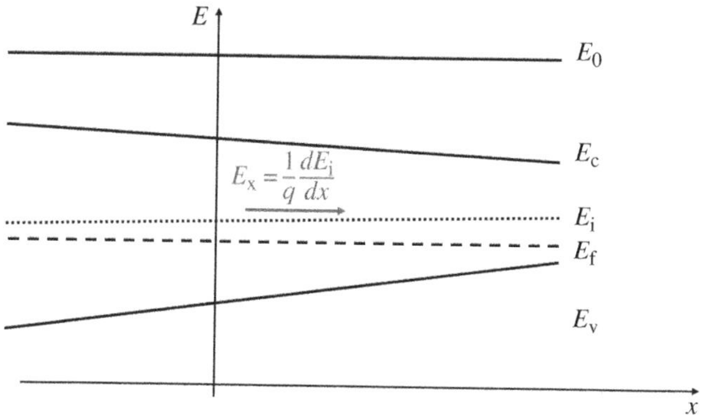

FIGURE 2.18 Schematic explanation of the induction of a built-in electric field in a heterogeneous semiconductor.

and

$$E_x = \frac{kT}{qN_a}\frac{dN_a}{dx}. \tag{2.47}$$

An internally induced electric field can also be formed when there is a gradual change in semiconductor properties such as *bandgap* and *electron affinity* where the *doping* is constant. An example of this situation is depicted in Figure 2.18. Such materials, which are known as *heterogeneous*, can nowadays be produced using *epitaxial* growth techniques such as *MBE* and *MOCVD* with monolayer control over the growth condition.

The aforementioned *quasi-neutrality* approximation is, for example, useful in the case of an exponential *dopant* distribution profile such as $N_d = N_0 \exp(-x/\lambda)$, with proper values of λ (describing the decay in this profile). This kind of profile is frequently encountered in semiconductor processing. As a result of the approximation expressed in (2.46), the internally induced electric field will be a constant ($kT/q\lambda$), throughout the region of exponential decay of the *donor* profile. Only for large values of λ will this electric field be weak enough to validate the assumption of *quasi-neutrality*.

According to (2.45), the potential difference between any two points can be evaluated in terms of the difference in electron concentration (or *hole* concentration):

$$\phi_3 - \phi_2 = \frac{kT}{q} Ln\left(\frac{n_3}{n_2}\right). \tag{2.48}$$

Inversely, the ratio between the charge concentrations can be evaluated in terms of the difference in potential[13]:

$$\frac{n_3}{n_2} = \exp\left[\frac{(\phi_3 - \phi_2)}{(kT)/q}\right]. \tag{2.49}$$

[13] Equivalently for *holes*, $\frac{p_2}{p_3} = \exp\left[\frac{(\phi_3 - \phi_2)}{(kT)/q}\right]$.

In this evaluation, we use the definition of potential at different points in the semiconductor as

$$\phi(x)=\frac{1}{q}(E_\mathrm{f}-E_\mathrm{i}(x)) \tag{2.50}$$

in which E_i/q is used as the reference potential.

For a general case, without the assumption of *SDA* and *quasi-neutrality*, using *Poisson's equation* for a given charge profile $\rho(x)$ (omitting argument x in the potential and ρ), we have

$$\frac{d^2\phi}{dx^2}=-\frac{\rho}{\epsilon_\mathrm{s}}=-\frac{q}{\epsilon_\mathrm{s}}(p-n+N_\mathrm{d}-N_\mathrm{a}). \tag{2.51}$$

In (2.51) full *activation* of *dopants* is assumed.

According to *Maxwell–Boltzmann* statistics,

$$n=n_\mathrm{i}\exp\left(\frac{E_\mathrm{f}-E_\mathrm{i}}{kT}\right)=n_\mathrm{i}\exp\left(\frac{q\phi}{kT}\right) \tag{2.52}$$

and

$$p=n_\mathrm{i}\exp\left(-\frac{E_\mathrm{f}-E_\mathrm{i}}{kT}\right)=n_\mathrm{i}\exp\left(-\frac{q\phi}{kT}\right). \tag{2.53}$$

The Equation (2.51) can therefore be presented as

$$\frac{d^2\phi}{dx^2}=\frac{q}{\epsilon_\mathrm{s}}\left(2n_\mathrm{i}\sinh\left(\frac{q\phi}{kT}\right)+N_\mathrm{a}-N_\mathrm{d}\right) \tag{2.54}$$

Whereas (2.54) provides the generic foundation for calculating the potential profile from any arbitrary distribution of *dopants*, it does not often yield an analytical solution. For that reason, instead of employing this equation, junctions are often analyzed on the simplified foundations provided through *SDA* and *quasi-neutrality* approximations (i.e., at least in the first-order analyses).

According to these approximations, a *p–n* junction is divided between three regions: two *neutral* regions, across which charge neutrality (or *quasi-neutrality*) is preserved and electric field is 0 (or very small), and a third region sandwiched between these two, which is fully devoid of *charge carriers*. This third region is referred to as the *depletion* (or space-charge) region.

2.3.2.1 **P–N *Step Junctions*** According to the first-order analyses, upon establishing the state of *thermal equilibrium* between the *p*- and *n*-sides of a *p–n* junction,

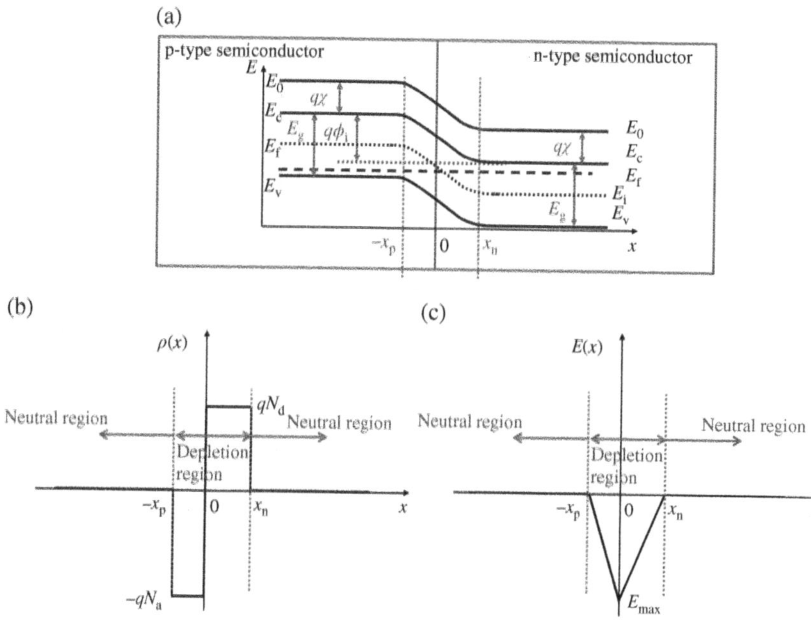

FIGURE 2.19 (a) Thermal-equilibrium band diagram of a homogenous *p–n* junction. (b) Charge concentration and (c) electric-field profiles across the depletion and the neutral regions of the uniformly doped contact depicted in (a).

the *diffusion current*, which is induced by the sharp variations of electron and *hole* concentration profiles, should come into balance with a *drift-current* term. This *drift current* is driven by the *electric field* created by the *ionized impurities* left behind in the *depletion region*. This region is formed on both sides of the *metallurgical junction*. Beyond the boundaries of the *depletion region*, *zero* electric field and charge neutrality is assumed. According to this approximation for the step *homojunction* depicted in Figure 2.19, we will have

$$\nabla \cdot E = \frac{\rho}{\epsilon_s} \Rightarrow \begin{cases} 0 < x < x_n : E(x) = \dfrac{qN_d}{\epsilon_s}(x - x_n) \Rightarrow \phi(x) = -\dfrac{qN_d}{2\epsilon_s}(x - x_n)^2 + \dfrac{E_f - E_{in}}{q} \\ -x_p < x < 0 : E(x) = -\dfrac{qN_a}{\epsilon_s}(x + x_p) \Rightarrow \phi(x) = \dfrac{qN_a}{2\epsilon_s}(x + x_p)^2 + \dfrac{E_f - E_{ip}}{q} \end{cases}. \tag{2.55}$$

In the junction depicted in Figure 2.19, *p*- and *n*-side of the junction are uniformly *doped* (hence the name step junction). In this figure, x_n and $-x_p$ are identified as the boundaries of the *depletion region*.

Through enforcing the continuity of the electric field at the *metallurgical junction*,[14] we have

$$N_a x_p = N_d x_n. \tag{2.56}$$

According to (2.55), electric field maximizes at the *metallurgical junction* where

$$E_{max} = E(x=0) = -\frac{qN_a x_p}{\epsilon_s} = -\frac{qN_d x_n}{\epsilon_s}. \tag{2.57}$$

The regions developed on the two sides of the *depletion region* are referred to as *neutral regions*. In these two regions,

$$\begin{cases} x < -x_p : p = N_a = n_i \exp\left(\frac{E_{ip} - E_f}{kT}\right) \Rightarrow E_{ip} - E_f = kT \cdot Ln\left(\frac{N_a}{n_i}\right) \\ x_n < x : n = N_d = n_i \exp\left(\frac{E_f - E_{in}}{kT}\right) \Rightarrow E_f - E_{in} = kT \cdot Ln\left(\frac{N_d}{n_i}\right) \end{cases}. \tag{2.58}$$

As a result, the difference in potential energy between the two *neutral regions* will be given by

$$q\phi_i = E_{ip} - E_{in} = kT \cdot Ln\left(\frac{N_a N_d}{n_i^2}\right). \tag{2.59}$$

The quantity ϕ_i is known as the built-in potential.

The total width of the *depletion region*, on the basis of this model, is given by

$$x_d = x_p + x_n = \sqrt{\frac{2\epsilon_s \phi_i}{q}\left(\frac{1}{N_d} + \frac{1}{N_a}\right)} \tag{2.60}$$

This equation is derived on the knowledge that potential across a *homojunction* is single valued.

As expressed in (2.55) and (2.56) and illustrated in Figure 2.19, most of the built-in potential and a proportionally larger part of the width of the *depletion region* are dropped across the side of the *p–n* junction with a lower *dopant* concentration. In addition, it is worth noting that the potential at the *metallurgical junction* is not 0 unless $N_a = N_d$.

[14] To be more accurate, it is the *electric-flux density* (i.e., *D*) that we are enforcing to be continuous. However, since in a *homojunction relative dielectric constants* of the *p*- and *n*-side are identical, this can be equivalently presented as continuity of electric-field intensity (i.e., *E*, defined as D/ϵ_s). As will be observed later in this section, this approach should be modified in the case of *heterojunctions*.

Not unlike the *blocking MS* junction, the potential and electric-field profiles for a *p–n* step junction, which were derived in this section on the basis of *SDA* approximations, are valid throughout the *depletion region* with the exception of a few length scales from the boundaries of this region (i.e., x_n and $-x_p$). This length scale is known as *Debye length.* In order to evaluate the spatial extent over which *SDA* is only loosely applicable, (2.54) can be used to more accurately determine the potential profile across the junction.

By neglecting the concentration of *holes* in the vicinity of $x = x_n$, (2.54) yields

$$\frac{d^2\phi'}{dx^2} = \frac{q}{\epsilon_s}(N_d - n) = \frac{q}{\epsilon_s}\left\{N_d - n_i \exp\left[\frac{q(\phi_n - \phi')}{kT}\right]\right\} = \frac{q}{\epsilon_s}N_d\left[1 - \exp\left(-\frac{q\phi'}{kT}\right)\right]. \quad (2.61)$$

In this equation, we have assumed that the potential at $x = x_n$ is not exactly equal to the potential in the *neutral region* of the *n-side* (i.e., $\phi_n = (E_f - E_{in})/q$). The difference is indicated by ϕ'. The semiconductor is assumed to be *nondegenerate.* Assuming ϕ' to be relatively small, we can use the first-order *Taylor* series expansion to rewrite (2.61)

$$\frac{d^2\phi'}{dx^2} = \frac{q}{\epsilon_s}N_d\frac{q\phi'}{kT} = \frac{\phi'}{L_D^2} \quad (2.62)$$

in which L_D is the *extrinsic Debye length* given by

$$L_D = \sqrt{\frac{\epsilon_s kT}{q^2 N_d}}. \quad (2.63)$$

According to Equation (2.62), near the edges of the *depletion region,* potential varies exponentially, with a rate identified in terms of L_D. This is depicted in Figure 2.20. Based on this explanation, the shorter the *Debye length,* the more accurate the *SDA* assumption of a sharp reduction (i.e., to 0) of *carrier* concentration at the boundaries of *depletion* and *neutral* regions is.

To be more accurate, it should be indicated that the *Debye length* describes the *screening* of electric field by the *charge carriers.* As a result, in its definition both electron and *hole* concentration and not just $n = N_d$ are present, yielding

$$L_D = \sqrt{\frac{\epsilon_s kT}{q^2(n+p)}}. \quad (2.64)$$

As expected from (2.64), the maximum value of *Debye length* happens in an *intrinsic* semiconductor, which yields

$$L_{Di\text{-}max} = \sqrt{\frac{\epsilon_s kT}{2q^2 n_i}}. \quad (2.65)$$

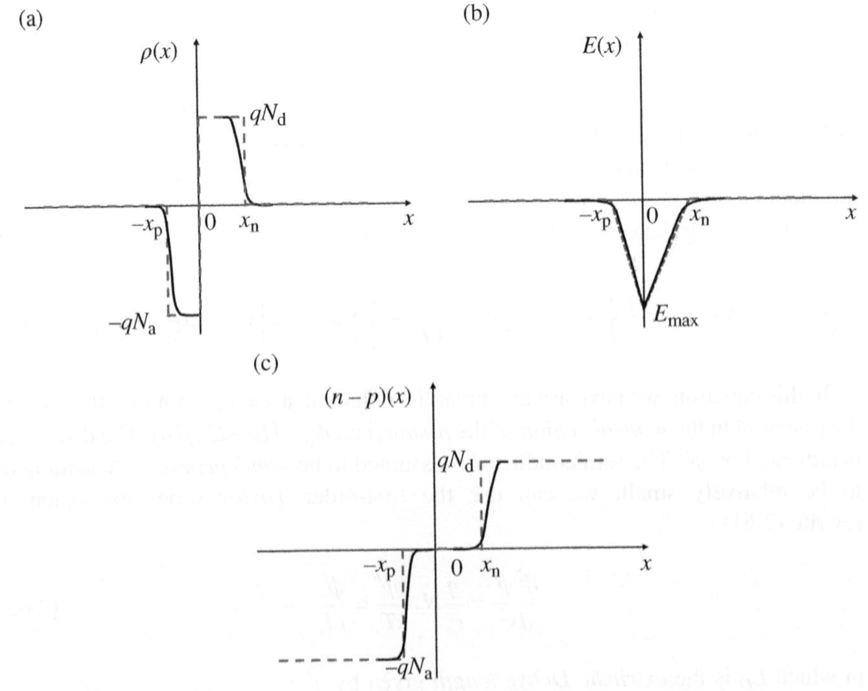

FIGURE 2.20 Charge concentration profile (a), variation of electric field (b), and profiles of electron and hole concentration across a *p–n* step junction (c). Predictions of *SDA* are shown in dashed gray lines, while the difference between the predictions of *SDA* and the reality are highlighted in solid curves. Transition regions are expressed in terms of the Debye length.

2.3.2.1.1 Application of SDA *to* P–N *Step Heterojunctions* In the present discussion of *heterojunctions*, for the sake of brevity, only crystalline semiconductor–semiconductor junctions in which *coherent* crystalline overgrowth has been achieved between two semiconductors of identical *lattice constant* and *type* are presented. In other words, in here a discussion of *strain* and formation of *interface states* will not take place. In addition, in this section the *quantum phenomena* linked to the formation of *quantum wells* at the *band*-edge discontinuities are not addressed. Later on in this chapter, and especially in Chapter 5, these issues are dealt with in detail.

Description of *p–n step heterojunctions* in terms of *SDA*, although generally following the explanations thus far presented in this section, requires a few major modifications. Those modifications are due to the differences between the *relative dielectric constants*, *effective carrier masses*, *effective density of states*, *bandgaps*, and *electron affinities* at the two opposite sides of the *heterojunction*. These differences command us to modify our presentation of the electric field in terms of the *electric-flux* density and its continuity. Due to the aforementioned differences

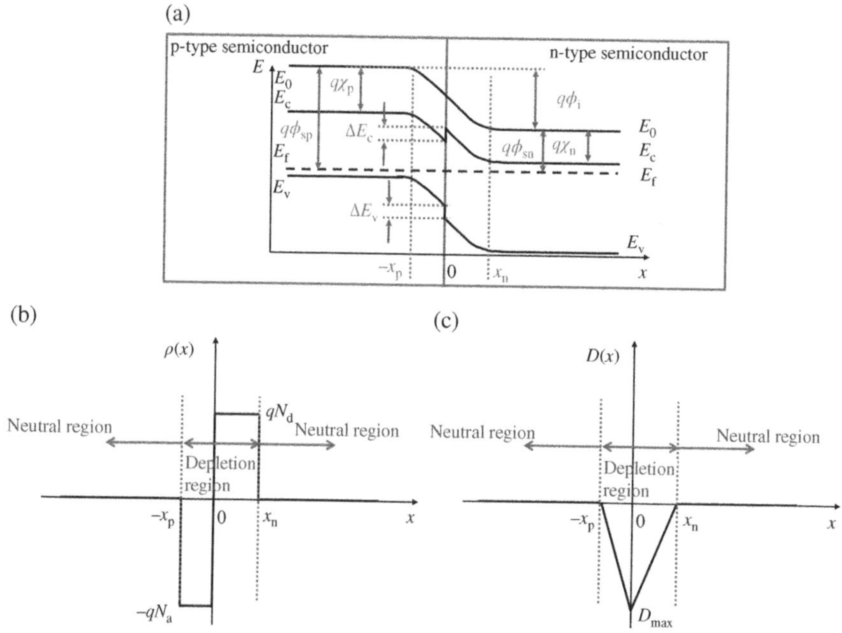

FIGURE 2.21 (a) Thermal-equilibrium band diagram of a *p–n* step heterojunction formed between a smaller-bandgap *p*- and a wider-bandgap *n*-side of *Type-I band lineup*. In this case, the size of the barrier imposed on *p*-side holes is larger than the one imposed on *n*-side electrons. (b) Charge concentration and (c) electric-flux density profiles across the depletion and the neutral regions of the step heterojunction depicted in (a).

between the *n*- and *p*-side of the *heterojunctions*, according to Figure 2.21,[15] (2.55)–(2.60) need to be rewritten based on the following foundation:

$$\nabla\cdot D=\rho\Rightarrow\begin{cases}0<x<x_n: D(x)=qN_d(x-x_n)\Rightarrow\phi(x)=-\dfrac{qN_d}{2\epsilon_{sn}}(x-x_n)^2+\dfrac{E_f-E_{in}}{q}\\ -x_p<x<0: D(x)=-qN_a\left(x+x_p\right)\Rightarrow\phi(x)=\dfrac{qN_a}{2\epsilon_{sp}}\left(x+x_p\right)^2+\dfrac{E_f-E_{ip}}{q}\end{cases}. \tag{2.66}$$

Here, ϵ_{sn} and ϵ_{sp} are the *dielectric constants* of the *n*- and *p*-side, respectively. In this section, we will maintain using the subscripts in this order.

Continuity of *electric-flux* density, in agreement with (2.56), maintains that $N_a x_p = N_d x_n$, where the *electric-flux* density is maximized at the *heterojunction*.

[15] As expressed earlier, *heterojunctions* with a variety of *band diagrams* are realizable. Figure 2.21 depicts only one of these.

In terms of the *intrinsic carrier* concentrations at the *n*- and *p*-side of the *heterojunction* (i.e., n_{in} and n_{ip}, respectively), across the *neutral* regions, which are developed away from the junction on its two sides, we can write

$$\begin{cases} x<-x_p : p=N_a=n_{ip}\exp\left(\frac{E_{ip}-E_f}{kT}\right) \Rightarrow E_{ip}-E_f=kT\cdot Ln\left(\frac{N_a}{n_{ip}}\right) \\ x_n<x : n=N_d=n_{in}\exp\left(\frac{E_f-E_{in}}{kT}\right) \Rightarrow E_f-E_{in}=kT\cdot Ln\left(\frac{N_d}{n_{in}}\right) \end{cases}. \quad (2.67)$$

As depicted in Figure 2.21, due to the presence of a mismatch between the *electron affinities* and/or *bandgaps* at the two sides of the *heterojunction*, generally speaking the difference between the *intrinsic Fermi levels* of the *n*- and *p*-side is not representative of the potential barrier height observed by majority *carriers* from one or both sides of the *heterojunction*.

Gaining insight into the operation of *heterojunctions* requires knowledge of the amount of discontinuity of the *conduction*- and *valence-band* edges (i.e., ΔE_c and ΔE_v, respectively). As shown in Figure 2.21, since *vacuum level* is continuous, the discontinuity of the *conduction-band* edge can be expressed in terms of the difference between the *electron affinities*,

$$\Delta E_c=|q\chi_n-q\chi_p|. \quad (2.68)$$

The *valence-band* edge discontinuity, or ΔE_v, is the complementary of this value with respect to *bandgap* discontinuity (i.e., ΔE_g). However, in spite of its agreement with Figure 2.21, Equation (2.68) presents only a first-order insight into the *band discontinuity*. This is rooted in the definition of *electron affinity*, which as pointed out in Section 1.1.6 is not clear-cut.

In *heterojunctions*, built-in potential is evaluated in terms of the bending of the *vacuum level* between the two sides. According to Figure 2.21, in terms of energy, this built-in potential can be expressed by the difference between the *work functions* of the two sides:

$$q\phi_i=|q\phi_{Sp}-q\phi_{Sn}|=|(q\chi_p+E_{cp}-E_f)-(q\chi_n+E_{cn}-E_f)|. \quad (2.69)$$

We should be warned that due to the presence of *conduction*- and *valence-band* edge discontinuities, this definition is not very descriptive of the behavior of the junction. The difference in potential energies of electrons and *holes* from the two sides of the junction (which are indicated by the difference in energy of the *conduction*- and *valence-band* edges between the two *neutral* regions, respectively) depending on the type of *band lineup* can be smaller or larger than this amount by ΔE_c and ΔE_v, respectively.

The presence of *band-edge* discontinuities is often problematic to *charge-carrier transport* across a *heterointerface*. It is worth mentioning that through adoption of appropriate techniques of *heterogeneous* growth of *heterostructures* such as tapering of the *bandgaps* and *dopant* concentrations, the discontinuities in the *conduction*- and *valence-band* edges can be prevented. Although in these structures the *spikes* and *notches* are eliminated, the *heterostructure* still offers an asymmetry between the potential barrier sizes observed by electrons and *holes*.

Using *Maxwell–Boltzmann* statistics, in terms of the concentration of *dopants* and the *effective density of states* at the *n*- and *p*-sides of the junction, (2.69) yields an equivalent form for calculating the built-in potential

$$\phi_i = \chi_p - \chi_n + \frac{E_{gp}}{q} - \frac{kT}{q} Ln\left(\frac{N_{cn}N_{vp}}{N_d N_a}\right). \tag{2.70}$$

In terms of this built-in potential, which as shown in Figure 2.21 is defined by the addition of the amounts of *band bending* on the two sides of the junction, the total *depletion-region width* is given by

$$x_d = x_p + x_n = \sqrt{\frac{2\phi_i}{qN_a}\frac{\epsilon_{sn}\epsilon_{sp}N_d}{N_d\epsilon_{sn} + N_a\epsilon_{sp}}} + \sqrt{\frac{2\phi_i}{qN_d}\frac{\epsilon_{sn}\epsilon_{sp}N_a}{N_d\epsilon_{sn} + N_a\epsilon_{sp}}}. \tag{2.71}$$

In this equation, respective values of x_n and x_p are individually distinguished.

Example

Consider a *GaAs p–n* step junction where the *p*- and *n*-side are *doped* to the level of 10^{15} and 10^{16} cm^{-3}, respectively. Calculate the built-in potential, width of the *depletion region*, and maximum value of the electric field. Perform the calculations at room temperature.

According to (2.59) and the numbers provided in Table 1.4,

$$\phi_i = \frac{kT}{q} Ln\left(\frac{N_a N_d}{n_i^2}\right) = 0.0258 Ln\left[\frac{10^{15} \times 10^{16}}{(9 \times 10^6)^2}\right] \cong 1.01 \text{ V}.$$

Based on (2.60),

$$x_d = \sqrt{\frac{2\epsilon_s\phi_i}{q}\left(\frac{1}{N_d} + \frac{1}{N_a}\right)} = \sqrt{\frac{2 \times 12.9 \times 8.85 \times 10^{-14} \times 1.01}{1.6 \times 10^{-19}}\left(\frac{1}{10^{15}} + \frac{1}{10^{16}}\right)}$$

$$\cong 1.26 \times 10^{-4} \text{ cm} = 1.26\ \mu\text{m}.$$

Knowing that $N_a x_p = N_d x_n$, $x_n \cong \dfrac{1.26 \times 10^{-4}}{11} = 1.14 \times 10^{-5}$ cm.

Hence, based on (2.57),

$$|E_{max}| = \frac{1.6 \times 10^{-19} \times 10^{16} \times 1.14 \times 10^{-5}}{12.9 \times 8.85 \times 10^{-14}} \cong 1.6 \times 10^4 \text{ V/cm}.$$

2.3.2.1.2 Biasing a P–N *Step Junction* According to *SDA*, *biasing* a *p–n* junction can be treated by replacing ϕ_i with $\phi_i - V_a$ in (2.60), which was developed for a *p–n homojunction.*[16] V_a refers to the external potential applied between the *p*- and *n*-side, respectively. In terms of this approximation, the *band diagram* of a *biased p–n* junction is developed by considering two separate states of *thermal equilibrium* in the two originally *neutral* regions. As long as the *bias* is not very strong,[17] this is a valid approximation for which case, these regions are referred to as *quasi-neutral* regions. These two states of *thermal equilibrium* are expressed in terms of two constant *Fermi levels*. Whereas the grounded side has its *Fermi level* fixed to its *unbiased* position, the *Fermi level* of the *positively biased* side drops by a certain amount (i.e., qV_a). The two *Fermi levels* are thereby separated by an amount of qV_a.

As shown in Figure 2.22 and Equations (2.55)–(2.60), when V_a is positive, reduction of the size of the potential barrier and retraction of the boundaries of the *depletion region* occur. In addition, the maximum electric field reduces. This is referred to as *forward bias*. Under *reverse bias* (negative V_a), the boundaries of the *depletion region* expand further into the *p*- and *n-side* of the junction, and the size of the potential barrier grows. According to this explanation, majority *carriers* will move more easily past the junction under *forward bias*. *Reverse bias*, however, impedes this *carrier transport*. This is the foundation that gives these junctions a rectifying characteristic. Here, it is assumed that the *bias* is applied to each region through an *Ohmic* contact.

In a *p–n homojunction*, such as the one depicted in Figures 2.19 and 2.22, the sizes of the potential barriers observed by majority *carriers* on both sides of the junction are equivalent. *Biasing* the junction has an equivalent impact on these two potential barriers. Figure 2.23 depicts a *biased p–n heterojunction*.

2.3.2.1.3 C–V Characteristics of P–N *Step Junctions* In the discussions of *blocking MS* junctions, we noticed that the study of the variation of the *small-signal depletion capacitance* with *reverse bias* voltage can provide meaningful insight into the nature of the semiconductor. In this section, we will study the foundation of this characterization technique applied to a *p–n* step junction.

For the case of a *step homojunction*, the *small-signal* per unit area *depletion capacitance* can be calculated in terms of the per unit area charge (Q_s), which is stored in the *space-charge* region

$$Q_s = qN_d x_n = qN_a x_p. \tag{2.72}$$

Accordingly, the *small-signal capacitance* of the *depletion* region (i.e., *junction* or *depletion capacitance*) is given by

$$C_j = \frac{d|Q_s|}{dV_a} = qN_d \frac{dx_n}{dV_a} = qN_a \frac{dx_p}{dV_a} \tag{2.73}$$

[16] Or, equivalent relationship developed for *p–n heterojunctions* in that section.

[17] That is, in case of positive V_a much smaller than the built-in potential and in case of negative V_a small enough not to cause excessive current (i.e., *breakdown*). *Breakdown* will be addressed in Section 2.3.2.1.5

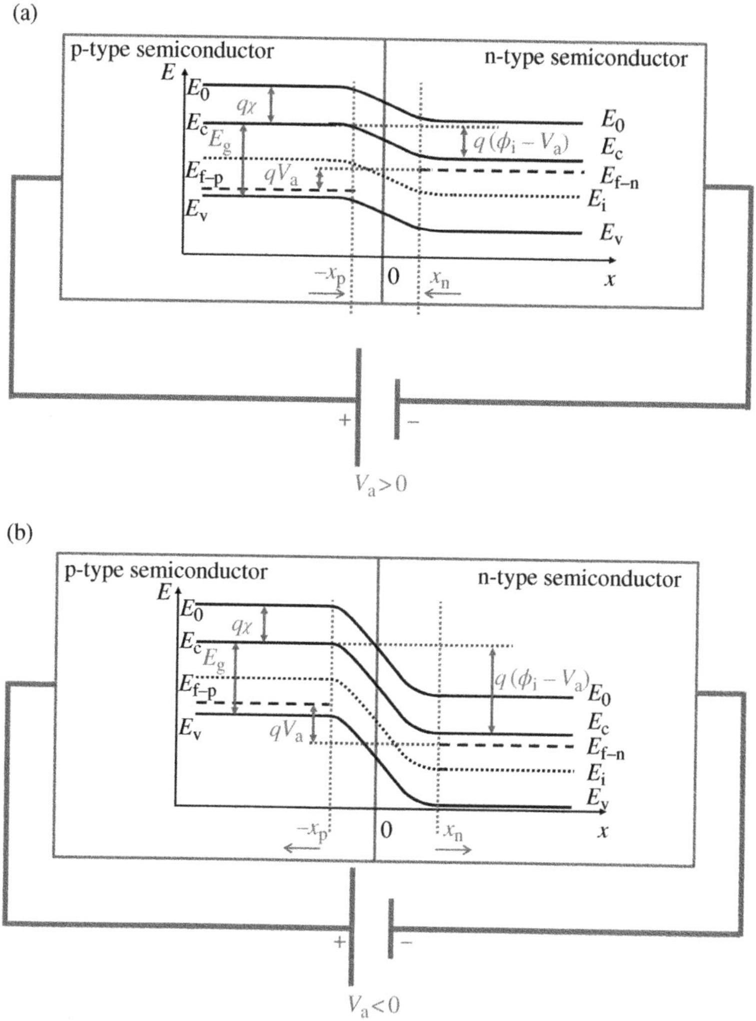

FIGURE 2.22 Band diagrams of a *p–n* step homojunction under (a) forward bias and (b) under reverse bias.

where

$$x_n = \frac{N_a}{N_d} x_p \text{ and } x_d = x_n\left(1 + \frac{N_d}{N_a}\right) = \sqrt{\frac{2\epsilon_s}{q}\left(\frac{1}{N_d} + \frac{1}{N_a}\right)(\phi_i - V_a)}. \tag{2.74}$$

As a result,

$$C_j = \frac{qN_d}{1 + (N_d/N_a)} \times \frac{1}{2\sqrt{(\phi_i - V_a)}} \sqrt{\frac{2\epsilon_s}{q}\left(\frac{1}{N_d} + \frac{1}{N_a}\right)} = \frac{\epsilon_s}{x_d}. \tag{2.75}$$

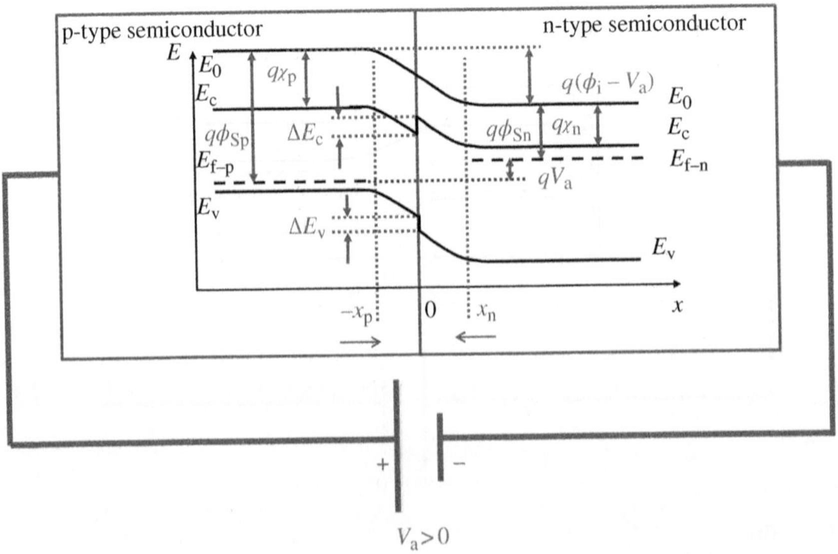

FIGURE 2.23 Band diagrams of a *p–n* step heterojunction under forward bias.

Ideally, the *small-signal C–V* characterization of *reverse-biased p–n* junctions can be used in the characterization of the *dopant* concentration profile. However, due to the two-sided expansion of the *depletion* region, this characterization is not as useful and straightforward. The following equations indicate the foundation of these measurements:

$$\frac{dC_j}{dx_n} = -\frac{\epsilon_s}{x_d^2}\left(1 + \frac{dx_p}{dx_n}\right) \tag{2.76}$$

$$|dQ| = \left|qN\left(-x_p\right)\right| dx_p = qN(x_n)dx_n \Rightarrow \frac{dC_j}{dx_n} = -\frac{C_j^2}{\epsilon_s}\left(1 + \frac{N(x_n)}{N\left(-x_p\right)}\right). \tag{2.77}$$

While $\frac{dC_j}{dx_n} = \frac{dC_j = d(Q/V)}{dQ/(qN(x_n))} = qN(x_n)\left(dC_j/dQ\right)$, (2.77) yields

$$N(x_n) = \frac{-C_j^3}{\epsilon_s q\left(dC_j/dV_a\right)}\left(1 + \frac{N(x_n)}{N\left(-x_p\right)}\right). \tag{2.78}$$

The presence of the dC_j/dV_a in (2.78) makes it difficult to apply this technique of characterization. Especially in the presence of noisy *signals*, this differentiation becomes quite limiting.

In the case of a *step heterojunction*, due to the difference in the *relative dielectric constants* at the two sides of the junction, the *depletion capacitance* modifies to

$$\frac{1}{C_\mathrm{j}} = \frac{x_\mathrm{n}}{\epsilon_\mathrm{sn}} + \frac{x_\mathrm{p}}{\epsilon_\mathrm{sp}} = \sqrt{\frac{2\phi_\mathrm{i}}{q}\frac{\epsilon_\mathrm{sn}N_\mathrm{d} + \epsilon_\mathrm{sp}N_\mathrm{a}}{\epsilon_\mathrm{sn}\epsilon_\mathrm{sp}N_\mathrm{a}N_\mathrm{d}}}. \tag{2.79}$$

This *depletion capacitance* is expressed in terms of a series combination of two capacitors: one defined on the *p*-side and one on the *n*-side of the *heterojunction*. Equations for x_n and x_p are provided in (2.71).

2.3.2.1.4 Current–Voltage Characteristic of P–N *Junctions* In describing the current–voltage characteristic of the junctions, we have so far neglected a few relevant processes including *carrier generation* and *recombination*. Whereas the single-sided contacts of *MS* junctions are much more forgiving to this simplification, *p–n* junctions cannot be treated without careful attention to these phenomena.

The reason for this fundamental difference between *MS* and *p–n* junctions is that in the former, as expressed in (2.31), current is carried by the majority *carriers* of the semiconductor (since electrons from the metal see a larger and almost *bias-independent* potential barrier). However, in the latter case *forward biasing* causes a rush of *carriers* across the interface and results in an excess population of minority *carriers* on each side. As will be seen in this section, it is these minority *carriers* that will essentially carry current across a *forward-biased p–n* junction. Under moderate *bias* or *low-level injection* conditions (i.e., *forward bias* much smaller than the built-in potential), majority *carrier* concentration is predetermined by the *dopant* concentration, while minority *carrier* concentration exceeds that of the *thermal equilibrium*. Such a condition, as will be observed shortly, is a cause for the *G–R* rate to become significant.

In this section, based on the discussions on the *current-continuity* equation from Section 1.4.1, a more or less accurate presentation of the current–voltage characteristic of *p–n* junctions will be provided. The model resulting from these discussions is relevant as long as the *drift–diffusion* formalism of *transport* is applicable. While both *Shockley–Hall–Read*[18] (also known as *SHR*) and *Auger G–R* models are visited, *SHR*, which is more relevant to *low-level injection* situations in *nondegenerate* environments, will receive a more prominent spotlight.

Current Continuity: Revisited Since the *current-continuity* equations are among the foundation of this model, here we repeat those equations from Section 1.4.1. As shown in the following, these equations quantify the temporal gradient of the electron and *hole* concentrations in terms of the spatial gradient of electron and *hole*

[18] Named after *William Shockley*, *Robert Hall*, and *W. Read*.

current densities in addition to *generation* (i.e., G) and *recombination* rates (i.e., R) of both *charge carriers*[19]:

$$\frac{\partial n}{\partial t} \cong \frac{1}{q}\frac{\partial J_n}{\partial x} + (G_n - R_n) \quad (2.80)$$

$$\frac{\partial p}{\partial t} \cong -\frac{1}{q}\frac{\partial J_p}{\partial x} + (G_p - R_p). \quad (2.81)$$

As long as electrons and *holes* are *generated*, or *recombined*, in terms of *electron–hole pairs*, the *generation* rate of electrons and *holes* (and also their *recombination* rates) will be equal. Such a situation can be envisioned in the form of a *direct band-to-band* transfer (i.e., between E_c and E_v) in *direct-bandgap* semiconductors. The momentum difference of the *band* edges in *indirect-bandgap* semiconductors does not allow this to happen, at least in the absence of another momentum-conserving interaction.

During the transition of a *charge carrier* between the two *band* edges, both momentum and energy should be conserved. In *indirect-bandgap* semiconductors interaction with a third particle (such another *carrier* or a *phonon*) is needed to conserve the momentum. As expected, however, the probability for three-particle interactions is much smaller than that of two-particle interactions. The probability of the former becomes considerable only when the concentration of *carriers* and *phonons* is extremely large. This happens in *degenerately doped* semiconductors under *high-level injection* and/or under very high temperatures, which are not usually of our primary interest.

In spite of this limitation, even in *indirect-bandgap* semiconductors (such as *Si* and *Ge*), *generation* and *recombination* processes of *charge carriers* are quite possible. This is due to the presence of nonideal *localized states* in the *forbidden bandgap* of the semiconductor. As will be expressed by the *SHR G–R* model, these *states* serve as stepping-stones for *charge carriers* to move between the *conduction-* and *valence-band* edges. According to *Heisenberg's uncertainty principles*, the *localized* nature of these *states* renders the momentum (i.e., the *conjugate variable* of distance) completely *uncertain*. While taking the *localized states* as stepping-stones, these *states* absorb the momentum difference between the *conduction-* and *valence-band* edges. The presence of a sizeable concentration of these *states* in *indirect-bandgap* semiconductors such as *Si* and *Ge* makes this a possible path for *generation* and *recombination* of *charge carriers*.

In terms of aforementioned possibilities for *G–R* processes in both *direct-* and *indirect-bandgap* semiconductors, we can apply (2.80) and (2.81) to the calculation of current density across the *quasi-neutral* regions of a *forward-biased p–n* junction. Since the electric field in these regions is small, we can use the definitions of *carrier*

[19] Subscripts *p* and *n* refer the *generation* and *recombination* rates of *holes* and electrons, respectively.

mobility and *diffusion constant* and assume these two *transport* variables to be position independent. On this basis,

$$\frac{\partial n}{\partial t} \cong \mu_n n(x)\frac{\partial E(x)}{\partial x} + \mu_n E(x)\frac{\partial n(x)}{\partial x} + D_n \frac{\partial^2 n(x)}{\partial x^2} + (G_n - R_n) \tag{2.82}$$

$$\frac{\partial p}{\partial t} \cong -\mu_p p(x)\frac{\partial E(x)}{\partial x} - \mu_p E(x)\frac{\partial p(x)}{\partial x} + D_p \frac{\partial^2 p(x)}{\partial x^2} + (G_p - R_p). \tag{2.83}$$

In this section our aim is to quantify the current density under steady-state conditions. As a result, the left side of (2.82) and (2.83) will be equal to 0. Also, since in the *quasi-neutral* regions the electric field and its gradient are small, we can neglect the first two terms from the right side of these equations. These approximations in the *quasi-neutral* regions yield

$$D_n \frac{\partial^2 n(x)}{\partial x^2} + (G_n - R_n) \cong 0 \tag{2.84}$$

$$D_p \frac{\partial^2 p(x)}{\partial x^2} + (G_p - R_p) \cong 0. \tag{2.85}$$

In order to proceed further in analyzing the current–voltage characteristic of the *p–n* junction, the *G–R* rate should be quantified. As mentioned earlier, we will do that according to the *SHR* model.

SHOCKLEY–HALL–READ *G–R* MODEL Figure 2.24 illustrates the foundation of the *SHR G–R* model and its involved processes. As identified in this figure, four individual processes are involved: *capture* of electrons from the *conduction band*, *emission* of electrons from the *localized state* to the *conduction band*, *capture* of *holes* from the *valence band*, and *emission* of *holes* from the *localized state* to the *valence band*. The rates of these four processes are identified by r_1, r_2, r_3, and r_4, respectively.

According to this picture, the *capture* and *emission* of *charge carriers* to and from the *localized states* can happen between these *states* and either the *valence* or the *conduction band*. The charge nature of the *localized states* (i.e., *acceptor* or *donor*)[20] is also important for understanding the balance between the four processes of *electron–hole capture* and *emission*.

In evaluation of the rates of electron- and *hole-capture* processes to each type of *charge carriers* and the *localized states*, a quantity known as *capture cross-section* is attributed (i.e., σ_n and σ_p, for electrons and *holes*, respectively).Typically an effective *recombination center*[21] has a *capture cross section* of about 10^{-15} cm^2 (i.e., observed

[20] We defined these earlier in Chapter 1 (i.e., Section 1.2.2).

[21] Earlier in this chapter (i.e., Section 2.2.2.1.4), the differences between the two types of characters of the *localized states* (i.e., *recombination center* vs. *trap*) were described.

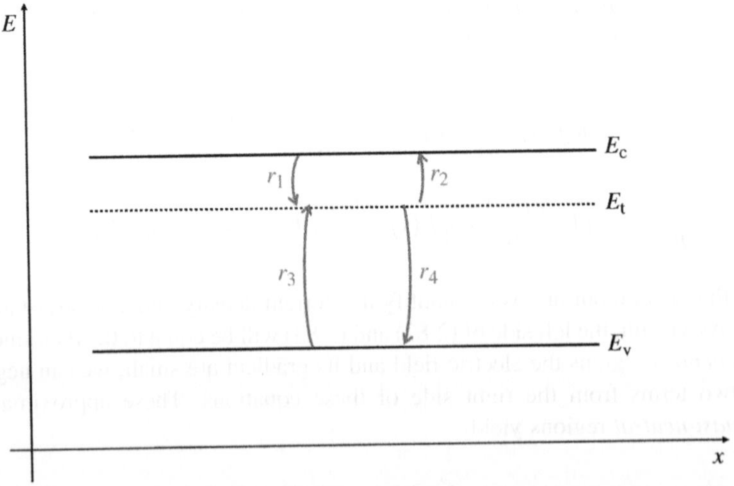

FIGURE 2.24 Band diagram depiction of the foundation of the *SHR G–R* model on the basis of four processes of electron capture (of rate r_1), electron emission (of rate r_2), hole capture (of rate r_3), and hole emission (of rate r_4) in/out of a recombination center E_t.

for *gold* and *iron*). Among the most detested impurities of semiconductor technology is *beryllium* with a very large *capture cross-section* of about 10^{-10} cm^2. As we will see later on, this large *capture cross-section* results in ultrafast *recombination* of minority *carriers* and degradation of device performance in the cases when minority *carrier* population is important to the device operation.

The larger the *capture cross-section* of an *energy state* for a particle, the higher the *capturing* probability of that *charge carrier* while randomly moving in the crystal. In this description, instead of envisioning electrons and *holes* as average moving particles (with their velocity equal to *thermal velocity*), we can imagine them as stationary particles in space. *Traps* (i.e., *localized states*) instead can be seen as butterfly nets with opening area equal to *capture cross-section*, which are moving at random *thermal velocity*. Since *holes* and electrons are *Fermions*, in evaluation of their *capture* rates by the *localized states* in addition to the information about the volume traversed *per unit time* by the "butterfly net" (i.e., $v_{th}\sigma_n$ and $v_{th}\sigma_p$, for electrons and *holes*, respectively), we need to know of the density of electrons (and *holes*) in the *conduction* (and *valence bands*) and also of the concentration of empty *trap sites*. In quantifying the *capture* rates, these three quantities are multiplied together. As a result, r_1 and r_3 (i.e., electron and *hole capture* rates) are written as

$$r_1 = nN_t[1-f(E_t)]v_{th}\sigma_n \tag{2.86}$$

$$r_3 = pN_t f(E_t)v_{th}\sigma_p \tag{2.87}$$

where N_t describes the *density of states* of the *trap sites* and the function *f*(*E*) is their occupation probability.

The *emission* rates of *charge carriers* are expressed in terms of *emission coefficients* (e_n for electrons and e_p for *holes*) and the concentration of *charge carriers* found at the *localized states*. The *emission* rates of electrons and *holes* (r_2 and r_4, respectively) are calculated as

$$r_2 = e_n N_t f(E_t) \quad (2.88)$$

$$r_4 = e_p N_t [1 - f(E_t)]. \quad (2.89)$$

In this formalism, the *localized states* are assumed to be of energy E_t. In terms of (2.86)–(2.89), we still need to quantify the *generation* and the *recombination* rate. Toward that end, we need to develop tangible definitions for the components used in these rate equations. Such a task will be more at hand when we consider *thermal equilibrium*, or if not possible a *quasi-thermal equilibrium*, state.

We can imagine the nonidealities of the crystal (i.e., *traps*) as a separate system from the semiconductor. In situations where the semiconductor and the *trap* systems have not achieved *thermal equilibrium*, the two systems will not have distribution functions (i.e., $f(E)$) in the form of a *Fermi–Dirac* function (i.e., f is not equal to f_D). Also, a single *Fermi level* between the two systems cannot be established. Since any situation concerning a *net* flow of current, like the one at hand, is contradictory to *thermal equilibrium*, we should indicate that only for low values of current we can imagine *quasi-thermal equilibrium* between the semiconductor and the *trap* system. Under such a state of quasi-*thermal equilibrium*, which can be applied to slightly *forward-biased p–n* junctions, we can still take advantage of some of the properties of *thermal equilibrium* (i.e., $r_1 = r_2$, $r_3 = r_4$, and $f(E) = f_D(E)$).

Assuming these properties and *Maxwell–Boltzmann* statistics, in a *nondegenerately doped* semiconductor, we will have

$$e_n = v_{th}\sigma_n n_i \exp\left(\frac{E_t - E_i}{kT}\right) \quad (2.90)$$

$$e_p = v_{th}\sigma_p n_i \exp\left(\frac{E_i - E_t}{kT}\right). \quad (2.91)$$

According to (2.90) and (2.91), e_n will not be large if $E_t < E_i$. Also e_p will not be large if $E_i < E_t$. These are physically expected facts. As a result of these observations, a *localized state* is an effective G–*R center* (and not a *trap site*) especially if it is located near the middle of the *bandgap* (i.e., E_i: *intrinsic Fermi level*).

The *SHR G–R* formalism considers *G–R* under *low-level injection* (i.e., *LLI*). Under such circumstances, the concentration of *charge carriers* in the *recombination centers* remains close to the values at *thermal equilibrium*. This is because under *LLI*, *recombination centers* quickly *capture* majority *carriers*. In contrast, due to the lower frequency in encountering minority *carriers*, they must wait for minority *carriers* to arrive. Essentially *recombination centers* can be seen to be predominantly occupied by majority *carriers*. As the majority *carrier* concentration remains almost intact under *LLI* conditions, the charge concentration of these *localized states* also remains intact (i.e., as a result for those states: $f=f_D$). Under such circumstances the *net* rate of electron *capture* (i.e., r_1-r_2) should be equal to the *net* rate of *hole capture*, which are equal to the *net* rate of *recombination*

$$U \equiv R_{sp}-G_{sp}=r_1-r_2=r_3-r_4. \tag{2.92}$$

This *G–R* process is a spontaneous process. *G–R* processes such as those used in *lasers* are, however, stimulated.

According to (2.92), by eliminating the function *f*(*E*), we will have

$$U=\frac{pn-n_i^2}{\tau_{p_0}[n+n_i\exp((E_t-E_i)/kT)]+\tau_{n_0}[p+n_i\exp((E_i-E_t)/kT)]} \tag{2.93}$$

where $\tau_{n_0}=1/(N_t v_{th}\sigma_n)$ and $\tau_{p_0}=1/(N_t v_{th}\sigma_p)$. This rate equation is now in a form that can be plugged into (2.84) and (2.85) to calculate the steady-state current in the *quasi-neutral* regions of a slightly *forward-biased p–n* junction. However, before doing that we will first try to simplify this equation a little further in order to gain insight into the *lifetime* of excess minority *carriers*. In addition, we will briefly talk about another *G–R* process.

In (2.93), the numerator, which describes the deviation from *thermal equilibrium* in terms of the *mass action law*, acts as a restoring force toward *thermal equilibrium*. For a positive value of this factor, in the presence of excess concentration of *carriers*, *recombination* should take over the *generation* (i.e., $R>G$) to bring the system back to *thermal equilibrium*, and vice versa.

When the *capture cross-sections* are equal (i.e., $\sigma_n=\sigma_p$), (2.93) yields a simpler form

$$U=\frac{pn-n_i^2}{\tau_0[p+n+2n_i\cosh((E_t-E_i)/kT)]}. \tag{2.94}$$

As alluded to earlier, this equation positively shows that due to the presence of the exponential term in the denominator, only the *localized states* at or in the vicinity of E_i are effective *G–R* centers. As impurities in *silicon*, *gold* and *copper* give rise to two effective *recombination centers* at about 0.03 and 0.01 eV above E_i, respectively.

Under the assumed *low-level injection* and paired *G–R* conditions, (2.94) further simplifies to

$$U\cong\frac{(p_0+n')(n_0+n')-n_i^2\cong n'(p_0+n_0)}{\tau_0[p_0+n_0+2n_i\ \cosh((E_t-E_i)/kT)]}. \tag{2.95}$$

As a result of this *G–R* rate equation and the continuity equations, in the absence of current, the excess *carrier* concentration will exponentially decay with time with a time constant of

$$\tau=\tau_0\frac{[p_0+n_0+2n_i\cosh((E_t-E_i)/kT)]}{(p_0+n_0)}. \tag{2.96}$$

This time constant is referred to as the *excess carrier lifetime*. For effective recombination centers (i.e., small $E_t - E_i$), $\tau \cong \tau_0$. In other words, for these conditions, the *excess carrier lifetime* is independent of the majority *carrier* concentration.

At this point in our description of *G–R* processes, it is worth mentioning that *surface* and *interface states*, which we spoke of earlier in Section 2.2.2.1.4, are also quite involved in *G–R* processes. Although less important to current–voltage characteristics of *p–n homojunctions*, the presence of these *states* at some *heterointerfaces* and at semiconductor/insulator interfaces present in the vicinity of the *channel* of many *field effect transistors* makes them quite relevant to the current description of these *channels*.

In analogy with the *bulk* (i.e., Equation 2.93), the *net recombination* rate through the *surface states* can be written as

$$U_s = \frac{N_{\mathrm{st}} v_{\mathrm{th}} \sigma_{\mathrm{n}} \sigma_{\mathrm{p}} \left(p_{\mathrm{s}} n_{\mathrm{s}} - n_{\mathrm{i}}^2\right)}{\sigma_{\mathrm{p}}\left[p_{\mathrm{s}} + n_{\mathrm{i}} \exp((E_{\mathrm{i}} - E_{\mathrm{st}})/kT)\right] + \sigma_{\mathrm{n}}\left[n_{\mathrm{s}} + n_{\mathrm{i}} \exp((E_{\mathrm{st}} - E_{\mathrm{i}})/kT)\right]}. \tag{2.97}$$

In Equation (2.97), the subscript s is referring to the surface. The dimension of N_{st} is per square centimeter.

Due to *Fermi-level pinning*, the surface is often at a different potential than the bulk of the semiconductor. As a result, a *depletion region* is formed at the surface.

Auger Recombination Process As discussed earlier, with the increase in *carrier* concentration, *Auger recombination*, which is a three-, rather than a two-, particle *recombination* process, becomes more plausible. This can happen in cases of *degenerately doped* semiconductors or when *high level-injection* is taking place. This is because the frequency of mobile charges encountering one another surpasses the frequency of encounter of a *carrier* with a *localized state*. *Auger recombination* in an n-type semiconductor involves the interaction of two electrons and one *hole*. For p-type semiconductors, it involves the interaction of two *holes* and an electron.

In terms of this three-particle interaction, the *Auger recombination* rate is formulated as

$$U_{\mathrm{A}} = R_{\mathrm{A}} - G_{\mathrm{A}} = c_{\mathrm{n}} n\left(pn - n_{\mathrm{i}}^2\right) + c_{\mathrm{p}} p\left(pn - n_{\mathrm{i}}^2\right) \tag{2.98}$$

where c_{n} and c_{p} are known as the *Auger recombination* coefficients. c_{n} for electrons in *silicon* is about 1×10^{-31} cm^6/s. The *Auger lifetime* for *holes* in n-type semiconductors is about *half* to a *third* of that of electrons in p-type semiconductors. This can be described in terms of the differences in electron and *hole effective masses* and *mobilities*.

In the presence of both *SHR* and *Auger recombination* processes, the *effective lifetime* (i.e., τ) can be evaluated as a combination of the two:

$$\frac{1}{\tau} = \frac{1}{\tau_{\mathrm{SHR}}} + \frac{1}{\tau_{\mathrm{A}}}. \tag{2.99}$$

Auger recombination process is the inverse of the *avalanche generation* process.

QUASI-NEUTRAL REGION CURRENT COMPONENTS OF A SLIGHTLY FORWARD-BIASED *P–N* STEP HOMOJUNCTION With the use of the *continuity equations* (2.84)–(2.85) and the adoption of the appropriate *G–R* model of (2.95), the current–voltage characteristic of a *p–n* junction is developed.

At this point, it is quite informative to recite our list of assumptions in the current treatment of *transport* in a *p–n* junction:

- Dimensions of *p* and *n* regions are much larger than the *Debye length*[22] so that *SDA* remains valid throughout most of the device length. As a result of this, *electric field* in the *neutral* or *quasi-neutral* regions of the *p*- and *n*-side is assumed to be 0, or negligible, and *diffusion constant* is assumed to be position independent.
- *Carrier* concentrations and *lattice temperature* are low enough so that *Maxwell–Boltzmann* statistics remains a valid approximation to *Fermi–Dirac* statistics.
- *G–R* can be accurately expressed in terms of the chosen *G–R* model (i.e., *SHR*) throughout the length of the *quasi-neutral* regions.
- Whereas in the present treatment a planar interface between the *p*- and *n-side* is assumed, the presence of corners in actual *p–n* junctions can result in local strengthening of the electric field and current (known as *current crowding*). This latter effect can cause localized heating and a temperature gradient across the junction. These secondary effects are also not considered.

Considering these assumptions, based on (2.49) and its counterpart for *hole* concentration, minority *carrier* concentrations at the edges of the *neutral regions* of a *slightly forward-biased p–n* junction are given by[23]

$$n_p(-x_p) = N_d(x_n)\exp\left(-\frac{q(\phi_i - V_a)}{kT}\right) \tag{2.100}$$

$$p_n(x_n) = N_a(-x_p)\exp\left(-\frac{q(\phi_i - V_a)}{kT}\right). \tag{2.101}$$

Since the *forward bias* has been limited, the boundaries of the *depletion* region are assumed to have remained intact. As a result, the *excess* minority *carrier* concentration at the inner boundaries of the *quasi-neutral* regions can be written as[24]

$$n'_p(-x_p) = n_{p0}(-x_p)\left[\exp\left(\frac{qV_a}{kT}\right) - 1\right] \tag{2.102}$$

[22] We have previously described this length scale in the case of an *MS* barrier. For *p–n* junctions, also, a *Debye length* can be defined following the same principles.

[23] n_p denotes the electron concentration of the *p-side* and p_n is representative of the *hole* concentration of the *n-side* of the junction.

[24] As indicated before, subscript 0 refers to *thermal-equilibrium* concentrations.

$$p'_{\rm n}(x_{\rm n}) = p_{n_0}(x_{\rm n})\left[\exp\left(\frac{qV_{\rm a}}{kT}\right) - 1\right]. \tag{2.103}$$

With the aid of these boundary conditions, as shown later, (2.84)–(2.85) yield the current–voltage characteristic of the junction (at least with the consideration of the *quasi-neutral* regions).

According to (2.85) and (2.95),

$$D_{\rm p}\frac{\partial^2 p_{\rm n}}{\partial x^2} - \frac{p_{\rm n} - p_{n_0}}{\tau_{\rm p}} = 0, \tag{2.104}$$

which yields the solution of

$$p'_{\rm n}(x) = A\exp\left(-\frac{x - x_{\rm n}}{\sqrt{D_{\rm p}\tau_{\rm p}}}\right) + B\exp\left(\frac{x - x_{\rm n}}{\sqrt{D_{\rm p}\tau_{\rm p}}}\right). \tag{2.105}$$

Since minority *carriers* are provided from the opposite side of the junction, the only admissible term in this solution is the decaying term. The boundary condition presented in (2.103) yields

$$A = p'_{\rm n}(x = x_{\rm n}) = p_{n_0}\left[\exp\left(\frac{qV_{\rm a}}{kT}\right) - 1\right] \tag{2.106}$$

where $B = 0$.

The decaying length scale of the exponential presented in (2.105) is known as the *diffusion length*:

$$L_{\rm p} = \sqrt{D_{\rm p}\tau_{\rm p}}. \tag{2.107}$$

Application of the same procedure to the *p-side* of the junction yields a decaying exponential into the *quasi-neutral* region for the excess electron concentration with a *diffusion length* defined as

$$L_{\rm n} = \sqrt{D_{\rm n}\tau_{\rm n}}. \tag{2.108}$$

According to these minority *carrier* profiles, it is observed that if the length of a *quasi-neutral* region was to be much longer than the *diffusion length* of minority *carriers* in that region, full recombination of excess *carriers*, with the majority *carriers* injected from the *Ohmic* contact, will happen in this region before arriving at the *Ohmic* contact. Such a case is referred to as the *long-base diode*. However, if in one or both sides of the *p–n* junction the length of the *quasi-neutral* region was to be shorter than the *diffusion length* of minority *carriers*, excess minority *carriers* will arrive at the *Ohmic* contact with effectively no intermediate *recombination*. In this

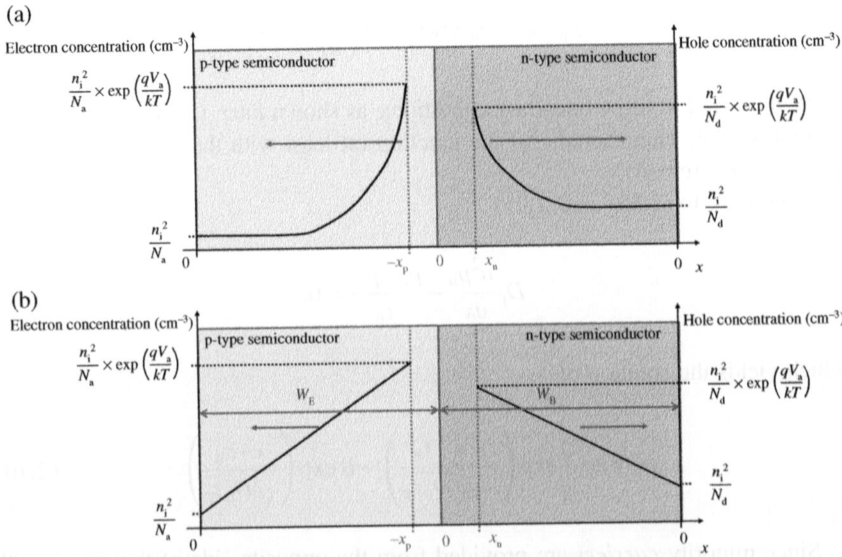

FIGURE 2.25 Minority carrier concentration profiles in the quasi-neutral regions of a forward-biased step *p–n* junction with bias V_a, when both sides of the junction are long (a), when both sides of the junction are short (b). N_a and N_d are the acceptor and the donor concentrations across the junction, respectively.

case, due to the abundance of majority *carriers* at the *Ohmic* contact, at the site of this contact, excess minority *carrier* concentration is forced to vanish. This yields a linearly, rather than exponentially, decaying profile to the excess minority *carriers* in the *quasi-neutral* region. This case is referred to as the *short-base diode*. This can be the case in one, or both, sides of the *p–n* junction. Figure 2.25 depicts the minority *carrier* concentration profile both in a *short-base* and a *long-base p–n* junction.

In the absence of a sizeable electric field and *drift current*, the *diffusion-only current* for the *short-* and *long-base diodes* can be evaluated.

Equations (2.109) and (2.110) present the *diffusion current* of minority *carriers* in a long-base *diode* in the *quasi-neutral* regions of the *n-* and *p-side*, respectively;

$$J_p(x) = -qD_p\frac{\partial p_n}{\partial x} = qD_p\frac{p_{n_0}}{L_p}\left[\exp\left(\frac{qV_a}{kT}\right)-1\right]\exp\left(-\frac{x-x_n}{L_p}\right)$$
$$= qD_p\frac{n_i^2}{N_dL_p}\left[\exp\left(\frac{qV_a}{kT}\right)-1\right]\exp\left(-\frac{x-x_n}{L_p}\right) \tag{2.109}$$

$$J_n(x) = qD_n\frac{n_i^2}{N_aL_n}\left[\exp\left(\frac{qV_a}{kT}\right)-1\right]\exp\left(\frac{x+x_p}{L_n}\right). \tag{2.110}$$

Since current density at any point along the length of the planar *p–n* junction is *constant*, the total current density can be calculated by adding up the *diffusion-current* densities calculated at the boundaries of the *depletion region*. As a result, in a *p–n* junction, which is *long base* on both sides, the total current density in the *quasi-neutral* regions is given by

$$J_t = J_p(-x_p) + J_n(x_n) = qn_i^2\left(\frac{D_n}{N_a L_n} + \frac{D_p}{N_d L_p}\right)\left[\exp\left(\frac{qV_a}{kT}\right) - 1\right] = J_0\left[\exp\left(\frac{qV_a}{kT}\right) - 1\right]. \tag{2.111}$$

Assuming a *short-base diode* on the *n-side* of the junction,

$$J_p = -qD_p\frac{\partial p_n}{\partial x} = qD_p\frac{p_{n0}}{W_B'}\left[\exp\left(\frac{qV_a}{kT}\right) - 1\right] = qD_p\frac{n_i^2}{N_d W_B'}\left[\exp\left(\frac{qV_a}{kT}\right) - 1\right] \tag{2.112}$$

where

$$p_n'(x) = p_{n0}\left[\exp\left(\frac{qV_a}{kT}\right) - 1\right]\left(1 - \frac{x - x_n}{W_B'}\right). \tag{2.113}$$

In these equations, $W_B' = W_B - x_n$. If the junction was to be *short base* on both sides,

$$J_t = J_p(-x_p) + J_n(x_n) = qn_i^2\left(\frac{D_n}{N_a W_E'} + \frac{D_p}{N_d W_B'}\right)\left[\exp\left(\frac{qV_a}{kT}\right) - 1\right] \tag{2.114}$$

where $W_E' = W_E - x_p$. As shown in Figure 2.25, W_B and W_E are the total width of the *n-* and *p-side* of the junction.

Figure 2.26 depicts the components of current across a *long-base p–n* junction. While so far we have visited only those components that are induced due to *G–R* across (or at the end) of the *quasi-neutral* regions, the remaining component of current, which is depicted in this figure, is developed across the *depletion region*. We will deal with this component in the next section. Among the components of current evolved in the *quasi-neutral* regions, as shown in Figure 2.26, the total current is entirely carried by the majority *carrier* current being supplied at the *Ohmic* contacts. As shown on the figure, this majority current is complementary of the minority *carrier* current terms calculated in (2.109), (2.110), and (2.112).

Depletion-Region Current Components of a Slightly Forward- or Reverse-Biased *P–N* Step Homojunction So far in our treatment of the current–voltage characteristic of *p–n* junctions, we have solely viewed the *depletion region* as a barrier against the flow of the majority *carriers*. In this part of the discussions, through studying the *G–R* processes in this region, we will evaluate the current components of the *depletion region*. This is a first-order departure from the notion of an *ideal diode*. This

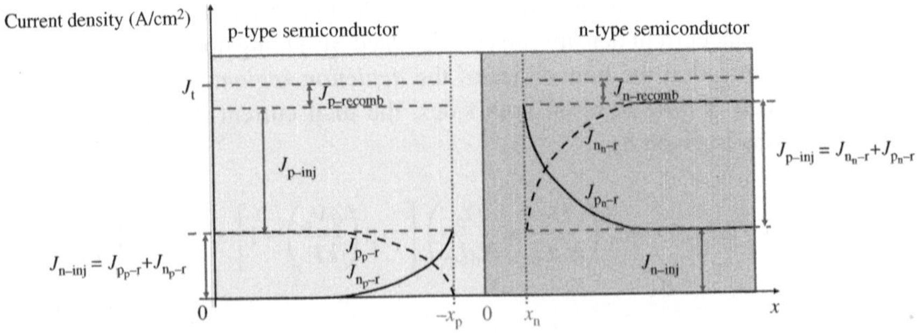

FIGURE 2.26 Current components across a long-base *p–n* junction. J_{n_p-r} and J_{p_n-r} represent the diffusion current components carried by the excess minority carriers in the quasi-neutral regions of the *p*- and *n*-side, respectively. J_{p_p-r} and J_{n_n-r} represent the diffusion-current components carried by the majority carriers recombining with the excess minority carriers in the quasi-neutral regions of the *p*- and *n*-side, respectively. $J_{\text{p-recomb}}$ and $J_{\text{n-recomb}}$ highlight the current components induced by the recombination of injected carriers across the depletion region. These two components are not so far considered in our treatment of the current of a *p–n* junction.

component of current is especially important in evaluation of *reverse bias* current of a *p–n* junction. Unlike the *forward bias* current, due to the increase in the potential barrier height, the leakage current under *reverse bias* is carried only by the flow of minority *carriers* across the *depletion region*.

Using the simplified *SHR* theory in this region (i.e., assuming equal *hole* and electron *capture cross-sections*) and taking $pn = n_i^2 \exp(qV_a/kT)$ on the basis of (1.43)–(1.44), (2.94) yields

$$U = -\frac{dn}{dt} = -\frac{dp}{dt} = \frac{n_i^2[\exp(qV_a/kT)-1]}{\tau_0[p+n+2n_i\cosh((E_t-E_i)/kT)]}. \tag{2.115}$$

The *G–R* current of the *depletion region* can then be evaluated by integrating *U* over the width of the *depletion region*:

$$J = q\int_{-x_p}^{x_n} U dx. \tag{2.116}$$

In spite of the difficulty in evaluating this integral in general, a quantitative discussion of this process for a few simple cases allows us to gain insight into its current versus voltage dependence.

Considering the case of a *forward-biased p–n* junction, where only the most effective of *recombination centers* (i.e., $E_t \sim E_i$) take part in the *G–R* process, it can be concluded that *J* will be maximized when $p+n$ in the denominator of *U* is minimized.

Knowing that $pn = n_i^2 \exp(qV_a/kT)$ and assuming that all the potential drop is happening across the *depletion region*, $p + n$ will be minimum when $p = n = n_i \exp(qV_a/2kT)$. In reality, due to the *forward biasing* of the junction and the existence of a large gradient of *charge-carrier* concentrations across the *depletion region*, this minimum value of $p + n$ will not provide an accurate representation of $p + n$ across the full width of the *depletion region*. However, assuming that this minimum $p + n$ extends across a small portion of the *depletion region* identified by x' and taking into account the dominance of U over this region, we can evaluate J as

$$J = \frac{qx'n_i^2[\exp(qV_a/kT) - 1]}{2n_i\tau_0[\exp((qV_a/2kT) + 1)]}. \tag{2.117}$$

Since under *forward bias carriers* are being injected across the *depletion region*, the presence of excess *carrier* concentrations makes this a *recombination* and not a *generation* current.

In this equation, with the valid assumption of the dominance of both exponential terms under *forward bias*, the *recombination current* will be exponentially dependent on $qV_a/2kT$. This is in contrast to the $\exp(qV_a/kT)$ dependence of the *ideal diode* equation (i.e., (2.111) and (2.114)). This is used in explaining the *diode's* nonideality factor. This weaker voltage dependence of current is observed especially under low current levels.

It is clear that for a *reverse-biased p–n* junction (i.e., $V_a < 0$), both of the exponential terms in (2.117) will lose dominance and the so-called *generation* (and not *recombination*) *current* calculated under this regime will be *bias* independent to the first order. In reality, since x_d and as a result x' become larger with the *reverse bias*, the *reverse* current also increases with voltage.

As mentioned earlier, in these equations $\tau_0 = 1/(N_t v_{th}\sigma)$ is the time constant associated with the *carrier generation* process and also the *recombination* of excess *carriers*, in a region with a density N_t of *recombination centers* of *capture cross-section* σ. Complexities in accurate evaluation of τ_0 and x' often make the accurate evaluation of the *G–R* current impossible. In real devices, the situation becomes even more complicated because of the presence of nonuniformity in distribution of *G–R centers* at the surfaces and interfaces.

Aside from this complexity, the model presented in this section also overlooks the presence of *interface states* at the *p–n* junction. This is especially important in *lattice-mismatched heterojunctions*. Usually, empirical models, with inclusion of an experimental nonideality factor n, are used in cases in which these effects are more predominant, yielding

$$J = J_0\left[\exp\left(\frac{qV_a}{nkT}\right) - 1\right]. \tag{2.118}$$

Also as alluded to earlier, due to the presence of *band*-edge discontinuities, other modes of *transport* such as *thermionic emission*, *field emission*, and *thermionic field emission* have to be considered in evaluation of the current–voltage characteristic of some *p–n heterojunctions*. This is of course dependent on the specifics of the *band lineup*. We will pay further attention to these topics in Section 2.5.

2.3.2.1.5 Breakdown Characteristics of P–N *Step Junctions* In Section 2.3.2.1.2, we observed that application of a *reverse bias* across a *p–n* junction results in expansion of the *depletion region* and increase in the maximum electric field at the *metallurgical junction*. However, as noted in Section 1.5, since for each of the materials of interest in semiconductor technology there is a maximum applicable electric field, such a framework has a limited range of applications. For semiconductors, beyond this critical electric field, *impact ionization* results in excessive increase in *carrier* concentration in both *conduction* and *valence bands*. For the case of other materials employed in semiconductor *chips* (e.g., SiO_2 dielectric and the plastic used in the packages of semiconductor *chips*), exceeding a critical electric field results in irreversible damage due to rupture of the insulator and its physical decomposition. Previously in Section 1.5, *avalanche breakdown* in semiconductors was described. In this section both *avalanche* and *Zener breakdown* processes in *p–n* junctions are addressed in further details.

Prompted by the presence of a sizeable electric field, *breakdown* of a semiconductor junction is indicated in terms of an excessive increase in the current passing through the junction. This is to various degrees uncontrollable by the potential drop across the junction. In other words, a junction breaks down when excessive increase in current is observed with very little or no increase in *bias*. Junction *breakdown* can be destructive or nondestructive. The rush of current (and the *Joule heating* inflicted on the junction and possibly melting/decomposition of the junction), in addition to the movement of ions across the junction, can have adverse effects on its structure. As a result, *breakdown* can leave irreversible impacts on the current–voltage and *C–V* characteristics of the junction. These cases are examples of destructive *breakdown*. Junction *breakdown* can, however, be nondestructive if the irreversible impacts are eliminated (e.g., by limiting the current and *Joule heat*).

Junction *breakdown* across various types of semiconductor systems including *MS* contacts, semiconductor–semiconductor *homo-* and *heterojunctions*, and metal–oxide–semiconductor junctions takes different forms. In this section we will focus on the *breakdown* processes of *p–n homojunctions*. *Breakdown* processes of other junctions will be addressed elsewhere.

In *p–n* junctions, the *breakdown* mechanism is explained though *Zener breakdown*, *avalanche breakdown*, or a combination of the two. As observed in Section 1.5, while *avalanche breakdown* is the result of the *impact-ionization* process (and *avalanching*) in multiplication of the number of *carriers* across a relatively long high electric-field region, *Zener breakdown* is the product of *carrier tunneling* across a very thin potential barrier. Due to this, while *avalanche breakdown* is seen in *lightly doped* samples, *Zener breakdown* is dominant in *heavily doped* samples (where according to (2.57) and (2.60) the electric field is stronger and the *depletion-region width* is smaller). As will be shown in this section, these two processes have two opposite trends of temperature dependence.

According to this description, the first step in evaluation of the *breakdown voltage* of a junction is to decide whether *avalanche* or *Zener breakdown* is the dominant process. For both of these two processes, *breakdown voltage* can be evaluated fairly well with the knowledge of geometry and *dopant* composition of the junction.

AVALANCHE BREAKDOWN IN A *P–N* STEP JUNCTION In this description, the *depletion region* of a *reverse-biased p–n* junction is dealt with. As a first-order approximation, it will be assumed that all *scattering* processes happening on average after a free flight (which is expressed by the length of the *mean free path*: λ) are *inelastic*. The maximum kinetic energy gained from the electric field $\vec{E}$, over this *mean free path*, can be evaluated as

$$\Delta E = q\int_0^{\lambda} \vec{E}\cdot d\vec{x}. \tag{2.119}$$

This amount of energy will be lost to *inelastic scattering*.

Depending on the amount of this energy, *scattering* in the form of collision with a covalent bond can result in breakage of the bond and *generation* of an *electron–hole pair*. Based on the length of the high electric-field region, in comparison to the *mean free path*, this process of producing *three charge carriers* from *one* can take place enough times to prompt an *avalanche*. However, if the middle part of the *depletion region* (i.e., the only region of high electric field across a *p–n* junction) is just a few times λ in length, *avalanche* will not be prompted.

Considering the semiclassical theory of quantum mechanics and observing electrons and *holes* as particles, we can study this process in terms of conservation of momentum and conservation of energy in the three-particle system. In order to simplify the conclusions, we will assume that the *effective masses* of *holes* and electrons are equal. According to this assumption, conservation of momentum and energy enforces that

$$\begin{cases} m^*(v_1 - v_2) = m^* v_2 + m^* v_2 \\ \frac{1}{2}m^* v_1^2 = E_g + \frac{3}{2}m^* v_2^2 \end{cases} \Rightarrow v_2 = \frac{v_1}{3} \Rightarrow \frac{1}{2}m^* v_1^2 = \frac{3}{2}E_g \tag{2.120}$$

where v_1 is the velocity of the original electron and v_2 is the velocity of any of the three *charge carriers* resulting from breaking the covalent bond.

For this simple case, according to (2.120), in order to produce the first *electron–hole pair*, the original electron should have acquired a kinetic energy of at least $\Delta E = 3/2E_g$.

For the discussion in this section, we will assume a p–n^+ junction. While *reverse biasing* this junction expands the predominantly one-sided *depletion region* on the *p-side*, it facilitates the injection of minority *carriers* across the junction. As the *hole* concentration on the *highly doped n*-side is very limited, it can be imagined that if we were to have any *avalanche breakdown* in this junction, it has to be instigated by the

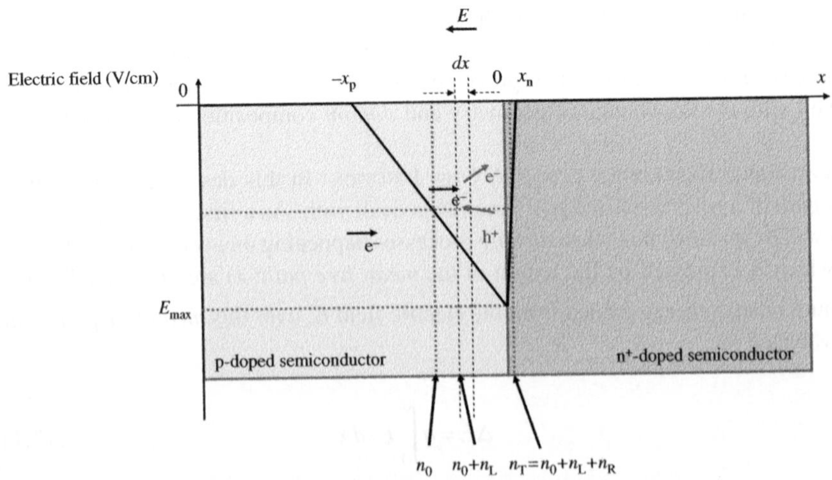

FIGURE 2.27 Schematic depiction of the process of avalanche breakdown within the high electric-field part of the depletion region of an p–n^+ junction. The indicated electron concentrations are defined in the text.

electrons injected into the middle part of the *depletion region* from the *lightly doped p*-side. The closer these electrons fly to the *metallurgical junction*, as expected from (2.119), the larger the ΔE and the higher the probability for *electron–hole pair generation*. Figure 2.27 provides a schematic description of *avalanche breakdown* in this junction.

To quantify *impact ionization*, we assume the aforementioned p–n^+ junction of Figure 2.27 to have a uniform geometry of cross-sectional area A, running parallel to the *interface*. According to Figure 2.27, we look into the additional concentration of *carriers* produced by the *impact-ionization* process in a small volume Adx. As it has been indicated already to maximize the probability of *impact ionization*, this volume is envisioned of being close to the *metallurgical junction*. This elemental volume is encompassed within the boundaries of the high electric-field region identified by dashed lines in Figure 2.27. These boundaries are determined by the required minimum electric field calculated on the basis of (2.119) and (2.120).

Assume n_0 to represent the concentration of electrons entering the high electric-field region from the left boundary (i.e., moving opposite to the direction of the electric field). In addition, assume that between the left-side boundary of this region and the left-side boundary of the aforementioned elemental volume Adx, n_L *electron–hole pairs* per *cubic centimeters* are produced by *impact ionization*. Out of the n_L *electron–hole pairs generated* per unit volume between these two boundaries, assuming that the distance is short so that *scattering* can be neglected, the direction of the electric field moves the electrons toward the volume Adx while pushing the *holes* away from this volume. According to these assumptions, in the absence of *carrier recombination*, the electron concentration at the left-side boundary of the volume Adx would be equal to $n_0 + n_L$. In terms of the definition of *ionization rate*, which was provided earlier in

Section 1.5, the probability of instigation of *impact ionization* by these electrons within the boundaries of this volume is formulated in terms of the product of the *ionization rate* of electrons α_n and the width of the elemental volume dx. As mentioned in Section 1.5, α_n is exponentially dependent on the electric field.

In terms of this probability and assuming dx to be very small, the additional concentration of electrons and *holes* created within the elemental volume Adx is given by

$$dn' = dp = \alpha_n n dx = \alpha_n (n_0 + n_L) dx. \tag{2.121}$$

In addition, *holes* entering this volume from the right-side boundary can also cause *impact ionization*. As we have already assumed that no *holes* enter the high electric-field region from the n^+ side of the junction, the only *holes* arriving at the right-side boundary of the elemental volume Adx would be those created between the right-side boundaries of this small volume and the high electric-field region. Let us assume the concentration of those *holes* at the right-side boundary of elemental volume Adx to be p_R. As a result, the additional concentration of electrons in the elemental volume delivered by the *impact-ionization* process of these *holes* is given by

$$dn'' = dp = \alpha_p p dx = \alpha_p p_R dx \tag{2.122}$$

where α_p denotes the *ionization rate* of the *holes*.

Therefore, the total addition of *impact ionization* to the electron concentration in the elemental volume Adx is

$$dn = dn' + dn'' = \alpha_n (n_0 + n_L) dx + \alpha_p p_R dx. \tag{2.123}$$

In terms of these quantities, the concentration of electrons at the right-side boundary of the high electric-field region would be equal to

$$n_T = n_0 + n_L + n_R \tag{2.124}$$

where

$$n_R = p_R. \tag{2.125}$$

Using n_T we can rewrite (2.123)

$$\frac{dn}{dx} = \alpha_n (n_0 + n_L) + \alpha_p n_R = (\alpha_n - \alpha_p)(n_0 + n_L) + \alpha_p n_T. \tag{2.126}$$

If we assume the *ionization rates* of electrons and *holes* to be equal to one another (i.e., $\alpha_n = \alpha_p = \alpha$), integration of (2.126) between the boundaries of the high electric-field region results in

$$n_T - n_0 = n_T \int_{x_L}^{x_R} \alpha dx \Rightarrow \text{Multiplication factor} = \frac{n_T}{n_0} = \frac{1}{1 - \int_{x_L}^{x_R} \alpha dx}. \tag{2.127}$$

It can be conferred from (2.127) that the *multiplication factor* becomes infinity when $\int_{x_L}^{x_R} \alpha dx = 1$. This condition is defined as the onset of *avalanche breakdown.*

In Section 1.5, we have already suggested equations for calculating the *ionization rate* in terms of the electric field. In this section the foundations of those equations are further elaborated.

As we already know, for *impact ionization* to occur, we have to have *carriers* with sufficient kinetic energy to break a covalent bond. The kinetic energy of electrons is denoted in terms of the amount of energy they have in excess of E_c.[25] As a result of *Maxwell–Boltzmann* statistics, the deeper a *state* is in the *conduction band*, the probability of finding an electron in that *state* is exponentially lower (i.e., a kinetically energized electron). From our discussion in Chapter 1, we already know that while an average electron *scatters inelastically* after flying for a *mean free path* of λ (and drops to E_c), there are always electrons that encounter these *scattering* processes more or less often than average. If an electron travels for distance d (longer than the *mean free path*) before the *inelastic scattering* robs it of the kinetic energy gained from the electric field, its energy can exceed the values indicated by *thermal equilibrium.* An indication of this energy excess is given in the form of the d/λ ratio. Resulting from *Maxwell–Boltzmann* statistics, the concentration of these *hot* electrons can be evaluated in terms of electron concentration under *quasi-thermal equilibrium* (i.e., n) as

$$n^* = n \cdot \exp\left(-\frac{d}{\lambda}\right). \tag{2.128}$$

This electron density helps in determining the density of *ionizing* collisions.

As long as the electric field is changing slowly, we can avoid the complexities created by *energy relaxation time* and define the energy of an electron at any position in terms of electric field at the same position in space.

As we have seen before, *impact ionization* requires a minimum amount of kinetic energy (denoted here as $\mathrm{E_{min}}$). For our simplified situation, we can define this amount of energy in terms of a minimum distance d, which a particle needs to travel freely within a region of constant electric field E:

$$d = \frac{\mathrm{E_{min}}}{qE}. \tag{2.129}$$

With this background, assuming that an electron immediately after gaining the minimum energy E_{min} breaks a covalent bond, the concentration of *ionizing* collisions in the distance dx can be written in proportionality to dx/d as

$$dn \propto n^* \frac{dx}{d} = \frac{qE}{\mathrm{E_{min}}}\left[\exp\left(-\frac{\mathrm{E_{min}}}{qE\lambda}\right)\right] ndx. \tag{2.130}$$

[25] Similarly, for *holes* this energy is calculated with regards to E_v.

Comparing this to Equation (1.96), we will have a first-order definition for the *ionization rate*:

$$\alpha = A \cdot E \exp\left(-\frac{B}{E}\right) \qquad (2.131)$$

where A and B are constants.

As we have already taken note, a large number of assumptions have been incorporated into this definition. In evaluation of the *breakdown voltage* due to the *avalanche* process, it should be emphasized that *reverse bias* not only increases the maximum electric field (which is further removed from the assumption of constant electric field) but also expands the boundaries of the *depletion region*. These impacts result in variation of the *ionization rate* throughout the *depletion region*. As a result of these complexities, oftentimes empirical equations are used instead of Equation (2.127). For the case of a *p–n* junction, we often encounter the following empirical relationship:

$$\text{Multiplication factor} = \frac{1}{1 - \left(\frac{|V_R|}{BV}\right)^n} \quad 2 < n < 6 \qquad (2.132)$$

in which $V_R < 0$ is the applied *reverse bias* and BV is the *breakdown voltage*.

This *breakdown voltage* often varies with *dopant* concentration as $N^{-2/3}$.

In the *avalanche breakdown* process, increasing the temperature results in an increase in the *breakdown voltage*. Especially at higher temperatures, such a thermal dependence is observed. This is because at high *lattice temperatures*, the rise in temperature results in reduction of the *mean scattering time* and effective reduction of the chance of encountering *hot carriers*.

Neglecting the *energy relaxation time* and the lag between energy and electric field often results in underestimation of the *avalanche breakdown voltage* of modern devices in which electric field is rapidly changing.[26]

ZENER BREAKDOWN IN *P–N* JUNCTIONS For potential barriers of small *width*, quantum *tunneling* can also serve as a competing *breakdown* process. In *p–n* junctions, this situation is encountered when the junction is formed between two *highly doped* semiconductors. The increase in *dopant* concentration, through reduction of the *depletion-region width*, postpones the *avalanche breakdown* to higher *reverse biases*. It also increases the chance of quantum *tunneling*. *Tunneling* takes place directly between the *conduction band* on the *n*-side of the junction and *valence band* on the *p*-side. Due to the large concentration of the majority *carriers* on these two sides, an improved

[26] J. Slotboom, G. Streutker, M. van Dort, P. Woerlee, A. Pruijmboom, and D. J. Gravesteijn, *Non-Local Impact Ionization in Silicon Devices*, in Proceedings of the IEEE Electron Devices Meeting, December 8–11, 1991, Washington, DC, USA. IEEE, Piscataway, NJ, pp. 127–130, 1991.

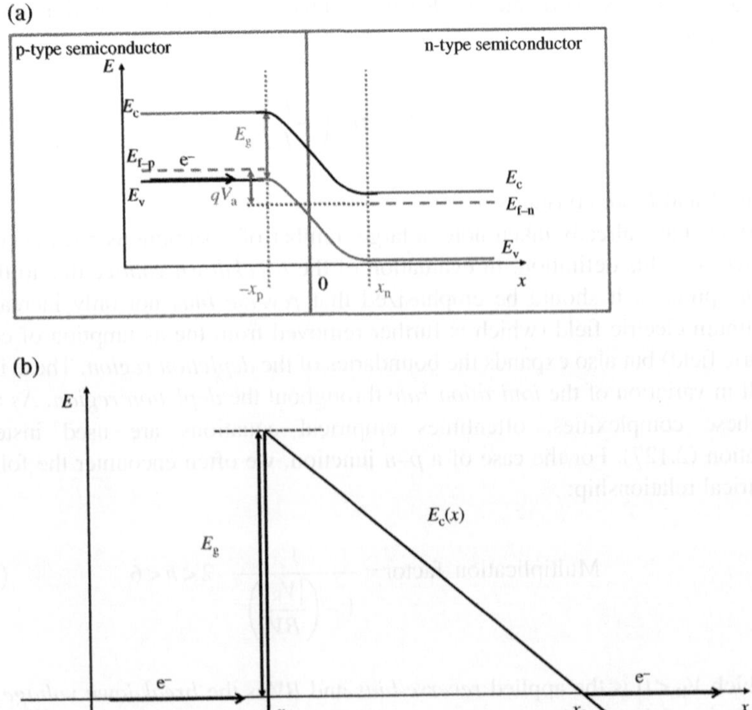

FIGURE 2.28 (a) Electron tunneling across a reverse-biased *p–n* junction (with bias V_a). (b) The triangular approximation of the band bending across a *p–n* junction used in formulating the tunneling probability according to the *WKB* approximation.

probability of *tunneling* causes a rush of current and hence a *breakdown* process known as *Zener breakdown*.

According to Figure 2.28, assuming a triangular potential barrier and using the *zero-scattering WKB* approximation, the *tunneling* probability can be approximated by

$$T \approx \exp\left[-2\int_0^d \sqrt{\frac{2m^* E_c(x)}{\hbar^2}}dx\right] \tag{2.133}$$

in which $E_c(x)$ is the linearly varying potential barrier height of the triangular barrier.

Integration results in

$$T = \exp\left(-\frac{K}{E}\right) = \exp\left(\frac{-qdK}{E_g}\right) \tag{2.134}$$

where $K = 4\left(\sqrt{2m^*}/3q\hbar\right)E_g{}^{3/2}$.

According to (2.133), the *tunneling* probability decreases rapidly with the reduction in the slope of $E_c(x)$ (i.e., the electric field) or increase of the tunneling distance d.

Zener breakdown voltage reduces with increased *lattice temperature*. This is because with an increase in *thermal velocity*, the number of *carriers* impinging on the *potential barrier* per unit time also increases. Consequently, for the same *tunneling* probability, the current through the barrier becomes larger. The distinctive temperature signatures between the two *breakdown* mechanisms of *p–n* junctions can be used to identify the dominant *breakdown* mechanism and also to design temperature-insensitive circuits.

2.3.2.2 Linearly Graded* P–N *Junctions: A Broader Perspective While a step junction can be realized with the use of *in situ doping* in *epitaxial* growth techniques, it is not encountered in the case of traditional *ion-implanted*, or *diffusely doped* profiles. The *diffused p–n* junctions to different degrees of approximation, are either treated as exponentially graded or linearly graded profiles. In this section, through applying *SDA*, the case of linearly graded *p–n* junctions is considered. As shown in Figure 2.29, in this junction we assume the *dopant* concentrations to vary linearly along the length of the *diode*. This variation is expressed by

$$N_d - N_a = ax \tag{2.135}$$

in which a has a cm^{-4} dimension.

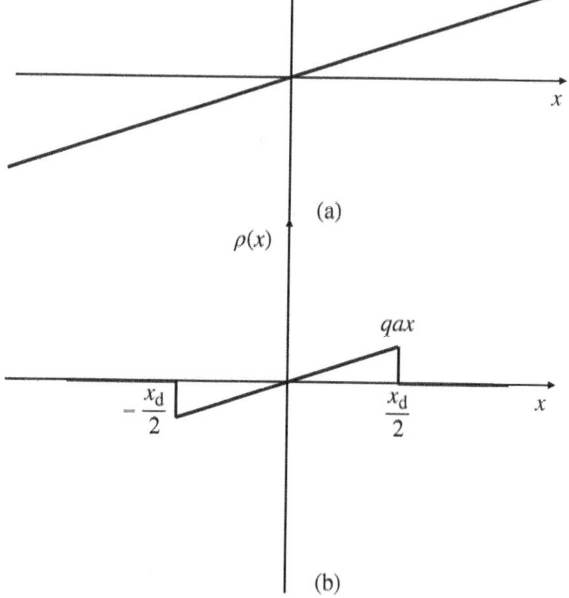

FIGURE 2.29 (a) The dopant concentration profile across a linearly graded *p–n* junction. (b) Charge concentration profile across a linearly graded junction according to *SDA*.

For positive values of x, the semiconductor is n-type *doped*, whereas when $x<0$ it is p-type *doped*. Following the same principles as those applied to a step junction, *SDA* assumes a *depletion region*, which is partly extended to *p-* and *n-side* of the junction (i.e., $-x_p<x<x_n$), while beyond this region two *neutral regions* with 0 electric field are considered.

The charge concentration profile is also depicted in Figure 2.29 where

$$\rho(x)=qax. \tag{2.136}$$

As shown in the following, with the use of *Gauss*'s law, this profile yields the spatial variation of electric field and as a result potential across the junction:

$$\nabla\cdot E=\frac{\rho}{\epsilon_s}\Rightarrow\left|\begin{array}{l} 0<x<x_n: E(x)=\dfrac{qa\left(x^2-x_n^2\right)}{2\epsilon_s}\Rightarrow\phi(x)=-\dfrac{qa}{2\epsilon_s}\left(\dfrac{1}{3}x^3-xx_n^2+c_1\right)\\ -x_p<x<0: E(x)=\dfrac{qa\left(x^2-x_p^2\right)}{2\epsilon_s}\Rightarrow\phi(x)=-\dfrac{qa}{2\epsilon_s}\left(\dfrac{1}{3}x^3-xx_p^2+c_2\right)\end{array}\right. \tag{2.137}$$

Requirement of continuity of electric field at the *metallurgical junction* enforces that

$$x_n=x_p=\frac{x_d}{2} \tag{2.138}$$

where x_d is the *depletion-region width*. Continuity also enforces c_1 to be equal to c_2.

Accordingly, the built-in potential, which is the potential difference between the two *neutral regions*, is calculated as

$$\phi_i=\phi(x_n)-\phi(-x_p)=\frac{qa}{2\epsilon_s}\cdot\frac{4}{3}x_n^3=\frac{qa}{12\epsilon_s}x_d^3. \tag{2.139}$$

As a result, the *depletion-region width* in the general form (i.e., under *bias*) is given by

$$x_d=\sqrt[3]{\frac{12\epsilon_s(\phi_i-V_a)}{qa}}. \tag{2.140}$$

In addition, since according to *Maxwell–Boltzmann* statistics

$$n(x_n)=N_d-N_a=\frac{ax_d}{2}=n_i\exp\left(\frac{E_f-E_{in}}{kT}\right) \tag{2.141}$$

and

$$p(-x_p) = N_a - N_d = -a\left(-\frac{x_d}{2}\right) = n_i \exp\left(\frac{E_{ip} - E_f}{kT}\right) \tag{2.142}$$

the built-in potential can also be expressed by

$$\phi_i = \frac{1}{q}(E_{ip} - E_{in}) = 2\frac{kT}{q} Ln\left(\frac{ax_d}{2n_i}\right). \tag{2.143}$$

According to (2.140), the *small-signal depletion capacitance* is calculated as

$$C_j = \frac{\epsilon_s}{x_d} = \frac{\epsilon_s}{\sqrt[3]{\dfrac{12\epsilon_s(\phi_i - V_a)}{qa}}}. \tag{2.144}$$

In terms of (2.137) and (2.140), the dependence of the maximum electric field developed across a linearly graded *p–n* junction under *bias* is expressed by

$$E_{max} = \frac{3(\phi_i - V_a)}{2x_d}. \tag{2.145}$$

In many circuit applications, as a result of its voltage dependence, the *depletion capacitance* of a *reverse-biased p–n* junction is employed as a *variable reactor* (also known as *varactor*). As indicated in (2.75) and (2.144), the *depletion capacitance* of *reverse-biased* step and linearly graded *p–n* junctions, in terms of the *reverse bias* V_R, can be expressed by

$$C_j \propto (\phi_i + |V_R|)^{-n} \tag{2.146}$$

or approximately as

$$C_j \propto (|V_R|)^{-n} \text{ for } |V_R| \gg \phi_i \tag{2.147}$$

in which $n = 1/3$ for a linearly graded junction and $n = 1/2$ for a step junction.

In terms of this comparison, the voltage sensitivity of a *p–n* step junction is greater than that of a linearly graded junction. In the case of a hyperabrupt junction, this sensitivity can be even further improved by making n larger than 1/2.

A hyperabrupt p^+–n junction has an exponentially decaying n-type *dopant* distribution in the *n-side* such that $N_d - N_a = Bx^{-m}$ where $m > 1$. Techniques of *epitaxy* can be used in realization of these junctions. For a *dopant* profile like this, we can write the *Poisson* equation as

$$\frac{d^2\phi}{dx^2} = -Bx^{-m}. \tag{2.148}$$

According to *SDA*, (2.148) results in the following relationship for the *depletion-region width*:

$$x_d \propto |V_R|^{1/-m+2}, \tag{2.149}$$

which yields

$$C_j = \frac{\epsilon_s}{x_d} \propto V_R{}^{1/m-2}. \tag{2.150}$$

According to (2.150), with the appropriate choice of *m*, *varactors* with different voltage sensitivities are realizable.

2.4 METAL–INSULATOR–SEMICONDUCTOR SYSTEM

In the implementation of different varieties of *field effect transistors*, the modulating effect of the electric field, defined by a *gate* terminal, on the *conductivity* of the *channel* is at the heart of the device design. This modulating electric field is exerted on the *channel* through a variety of forms including:

- Expansion and retraction of the boundary of the *depletion region* of a *blocking MS* junction through changing the *reverse bias* (in the case of *MESFETs*[27])
- Expansion and retraction of the boundaries of the *depletion region* of a *p–n homojunction* through changing the *reverse bias* (in the case of *JFETs*[28])
- The capacitive effect of the *gate* electrode on the *carrier* population of the semiconductor *channel* while an insulator (such as *silicon dioxide*) is sandwiched between the two (in the case of *MOSFETs*[29] or generally speaking *MISFETs*[30])
- The capacitive effect of the *gate* electrode on the *carrier* population of an *undoped* semiconductor *channel* while another often *doped* semiconductor, either sandwiched between the two and/or placed below the semiconductor *channel* layer, confines the electron population of the *channel* at the *heterointerface* of the two semiconductors (in case of *HFETs*[31])

What is common between all these *field effect transistors* is that the *gate* terminal is ideally expected to have *zero* direct current. Due to that reason, the general name for these devices is *insulated gate field effect transistor* (i.e., *IGFET*).

The most common *field effect transistor* is the *MOSFET*. The popularity of *silicon* technology, especially in the form of *silicon MOSFETs*, has been mostly due to the

[27] *Metal–semiconductor field effect transistor.*
[28] *Junction field effect transistor.*
[29] *Metal–oxide–semiconductor field effect transistor.*
[30] *Metal–insulator–semiconductor field effect transistor.*
[31] *Heterojunction field effect transistor.*

very special material properties of the interface of *monocrystalline silicon* and *amorphous* SiO_2 grown by *thermal oxidation*. The interface between these two widely different materials provides one of the best of *heterointerfaces* in terms of low *trap density* and superb mechanical strength. The possibility of forming this *heterointerface* played a tremendous role in shaping up the *planar* technology. For this reason, in this part of Chapter 2, we delve into drawing the *band diagram* of an *MOS* system and in light of that examine its properties. In the later chapters, on the basis of this *band diagram* and other information given in this part of the chapter, we will study a large variety of *field effect transistors*.

As will be seen later in this volume, whereas enhancing the speed of *silicon MOSFETs* requires the thickness of SiO_2 to be reduced, a substantial reduction in the *gate* insulator's thickness renders some of the findings of this section irrelevant. These issues will be dealt with in appropriate places of Chapters 3 and 4. In the meanwhile, issues raised by the reduced thickness of gate SiO_2 of *MOSFETs* have caused SiO_2 to be replaced by oxides of higher *dielectric constants* such as *silicon oxi-nitrides*, *tantalum pentoxide*, and *hafnium*-based oxides. While the insulator/semiconductor interfaces realized with these dielectrics are much less perfect than the Si/SiO_2 interface, most of this section's descriptions of *MOS* systems are readily extendable to these structures.

2.4.1 Thermal-Equilibrium Band Diagram of *MOS* System

In our treatment of the *MOS* system, we start from a system under *thermal equilibrium* (i.e., constant *Fermi level* throughout the system). If the metal and the semiconductor were to be isolated from one another (i.e., in the ideal classical limit of *zero* current across the insulator and absence of a conductive path between metal and semiconductor), reaching this state will be impossible. Under these conditions, the *Fermi levels* of the metal and semiconductor will remain unequal even within the context of the contact. This preservation of *nonequilibrium* status is often used in *nonvolatile* memory cells.

Imperfections of the insulator and the resulting presence of a finite but limited current across this layer facilitate the establishment of *thermal equilibrium* across the *MOS* system even in the absence of a direct conductive path between metal and semiconductor. Whereas this statement depicts a correct picture within the classical realm, exponential improvement of the probability of *quantum tunneling* with reduction of the barrier thickness provides a path for establishing *thermal equilibrium* between metal and semiconductor when the oxide is ideal.

In the present treatment of *silicon MOS* systems, we will start at an ideal limit where the charge concentration within the oxide layer, and also at the two interfaces of the system, is assumed to be 0. Following this ideal treatment, impacts of the presence of charge in these regions will be dealt with.

In drawing the *thermal-equilibrium band diagram* of this *MOS* system, the notions of *bandgap* and *electron affinity*, which were earlier developed for crystalline semiconductors, will also be attributed to the *amorphous* SiO_2. This is a stretch from reality. However, this stretch has been proven affordable and useful.

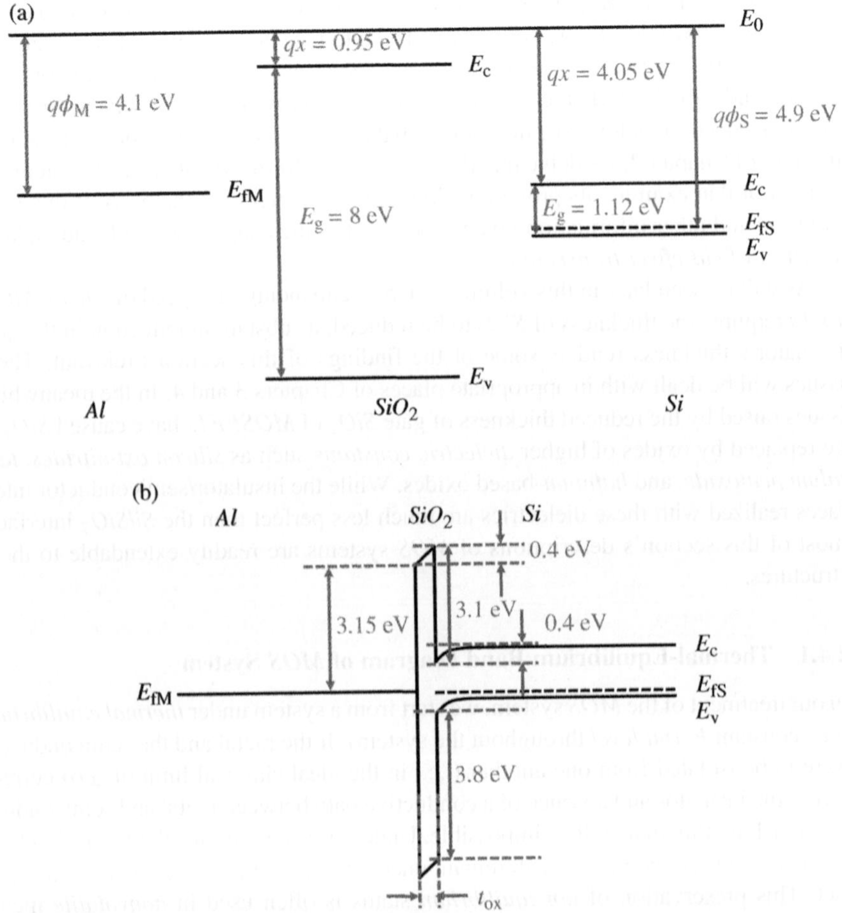

FIGURE 2.30 (a) Bandgap information of the constituent parts of the *MOS* system described in the text. (b) Thermal-equilibrium band diagram of the aforementioned *MOS* system.

Figure 2.30a illustrates the *electron affinity*, *work function*, and *bandgap* values for the three constituents of an *MOS* system at room temperature. The depicted metal is *aluminum* and the *silicon* is uniformly p-type *doped* with $N_a \sim 10^{15}$ cm^{-3}. In the discussions of Section 2.4, we assume a p-type semiconductor. The findings of this section, however, are readily extendable to *MOS* systems where semiconductor is n-type *doped*.

In the presence of a conductive path between metal and semiconductor, when the three materials depicted in Figure 2.30a are brought together, spontaneous electron transfer from aluminum to p-type *silicon* is prompted to establish *thermal equilibrium*. This is because, as shown in Figure 2.30a, the semiconductor *work function* is 0.8 eV larger than that of the metal. This electron transfer results in *electron–hole pair recombination* at the *Si/SiO₂* interface and formation of a negatively charged *space-charge* region at this interface. Assuming that the insulator and the interfaces are devoid of any charge concentration, the resulting formation of the *space-charge* region will be balanced by a positive-charge profile formed at the metal side of the metal/insulator interface.

As a result of the evolution of this *space-charge* region, the 0.8 eV *work-function* difference between the metal and the semiconductor will drop across the oxide and the *space-charge* region. The *space-charge* region and the electric field developed across it can be simply treated on the basis of our discussions in Section 2.2. On that basis, while in the absence of charge concentration within the oxide, ideally the potential drop V_{ox} across the oxide induces a constant electric field across this layer, continuity of *electric-flux density* at the interface between the oxide (i.e., D_{ox}) and the semiconductor (i.e., D_s defined at $x=0$) demands that

$$\epsilon_{ox}E_{ox} = \epsilon_{ox}\frac{V_{ox}}{t_{ox}} = \epsilon_s E_s(x=0) \tag{2.151}$$

where t_{ox}, ϵ_{ox}, and ϵ_s are the thickness of the oxide, *permittivity* of the oxide, and *permittivity* of the semiconductor, respectively. According to (2.9) and (2.13), for a *band bending* of $q\phi_i$ across the semiconductor,

$$E_s(x=0) = \sqrt{\frac{2qN_a\phi_i}{\epsilon_s}}. \tag{2.152}$$

For an oxide thickness of about 110 nm (and considering the *relative permittivity* of *silicon* and *SiO*$_2$ of 11.7 and 3.9, respectively), half of the *work-function* difference will drop across the oxide (i.e., $V_{ox}=0.4$ V), while the other half produces the *band bending* depicted in Figure 2.30b across the semiconductor layer (i.e., $\phi_i=0.4$ V). As observed through these calculations, in comparison to a *blocking MS* junction, the presence of the oxide layer in the *MOS* system reduces the surface electric field through distancing the two charge profiles.

Example

For the *MOS* system described in Figure 2.30, evaluate the proportion of the *work-function* difference dropping across the oxide if only *aluminum* were to be replaced by another metal of *work function* 4.5 eV.

The *work-function* difference in this case is equal to 0.3 eV. Assuming ϕ_i to be the amount of *band bending* on the semiconductor side, according to (2.152) the maximum electric field developed at the *SiO*$_2$/*Si* interface is given by

$$E_s(x=0) = q\frac{N_a x_d}{\epsilon_s} = \sqrt{\frac{2qN_a\phi_i}{\epsilon_s}} = \sqrt{\frac{2\times 1.6\times 10^{-19}\times 10^{15}}{11.7\times 8.85\times 10^{-14}}\phi_i}.$$

Through enforcing the continuity of the electric field across the *SiO*$_2$/*Si* interface, (2.151) yields

$$3.9\times\frac{V_{ox}}{110\times 10^{-7}} = 11.7\times\sqrt{\frac{2\times 1.6\times 10^{-19}\times 10^{15}}{11.7\times 8.85\times 10^{-14}}\phi_i}.$$

Having $V_{ox}+\phi_i=0.3$ V yields V_{ox} of almost 0.2 V.

2.4.2 Biased *MOS* System

In the description of a *biased MOS* system, we continue to assume that the oxide layer, like a classical insulator, is not passing any *DC* current across it. When *biasing* the *MOS* system of Figure 2.30 (i.e., through applying a potential difference V_{MS} between metal and semiconductor), *thermal equilibrium* between these two constituents of the system will be broken. In drawing the *band diagram* of the *biased* system, whereas the *conduction-* and *valence-band* discontinuities (i.e., ΔE_c and ΔE_v, respectively) between oxide and semiconductor and the potential barrier height at the metal/oxide interface remain unchanged, the amount of the applied *bias* voltage will be divided between changes in the *band bending* across the oxide and the surface of the semiconductor. As noted in Section 2.4.1, the proportions of this division are calculated through invoking the continuity of *electric-flux density* at the insulator/semiconductor interface. So far, this description is not very different from what we saw earlier in Section 2.2 in terms of *biasing* a *blocking MS* junction. What is different between an *MOS* system and a *blocking MS* system is that in the former, due to lack of charge transfer across the oxide, *equilibrium* between the *surface* and *bulk* of the semiconductor often persists even after applying a voltage between the metal and the semiconductor. According to this description, whereas applying a *bias* (i.e., V_{MS}) between the metal and semiconductor dissociates the *Fermi levels* of the metal and the semiconductor with an energy separation defined by qV_{MS}, a constant semiconductor *Fermi level* is often defined all the way to the surface of the semiconductor.

According to this description, as illustrated in Figure 2.31a, whereas application of an MS *bias* of −0.8 V produces *zero band banding* across the oxide and the semiconductor (a condition that is referred to as *flat band*), applying a more negative *bias* produces an excess concentration of *holes* at the surface of the p-type semiconductor (this condition of *bias* is known as *accumulation*). The required *bias* to produce flat *bands* in oxide and semiconductor is known as *flat-band* voltage (i.e., ϕ_{FB}). If the semiconductor is p-type *doped*, a sufficient source of *holes* is present to yield this state of *equilibrium*. The presence of this surface *accumulation layer*, which is confined in an almost triangular *quantum well* formed in the *valence band* of insulator/semiconductor interface, is often expressed in terms of a parallel-plate capacitor between metal and semiconductor. In other words, the thickness of this layer is approximated by 0 (i.e., *accumulation* charge is presented as a sheet of positive *charge carriers*). Although often a valid approximation, this is not quantum mechanically correct. According to the confinement of *holes* indicated in Figure 2.31b, quantum mechanics predicts the *hole wave function* to maximize at a certain distance from the insulator/semiconductor interface (i.e., Δ). In terms of this separation and the finite width of the *accumulation layer*, a *quantum capacitance* (i.e., ϵ_s/Δ in F cm^{-2}) should be considered in series with the *oxide capacitance* (i.e., ϵ_{ox}/t_{ox} in F cm^{-2}).[32]

Following this procedure for drawing the *band diagram* when a *bias* is more positive than the *flat-band* voltage is applied, however, runs into a certain conundrum. As shown in Figure 2.31c, under this *bias* condition, for smaller *bias* voltages a region of

[32] More will be said on this topic in Section 3.3.

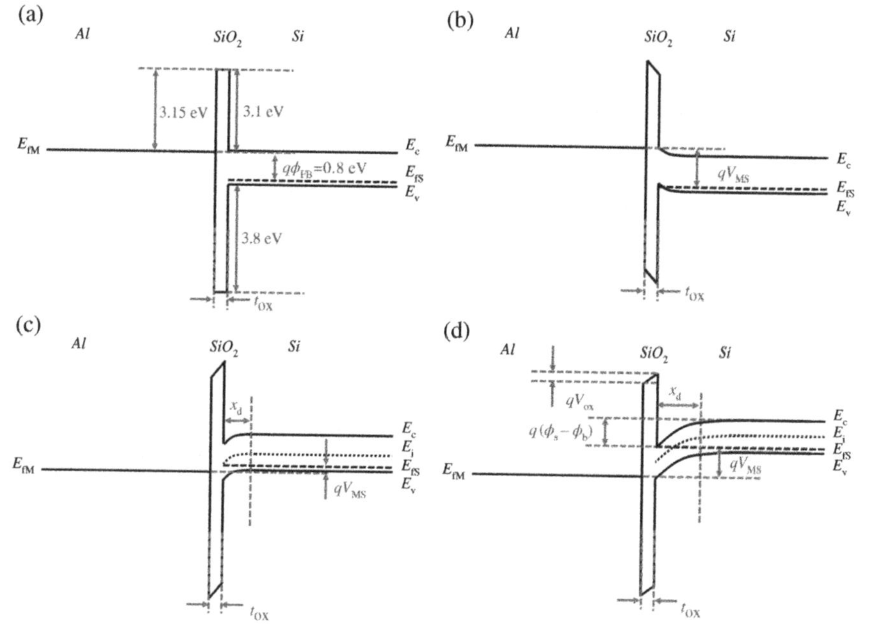

FIGURE 2.31 (a) Flat-band band diagram of the *MOS* system of Figure 2.30b when semiconductor is maintained under thermal equilibrium. (b) Band diagram of the *MOS* system of Figure 2.30b when semiconductor is maintained under thermal equilibrium and biased in accumulation. (c) Band diagram of the *MOS* system of Figure 2.30b when semiconductor is maintained under thermal equilibrium and biased in depletion. (d) Band diagram of the *MOS* system of Figure 2.30b when semiconductor is maintained under thermal equilibrium and biased in inversion.

reduced *hole* concentration will evolve at the surface of the semiconductor (i.e., *depletion* region). This *bias* regime is referred to as *depletion*. Continuing to increase the *bias* voltage, however, results in elevation of the *Fermi level* above the *intrinsic Fermi level*, which indicates an n-type character for the *surface* of the semiconductor (Fig. 2.31d). For that reason, this *bias* condition, which causes the induction of a majority electron population at the *p-doped surface*, is known as *inversion*. The electron population of this *inversion layer* as shown in Figure 2.31d is confined to an almost triangular *quantum well* formed in the *conduction band* of the insulator/semiconductor interface.

As expressed in Section 1.3, electron concentration of this *inversion layer* will be exponentially dependent on potential drop across the semiconductor. In the presence of a highly populated *inversion layer*, not unlike the *accumulation layer*, this negatively charged layer can be approximately treated through the assumption of the *oxide capacitance*. However, as shown in Figure 2.31d, electrons are not the only negative charges on the semiconductor side of the *MOS capacitance*. Uncompensated negatively charged *ionized impurities* of the *depletion* region provide an additional profile of negative charges in the semiconductor.

The *bias* condition for which the same (but opposite) concentration of *carriers* in the *inversion layer*, as that of the *bulk* semiconductor, is induced is referred to as the *onset of strong inversion*. Due to the exponential dependence of the *inversion-layer* electron concentration on the position of the *Fermi level*, the position of this *level* at the surface of the semiconductor can be safely assumed to be quite constant beyond the *onset of strong inversion*. As a result, beyond this limit, an increase in the *bias* tends to have negligible impact on the size of the *depletion region* since any voltage in excess of this value is dropped across the insulator. Increase in the voltage drop across the oxide, while ideally increasing the electron population of the *inversion layer*, due to the exponential dependence of electron concentration on *bias*, changes the *surface* potential only negligibly from its value at the *onset of strong inversion*.

Whereas assumption of *equilibrium* across the semiconductor proposes the evolution of the *inversion layer*, the p-type nature of the semiconductor poses a certain difficulty in sourcing sufficient concentration of electrons to this layer. This is the cause of the aforementioned conundrum. While the generation process can be envisioned as such a source, the slow *rate* of this process prevents it from being efficiently capable of reestablishing *equilibrium* between the semiconductor's *bulk* and *surface* in practical *MOS* systems. It is only in the presence of a *highly doped* n^+ region in the *unblocked* vicinity of the *MOS* system that the required number of electrons for forming the *inversion layer* can be sourced into the semiconductor *surface*. This is especially true for *strong inversion*. Without the presence of this source of minority *carriers*, *equilibrium* between *bulk* and *surface* of the semiconductor is not established. Under this latter condition, the negative charge concentration of the *MOS capacitance* is purely provided in terms of the *depletion*-region charge. As a result, *MOS capacitance*, given by the series combination of the *oxide* and the *depletion capacitances*, is primarily defined by the latter, which is of reducing orders with increase in the positive *bias* across the *MOS* system. This *bias* condition is known as *deep depletion*.

According to the explanations presented in this section, in the presence of an *unblocked* source of minority *carriers*, depending on the polarity and the amount of potential difference applied between the metal and the semiconductor, the *MOS* system can be *biased* in three different regimes:

1. *Accumulation*
2. *Depletion*
3. *Inversion*

2.4.2.1 Biased* MOS *System: Semiconductor under Equilibrium Table 2.1 summarizes the aforementioned *biasing* scenarios. As will be shown in Chapter 3, in a *MOSFET* an *unblocked* source of minority *carriers* is present. In the rest of the discussions of this section, we will assume this presence as granted.

As indicated in the aforementioned qualitative description of the three regimes of operation of an *MOS* capacitor, we can quantify the operation of an *MOS* system under

TABLE 2.1 Summary of the Biasing Scenarios of an *MIS* System Made to a p-Type Semiconductor

Bias Condition in Terms of: $V_{MS} - \phi_{FB}$	Polarity of the Surface Potential ϕ_S	Charge Condition of the Surface	Majority Carrier Concentration and Type at the Surface
Negative	Negative	Accumulation	$p_s > N_a$ p-Type surface
0	Negative	Neutral	$p_s = N_a$ p-Type surface
Positive (small)	Negative	Depletion	$n_i < p_s < N_a$ p-Type surface
Positive (larger)	0	Intrinsic	$p_s = n_s = n_i$ Intrinsic surface
Positive (larger)	Positive	Week inversion	$n_i < n_s < N_a$ n-Type surface
Positive (larger)	Positive	Onset of strong inversion	$n_s = N_a$ n-Type surface
Positive (larger)	Positive	Strong inversion	$n_s > N_a$ n-Type surface

the *accumulation*, *depletion*, and *inversion* modes through tracking the *surface* versus *bulk* potential in the semiconductor.

Where potential is defined as

$$\phi(x) = \frac{1}{q}[E_f - E_i(x)], \tag{2.153}$$

we can identify *surface* potential ϕ_s as

$$\phi_s = \phi(0) = \frac{1}{q}[E_f - E_i(0)]. \tag{2.154}$$

Likewise, the *bulk* potential is given by

$$\phi_b = \phi(\infty) = \frac{1}{q}[E_f - E_i(\infty)]. \tag{2.155}$$

In the case of the *MOS* system made to a p-type semiconductor, the value of ϕ_b (which is the potential in the neutral *bulk* of the semiconductor) is negative, while it takes positive values for an n-type *bulk*. Obviously under *inversion*, ϕ_s will have the opposite polarity of ϕ_b.

As a result of these definitions, in terms of the *acceptor* concentration of the *bulk* (i.e., N_a), the *surface carrier* concentrations can be calculated as

$$p_s = N_a \exp\left[\frac{q(\phi_b - \phi_s)}{kT}\right] \tag{2.156}$$

and

$$n_s = \frac{n_i^2}{N_a} \exp\left[\frac{q(\phi_s - \phi_b)}{kT}\right]. \tag{2.157}$$

As it was qualitatively described earlier, the exponential behaviors expressed in (2.156) and (2.157) reveal that under *strong inversion* or *accumulation*, *surface* potential remains relatively constant. In the case of *strong inversion*, this *surface* potential is quite often indicated by the *onset of strong inversion* (i.e., $\phi_s = -\phi_b$).

In quantifying the *depletion* regime, as a crude approximation, we often assume this mode to be dominant for all values of *bias* between the *onset of strong inversion* and *accumulation*. In this approximation of the *depletion* regime of operation of an *MOS* system, the only *space charge* assumed in this *bias* range is the *depletion* charge of *ionized impurities*. In other words, before the *onset of strong inversion*, excess electron population at the *surface* of the semiconductor is overlooked in first-order approximations. In light of the interest in the operation of *MOSFETs* between *inversion* and *depletion* modes, however, a more accurate approximation is often called for. This need is especially further pronounced by the popularity of the so-called *subthreshold*[33] operation of *MOSFETs*. Based on Figure 2.31, it is clear that

$$V_{MS} - \phi_{FB} = V_{ox} + \phi_s - \phi_b \tag{2.158}$$

where V_{ox} is the potential drop across the oxide and $\phi_s - \phi_b$ represents the potential difference between *bulk* and *surface* of the semiconductor.

Assuming the oxide to be a perfect insulator, the constant electric field across this layer is given by

$$E_{ox} = \frac{V_{ox}}{t_{ox}} = \frac{V_{MS} - \phi_{FB} + \phi_b - \phi_s}{t_{ox}}. \tag{2.159}$$

As a result, based on the continuity of *electric-flux density*, the electric field right at the semiconductor side of the interface will be

$$E_s(x=0) = \frac{\epsilon_{ox} E_{ox}}{\epsilon_s}. \tag{2.160}$$

Using the definition of *oxide capacitance* as

$$C_{ox} = \frac{\epsilon_{ox}}{t_{ox}}, \tag{2.161}$$

we will have

$$\epsilon_s E_s(x=0) = C_{ox}[V_{MS} - \phi_{FB} + \phi_b - \phi_s]. \tag{2.162}$$

[33] This will be defined in Chapter 3.

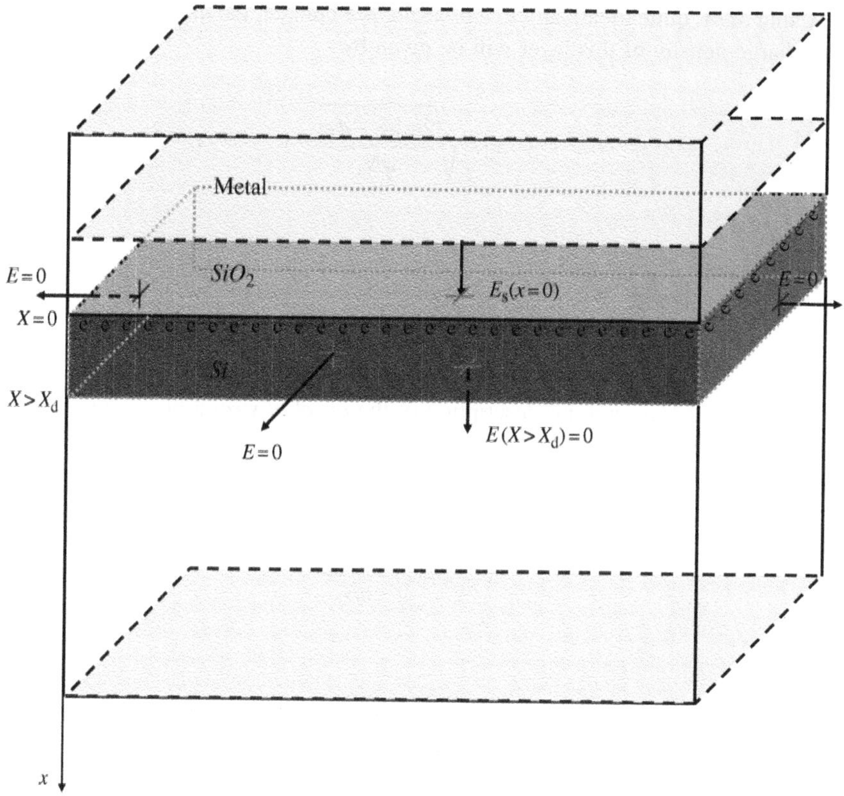

FIGURE 2.32 Schematic depiction of the *Gaussian* box referred to in the text in exploring the inversion characteristics of an *MOS* system. The boundaries of this box are indicated in gray. The box is defined within the semiconductor part of the *MOS* system. The inversion-layer charge is indicated by e^-'s. According to *SDA*, in this case a depletion region of charge concentration qN_a is assumed to extend from $x=0$ to $x=x_d$.

Imagining a *Gaussian* box encompassing the total *space-charge* profile (i.e., *depletion* and *inversion* profiles), where the *zero* electric field at the bottom of this box serves as a boundary condition in applying *Gauss*'s law, *Gauss*'s law yields

$$-\epsilon_s E_s(x=0) = Q_n + Q_d \tag{2.163}$$

where Q_n and Q_d stand for the per unit area charge density of the *inversion* and the *depletion* charge profiles, respectively. This *Gaussian* box is shown in Figure 2.32.

According to (2.162) and (2.163),

$$Q_n = -C_{ox}[V_{MS} - \phi_{FB} + \phi_b - \phi_s] - Q_d. \tag{2.164}$$

Using *SDA*, quite identically to a *blocking MS* contact, the *depletion-region width* and charge density of this layer will be given by

$$x_{\mathrm{d}} = \sqrt{\frac{2\epsilon_{\mathrm{s}}|\phi_{\mathrm{s}} - \phi_{\mathrm{b}}|}{qN_{\mathrm{a}}}} \tag{2.165}$$

and

$$Q_{\mathrm{d}} = -\sqrt{2\epsilon_{\mathrm{s}} qN_{\mathrm{a}}|\phi_{\mathrm{s}} - \phi_{\mathrm{b}}|} \tag{2.166}$$

where the substrate is simply assumed to be uniformly *doped.*

Based on (2.165) and our description of the *onset of strong inversion* as the *bias* condition beyond which the *surface* potential remains constant, the maximum value of the *depletion-region width* is given by

$$x_{\mathrm{d\text{-}max}} = \sqrt{\frac{4\epsilon_{\mathrm{s}}|\phi_{\mathrm{b}}|}{qN_{\mathrm{a}}}} \tag{2.167}$$

for which

$$Q_{\mathrm{d\text{-}max}} = -\sqrt{4\epsilon_{\mathrm{s}} qN_{\mathrm{a}}|\phi_{\mathrm{b}}|}. \tag{2.168}$$

Beyond the *onset of strong inversion*, while Q_{d} in (2.164) is replaced by $Q_{\mathrm{d\text{-}max}}$, increase in V_{MS} results in a linear increase of Q_{n}. This framework lacks any quantum mechanical insight into the evolution of *quantum capacitance* and *subbands* at the insulator/semiconductor interface. Although a good approximation in many applications (and for an acceptable range of values of V_{MS}), the lack of representation of these factors limits the applicable range of (2.164).

2.4.2.2 ***Biased* MOS *System: Semiconductor under Nonequilibrium*** In the presence of the aforementioned source of minority *carriers*, which essentially yields the possibility to form an *inversion* layer at the insulator/semiconductor interface, one can also envision the opportunity to disturb the *equilibrium* state of the semiconductor by making an *Ohmic* contact to this charge reservoir. As shown in Figure 2.33, such a *nonthermal-equilibrium* state is established by *biasing* the *p–n* junction between the charge reservoir and the p-type *bulk* semiconductor. Since the *carrier* concentration of the *inversion* layer originates from the charge reservoir, these two high *carrier* concentration regions are effectively *shorted* together. As a result, *reverse biasing* the *p–n* junction formed between the charge reservoir and the *bulk* semiconductor removes the *equilibrium* status of the semiconductor through depleting the *excess* electrons required for forming the *inversion* layer. This *carrier* depletion essentially

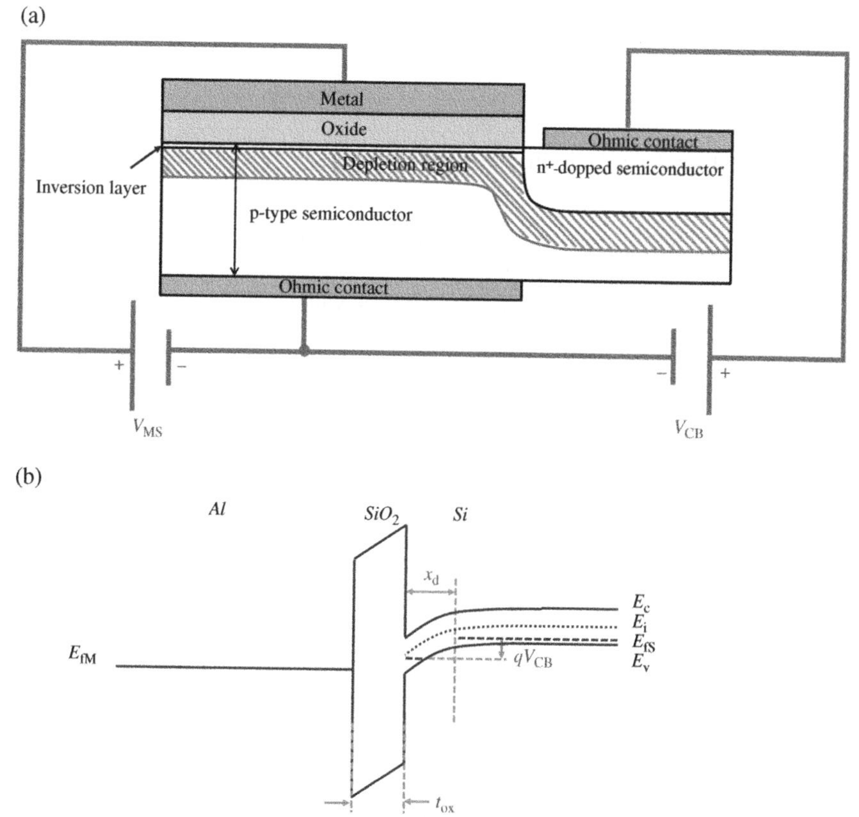

FIGURE 2.33 (a) Schematic depiction of the bias configuration between the charge reservoir and the *MOS* capacitance expressed in the text. (b) Resulting band diagram from the bias configuration indicated in (a). As indicated by the discontinuity of Fermi level across the semiconductor, the state of thermal equilibrium across this layer is disrupted upon application of V_{CB}.

postpones the *inversion* to a higher positive[34] potential difference between metal and semiconductor (i.e., V_{MS}) in the *MOS* capacitor. In terms of the *reverse bias* V_{CB} (i.e., between the *carrier* reservoir and the semiconductor *bulk*), the maximum *depletion-region width* will become larger than the value calculated at *equilibrium* (i.e., the value expressed by (2.167)),

$$x_d = \sqrt{\frac{2\epsilon_s(2|\phi_b| + V_{CB})}{qN_a}}. \tag{2.169}$$

[34] This is obviously in reference to the p-type *bulk*.

For this *depletion-region width,* (2.168) modifies to

$$Q_d = -\sqrt{2\epsilon_s q N_a(2|\phi_b| + V_{CB})}. \tag{2.170}$$

As a result, under *strong inversion* (i.e., $\phi_s = |\phi_b| + V_{CB}$), based on (2.164),

$$Q_n = -C_{ox}[V_{MS} - \phi_{FB} - V_{CB} - 2|\phi_b|] + \sqrt{2\epsilon_s q N_a(2|\phi_b| + V_{CB})}. \tag{2.171}$$

Equation (2.171) allows the calculation of the value of V_{MS} (also referred to as the *gate* voltage V_G in the terminology of *MOS* systems) necessary for inducing the *inversion* layer. This value, which is known as the *threshold voltage* V_T, is identified by the *gate* voltage yielding $Q_n = 0$,

$$V_T = \phi_{FB} + V_{CB} + 2|\phi_b| + \frac{1}{C_{ox}}\sqrt{2\epsilon_s q N_a(2|\phi_b| + V_{CB})}. \tag{2.172}$$

In terms of this definition, under the regime of *strong inversion*, the *inversion* charge density can be simply defined as

$$Q_n = -C_{ox}(V_G - V_T). \tag{2.173}$$

This is a valid approximation especially when V_G is not too close to the *threshold voltage*. For small positive values of $V_G - V_T$, and even more so for negative values of this quantity (i.e., a condition referred to as *subthreshold*), this approximation fails the test of accuracy. This inaccuracy is caused by the almost arbitrary definition of *strong inversion*. As we took note earlier, while in reality the formation of the *inversion* layer happens gradually with the increase of the *gate* voltage, we have assumed that there are no electrons at the insulator/semiconductor interface before the *onset of strong inversion*. Alleviation of this problem requires an exact solution of the *Poisson* equation. Nowadays, the requirement of the reduction of power dissipation of *ULSI*[35] circuits is driving circuit designers to operate transistors under *subthreshold* conditions. In this regime of operation, relationships such as (2.172) and (2.173) are not sufficient.

2.4.3 Threshold-Voltage Adjustment and Calculations

The *threshold voltage* of an *MOS capacitor* is a very significant factor in the operation of a *MOSFET*. As a result, accurate evaluation of this parameter in terms of the process information is pivotal to successful design of a process flow for realization of *MOSFETs*. Whereas so far we have assumed only a uniform *dopant* concentration across the semiconductor, during the *microfabrication* process we are often in need of *ion implanting* additional *doping* profiles close to the insulator/semiconductor

[35] Stands for *ultra large-scale integrated.*

interface to adjust the value of the *threshold voltage*. According to (2.172), besides the dependence of Q_d on the *dopant* concentration, other factors on which the *threshold voltage* is dependent include metal *work function*, *oxide capacitance*, and potential of the neutral *bulk*. Among these factors, while the *bulk* potential is only logarithmically variable with the *dopant* concentration, choices of insulator thickness/*permittivity* and metal *work function* are predetermined by other requirements of the technology (such as the possibility to yield a low-*defect* insulator/semiconductor interface, high-temperature stability, and high *conductivity* of the *gate* contact).

In this section, based on the descriptions presented in Section 2.4.2.2, a framework for calculation of the *threshold voltage* on the basis of *ion-implanted threshold-voltage adjustment dopant* distribution profiles is provided.

We start the presentation of this framework with an overview of the process of *ion-implantation doping*. In this process, the surface of the *solid* is bombarded by accelerated ions of an ion beam. While penetrating the surface of the *solid*, the energetic ions, through interactions with the *solid*, eventually lose their kinetic energy and stop unless they are redistributed through a *thermal-diffusion* process. The profile of *dopant* distribution according to this technique is quite realistically presented in the form of a *Gaussian* distribution,

$$N(x) = N_p \cdot \exp\left(-\frac{(x-R_p)^2}{2\Delta R_p^{\,2}}\right) \tag{2.174}$$

where N_p, R_p, and ΔR_p are the peak, mean depth (known as *projected range*), and standard deviation (known as vertical *straggle*) of the distribution, respectively. The *projected range* and *straggle* can be controlled through the acceleration energy of the implanted ions, while N_p is monitored by measuring the ionic current of the implanter beam. According to this distribution, for a sufficiently energetic ion beam where R_p is larger than a few times ΔR_p plus the thickness of the insulator through which oftentimes implantation is taking place (and hence both of the tails of the distribution profile of (2.174) fall inside the semiconductor), the value of per unit area density of *implanted* ions (also known as *dose*) is given by

$$Q_{II} = \sqrt{2\pi}\Delta R_p N_p. \tag{2.175}$$

In describing the *threshold-voltage adjustment* calculation framework, the shallow[36] implant seen in (2.174) is often expressed by a rectangular distribution profile (Fig. 2.34). For the conditions expressed previously, in this piecewise distribution profile,

$$N_s = \frac{1}{2}N_p \tag{2.176}$$

[36] Since the implant is done in the vicinity of the insulator/semiconductor interface.

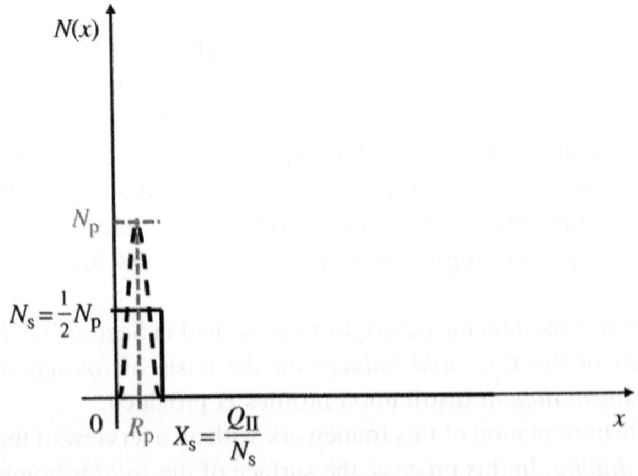

FIGURE 2.34 Piecewise linear approximation of the ion-implanted threshold-voltage adjustment dopant distribution profile.

and

$$X_s = \frac{Q_{II}}{N_s}. \tag{2.177}$$

In order to induce a meaningful change to the *threshold voltage*, N_s must be considerably larger than the background *dopant* concentration (i.e., N_a).

In terms of the rectangular distribution of Figure 2.34, three cases are identified in the evaluation of *threshold-voltage* adjustment:

CASE (1):

When $X_s \ll x_{d\text{-max}}$, the *ion-implanted threshold-voltage adjustment* profile can be seen as a *delta-doped*[37] layer at the insulator/semiconductor interface. In this case, the predominant change in the value of V_T is determined by $\pm q(Q_{II}/C_{ox})$. The sign of this adjustment is determined by the *type* of *dopants*. In the case of a p-type *bulk*, *acceptor* implant yields a positive shift in *threshold voltage*. In the case of *donor* implant, the shift is negative. Evidently, the trends are opposite in the case of an n-type *bulk*.

[37] That is, a very thin layer of *dopants*, resembling a delta-function distribution profile.

CASE (2):

When $X_s < X_{d\text{-test}}$, where

$$X_{d\text{-test}} = \sqrt{\frac{4\epsilon_s|\phi_{test}|}{qN_s}} \text{ and } |\phi_{test}| = \frac{kT}{q} Ln\left(\frac{N_s}{n_i}\right). \tag{2.178}$$

A comparison to test values $X_{d\text{-test}}$ and ϕ_{test} is aimed at assessing the possibility of the *ion-implanted threshold-voltage adjustment* profile to overwhelm the *bulk dopant* concentration throughout the width of the *depletion* region. As illustrated in Figure 2.35, this is evidently not the case. According to Figure 2.35, the *Poisson* equation is used to calculate the surface potential ϕ_s

$$\begin{cases} E(x) = \dfrac{qN_sX_s}{\epsilon_s} + \dfrac{qN_ax_d}{\epsilon_s} - \dfrac{q(N_s+N_a)x}{\epsilon_s} & \text{for } 0<x<X_s \\ E(x) = \dfrac{qN_a(x_d-x)}{\epsilon_s} & \text{for } X_s<x<x_d \end{cases}. \tag{2.179}$$

Using these electric-field profiles, the variation of potential can be formulated as

$$\begin{cases} \phi(x) = -\dfrac{q}{\epsilon_s}(N_sX_s + N_ax_d)x + \dfrac{1}{2}q\left(\dfrac{N_s+N_a}{\epsilon_s}\right)x^2 + \phi_s & \text{for } 0<x<X_s \\ \phi(x) = \dfrac{qN_a}{\epsilon_s}\left(\dfrac{1}{2}x^2 - x\cdot x_d + \dfrac{1}{2}x_d^2\right) + \phi_b & \text{for } X_s<x<x_d \end{cases}. \tag{2.180}$$

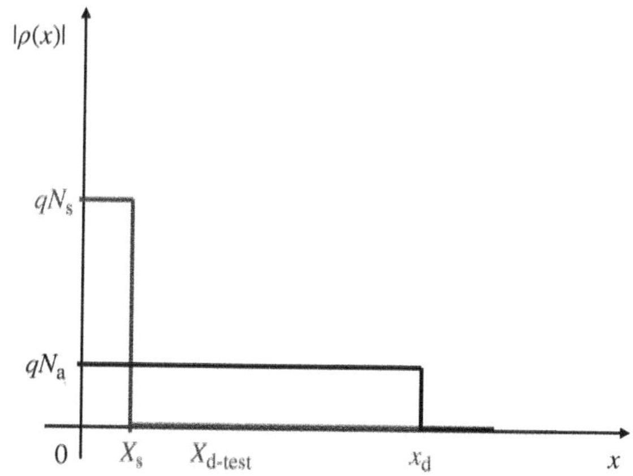

FIGURE 2.35 Charge concentration profile governing the circumstances in evaluating the threshold voltage under Case (2).

According to (2.180),

$$\phi_s = \frac{qN_a}{2\epsilon_s}x_d^2 + \frac{qN_s}{2\epsilon_s}X_s^2 + \phi_b. \tag{2.181}$$

By comparing (2.181) to Equation (2.165), the second term on the right-hand side of (2.181) is identified as the contribution of the shallow *threshold-voltage adjustment implant* to the *surface* potential. According to this insight, at the *onset of strong inversion* (i.e., $\phi_s = -\phi_b$),

$$x_{d\text{-max}} = \sqrt{\frac{2\epsilon_s}{qN_a}\left(-\frac{qN_s}{2\epsilon_s}X_s^2 - 2\phi_b\right)}. \tag{2.182}$$

Accordingly, we will have

$$Q_n = -C_{ox}[V_{MS} - \phi_{FB} + \phi_b - \phi_s] - Q_d + qQ_{II} \tag{2.183}$$

where Q_d still represents the *depletion-layer* charge density calculated purely in terms of the *bulk dopant* concentration (i.e., N_a) and Q_{II} is the implanted *dose*. Incorporating (2.166), (2.181), and (2.182) into (2.183) yields

$$Q_n = -C_{ox}[V_{MS} - \phi_{FB} + 2\phi_b] + \sqrt{2\epsilon_s qN_a\left(-\frac{qN_s}{2\epsilon_s}X_s^2 - 2\phi_b\right)} + qQ_{II}. \tag{2.184}$$

Setting $Q_n = 0$ results in the definition of the *threshold voltage*

$$V_T = \phi_{FB} - 2\phi_b + \frac{1}{C_{ox}}\sqrt{2\epsilon_s qN_a\left(-\frac{qN_s}{2\epsilon_s}X_s^2 - 2\phi_b\right)} + \frac{q}{C_{ox}}Q_{II}. \tag{2.185}$$

Equivalently, in the presence of *bias* V_{CB} between the charge reservoir and the semiconductor *bulk*,

$$V_T = \phi_{FB} - 2\phi_b + V_{CB} + \frac{1}{C_{ox}}\sqrt{2\epsilon_s qN_a\left(-\frac{qQ_{II}}{2\epsilon_s}X_s - 2\phi_b + V_{CB}\right)} + \frac{q}{C_{ox}}Q_{II}. \tag{2.186}$$

Comparing this definition to that of (2.172), the modification embodied by the last term of the expression is often dominant. As a result, *threshold-voltage* adjustment under Case (2) is almost linearly dependent on the *dose* of the shallow implant. The difference between Case (1) and Case (2) is only in terms of the fourth term of (2.186), which as mentioned earlier is often negligible. In this derivation, we have implicitly assumed that the *flat-band voltage* is quite independent of the *adjustment implant*. Since the position of the *Fermi level* with respect to E_c is only logarithmically dependent on the *dopant* concentration and in addition as the width of the *ion-implanted* profile is relatively small, the assumption of invariability of semiconductor *work function* is quite acceptable.

CASE (3):

When $X_s > X_{d\text{-test}}$, in this case, throughout the *depletion-region width*, the *threshold-voltage adjustment implant* is dominating the *dopant* concentration. As a result, according to (2.172),

$$V_T = \phi_{FB}(N_{tot}) + V_{CB} + 2|\phi_{test}| + \frac{1}{C_{ox}}\sqrt{2\epsilon_s q(N_a + N_s)(2|\phi_{test}| + V_{CB})} \quad (2.187)$$

or equivalently

$$V_T = \phi_{FB}(N_{tot}) + V_{CB} + 2|\phi_{test}| + \frac{1}{C_{ox}}\sqrt{2\epsilon_s q\left(N_a + \frac{Q_{II}}{X_s}\right)(2|\phi_{test}| + V_{CB})}. \quad (2.188)$$

In (2.187) and (2.188) the *flat-band voltage* is evaluated in terms of the *work-function* difference between metal and semiconductor where semiconductor *doping* is determined by the algebraic sum of N_a and N_s (i.e., N_{tot}). In terms of *Maxwell–Boltzmann* statistics, *metal work function* ($q\phi_M$), and semiconductor *electron affinity* ($q\chi$),

$$\phi_{FB}(N_{tot}) = \phi_M - \left[\chi + \frac{E_g}{2q} \pm \frac{kT}{q} Ln\left(\frac{\left|N_a \pm \frac{Q_{II}}{X_s}\right|}{n_i}\right)\right] \quad (2.189)$$

where the positive signs refer to an *acceptor-type ion-implanted* profile and negative signs refer to a *donor-type* profile of *adjustment implant*. It should be reemphasized that N_s is assumed to be greater than N_a.

Since ϕ_{FB} and ϕ_{test} are only logarithmically dependent on the *dose* of the *ion-implanted threshold-voltage adjustment* profile, under Case (3) the *threshold voltage* will be approximately changing in proportion to the square root of the *implanted dose*.

Whereas throughout the discussions of this section, we assumed the shallow *threshold-voltage adjustment implant* to reside fully within the semiconductor, *implants* of small energies are also sometimes used, which result in only a partial incorporation of the *adjustment dose* in the semiconductor. Since *implantation* takes place through the oxide layer, the rest of the *implanted dose* resides within this oxide film. In addition to this, due to exposure of the *ion-implanted* profile to high-temperature environments, throughout the course of the succeeding *microfabrication* steps, the *straggle* of the *ion-implanted* profile undergoes certain variations that are not yet incorporated into our calculation framework.

Under these circumstances, Equations (2.176) and (2.177) need to be modified. The extraction of the values of N_s and X_s, in terms of the *ion implantation* and all of the succeeding *thermal-diffusion* processes (which are also represented by *normal* distribution functions), is presented based on the assumption of these processes being independent. The overall *standard deviation* of the *Gaussian* distribution will therefore be given by

$$\sigma = \sqrt{\Delta R_p{}^2 + \sum_i 2D_i t_i} \tag{2.190}$$

where $D_i t_i$ is the product of the *diffusion constant* of the specific *implanted* impurity at the given temperature of the processing step i (which is succeeding the *ion implantation*) and the duration of that process.

When the value of the *projected range* (i.e., R_p) is larger than the thickness of the oxide through which *ion implantation* is carried out by approximately three times the standard deviation (i.e., σ), since approximately all of the *implanted* ions are residing within the semiconductor, equations quite identical to (2.176) and (2.177) can be used. The only change would be replacing the *straggle* in (2.175) with the standard deviation calculated according to (2.190) to identify the value of N_p.

In contrast to this case, when the *projection range* is equal to the thickness of oxide, while everything else remains the same, (2.177) modifies to

$$X_s = \frac{Q_{II}}{2N_s}. \tag{2.191}$$

The factor 2 in the denominator indicates that only half of the *implanted dose* is contributing to the adjustment of *threshold voltage*.

Obviously, as a generic case between these two situations (where either more or less than half of the *implanted dose* is residing within the semiconductor), we will have

$$N_0 = \frac{Q_{II}}{\sqrt{2\pi}\sigma} \exp\left(-\frac{(x_0 - R_p)^2}{2\sigma^2} \right) \tag{2.192}$$

where x_0 is the thickness of the oxide. According to N_0,

$$N_s = \frac{1}{2} N_0 \tag{2.193}$$

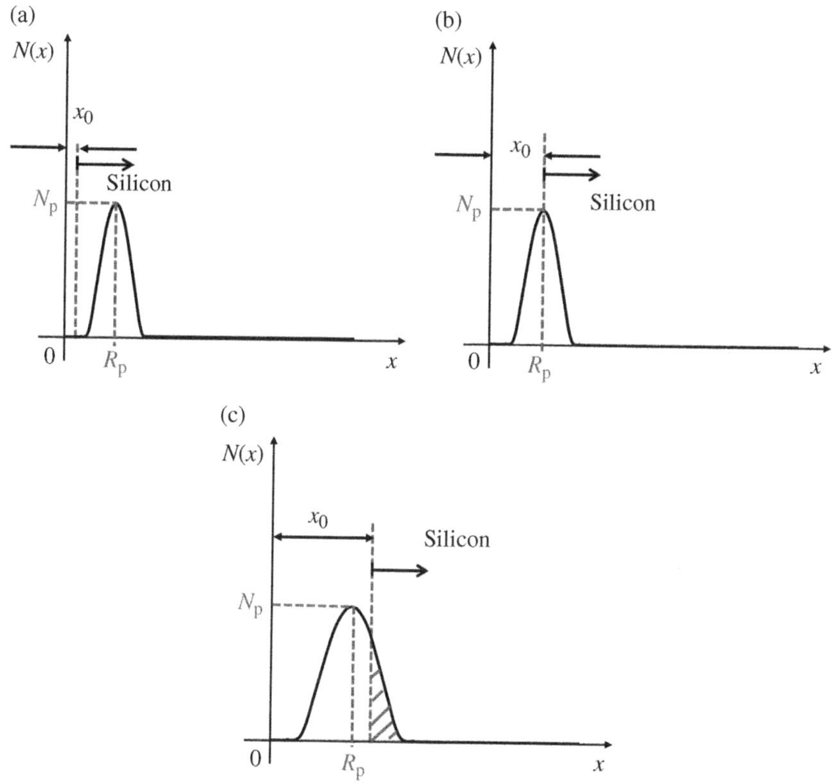

FIGURE 2.36 The three identified situations in calculating N_s and X_s: (a) when R_p is much larger than x_0, (b) when R_p is equal to x_0, and (c) when R_p is smaller than x_0. In (c) the portion of the dose implanted into silicon is indicated by the hash-marked area.

and

$$X_s = \frac{\displaystyle\int_{x_0}^{\infty} \frac{Q_{\text{II}}}{\sqrt{2\pi}\sigma} \exp\left(-\frac{(x-R_p)^2}{2\sigma^2}\right) dx}{N_s}. \tag{2.194}$$

The scenarios explained for the calculation of X_s and N_s are depicted in Figure 2.36. Equations (2.186) and (2.188), in terms of the *implanted dose*, need to be modified according to the assessment of the *ion-implanted* profile versus the situations presented in this figure.

2.4.4 *C–V* Characteristic of *MOS* Systems

Measurement and analysis of *small-signal capacitance* of *MOS* structures have been vital to understanding the physics of these structures and crucial in improving their fabrication technology. So far within the discussions of Section 2.4.2 regarding the modes of operation of *MOS* capacitors, we have made certain references to *bias* dependence of these capacitors. In this section we will delve into this topic at a greater depth. First, the *C–V* behavior of an ideal *MOS* system is examined. Following this discussion, a brief overview of different forms of nonidealities in *MOS* systems and their correlation with *C–V* characteristics is presented.

In the analysis of the *C–V* characteristics of *MOS* systems, we follow the same description as the one observed in *MS* and *p–n* junctions for the *small-signal capacitance*:

$$C = \frac{dQ}{dV} \tag{2.195}$$

where Q is the charge stored at the junctions of the *MOS* system and V represents the potential difference between metal and semiconductor. In the discussions of *C–V* characterization of *MOS* structures, it should be kept in mind that a small-amplitude *AC* signal overrides the *DC bias* applied for *biasing* the structure into its three operation regimes (i.e., *depletion, accumulation*, and *inversion*).

The intricacies in the analysis of *C–V* characteristics of *MOS* systems are rooted in the possibility of inverting the *carrier* population in these systems and in the large minority *carrier dielectric relaxation time constant* of an ideal *MOS* structure. In the analysis of *MS* or *p–n* junctions, we assumed the charge concentrations to almost instantaneously (i.e., with *dielectric relaxation time constants* in the order of *picoseconds*) respond to *small-signal* variation of voltage. However, in *MOS* systems certain components of the *space-charge* profile can be incapable of responding quickly enough. This fact results in a significant dependence of the *C–V* characteristics on the frequency of the *small-signal* voltage stimulus. Resulting cases of *MOS C–V* characteristics are summarized in Table 2.2.

According to the information presented in Table 2.2, depending on the *bias* condition, *C–V* characterization of *MOS* structure is performed under three different regimes of operation:

1. Low frequency
2. High frequency
3. Deep depletion

The demarcation between these three different *bias* conditions is established through distinguishing the *space-charge* distribution to which *AC* or *DC* coupling has been made. As noted earlier in Section 2.4.2, two components of the *space-charge* distribution are present in *MOS* systems:

1. One in the *depletion region* formed in the *bulk* of the semiconductor
2. One in the form of the *surface* charge concentration materialized at the insulator/semiconductor interface (i.e., *accumulation* or *inversion*)

TABLE 2.2 Summary of the Observations Made on the *C–V* Characteristics of an *MIS* System for V_{MS} beyond the Onset of Strong Inversion

Bias Condition	Equivalent Capacitance Seen across the *MIS* System	Description of the Charge Equilibrium
Low-frequency *C–V*	$C_{eq} = C_{ox}$	Charge distribution across the semiconductor side of the system is under equilibrium both for the *DC* and the *AC* signal. As a result, the *AC* signal can be mirrored on the surface concentration, which bears a closer proximity to the metal interface
High-frequency *C–V*	$C_{eq}^{-1} = C_{ox}^{-1} + \frac{x_{d\text{-max}}}{\epsilon_s}$	Charge distribution across the semiconductor side of the system is under equilibrium for the *DC* but not the *AC* signal. As a result, the *AC* signal can be mirrored only on the bulk charge distribution
Deep depletion	$C_{eq}^{-1} = C_{ox}^{-1} + \frac{x_d > x_{d\text{-max}}}{\epsilon_s}$	Charge distribution across the semiconductor side of the system is not under equilibrium either for the *DC* or the *AC* signal. As a result, depletion-region width x_d surpasses the equilibrium calculated value $x_{d\text{-max}}$

The availability of sources of *space charge* in these two distribution profiles, for mirroring the charge variation at the metal side of the metal/insulator interface (i.e., due to *AC* and *DC bias*), is the key factor in determining the *bias* condition of *MOS* structures. In application of an overriding *small signal* on a positive V_{MS} *bias* applied to *MOS* structures with n-type semiconductor (or negative *bias* to *MOS* structures with p-type semiconductor), the very small *dielectric relaxation time constant* of majority *carriers* prompts an almost instantaneous response at the insulator/semiconductor interface in the form of variation of majority *carrier* concentration of the *accumulation* layer. Due to the scarceness of minority *carriers*, such a rapid response to potential variation is not normally envisioned at this interface when *bias* polarity is reversed.

Under the latter *bias* condition, variation of the charge concentration in the *inversion* layer is only possible when either the frequency of the *AC signal* used in *C–V* characterization is very low (with regards to the *dielectric relaxation time constant* of minority *carriers*) or when an abundant source of minority *carriers* is present (e.g., due to the presence of a *forward-biased p–n* junction, heating, or illumination). It is in reference to this frequency dependence that terms *low-* and *high-frequency C–V* are used among the entries of Table 2.2.

In the present explanation of *HF* and *LF C–V* characteristics, we have made the assumption that prior to measuring the *small-signal capacitance*, the *MOS* system has been left under *DC bias* long enough to establish the *equilibrium* state across the semiconductor (and as a result the *inversion* layer).

In the absence of an *unblocked* source of minority *carriers*, a *C–V* characterization done at high frequencies will result in recording a small capacitance when the *MOS* system is *biased* in *inversion* (i.e., smaller than the value recorded under *accumulation*-mode *bias*). This is because the variation of the *AC signal* applied to the metal is coupled to the charge concentration of the *depletion region* formed in the *bulk* (whose boundary is further distanced from the insulator/semiconductor interface, which as a result poses a smaller capacitive effect) and not the *inversion* layer. This is understandable, since the variation of charge concentration in the *depletion region* follows the *dielectric relaxation time constant* of majority *carriers* (because this profile is created by the *ionized impurities* left behind in the *depletion region* by the majority *carriers*). However, as expressed earlier, variation of charge concentration in the *inversion* layer follows, in this case, the much longer *dielectric relaxation time constant* of minority *carriers*. Of course, a normally *HF C–V* can be turned into an *LF C–V* when sources of minority *carriers*, in the aforementioned fashions, are present.

In the following, we summarize these *bias* conditions more quantitatively with the use of the *depletion capacitance* and *oxide capacitance* (or generally speaking *insulator capacitance*). Each of the three identified cases considers a certain *bias* condition for the *MOS* system:

1. *DC biased* in *accumulation*, or *inversion*, mode (when conditions for expressing *LF C–V* behavior are met), while semiconductor *bulk* and *surface* maintain an *equilibrium* status

As long as the variation of charge concentration on the metal is mirrored by the opposite charge polarity formed at the insulator/semiconductor interface (i.e., under *accumulation*, *low-frequency inversion*, or *inversion* in the presence of an *unblocked* source of minority *carriers*), the *small-signal capacitance* is approximately modeled by the *oxide capacitance*: $C_{ox} = \epsilon_{ox}/t_{ox}$. However, as indicated earlier in Section 2.4, in this expression the impacts of the quantum mechanical nature of electrons in the formation of a triangular *quantum well* and energy *subbands* are not taken into account. Such impacts can be approximately described by inclusion of a *quantum capacitance* in series with the *oxide capacitance*. This is often done simply by assuming a certain addition to the thickness of insulator layer (i.e., Δ). This additional thickness, which is traditionally seen as the width of the *quantum well*, is on the order of a few *nanometers*.

2. *DC biased* in *flat-band*, *depletion*, or *inversion* mode (when conditions for expressing *LF C–V* behavior are not met), while semiconductor *bulk* and *surface* maintain an *equilibrium* status

As expressed earlier, in the absence of coupling of *AC* or *DC bias* to the *space-charge* component formed at the insulator/semiconductor interface (i.e., under *flat band* and *depletion*-mode *bias* or when *MOS* system is *biased* in *inversion* and *C–V* characterization is done at high frequencies while an *unblocked* source of minority *carriers* is not present), any potential variation will be coupled to the *space-charge* profile of the *depletion region* formed in the *bulk* of the semiconductor. In such cases, the *MOS capacitance* is quantifiable in correlation with the *depletion capacitance*.

As it was observed in the cases of *MS* and *p–n* junctions, in the analysis of the *depletion capacitance* of *MOS* systems, the assumptions of the *depletion approximation* should be validated with regards to the *Debye length*. In this case, since through reduction in *surface carrier* concentration, while changing the *DC bias* from *accumulation* to *flat band* the *Debye length* is increasing, full solution of *Poisson's* equation with incorporation of not only *ionized impurities* but also *charge carriers* is needed in evaluation of the *small-signal MOS capacitance* under *flat band* or early *depletion* conditions. An approximate closed-form expression for calculating the *small-signal capacitance* at $V_{MS} = \phi_{FB}$ is given by

$$C_{FB} = \frac{1}{(1/C_{ox}) + (L_D/\epsilon_s)} \tag{2.196}$$

in which L_D is the *extrinsic Debye length*.

Equation (2.196), in agreement with the earlier discussions in this section, for gate voltages close to the *flat-band* voltage indicates that reduction of the *accumulation* charge concentration results in a decrease in the *small-signal MOS capacitance*. For *gate* voltages in the *depletion* regime of operation, this *small-signal capacitance* is expressed in terms of the series combination of the *oxide capacitance* and the *depletion capacitance*,

$$C = \frac{1}{(1/C_{ox}) + (1/C_s)} = \frac{1}{(1/C_{ox}) + (x_d/\epsilon_s)} \tag{2.197}$$

where x_d is the width of the *depletion region*.[38] In this case, due to the depletion of majority *carriers* from the *surface*, the *depletion approximation* is used without any need for involving the *Debye length*. Since x_d becomes larger with *bias*, for *DC biases* deeper into the *depletion* mode, the *depletion capacitance*, and essentially the *MOS capacitance*, reduce with *bias*.

This reduction in capacitance, however, is often bound by the *pinning* of the *surface* potential at the *onset of strong inversion*. According to our earlier discussions in Section 2.4.2, for *gate* voltages larger in magnitude than the *threshold voltage* (i.e., corresponding to the *onset of strong inversion*), as long as under *DC bias* the semiconductor *bulk* and *surface* are in *equilibrium*, the *depletion-region width* will arrive at a certain maximum value (i.e., $x_{d\text{-max}}$). As a result, under these conditions the *small-signal MOS capacitance* develops a minimum value indicated by

$$C = \frac{1}{(1/C_{ox}) + (x_{d\text{-max}}/\epsilon_s)}. \tag{2.198}$$

[38] Strictly speaking, since C_s is a nonlinear capacitor, this relationship is not right. However, linearization at a given *bias* point allows the use of (2.198).

3. *DC biased* at voltages normally representative of *inversion* mode, while the semiconductor *bulk* and *surface* are not in *equilibrium*

This happens in cases where the *DC bias* is also gradually changing with a rate faster than the rate of the *generation* process of minority *carriers*. Under this condition, the semiconductor *bulk* and *surface* fail to arrive at an *equilibrium* state. As a result, in the absence of an *inversion* layer, *pinning* of the *surface potential* will not take place. In this case, as the *bias* moves deeper into the range normally representative of the *inversion* mode, the *depletion-region width* keeps expanding. The result of this expansion beyond $x_{d\text{-max}}$ is that the recorded *small-signal capacitance* will become even smaller than the value predicted by (2.198). This case, as identified in Table 2.2, is referred to as *deep depletion*.

The variation of the *small-signal MOS capacitance* with *bias* for the three cases identified in this description is demonstrated in Figure 2.37.

So far in Section 2.4, we have not paid any attention to the presence of charges in the insulator or at the insulator/semiconductor interface. In reality these types of nonidealities are present. Charges in the insulator can be fixed or mobile. Depending on this possibility, different modifications manifest in the *small-signal C–V* characteristics of *MOS* structures. Figure 2.38 illustrates a few cases of the impacts of these nonidealities on the *C–V* characteristics of *MOS* structures. *Low-frequency C–V* measurements are often used in technology diagnosis and in extraction of information about interface charge and *trap* densities.

The immediate impact of the presence of a fixed charge concentration in the insulator is the induction of charge profiles at the metal/insulator and insulator/semiconductor interfaces. Obviously, the interface with the closest proximity to the fixed charge profile observes the most charge induction. The induction of this charge

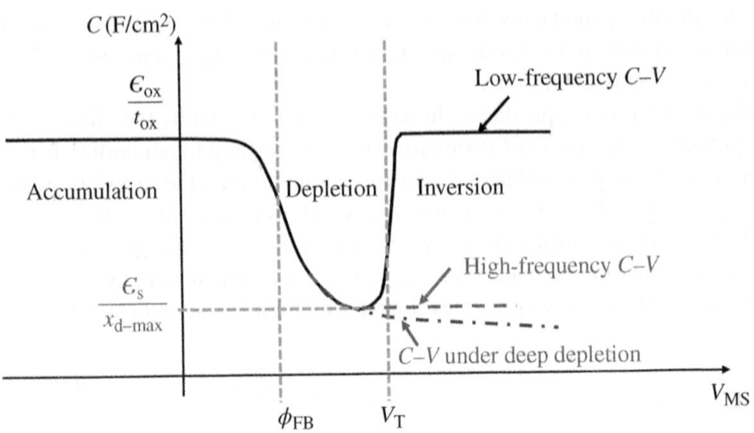

FIGURE 2.37 The three identified *C–V* profiles of an *MOS* capacitor made to a p-type semiconductor: low frequency, high frequency, and deep depletion.

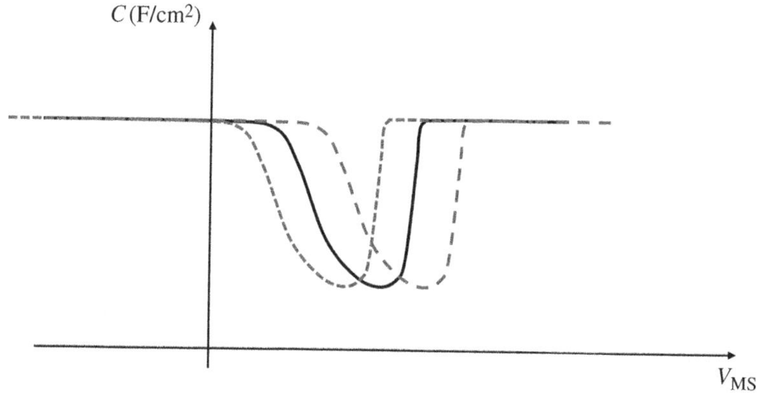

FIGURE 2.38 The low-frequency *C–V* characteristics of an ideal *MOS* capacitor shown in the continuous black curve. The dashed gray curves indicate the possibility of shifting this characteristic to the right (or to the left) due to semipermanent electron (or hole) trapping at the acceptor- (or donor-) type surface states. In the case of shallow states and mobile charged impurities, these transitions can present a hysteresis on the *C–V* characteristics.

concentration in the semiconductor clearly changes the *flat-band* and *threshold voltage* of an *MOS* system. *Gauss*'s law can be used to quantify the size of these changes. Figure 2.39 illustrates the impact of the presence of charge concentration within the insulator in modifying the previously assumed constant electric field across this layer.

The resulting expression for the new value of ϕ_{FB} in the presence of charge concentrations both within the insulator and at the insulator/semiconductor interface is provided by

$$\phi_{FB} = \phi_M - \phi_S - \frac{Q_f}{C_{ox}} - \frac{1}{C_{ox}} \int_0^{t_{ox}} \frac{x}{t_{ox}} \rho(x) dx \tag{2.199}$$

in which Q_f is the fixed insulator/semiconductor interface charge density and $\rho(x)$ represents the profile of the charge concentration within the insulator. In the case of the presence of mobile charges in the insulator layer, the amount of shift of the *flat-band* voltage and the resulting change in the *threshold voltage* vary with the *bias* condition. As shown in Figure 2.40, the mobile nature of insulator charges results in distorted *C–V* characteristics. In Figure 2.41, a schematic depiction of the four categories of oxide charges encountered in *silicon* technology is presented.

In terms of the descriptions presented in this section, Figure 2.42 presents the *C–V* measurement setups often used in characterizing *MOS* systems.

(a)

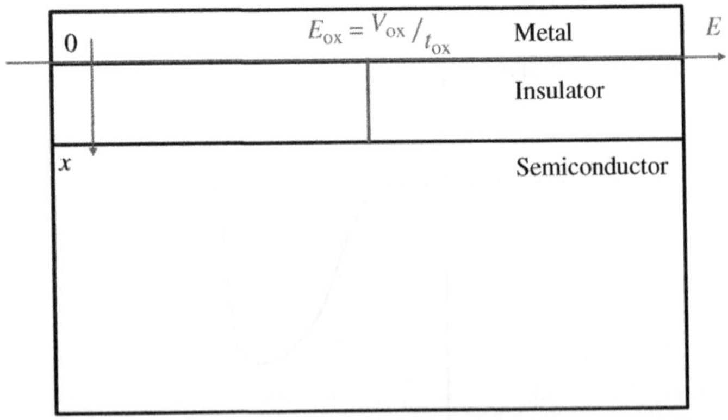

(b)

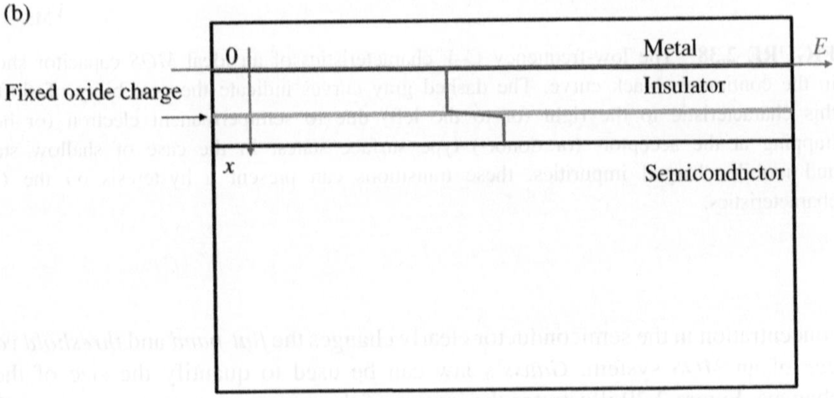

FIGURE 2.39 (a) Profile of electric field across the ideal insulator of an *MOS* capacitor. (b) Profile of electric field across a nonideal insulator with the presence of a layer of positive fixed oxide charge within the oxide (indicated in gray).

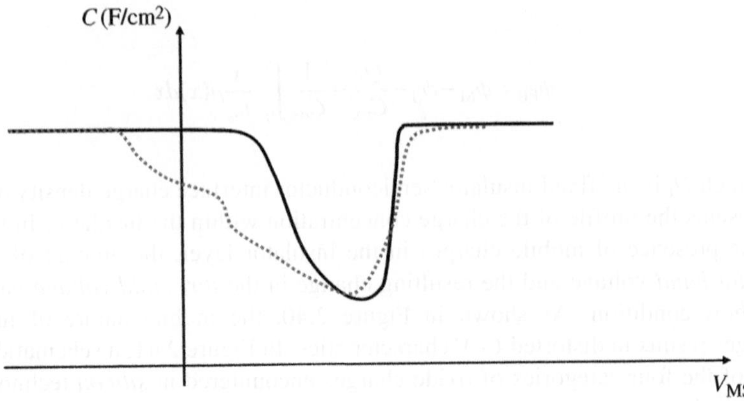

FIGURE 2.40 The low-frequency *C–V* characteristics of an ideal *MOS* capacitor shown in the continuous black curve. Depending on the type of nonideality in the oxide, a distortion in the *LF C–V* can present itself. An example of this distorted characteristic is shown by the dotted curve.

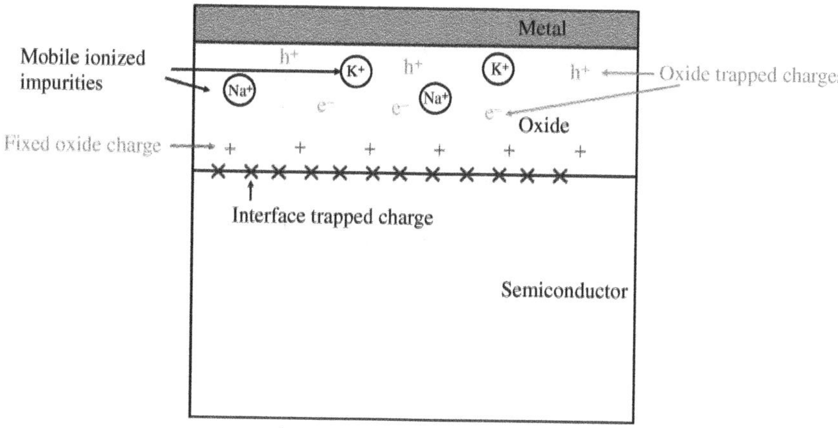

FIGURE 2.41 Four categories of nonidealities within or at the semiconductor interface of the oxide are as follows: interface trapped charges, fixed oxide charge in the vicinity of interface, mobile ionized impurities in the oxide, and trapped charges within the insulator.

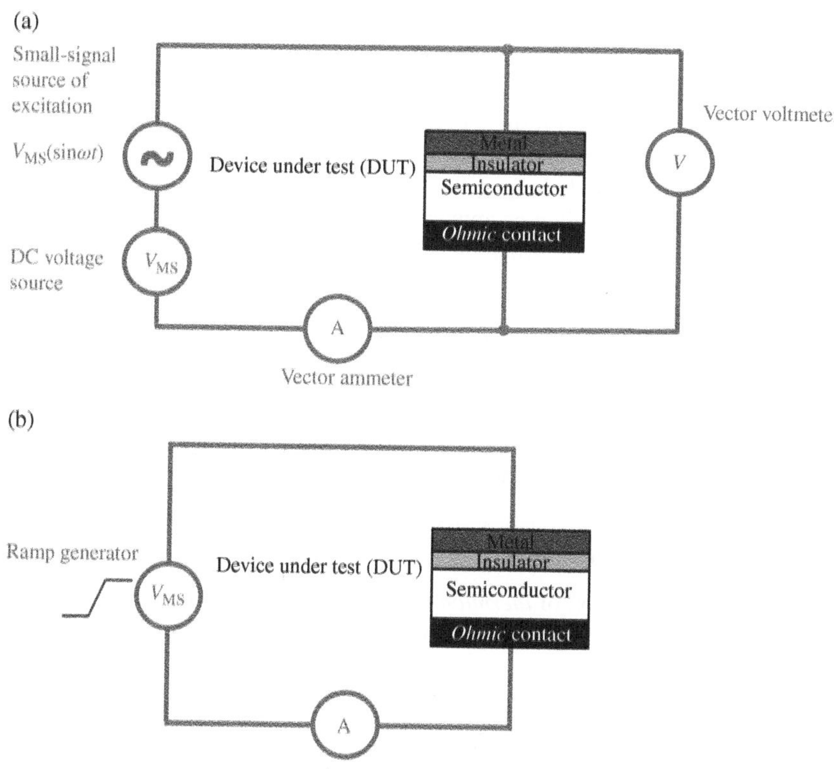

FIGURE 2.42 (a) A typically used *C–V* measurement setup for characterizing an *MOS* system. (b) A quasi-static *LF C–V* measurement setup.

2.5 CURRENT CONDUCTION IN THE PRESENCE OF BAND DISCONTINUITIES IN JUNCTIONS

As it was indicated earlier in this chapter and also in Chapter 1 the presence of potential barriers in forms such as *band* discontinuities in *MS* and *MIS* systems and *spikes* in the *band diagram* of semiconductor *heterojunctions* demands the evaluation of processes such as *thermionic emission* and *field emission* in formulating the current–voltage characteristics. In this section, we will deal with the general formalisms of these processes. Although at times they are seemingly presented for a specific example, we bear no intention to limit the scope of coverage to those few cases.

This description follows the introduction presented in Section 1.4.5. According to that introduction *thermionic emission* is caused by the high kinetic energy of *hot carriers* capable of rising over the potential barrier, whereas *field emission* is a quantum phenomenon (i.e., *tunneling*).

In addition to *thermionic emission* over the potential barriers inside a semiconductor, this phenomenon is also observed in the emission of electrons to the *vacuum* at the semiconductor *surface*. In this context, it is obvious that only electrons with kinetic energies larger than the *electron affinity* at the *surface* can be *thermionically emitted* into *vacuum*.

As we took note in Section 2.4, many semiconductor devices are composed of structures that include an insulating layer. Although so far most of our discussions on *MOS* systems have been restricted to ideal insulators (i.e., devoid of any charge concentration), a real insulator is always host to a number of *bulk* and *surface/interface* mobile and fixed charges. As a result, in the presence of *charge carriers* within the insulator, we often encounter *carrier transport* through these insulating layers. This is especially true when the *lattice* and/or *carrier temperatures* or electric field is high. Whereas *field emission* in the form of almost direct *tunneling* across the insulator, also known as *Fowler–Nordheim* process,[39] is the most common mode of *transport* through an ideal insulator, *carrier transport* in real insulators is sometimes the outcome of different modes of *thermionic* and *field emission* processes. Table 2.3 presents the basic conduction processes encountered in insulators along with their temperature and voltage dependences. Discussions of this section, in due course, will decipher the exact meaning of the entries of this table.

2.5.1 Thermionic Emission

In the simplest description of *thermionic emission*, a few assumptions are taken into account. A short list of these assumptions includes the following:

- Electron distribution in energy domain is treated as that of an *ideal gas* following *Maxwell–Boltzmann* statistics.

[39] Named after *Ralph Fowler* and *Lothar Nordheim*.

TABLE 2.3 The Basic Conduction Processes Encountered across the Insulators along with Their Temperature and Voltage Dependences

Type of Conduction	Voltage and Temperature Dependence of the Current
Direct tunneling	$\propto V_{ox}{}^2.\exp\left[-\frac{a}{V_{ox}}\right]$
Thermionic emission	$\propto T^2.\exp\left[\frac{q}{kT}\left(\sqrt{\frac{q}{4\pi\epsilon_{ox}t_{ox}}}.\sqrt{V_{ox}}-\phi_B\right)\right]$
Frenkel–Poole emission	$\propto V_{ox}.\exp\left[\frac{q}{kT}\left(\sqrt{\frac{q}{\pi\epsilon_{ox}t_{ox}}}.\sqrt{V_{ox}}-\phi_B\right)\right]$
Ohmic conduction	$\propto V_{ox}.\exp\left[-\frac{c}{kT}\right]$
Ionic conduction	$\propto \frac{V_{ox}}{T}.\exp\left[-\frac{d}{kT}\right]$
Space-charge limited conduction	$\propto V_{ox}{}^2$

- No collisions in the *depletion region* formed at the *interface* or *surface* regions are assumed. As a result, only electrons originating from the *quasi-neutral* region are considered.
- Only the emission of electrons moving toward the barrier, with kinetic energies larger than the potential barrier, is considered.
- In the case of *multivalley* semiconductors of anisotropic *effective electron mass*, the sum of *emission* currents from each *valley* is considered as the total *thermionic emission* current.

2.5.1.1 Thermionic Emission in Blocking Schottky Contacts In this section, we will first focus on the description of *thermionic emission* from the most popular semiconductor surface (i.e., {*100*} *silicon* surface) when placed in an *MS blocking* contact. In the case of an arbitrary *silicon* surface, *Crowell* formulated the *thermionic emission* current.[40] This derivation happens to be in its simplest form for this most popular *silicon* surface of {*100*}. As for the last of the assumptions indicated previously, according to our earlier discussions in Chapter 1, the *conduction band* of *silicon* is composed of *six* identical *valleys* located at k_x-, k_y-, and k_z-*axes*. In the ellipsoids representing these *valleys*, we have already identified a *longitudinal effective mass* m_l of $0.92m_0$ and a *transverse effective mass* m_t of $0.19m_0$.

In order to study the *thermionic emission* current, the distribution function of electrons in the *quasi-neutral* region with respect to momentum and energy should be

[40] See C. R. Crowell, "The Richardson constant for thermionic emission in Schottky barrier diodes," *Solid State Electron*, vol. **8**, pp. 395–399, 1965; C. R. Crowell, "Richardson constant and tunneling effective mass for thermionic and thermionic-field emission in Schottky barrier diodes," *Solid State Electron*, vol. **12**, pp. 89–105, 1969.

evaluated. The first step of this evaluation is to consider that the number of electronic *states* per unit volume with momenta between p_x and $p_x + dp_x$, between p_y and $p_y + dp_y$, and between p_z and $p_z + dp_z$ in each of the *conduction-band valleys* is given by[41]

$$N(p_x, p_y, p_z)\,dp_x dp_y dp_z = \frac{2}{h^3} dp_x dp_y dp_z. \tag{2.200}$$

Using the *effective-mass* theory, the kinetic energy defined by $E - E_c$ is given by

$$E - E_c = \frac{p_x^2}{2m_x} + \frac{p_y^2}{2m_y} + \frac{p_z^2}{2m_z}. \tag{2.201}$$

Using *Maxwell–Boltzmann* statistics and this definition of kinetic energy, (2.200) can be used to evaluate the number of electrons per unit volume in each of the *conduction-band valleys* for the indicated range of momenta (i.e., between p_x and $p_x + dp_x$, p_y and $p_y + dp_y$, and p_z and $p_z + dp_z$):

$$\begin{aligned} dn(p_x, p_y, p_z) &= \frac{2}{h^3} \exp\left(-\frac{E - E_f}{kT}\right) dp_x dp_y dp_z \\ &= \frac{2}{h^3} \exp\left(-\frac{E_c - E_f}{kT}\right) \exp\left(-\frac{p_x^2}{2m_x kT}\right) \exp\left(-\frac{p_y^2}{2m_y kT}\right) \\ &\quad \times \exp\left(-\frac{p_z^2}{2m_z kT}\right) dp_x dp_y dp_z. \end{aligned} \tag{2.202}$$

However, in evaluation of *thermionic emission* current, we are only interested in the concentration of electrons moving normal to the surface. According to (2.202), the concentration of electrons of a certain range of momentum in only one of the given principal directions of the *Cartesian* coordinate system (e.g., between p_x and $p_x + dx$) can be evaluated by integrating this profile over all values of momenta along the other two axes:

$$\begin{aligned} dn(p_x) &= \frac{2}{h^3} \exp\left(-\frac{E_c - E_f}{kT}\right) \exp\left(-\frac{p_x^2}{2m_x kT}\right) dp_x \int_{-\infty}^{\infty} \exp\left(-\frac{p_y^2}{2m_y kT}\right) dp_y \\ &\quad \times \int_{-\infty}^{\infty} \exp\left(-\frac{p_z^2}{2m_z kT}\right) dp_z = \frac{2}{h^3} \exp\left(-\frac{E_c - E_f}{kT}\right) \left(\sqrt{\pi}\right)^2 \sqrt{2m_y kT}\sqrt{2m_z kT} \cdot \\ &\quad \times \exp\left(-\frac{p_x^2}{2m_x kT}\right) dp_x = \frac{4}{h^3} \pi kT \sqrt{m_y m_z} \exp\left(-\frac{E_c - E_f}{kT}\right) \cdot \exp\left(-\frac{p_x^2}{2m_x kT}\right) dp_x. \end{aligned} \tag{2.203}$$

[41] The origin of this relationship is in the following definition of summation over momentum *states*, where $\hbar k$ represents the momentum in quantum mechanics and the factor 2 is introduced because of the *Pauli's exclusion principle*: $\frac{1}{\text{vol}} \sum_k = 2(\text{Pauli}) \iiint \frac{dk}{(2\pi)^3}$.

In our first-order treatment, we will ignore the *Schottky barrier lowering* effect. For this case, according to our list of assumptions, in a *Schottky diode biased* to V_a, only those electrons in the *quasi-neutral* region with a kinetic energy of at least $q(\phi_i - V_a)$ (which are traveling normal to the contact) can contribute to *thermionic emission* current. In $\{100\}$ *silicon*, this emission surface is normal to the k_x-*axis*. As a result, the *thermionic emission* current density of electrons from one of the *conduction-band valleys* into the metal will be given by

$$\begin{aligned} J_{1,S-M}(V_a) &= q\int_{p_x=p_{x0}}^{\infty} v_x dn(p_x) \\ &= \frac{4q}{h^3}\pi kT\sqrt{m_y m_z}\exp\left(-\frac{E_c-E_f}{kT}\right)\int_{p_x=p_{x0}}^{\infty}\frac{p_x}{m_x}\exp\left(-\frac{p_x^2}{2m_x kT}\right)dp_x \quad (2.204) \\ &= \frac{4\pi q\sqrt{m_y m_z}}{h^3}(kT)^2\exp\left(-\frac{E_c-E_f}{kT}\right)\exp\left[-\frac{q(\phi_i-V_a)}{kT}\right] \end{aligned}$$

in which $v_x = (p_x/m_x)$ is the velocity of an electron traveling in the x-direction and p_{x0} is defined in terms of the minimum required kinetic energy as ${p_{x0}}^2/2m_x = q(\phi_i - V_a)$.

According to the *band diagram* of a *blocking MS* contact, we know that the size of the potential barrier on the metal side of the junction in terms of the built-in potential is given by

$$q\phi_B = q\phi_i + E_c - E_f. \qquad (2.205)$$

As a result, (2.204) transforms to

$$\begin{aligned} J_{1,S-M}(V_a) &= \frac{4\pi q\sqrt{m_y m_z}}{h^3}(kT)^2\exp\left(-\frac{q\phi_B}{kT}\right)\exp\left(\frac{qV_a}{kT}\right) \\ &= \frac{A\sqrt{m_y m_z}}{m_0}T^2\exp\left(-\frac{q\phi_B}{kT}\right)\exp\left(\frac{qV_a}{kT}\right) \end{aligned} \qquad (2.206)$$

where $A \equiv (4\pi q m_0/h^3)k^2$. This constant is referred to as the *Richardson constant* for free electrons (i.e., $A = 120\ \text{A/cm}^2/\text{K}^2$).

In order to evaluate the total *thermionic emission* current density from a semiconductor to a metal, we have to take care of the other five *valleys* of the *conduction band*. As shown in Chapter 1, in the case of $\{100\}$ *silicon*, for two *valleys* on the k_x-*axis*, we have $m_y = m_z = m_t$, and for the other four *valleys* on the k_y- and k_z-*axes*, we have $m_y = m_l$ and $m_z = m_t$. As a result, the total *thermionic emission* current density is given by

$$J_{n-Si\{100\},S-M}(V_a) = \sum_{i=1}^{6} J_{i-Si\{100\},S-M}(V_a)$$
$$= A\left(\frac{2m_t}{m_0} + \frac{4\sqrt{m_t m_l}}{m_0}\right)T^2 \exp\left(-\frac{q\phi_B}{kT}\right)\exp\left(\frac{qV_a}{kT}\right) \quad (2.207)$$
$$= A^*_{n-Si\{100\}} T^2 \exp\left(-\frac{q\phi_B}{kT}\right)\exp\left(\frac{qV_a}{kT}\right)$$

where

$$A^*_{n-Si\{100\}} = A\left(\frac{2m_t}{m_0} + \frac{4\sqrt{m_t m_l}}{m_0}\right) = 2.05\ \text{A}. \quad (2.208)$$

$A^*_{n-Si\{100\}}$ is the *Richardson constant* for n-type {100} *silicon*. In the case of *silicon*, the dependence on *surface* orientation is not very strong. For example, in {111} *silicon*, the *Richardson constant* is 2.15 A. In experimental studies, this constant is often used as a fitting parameter.

Assuming no *Schottky barrier lowering* effect, the metal to semiconductor emission current will be *bias* independent. This is because, as indicated in Section 2.2, the metal to semiconductor barrier height (i.e., ϕ_B) is *bias* independent. Under this assumption, the total current density in the case of a *Schottky* contact to n-type {100} is given by

$$J_{\text{thermionic}\ n-Si\{100\}}(V_a) = J_{n-Si\{100\},S-M}(V_a) - J_{n-Si\{100\},MS}(V_a)$$
$$= A^*_{n-Si\{100\}} T^2 \exp\left(-\frac{q\phi_B}{kT}\right)\left[\exp\left(\frac{qV_a}{kT}\right) - 1\right]. \quad (2.209)$$

More accurate theories, with smaller set of assumptions, have also been proposed.[42] For example, in the *diffusion emission theory*, the assumption of no collisions in the *depletion region* is omitted. These other theories result in relationships similar in form to Equation (2.209), while the preexponential factor is different. The reader is encouraged to compare (2.209) to (2.33), which was earlier developed for the current–voltage characteristic of a *Schottky blocking* contact purely on the basis of the *drift–diffusion* formalism.

As shown earlier in Section 2.2.2.1.4, with incorporation of *Schottky barrier lowering*, $q\phi_B$ should be replaced by $q(\phi_B - \Delta\phi)$. *Forward bias* reduces *Schottky barrier lowering*, while *reverse bias* increases the value of $\Delta\phi$.

2.5.1.2 *Thermionic Emission across Semiconductor Heterojunctions* As indicated earlier, the relevance of *thermionic emission* of electrons to current *transport* is not restricted to *MS* and *MOS* systems. In operation of semiconductor

[42] See C. R. Crowell and S. M. Sze, "Current transport in metal-semiconductor barriers," *Solid State Electron*, vol. **9**, pp. 1035–1048, 1966.

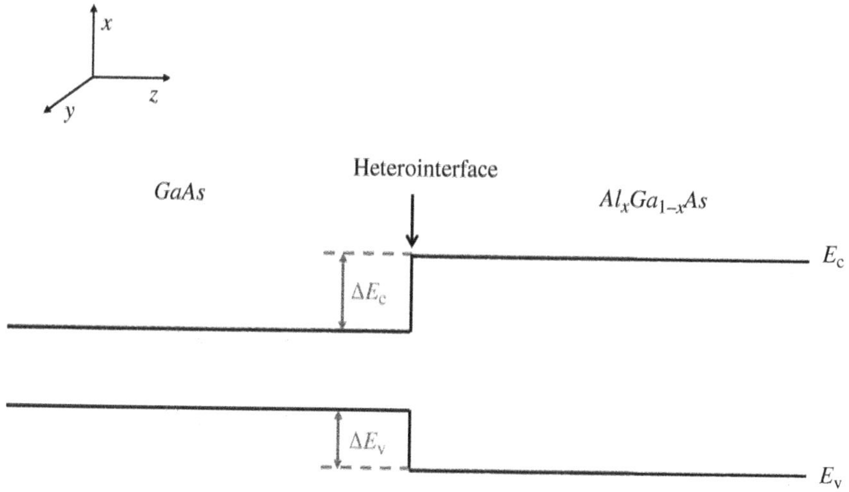

FIGURE 2.43 Approximate profile of band discontinuity at the *Type-I AlGaAs/GaAs* heterointerface, with assumption of no band bending.

heterostructure systems of interest to this volume, such as *$Al_xGa_{1-x}As$/GaAs high electron mobility transistors,*[43] such a process of current conduction is essential. As observed earlier in Section 2.3, the *bandgap* discontinuity at these *heterojunctions* imposes potential barriers on the *charge carriers.* In this section, for the specific example of an *$Al_xGa_{1-x}As$/GaAs heterojunction,* we will look into the evaluation of *thermionic emission* current density.

We will start this evaluation with assuming no *band bending* at the *heterointerface.* As a result of this discussion, we will reinvestigate the origins of *band bending.* The depiction of the *band diagram,* when the energy of the *band* edges is not changing with position, is presented in Figure 2.43. In this section, we will evaluate the *thermionic emission* current between *AlGaAs* and *GaAs* using the approximations employed in Section 2.5 (which are known as *Bethe*'s approximations, after *Hans Bethe*).

Now that we are acquainted with the assumptions of this framework, we will attempt to articulate its foundations in more details. In this approximation, *Bethe* assumed that the distribution function of electrons on each side of the *heterojunction* follows the *Fermi–Dirac* distribution or, where appropriate, the *Maxwell–Boltzmann* distribution. In this distribution, a *quasi-Fermi level* (i.e., E_{fn}) is assumed for electrons on each side of the *heterojunction.* These energy *levels,* although different on the two sides, are constant on each side of the junction. This model assumes a strong electron–electron interaction on each side of the junction. This results in a *Fermi–Dirac*-like distribution, even in the presence of the chance of electron emission to the opposite side.

In the case of a *Schottky* barrier, the presence of substantial electron emission to the opposite side of the junction makes it difficult to believe that *Fermi–Dirac* statistics

[43] Which will be dealt with in Chapter 5.

can still be maintained. However, for most practical cases, the assumptions of such distributions result in calculations that are only off by a factor of about 2 from the results of ensemble *Monte Carlo* treatment of the same problem. Hence, it is instructive to pay attention to this simple formalism.

According to the *band diagram* shown in Figure 2.43, we can indicate that all of the *GaAs* electrons of kinetic energy greater than the *conduction-band* discontinuity (i.e., ΔE_c), which are moving normal to *heterointerface* toward the $Al_xGa_{1-x}As$ layer, will propagate into this layer. Also, all the electrons in the *AlGaAs* layer with a velocity vector of the opposite direction will propagate into the *GaAs* layer. The situation depicted in Figure 2.43 is evidently a simplified situation in which we have not paid attention to the formation of energy *subbands*. In addition, proper definition of *effective mass* at higher energies and the process of redirectional electron collision before reaching the *heterointerface* is missing from this description. In this situation, the current density from *GaAs* to *AlGaAs* is given by

$$J_{GaAs-AlGaAs} = \frac{2q}{h^3} \iint_{p_x p_y} dp_x dp_y \int_{p_z > p_{z0}} dp_z v_z f_0(k) \tag{2.210}$$

where p_{z0} is the minimum momentum necessary to overcome the potential barrier and $f_0(k)$ is the *quasi-thermal-equilibrium* distribution function on the *GaAs side* of the *heterointerface*. As we saw in Section 2.5, p_{z0} can be classically evaluated through

$$\frac{m^* v_{z0}^2}{2} = \Delta E_c \tag{2.211}$$

where v_{z_0} is the minimum velocity necessary to overcome ΔE_c.

Considering $f_0(k)$ to follow a *Maxwell–Boltzmann* distribution function, we can rewrite (2.210) as

$$J_{GaAs-AlGaAs} = \left(\frac{m^*}{\hbar}\right)^3 \frac{q}{4\pi^3} \exp\left(\frac{E_{fn}^{GaAs} - E_c^{GaAs}}{kT}\right) \iint_{-\infty}^{\infty} dv_x dv_y \int_{v_{z0}}^{\infty} dv_z v_z \exp\left[-\frac{m^*}{2kT}\left(v_x^2 + v_y^2 + v_z^2\right)\right]. \tag{2.212}$$

In this equation, E_{fn}^{GaAs} is the position of the *quasi-Fermi level* of electrons on the *GaAs* side.

Now, the *x*–*y* integration in (2.212) readily yields

$$\iint_{-\infty}^{\infty} dv_x dv_y \exp\left[-\frac{m^*}{2kT}\left(v_x^2 + v_y^2\right)\right] = \frac{2\pi}{m^*} kT. \tag{2.213}$$

Based on (2.213), (2.212) can be simplified to

$$J_{GaAs-AlGaAs} = \frac{m^{*2}}{\hbar^3} \frac{q}{2\pi^2} kT \exp\left(\frac{E_{fn}^{GaAs} - E_c^{AlGaAs}}{kT}\right) \int_{v_{z0}}^{\infty} v_z \exp\left[-\frac{m^*}{2kT}\left(v_z^2\right)\right] dv_z, \tag{2.214}$$

which gives the *thermionic emission* current density from *GaAs* to *AlGaAs* as

$$J_{GaAs-AlGaAs} = A^* T^2 \exp\left(\frac{E_{\text{fn}}^{GaAs} - E_{\text{c}}^{AlGaAs} - |\Delta E_{\text{c}}|}{kT}\right) \tag{2.215}$$

in which $A^* \equiv (4\pi q m^* k^2)/h^3$ is the *effective Richardson constant*. If the electrons are energized (which, as noted in Section 1.4.4.1, is indicated by *electron temperature* T_e different from the *lattice temperature* T), then T_e takes the place of T in the argument of the exponential in (2.215).

The current in the opposite direction (i.e., from $Al_xGa_{1-x}As$ toward *GaAs*) in the absence of a potential barrier is given by

$$J_{AlGaAs-GaAs} = A^* T^2 \exp\left(\frac{E_{\text{fn}}^{AlGaAs} - E_{\text{c}}^{AlGaAs}}{kT}\right) \tag{2.216}$$

where E_{fn}^{AlGaAs} is the *quasi-Fermi level* of electrons on the $Al_xGa_{1-x}As$ side of the *heterointerface*.

According to (2.215) and (2.216), a net flow of *carriers* across the *heterointerface* is present. As a result of this, there is a charge transfer in the quest for establishing *thermal equilibrium* across the *heterointerface*. The *band diagram* of Figure 2.43 develops a *band bending* in the form discussed in Section 2.3. Establishing *thermal equilibrium* in this *Type-I heterostructure* results in transferring electrons from the *highly doped* wide-*bandgap* material to the *lightly doped* narrow-*bandgap* semiconductor and causes the formation of a triangular *potential well* at the *heterointerface*. Contingent upon sufficient *doping* in the wide-*bandgap* side and *carrier* transfer across the *heterointerface*, the *conduction-band* edge at this *quantum well* can dip below the unified *Fermi level*. This behavior is indicative of a *degenerate*-level concentration of electrons in the *quantum well*. Due to the small width of the *quantum well*, this electron profile takes on a two-dimensional character. We often refer to this electron concentration as *two-dimensional electron gas* or *2DEG* for short. The electron *mobility* of the *2DEG* created in this fashion is superior to the electron *mobility* of the *bulk*. The remoteness from *ionized impurities* is the key factor in enhancement of *mobility* of the *2DEG* formed at the *heterointerface*. In this fashion, in addition to confinement of *impurities*, *phonons* can also be separated from electrons. An example of this can be the transfer of *charge carriers* from the *polar* side of a *heterointerface* to its possibly *nonpolar* opposite side. In *carrier* transfer across the *heterointerface*, whereas *polar phonons* can still manage a long-range fringing field on the *nonpolar* side of the *heterointerface*, the *2DEG* electrons will be less affected by *polar phonons*.

The time constant of *charge-carrier* transfer from *AlGaAs* to *GaAs* is in the order of *picoseconds*. The charge transfer time constant is much longer in the opposite direction. The electron emission from *GaAs* to *AlGaAs* can be enhanced through energizing the *electron gas* (i.e., increasing T_e), by means such as applying an electric field parallel to the *heterointerface*. This scenario, which is referred to as *real-space transfer*, will be dealt with in Chapters 4 and 5.

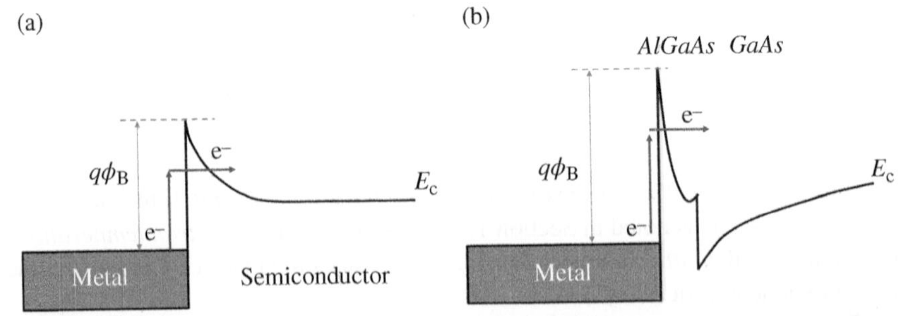

FIGURE 2.44 (a) Schematic depiction of thermionic field emission process across a blocking *Schottky barrier* made to a moderately doped semiconductor. (b) Schematic depiction of thermionic field emission process across a blocking *Schottky barrier* made to a *type-I heterostructure* (e.g., *AlGaAs/GaAs*).

2.5.2 Field Emission and Thermionic-Field Emission

As we have indicated earlier in Section 1.4.5, reduction in the thickness of the potential barrier formed at the semiconductor *surface* or at the *interfaces* can promote current conduction in the form of quantum *tunneling* (i.e., *field emission*) process. Such a situation can be encountered in *interfaces* with highly *doped* semiconductors. In conjunction with this mechanism, as shown in Figure 2.44, for moderate *doping* levels, a mixture of *thermionic-* and *field emission* processes can also be expected. This is known as *thermionic-field emission*. In these cases, even at low applied *biases*, a current larger than the predictions of *thermionic emission* flows through the junction (or a larger emission from the surface of a *field emitter* occurs).

In expressing the current through an *MOS* system, the often responsible *tunneling* mechanism is explained in terms of the *Fowler–Nordheim tunneling* formalism. In this formalism, electrons *tunnel* into the so-called conduction band of the oxide and *drift* through this layer. For the simplified case in which effects of temperature and *barrier lowering* are ignored, this *tunneling* current density is given by

$$J_{FN} = \frac{q^2 E_{ox}^2}{16\pi^2 \hbar \phi_{OS}} \exp\left(-\frac{4(2m^* q)^{1/2} \phi_{OS}^{3/2}}{3\hbar E_{ox}}\right) \tag{2.217}$$

in which case E_{ox} and ϕ_{OS} are the electric field across the oxide and the *conduction-band* discontinuity between oxide and semiconductor, respectively.[44] According to (2.217), the current is characterized by a straight line in the plot of $\log\left(J_{FN}/E_{ox}^2\right)$ versus $1/E_{ox}$. This behavior can be used in gauging the dominance of this mode of *transport*.

A schematic representation of this mechanism is depicted in Figure 2.45.

[44] See M. Lenzlinger and E. H. Snow, "Fowler-Nordheim tunneling into thermally grown SiO_2," *J Appl Phys*, vol. **40**, pp. 278–283, 1969.

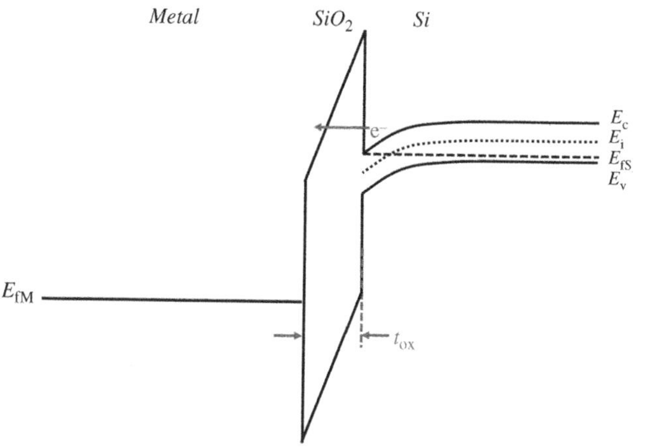

FIGURE 2.45 Schematic depiction of the *Fowler–Nordheim* tunneling process across the oxide layer in an *MOS* system.

In the case of a very thin oxide (i.e., <4 nm), the *MOS* leakage current instead of the *Fowler–Nordheim* mechanism is dominated by direct *tunneling* through the forbidden *energy gap* of the oxide. For this mechanism, the theory based on *WKB* is more complicated than the *Fowler–Nordheim tunneling*, and there is no simple analytical model describing the current–voltage characteristics.

The presence of *traps* inside the oxide or at the interfaces can also assist the *tunneling carrier transport*. This mode of *transport* is often referred to as *trap-assisted tunneling*. *Traps* serve as stepping stones in the *tunneling* process. While certain *trap* concentrations are present in as-grown oxide layers and at their *interfaces*, the *tunneling* of energetic *carriers* (i.e., *hot carriers*) through the oxide and the subsequent interactions with the oxide can also add to their concentrations. As a result, *trap-assisted tunneling* serves as a *positive feedback loop* in the degradation of the quality of *MOS* systems.

The empirical *lucky-electron tunneling* model, which was proposed by *Shockley*, is the simplest model used in the description of current injection through *MOS* junctions. As schematically depicted in Figure 2.46, the electron positioned at distance *d* from the *interface*, in the case of a *scatteringless drift* toward the *interface*, possesses sufficient amount of potential energy at the SiO_2/Si *interface* to surmount the potential barrier height $qV(d)$. Such an electron is referred to as a *lucky electron*. In this model, based on *Maxwell–Boltzmann* statistics, the probability of an electron at a distance *d* from the SiO_2/Si *interface* to be emitted into the oxide layer (i.e., to be dealt with as a *lucky electron*) is indicated through an exponential term,

$$P(d) \propto \exp\left(-\frac{d}{\lambda}\right). \qquad (2.218)$$

The factor λ (i.e., *mean free path*) in (2.218) is indicative of energy loss to the *inelastic* collisions in *silicon*.

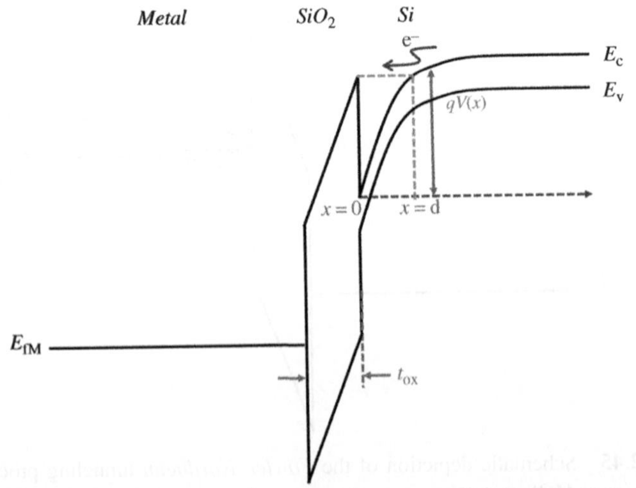

FIGURE 2.46 Schematic depiction of the lucky-electron tunneling process on the band diagram of an *MOS* system.

It has been empirically determined that the effective potential barrier height for electron emission in *MOS* systems can be written as

$$qV(d) = q\phi_{OS} - q\Delta\phi - \alpha E_{ox}^{2/3}. \tag{2.219}$$

The first two terms represent the potential barrier height at the SiO_2/*Si* interface and the *barrier lowering*, respectively. The third term represents the effective impact of *thermionic-field emission* on improving the chance of *tunneling* (where α is a fitting parameter). It has been shown that the effect of temperature on this process can be well expressed by the temperature dependence of the *effective mean free path*,[45]

$$\lambda(T) = \lambda_0 \tanh\left(\frac{E_{op}}{2kT}\right) \tag{2.220}$$

where E_{op} is the *optical-phonon energy* (which in the case of *silicon* is equal to 63 meV) and λ_0 is the low-temperature limit of λ. It has been found empirically that in case of *silicon* $\lambda_0 = 10.8$ nm.[46]

Another mode of electron *emission* through a potential barrier is known as *Frenkel–Poole* emission.[47] As shown in Figure 2.47, this *emission* is caused by the thermal

[45] See C. R. Crowell and S. M. Sze, "Temperature dependence of avalanche multiplications in semiconductors," *Appl Phys Lett*, vol. **9**, pp. 242–244, 1969; T. H. Ning, C. M. Osburn, and H. N. Yu, "Emission probability of hot electrons from silicon into silicon dioxide," *J Appl Phys*, vol. **48**, pp. 286–293, 1977.

[46] In the case of *MOS* structures incorporated in a *MOSFET*, we are in need of a more developed version of the *lucky-electron tunneling* model, which will be presented in Chapter 3.

[47] Named after *Yakov Frenkel* and *Horace Poole*.

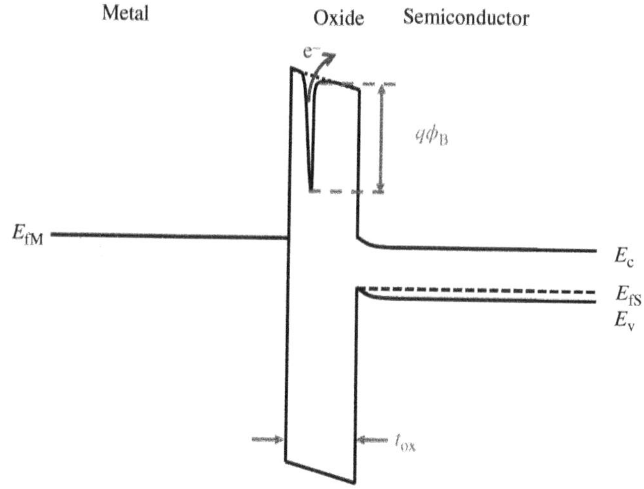

FIGURE 2.47 Schematic depiction of the *Frenkel–Poole* emission process through an unideal oxide layer of an *MOS* system. The modified barrier height in terms of the energy level of the defect within the oxide layer is indicated.

excitation of *trapped* electrons into the *conduction band*. The characteristics of *Frenkel–Poole emission* for *trap states* with *Coulomb* potentials are similar to that of *thermionic emission*. However, in this case the potential barrier height is defined by the depth of the *Coulombic trap potential well* where the *barrier lowering effect* is larger than the observed *lowering* in Section 2.2.2.1.4 by a factor of 2. This is because the positive charge is immobile and *barrier lowering* is dictated by both sides of the *Coulombic* well.

In the case of low voltages and high temperatures, emission occurs through the *hopping* of thermally excited electrons from one *isolated state* to the next. This *emission* process results in an *Ohmic* characteristic that is exponentially dependent on the temperature.

FURTHER READING

Following references have been used in preparation of the materials presented in this chapter. They are also suggested as sources for further reading.

PHYSICS OF SEMICONDUCTOR DEVICES

K. Hess, *Advanced Theory of Solid State Devices*, IEEE Press, Piscataway, NJ, 2000.

R. S. Muller, T. I. Kamins, and M. Chan, *Device Electronics for Integrated Circuits*, John Wiley & Sons, Inc., Hoboken, NJ, 2003.

R. F. Pierret, *Semiconductor Device Fundamentals*, Addison Wesley, Reading, MA, 1996.

B. G. Streetman and S. K. Banerjee, *Solid State Electronic Devices*, Prentice Hall, Englewood Cliffs, NJ, 2006.

S. M. Sze, *Semiconductor Device Physics and Technology*, John Wiley & Sons, Inc., Hoboken, NJ, 2002.

S. M. Sze and K. K. Ng, *Physics of Semiconductor Devices*, John Wiley & Sons, Inc., Hoboken, NJ, 2007.

Y. Taur and T. H. Ning, *Fundamentals of Modern VLSI Devices*, Cambridge University Press, Cambridge, UK, 2009.

S. Wang, *Fundamentals of Semiconductor Theory and Device Physics*, Prentice Hall, Englewood Cliffs, NJ, 1989.

PROBLEMS

2.1 Using *SDA* draw the thermal-equilibrium band diagram of an *MS* junction established between a phosphorus-doped slab of germanium and platinum at room temperature. Phosphorus dopants are distributed uniformly throughout the volume of the semiconductor slab with a concentration of 10^{16} cm^{-3}. Assume the surface of the semiconductor to be ideal.

2.2 Considering the *Schottky barrier* lowering, modify the band diagram produced in problem 1.

2.3 Produce the $1/C^2$ versus V_{MS} characteristics for blocking Schottky contacts made to:

a. A p-type semiconductor

b. An n-type semiconductor

- How would these diagrams ideally change if the metal were to be replaced with another one of a larger work function while still the blocking characteristic is maintained?
- How would these diagrams ideally change if the dopant concentration were to be increased while still the blocking characteristic is maintained?

2.4 Prove Equation (2.40).

2.5 Performing a literature search for the required constants, draw the room-temperature thermal-equilibrium band diagrams for the following heterojunctions and calculate the values of the built-in potential and the width of the depletion region. Indicate the sources used in extracting the required constants:

a. Si-doped $Al_{0.3}Ga_{0.7}As$/undoped *GaAs*, while the dopant distribution is uniform with a concentration equal to 10^{16} cm^{-3}.

b. Si-doped *GaAs*/undoped $In_{0.2}Ga_{0.8}As$, while the dopant distribution is uniform with a concentration equal to 10^{16} cm^{-3}.

c. B-doped $Si_{0.8}Ge_{0.2}$/P-doped *Ge*, while both dopant distributions are uniform. Boron concentration is equal to 10^{17} cm^{-3}, and the phosphorus concentration is 10^{15} cm^{-3}.

Assume all dopants activated and the interface to be ideal with no strain. Use Vegard's law wherever needed.

2.6 Prove Equation (2.60).

2.7 Prove Equations (2.70) and (2.71).

2.8 Try to prove (2.93). Toward that end carefully evaluate the following seminal article:

W. Shockley and W. Read, "Statistics of the recombinations of holes and electrons," *Phys Rev*, vol. **87**, no. 5, pp. 835–842, September 1952.

2.9 Present the complete proof of (2.114) without sparing any step.

2.10 Without sparing any step and with full description of the assumptions, prove Equation (2.134).

2.11 According to the procedure outlined in Figure 2.30, draw the thermal-equilibrium band diagram across an *MOS* system established between *Pt*, 30 nm thick SiO_2 layer, and uniformly n-doped *Si* slab doped to the level of $10^{17}\ \text{cm}^{-3}$.

2.12 Apply a voltage of +0.5 V between the metal and the semiconductor in the *MOS* system of Figure 2.30 and accordingly redraw the band diagram with the full indication of the amount of band bending across different layers.

2.13 For the *MOS* systems described in Figure 2.30, calculate the threshold voltage.

2.14 For the *MOS* system described in Figure 2.30, reevaluate the threshold voltage if boron adjustment implant with $N_p = 10^{18}\ \text{cm}^{-3}$, $R_p = 120$ nm, and $\Delta R_p = 3$ nm were to be used. Simplistically assume that all of the microfabrication steps succeeding the adjustment implant are performed at low enough temperatures so that no diffusion process is required to be considered.

APPENDIX 2.A LIMITATIONS OF *SDA* AND THE MEANING OF DEBYE LENGTH

Transition between the *depletion* and the *neutral* regions to the sides of a semiconductor junction happens within the *Debye length.*

Debye length, as indicated in Section 2.3.2, can be calculated from solving for potential in equations such as (2.54). It should be mentioned that this has its roots in the fact that the transition in electron concentration cannot be abrupt. Instead, based on thermal energy of kT, the *Debye length* identifies the characteristic length of spilling. As a result, the definition of a contact area in *SDA* is only possible within the uncertainty of this length scale.

As an example, consider a *highly doped* contact region formed to an *undoped* semiconductor in which *carriers* spill over to the *undoped* region. The length of the spilledover area, over which a potential drop of kT/q occurs, is the *Debye length.* In such a case, as previously indicated in (2.63), according to (2.54),

$$L_{\mathrm{D}}^2 \cong \frac{\epsilon_{\mathrm{s}} kT}{q^2 N_{\mathrm{c}}} \tag{2.A.1}$$

where N_c is the *dopant* concentration of the *highly doped* contact. According to *Poisson's equation*,

$$\frac{kT}{q} \cong \int_0^{L_{\mathrm{D}}} \frac{qN_{\mathrm{c}}}{\epsilon_{\mathrm{s}}} s(z) z dz. \tag{2.A.2}$$

In Equation (2.A.2), we are assuming that spillover happens in terms of the function $s(z)$, which is bound between 0 and 1. Equation (2.A.2) for an approximate constant *doping* N_c can be rearranged as

$$\frac{\epsilon_{\mathrm{s}} kT}{q^2 N_{\mathrm{c}}} \cong \int_0^{L_{\mathrm{D}}} s(z) z dz \tag{2.A.3}$$

where using the *mean-value theorem* of integral calculus, the integral can be assumed to be equal to L_{D}^2.

The aforementioned form of the *Debye length* offers only an approximation. Particularly in *nonequilibrium* situations, *Debye length* needs to be calculated numerically.

An interesting example with regards to the application of the concept of *Debye length* is encountered when we have a positively charged *donor* in a bath of negatively charged electrons. In this situation, electrons accumulate around the *donor* and *screen* its charge. The *charge screening* happens within the *Debye length*. As indicated in our discussion on *scattering processes* in Section 1.7, this *screening* also plays an important part in understanding the *scattering* of electrons by *impurities*.

3

TRADITIONAL PLANAR MOSFETs: OPERATION, MODELING, AND TECHNOLOGY SCALING

The 1930 patent of *Julius Edgar Lilienfeld*[1] provides the first concrete evidence of the inception of the idea of a *field effect transistor*. However, due to the lack of proper understanding of the physics and manufacturing of the insulator/semiconductor contacts, the proposed devices were not successfully realized for approximately another 30 years. In the meanwhile, *bipolar junction transistors* (i.e., *BJTs*), which normally operate on the basis of the proximity of two semiconductor–semiconductor *homojunctions*, were successfully implemented in the *planar silicon* technology. In spite of this original technological lag, the tremendous advantages of *field effect transistors*, especially when realized in the form of *silicon metal–oxide–semiconductor FETs* (i.e., *MOSFET*), made them capable of capturing most of the *BJT*'s market in less than three decades after their successful realization. In this chapter, discussions will start with a comparison between the operation principles of *BJTs* and *MOSFETs*. Following this discussion, behavior of earlier-generation *MOSFETs* (which are known as *long-channel MOSFETs*) will be analyzed. After providing analytical models for the behavior of *long-channel MOSFETs*, in a chronological order a few other device models describing the operation of *MOSFETs* of increasingly shorter *gate length* are presented. These models still adopt the *drift–diffusion transport* formalism, however, as *technology scaling* is reducing the *gate length* of *MOSFETs* to about 10 nm, considering *ballistic transport* is becoming increasingly relevant. This is because

[1] US patent #1,745,175. Method and apparatus for controlling electric currents.

Field Effect Transistors, A Comprehensive Overview: From Basic Concepts to Novel Technologies, First Edition. Pouya Valizadeh.

electrons in these short *channels* are suddenly introduced to a very high electric-field region without the possibility of reaching *steady state*. These non*steady-state* electrons are observed to be able to achieve *overshoot velocities* much higher than the *saturation velocity*. Due to the higher average *carrier* velocity along the *channel* of these devices, *current drive* exceeds the predictions of our *drift–diffusion*-based formalism. In the current state of the art of *MOSFET* technology, electrons fly *ballistically* through more than 40% of the *channel length*. Discussion of this new operation paradigm of *deeply scaled field effect transistors*, however, is left to Chapter 6.

3.1 BATTLE OF TRANSISTORS: MOSFET VERSUS BJT

Figure 3.1 illustrates the typical structure of an *n-channel planar MOSFET* (i.e., *NMOS*) next to an *n–p–n BJT*. Both of the depicted transistors are realizable in compliance with the restrictions of *planar silicon* technology. Although the discussions of this chapter are biased toward these transistors (unless otherwise indicated), appropriate modifications to the *doping* status of the semiconductor regions and the polarity of

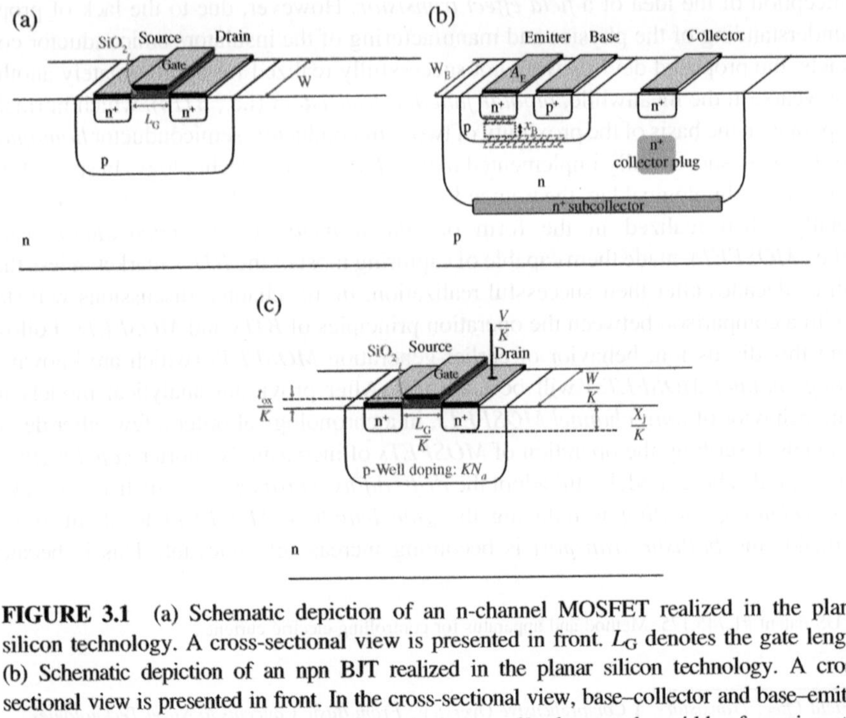

FIGURE 3.1 (a) Schematic depiction of an n-channel MOSFET realized in the planar silicon technology. A cross-sectional view is presented in front. L_G denotes the gate length. (b) Schematic depiction of an npn BJT realized in the planar silicon technology. A cross-sectional view is presented in front. In the cross-sectional view, base–collector and base–emitter depletion regions are indicated by hash-marked areas. X_B denotes the width of quasi-neutral base. (c) Constant-field scaling of MOSFET in terms of the scaling parameter K according to Dennard et al.

bias conditions yield the possibility of extending these discussions to *p-channel MOSFETs* (i.e., *PMOS*) and *p–n–p BJTs*.

Bipolar transistor action is defined through modulating the *reverse current* of the *base–collector p–n* junction through *forward biasing* the *base–emitter p–n* junction. Modulation of the *drain* current of the *MOSFET* depicted in Figure 3.1 is, however, carried out through the *field effect* of the *gate* electrode in controlling the electron population of the *inversion channel* formed between the *drain* and *source* electrodes. This *channel* is formed in the vicinity of the insulator/semiconductor *interface*. In terms of the differences in the modes of operation of the two transistors, we can put together a short list of eye-catching facts about the devices illustrated in Figure 3.1:

1. The expressed differences between the *transistor actions* in the *BJT* and the *MOSFET* grant the *planar MOSFET* a two-dimensional *carrier-transport channel* formed parallel to the surface of the *wafer*. In *BJTs*, *carrier transport* across the *quasi-neutral* region of the *base* takes on a three-dimensional character (while *transport* happens normal to the surface of the *wafer*). Caused by the nature of the *diffusion doping* process, since the *dopant* distribution profiles of the *emitter* and *base* decay with depth, the *emitter–base* junction will have its lowest built-in potential defined at the bottom of the *emitter* and not at its sides. Due to that reason, the aforementioned three-dimensional mode of *carrier transport* between *emitter* and *collector*, which is defined by *forward biasing* the *base–emitter* junction, takes place in the region marked by the dotted lines in Figure 3.1b.
2. The placement of the *emitter*, *base*, and *collector* electrodes of the *planar BJT* on a plane normal to the *carrier-transport channel* imposes a lateral extent on this device, which is much bigger than that of the actual surface area where the *transistor action* is taking place. This latter area is marked by the lateral dimensions of the n^+ *emitter*.
3. Due to lateral separation between where the actual *bipolar transistor action* is taking place (i.e., the so-called *intrinsic BJT*) and the metal–semiconductor contacts connecting this transistor to the outside world (i.e., *extrinsic base*, *emitter*, and *collector* terminals), *ion-implantation doping* is often required to reduce the parasitic resistance between the *extrinsic* and the *intrinsic collector*. This indicates the reason for incorporating the *ion-implanted* profiles of the *subcollector* and *collector plug* in Figure 3.1b.
4. In terms of the expressed differences between the operating principles of the two transistors, while the critical dimensions in determining the *current drive* and high-frequency operation of a *BJT* are defined by the length of the *quasi-neutral* region of the *base* (which, as shown in Fig. 3.1b, is defined vertically) and to some extent the surface area of the *emitter*, the operation of a *MOSFET*, according to Figure 3.1a, is primarily defined by the lateral dimensions of the *gate* electrode (i.e., especially the *gate length*).
5. Ideally, due to the presence of the *gate oxide*, the control terminal of the *MOSFET* (i.e., *gate*) operates at a *zero DC* current level. In the *BJT*, *base* current is identified by the *forward* current of the *base–emitter p–n* junction.

The development and success of *MOSFET* technology were primarily driven by the tremendous demands for digital circuits of higher functionality and complexity. In fulfilling these demands, *MOSFETs* enjoyed a three-fold advantage: they had a simpler fabrication technology, they were realizable over a smaller *chip* area, and due to the presence of an insulator (i.e., SiO_2) they drew much less input current at their control terminal. As a result of all these advantages, quite early-on *MOSFET* digital *chips* of higher packing density and better power efficiency than the contemporary *BJT chips* were realized. This happened after overcoming a few early setbacks.

The most important of these early technological setbacks were caused by the mind-sets inherited from *BJT* technology. Since the operation of a *BJT* relies on *carrier transport* in the *bulk* of the semiconductor and not in the vicinity of its *surface*, these devices were quite immune to the large concentration of *surface states* and dangling *bonds* at the surface of *(111) silicon*. This is how these substrates found their way into the *silicon BJT* technology. However, while inheriting *(111) silicon* for realization of *n-channel MOSFETs*, early technologists were faced with the unwanted formation of an *inversion channel* at the oxide/semiconductor *interface* (i.e., even in the absence of a *gate bias*). Such an observation was not made in manufacturing of *p-channel MOSFETs*. Understanding the link between the large number of dangling *bonds* at the *surface* of *(111) silicon* and the formation of a sizeable concentration of positive fixed charge in the thermally grown SiO_2 was pivotal to the success of *MOSFET* technology. As noted in Section 2.4.4, the presence of a positive fixed oxide charge can support the induction of an *inversion n-channel* on the semiconductor side of the *interface*. This kind of a characteristic grants the *n-channel MOSFET* a *channel* between the *drain* and *source* terminals, even under the standby *bias* condition (when *gate* electrode is at *zero bias*). To turn the transistor *off*, this *channel* needs to be *depleted* through the application of a negative voltage between the *gate* and the *channel*. This mode of operation is known as *depletion mode* (or *normally on*). In contrast to this mode of operation, we can imagine *n-channel MOSFETs* that require a positive *gate* voltage for induction of an *inversion channel* between the *source* and the *drain* n^+*-wells* (i.e., with positive *threshold voltage*). This latter mode of operation is known as *enhancement mode* (or *normally off*).

Depletion-mode MOSFETs, due to their standby power consumption, are less appealing to digital applications than *enhancement-mode MOSFETs*. Considerable reduction of the number of *surface* dangling *bonds* by switching from *(111) silicon* substrates to *(100) silicon* proved to be beneficial toward the effective reduction of the positive fixed oxide-charge concentration and hence realization of *enhancement-mode n-channel MOSFETs*.

The other *BJT*-inherited early technological difficulty in realization of *MOSFETs* was produced by the use of *aluminum* as the *gate* electrode. While *aluminum* serves as an appropriate metal in realization of the electrodes of *BJTs*, its low-melting temperature creates hassles for alignment between the inner sidewalls of the n^+*-wells* of *source* and *drain* and the *gate* electrode-oxide stack in *MOSFETs*. In the absence of such an alignment, as noticed in Section 2.4.4, the *inversion channel* of a *MOSFET* will not be able to respond to the *gate* potential quickly enough. In addition, caused by this misalignment and the resulting presence of a *p–n* junction, in conflict with the operation of a *MOSFET*, the *drain–source* conduction takes on a *blocking* character. While for achieving such a precise alignment, technologists require the use of a *gate* electrode-oxide stack as the masking material in realization of n^+*-wells* of *source* and *drain*,[2] the high-temperature requirements of the *doping* process (which are not compatible with the melting temperature of *aluminum*) render this choice of electrode incompatible with realization of *gate* in *MOSFET* technology. Replacing *aluminum* with a *highly doped* deposited film of *polycrystalline silicon* proved quite effective in alleviating this technological challenge. More recently, instigated by the need to improve the *conductivity* of *gate* electrode, *polysilicon gates* have been replaced by *copper gates*.

Complementary realization of *n*- and *p-channel enhancement-mode MOSFETs* in *silicon CMOS*[3] technology in 1963, and its role in effective reduction of the static power consumption of digital *cells*, is the other, even more, important factor that has strengthened the hegemony of *FET* technology particularly in digital applications. While realization of *CMOS* in other semiconductor technologies is being pursued, the lack of a high-quality thermally grown oxide (i.e., with an acceptable low *defect density interface* with the semiconductor) in semiconductor technologies other than *silicon* has caused those endeavors to be much less rewarding. Although the superb properties of thermally grown SiO_2 up until recently have served the *silicon CMOS* technology very successfully, the discussions of this chapter will show that *silicon* technology is now in dire needs of finding a substitute *gate* insulator.

Due to the aforementioned advantages of *MOSFET* technology, this technology has succeeded in replacing *BJT* technology in both memories and logic circuits. Another advantage is the simplicity of *MOSFET* structures (especially in earlier generations) resulting in their improved device reliability and fabrication *yield*.[4] These improvements opened more daring horizons to ultra large-scale integration (i.e., *ULSI*).

Despite these benefits, over the decades since the first realization of *MOSFETs*, *bipolar* transistors have managed to hold on to a number of more or less eroding advantages. These advantages are essentially linked to our earlier observation of the three-dimensional mode of *carrier transport* between the *emitter* and *collector* in a *BJT* compared to the two-dimensional *carrier transport* between the *drain* and *source* of a *MOSFET*. The three-dimensional *channel* of *bipolar* transistors offers these devices higher current, power, and frequency ratings, which are suitable characteristics for high-frequency power amplifiers in telecommunication systems. At least in the earlier years of technology, this advantage was even further boosted by the vertical definition of the *minimum feature size* of *BJTs* compared to the lateral definition of the *gate length* of *MOSFETs*. Whereas the vertical definition of the *quasi-neutral* region of the *base* in *BJTs* depends on the degree of accuracy of the *doping* process, reduction of the laterally defined *minimum feature size* of the *gate* in

[2] For obvious reasons, this is known as a *self-aligned technology*.

[3] Standing for *Complementary Metal Oxide Semiconductor*.

[4] *Yield* is a technology term representing the number of *chips* operating according to a certain standard of operation versus the total number of *chips* realized concurrently through the *planar* fabrication process.

MOSFETs is bound by the limits of the techniques of *lithography*. Over the past few decades, manyfold improvement on the resolution of the techniques of *lithography* has managed to reduce the *gate length* of *MOSFETs* from about 30 μm in the early generations to the contemporary values smaller than 20 nm. Consequently, the original advantage of *doping* techniques in realization of features in the range of tens of *nanometers* has been essentially eroded.

The aforementioned improvement in the resolution of techniques of *lithography* for realization of *MOSFETs* with smaller *gate length* (which as will be observed in this chapter demonstrate lower delay and higher switching speeds) has been the source of the technological trend known as *technology scaling*. *Technology scaling* is a concept representing reduction in *minimum feature size* of *MOSFETs* and also feature sizes of *passive* elements in *MOSFET* circuitry. As a result, this concept indicates both improvements in switching frequency and packing density[5] of a *MOSFET chip*. A *technology scaling* strategy serves as a guideline in predicting the possible improvements in *chip* performance offered by a certain advance in a specific unit of process technology such as *lithography*. While a number of *scaling* strategies have been proposed over the years, one of most influential strategies had been the *constant-field scaling* strategy, which was proposed by Dennard et al. in 1974.[6] According to this strategy, the voltages and *doping* levels are *scaled* so that the *scaling* of the physical dimensions will result in no specific alteration in the internal electric-field strength. Figure 3.1c illustrates a summary of the guidelines of the *constant-field scaling* strategy in terms of the *scaling* factor *K*. The *constant-field scaling* strategy has limited scope of applications. Dominance of *high-field carrier transport* (and for *FETs* of even shorter *gate length*, *ballistic transport*) in the so-called short-channel MOSFETs imposes a conceptually different mechanism of operation in *scaled* devices. As a result, according to the models presented later in this chapter, predictions of *constant-field scaling* strategy have become irrelevant.

The *International Technology Roadmap for Semiconductors*[7] (i.e., *ITRS*) frequently updates the roadmap for the development of semiconductor devices according to *scaling* strategies and technological breakthroughs. It also generates a list of short-term and long-term needs of technology according to the strategies identified for memories and logic circuits.

3.2 PRINCIPLES OF OPERATION OF MOSFETs AND DEVICE MODELING: FIRST-ORDER PRINCIPLES

As noted in Section 3.1, the operation principles of *MOSFETs* have their roots in the characteristics of *MOS* capacitors. Based on the discussions of Section 2.4.4, in a *MOSFET* such as the one depicted in Figure 3.1a, abundant availability of *minority*

[5] For example, represented by the number of transistors per unit area on a digital *chip*.

[6] R. H. Dennard, F. H. Gaensslen, H.-N. Yu, V. L. Rideout, E. Bassous, and A. R. LeBlanc, "design of ion-implanted MOSFETs with very small physical dimensions," *IEEE J. Solid-State Circuits*, vol. **9**, no. 5, pp. 256–268, 1974.

[7] Check out the ITRS website at: http://www.itrs.net/

carriers in the close *unblocked* proximity of the semiconductor side of the *MOS* capacitor gives the *gate* electrode the possibility to swiftly *bias* the *MOS* capacitor under each of its three regimes of operation (i.e., *depletion*, *accumulation*, and *inversion*), even for relatively high-frequency *gate signals*.

Biasing the *MOS* capacitor of these structures in *inversion* will subsequently result in formation of a conductive path between the two *minority carrier* reservoirs (i.e., n^+-*wells* of the *source* and *drain* in an *n-channel MOSFET*). The *carrier* population, and as a result the *conductivity* of this *inversion-induced* quasi two-dimensional *channel*, are controlled through the *field effect* of the *gate*. Ideally in presence of a thick enough insulator with sufficient *band discontinuity* to semiconductor, no *DC* current will be established between the *gate* and this *inversion channel*. The application of an unequal *bias* across the *Ohmic MS* contacts formed to the two n^+-*wells* prompts current conduction through the *inversion channel*. The effectiveness of the *gate* in controlling the *conductivity* of this *channel* is identified through evaluating the *gate transconductance*. Since in the absence of the *inversion channel* the *source* to *drain* path converts into a front-to-back series combination of two *p–n* junctions, under this *bias* condition the current established between *drain* and *source* will be limited to the *reverse current* of a *p–n* junction. It is in terms of these modes of operation (i.e., *inversion* and *depletion*) that, respectively, the *on* and *off* states of operation of an *enhancement-mode MOSFET* are identified.

According to this elementary explanation of the operation of *MOSFETs*, in order to offer the needed *signal amplification* in an analog application or a swift transition between the *on* and *off* states in a digital application, it is evident that the *gate transconductance* should be as large as possible. While improving the *oxide capacitance* through reducing the thickness of the *gate* insulator can certainly improve the *gate*-induced electric field normal to the *transport channel* of a *MOSFET*, awareness of the presence of a *drain*-induced electric field along the *channel* (among other factors that will be spoken of later on in this chapter) is definitely needed in evaluation of the resulting improvement in *gate transconductance*. As will be explored at a deeper level in Section 3.2.2, the charge control of the *inversion channel* and improvement in *gate transconductance* hang in the balance of these two electric-field components. While originally, due to the much smaller value of the *gate*-insulator thickness in comparison to the *gate length*, favorably the *gate*-induced electric field was predominant, deeply scaled *MOSFETs* have left very little to this presumption.

Considering the contact to the so-called *body* of a *MOSFET* (which is the *well* within the boundaries of which a *MOSFET* is realized), we should pay attention to the fact that a *MOSFET* is a four-terminal device. Later in this chapter in Section 3.2.1.3, while analyzing the behavior of a *MOSFET*, consistent with Equation (2.172), we will look into the role of this *body* contact (known as the *body effect*) in modifying the *threshold voltage*. Figure 3.2 shows the circuit symbols adopted for *enhancement* and *depletion-mode n-* and *p-channel MOSFETs*. On these symbols, the direction of the *diode* symbol shown between the *source* and the *body* terminals identifies whether the *MOSFET* is *n-* or *p-channel*.

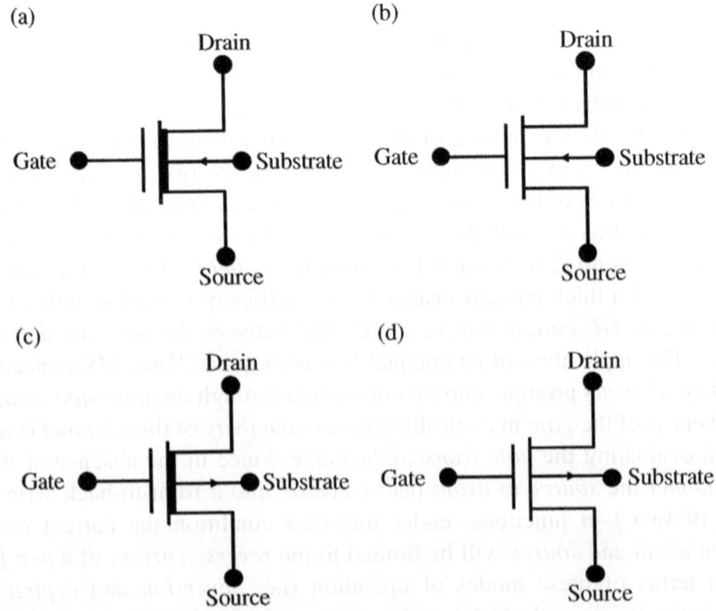

FIGURE 3.2 Circuits symbols for (a) a depletion-mode n-channel MOSFET, (b) an enhancement-mode n-channel MOSFET, (c) a depletion-mode p-channel MOSFET, and (d) an enhancement-mode p-channel MOSFET.

3.2.1 Modeling of the Operation of Long-Channel MOSFET

As a first step in producing a descriptive model for *MOSFETs*, oftentimes, undisputed by the *drain*-induced electric field, electrostatic dominance of the *gate* over the *channel* is assumed. In the first-order approximation, current across the *MOS* capacitor and the *reverse* current of *p–n* junctions are neglected. Although such models are quite descriptive of the behavior of early-generation *MOSFETs*, among other factors the so-called disputed *electrostatic integrity*[8] of the *gate* in *scaled MOSFETs* has left them quite insufficient. While these models are often referred to as *long-channel models*, assuming a literal correlation between this name and the *gate length* could be quite confusing. In Section 3.2.2, through the discussions of the so-called short-channel MOSFETs, we will attempt to identify a more precise metric for the applicability of the assumptions of the *long-channel models*. In terms of this description, in the present text whenever the terms *long channel* or *short channel* are used, it is in pure correlation with the applicability of the assumptions behind these models and not the actual *gate length*. Since *long-channel models* of *MOSFETs* provide a superb overview of the *technology scaling*, for gaining a historical insight into this trend (i.e., at least at the time of its inception), the present section provides a summary of these models.

[8] A quantitative definition of *electrostatic integrity* will be provided in Chapter 4.

In the presentation of a basic analytical model for the operation of a *MOSFET*, we first assume no potential difference between *source* and *body* terminals (while maintaining them at *zero* potential).[9] Later on, by assuming a V_{SB} different than *zero*, we will incorporate the so-called body effect into the model. In the following discussions we also assume that the *drain–source* distance (i.e., the *gate length*) is much longer than the *carriers' mean free path* and also longer than the width of the *depletion regions* of the two *p–n* junctions, which are formed between the *carrier* reservoirs and the *body*. As a result of these assumptions, a *drift–diffusion carrier transport* based on the notion of *low-field mobility* is first adopted. These stringent conditions will be somewhat loosened as we go further through the discussions of this chapter.

We start this treatment with the assumption of dominance of *drift transport* in the *channel* of the *MOSFET* when its *MOS* capacitor is *biased* under the condition of *strong inversion*.[10] Based on the discussions that are presented later in Section 3.2.2, this assumption can be substantiated by showing that the slow variation of the small *drain-induced* electric field along the length of the *linear channel* (i.e., in that part of the *channel* that is operating under the so-called *linear regime*[11]) and saturation of *carrier* velocity for *holes* and also electrons in many semiconductors in the high electric field part of the *channel* (i.e., in the rest of the *channel* that is operating under the so-called saturation regime[12]) limit the gradient of *carrier* concentration along the *channel*. As a result, this renders the *diffusion* current negligible. For semiconductors with *negative differential* electron *mobility*, due to the large value of potential drop in the part of the *channel* beyond the point of maximum *drift velocity*, this assumption may require a revisit.

According to the assumption of a *drift*-only *channel* in the *n-channel MOSFET* depicted in Figure 3.3, upon *biasing* the *MOS* capacitor into *inversion* and applying a *bias* between the *drain* and *source* terminals, the *drain–source* current passing through the *inversion channel* can be calculated as

$$I_D = WQ_n(y)v(y). \tag{3.1}$$

$Q_n(y)$ stands for the quasi two-dimensional *inversion* charge density per unit area at any arbitrary position y along the *channel*, $v(y)$ refers to the electron *drift velocity* at that point along the *channel*, and W stands for the *gate width* of the transistor (see Fig. 3.3). According to the assumption of *zero body*- and *gate*-leakage current, *current continuity* enforces current along the *channel* to be position independent. That is why no argument y has been indicated for I_D.

For small values of V_D, the *drain*-induced electric field along the *channel* is limited. As a result, the low-field electron *mobility* can be adopted in (3.1). Obviously, this is done through replacing $v(y)$ by $-\mu_n E(y)$, where μ_n is the low-field electron *mobility* in the *inversion channel* (i.e., under *strong inversion*). In the

[9] That is, the *built-in potential* of the *p–n* junction is neglected.

[10] See Section 2.4.4 for the definition of this condition.

[11] To be defined shortly.

[12] Also to be defined shortly.

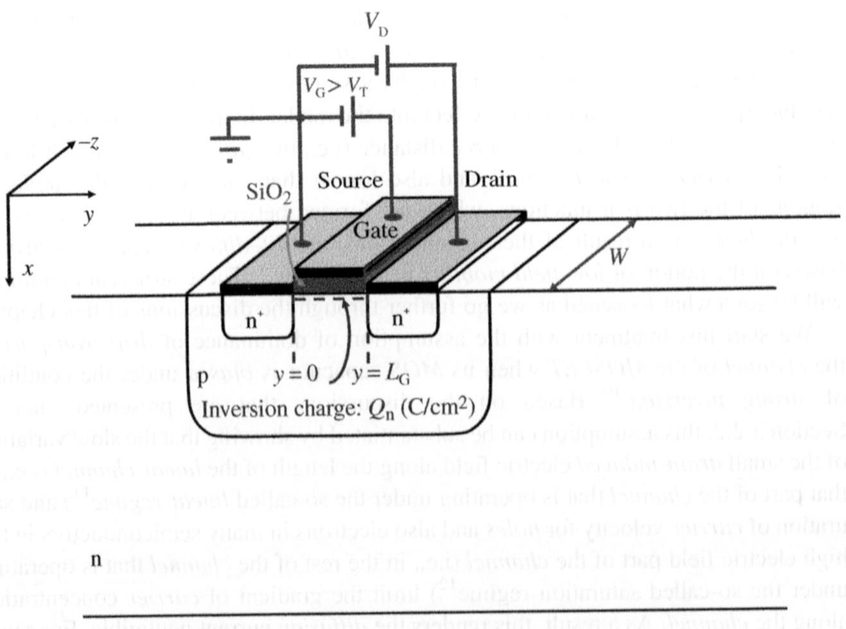

FIGURE 3.3 An n-channel MOSFET biased in inversion.

one-dimensional case, the *drain*-induced $E(y)$ is defined in terms of *channel* potential $V(y)$ as $-\partial V(y)/\partial y$.

In order to present a simple first-order analytical model for the *drain* current–voltage characteristics of *MOSFETs*, we also assume the *carrier mobility* (i.e., μ_n in case of *n-channel MOSFET*) to be position independent.

The conditions that we have so far considered in this model are often referred to as the *gradual channel approximation*. As stated earlier, according to these assumptions, both the *channel inversion* and *bulk depletion*[13] charge concentrations are fully under the control of the vertical *gate*-induced electric field. The passage of a current through the *inversion channel* and the resulting potential drop along the *channel*, according to (2.172), should prompt a nonlinear variation of the *threshold voltage* along the *channel*. By assuming the *carrier* concentration gradient in the *inversion channel* to be negligible, nonetheless we have implicitly assumed that the *threshold voltage* is position independent. The impact of such an assumption will be shortly evaluated within the context of the *long-channel MOSFET*.

With this set of assumptions and according to the descriptions provided for *MOS* capacitors in Section 2.4,

$$Q_n(y) = -C_{ox}[V_G - V_T - V(y)] \tag{3.2}$$

[13] See Section 2.4.2 for the correlation between the two charge profiles.

where C_{ox} is the per unit area *oxide capacitance* and V_T is the *threshold voltage*.

Substitution of (3.2) in (3.1) yields

$$I_D = WC_{ox}[V_G - V_T - V(y)]\mu_n \frac{\partial V(y)}{\partial y}. \tag{3.3}$$

Based on the earlier assumption of *current continuity* in the *channel*, (3.3) can be integrated along the length of the *channel* (i.e., L_G):

$$I_D \int_0^{L_G} dy = \mu_n W C_{ox} \int_0^{V_D} [V_G - V_T - V(y)] dV(y) \tag{3.4}$$

where *channel* potential at the *source* and *drain* sides are taken as *zero* and V_D, respectively.

According to (3.4),

$$I_D = \frac{\mu_n W C_{ox}}{L_G} \left[(V_G - V_T) V_D - \frac{1}{2} V_D^2 \right]. \tag{3.5}$$

The regime of operation that we have just explained is often referred to as the *linear regime*. Equation (3.5) is only applicable to small values of V_D, which are neither imposing a *channel* potential $V(y)$ larger than $V_G - V_T$ at any point along the channel nor a *drain*-induced electric field beyond the value corresponding to the onset of *drift-velocity saturation*. Failing to satisfy the first condition clashes with Equation (3.2) and electrostatic dominance of the *gate* over the *channel*. Failing to satisfy the latter contradicts the use of *low-field mobility* (i.e., μ_n).

As shown in (3.2), having a *drain bias* $V_D = V_G - V_T$ causes the *inversion channel* to be *pinched off* at its *drain* end (i.e., $y = L_G$).[14] Increasing V_D beyond this value results in expansion of the *depletion region* of the *body*-to-*drain* *p–n* junction and elimination of the *strong inversion* from the *channel* not just at its *drain* end but in the expanding proximity of the *depletion region* along the *channel*. Enforcing *current continuity* in this latter part of the *channel*, which has a very limited electron concentration, requires the electron population to achieve higher *drift velocities* (i.e., in inverse proportion to the reduced concentration of *inversion-layer* electrons). For a semiconductor like *silicon*, the need for a higher *drift velocity* demands the evolution of a higher *drain*-induced electric field in the *pinched-off* section of the *channel*. While speaking of *Schottky's depletion approximation* calls for an infinite *drift velocity* in the *pinched-off* section of the *channel* (which is certainly not obtainable), even without literally adopting the *SDA*, such a model, which lacks the notion of *drift-velocity saturation*, can still find itself restricted by the need for unphysical

[14] Again, the potential drop across the *source* and *drain* *p–n* junctions are neglected.

values of electric field or electric field discontinuity.[15] This is because the decaying electron concentration toward the *drain* end of the *channel* still demands an improved *drift velocity*. We will sometimes refer to the *pinched-off* section of the *channel* as *velocity-saturation region* or simply *saturation region*. The earlier part of the *channel* is called the *linear region*. Discussions in Section 3.2.2 will further clarify the origins of these names.

Due to the dependence of *carrier* population of the *inversion channel* on the *channel* potential as expressed in (3.2) and the nonlinear dependence of *drift velocity* on electric field, the *drain*-induced electric field along the *inversion channel* is not constant. However, the *gradual channel approximation* of the so-called long-channel MOSFETs justly assumes $E(y)$ to be approximately constant in the *linear region* but not at the vicinity of the *pinched-off* section. This assumption is justified by the much larger variation of the *drain*-induced electric field along the *pinched-off* section of the *channel* and its close proximity. *Long-channel models*, while maintaining a purely linear *drift-transport* characteristic, assume the electric field along this latter part of the *channel* to be not as strong as the electric field at the onset of *drift-velocity saturation*.

In terms of these approximations, while the *drain* current calculated according to Equation (3.5) is expected to increase with V_D, this increase will continue only to the onset of *pinch-off* at the *drain* end of the *channel*. According to (3.5), under this *bias* condition (i.e., $V_D = V_G - V_T$, which is known as the V_{Dsat}), I_D is evaluated by

$$I_D = \frac{1}{2}\mu_n C_{ox}\frac{W}{L_G}(V_G - V_T)^2. \tag{3.6}$$

Validating the long list of assumptions of the *long-channel model* certainly requires careful inspection of the *bias* and structural information of the device.

3.2.1.1 Channel-Length Modulation The regime of operation of a *MOSFET* when *drain* voltage is in excess of V_{Dsat} is known as *saturation*. Under this regime of operation, any increase in V_D in excess of $V_G - V_T$ is assumed by the *long-channel models* to drop across the *pinched-off* section of the *channel*. This potential drop, in terms of the assumptions of these models, induces no direct enhancement in the *drain* current by improving the *drain*-induced electric field and *drift velocity*. However, since increase in V_D beyond this limit signals the expansion of the *pinched-off* region in the direction of the *source* terminal, considering the constant potential drop of $V_G - V_T$ along the shrinking length of the *inversion channel*, a *drain* current dependence on V_D is expected. Such an effect, which is referred to as *channel-length modulation*, is taken into account by subtracting the length of *pinched-off* section of the *channel* (i.e., ΔL) from the total *gate length*. As a result, for V_D greater than $V_G - V_T$,

$$I_D = \frac{1}{2}\mu_n C_{ox}\frac{W}{L_G - \Delta L}(V_G - V_T)^2 \tag{3.7}$$

[15] This will be shown in Section 3.2.1.1.

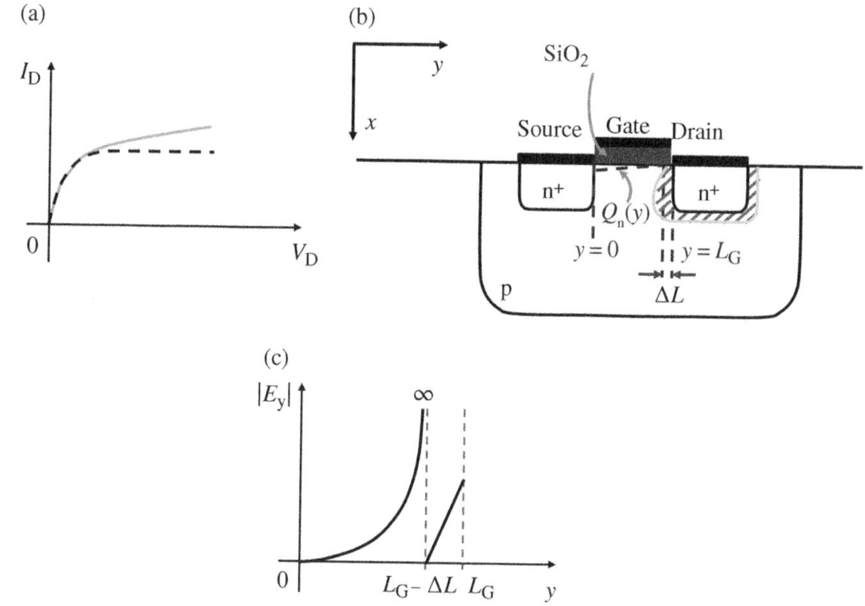

FIGURE 3.4 (a) Effect of channel-length modulation on a typical inversion-biased drain current–voltage characteristic of a MOSFET (which is shown in the dashed black curve) illustrated in gray. (b) Schematic depiction of the effect of channel-length modulation on the inversion charge concentration profile shown on the cross-sectional view of a MOSFET. The depletion region around the n^+-well of the drain is hash marked. (c) Prediction of gradual channel approximation for the variation of the electric field along the length of the inversion channel when channel-length modulation is in effect.

where ΔL can be evaluated through calculating the width of *depletion region* of a *p–n*$^+$ junction *reverse biased* at $V_D - V_G + V_T$.[16] Expansion of ΔL with V_D gives the *long-channel MOSFET* a finite output resistance under the *saturation regime* of operation (which is evaluated by $r_d^{-1} = \partial I_D / \partial V_D$). In evaluation of I_D on the basis of (3.7), applicable *drain biases* are limited by the *breakdown* processes of the *channel* and the *gate* contact (since these processes are not considered in the present model).

Figure 3.4 provides a qualitative indication of the effect of *channel-length modulation* on the *drain* current–voltage characteristics of a *MOSFET* (when V_G is increased beyond the *onset of strong inversion*: V_T). Figure 3.4b identifies ΔL on a cross-sectional view of a *MOSFET*.

[16] Which, in terms of the *dopant* concentration of the *bulk* (i.e., N_a), is approximately equal to $\sqrt{\frac{2\epsilon_s(V_D - V_G + V_T)}{qN_a}}$.

Combining (3.3) and (3.7) yields the following equation for the *drain*-induced electric field along the early part of the *channel* (i.e., $0<y<L_G-\Delta L$):

$$E(y)=\frac{V_G-V_T}{2(L_G-\Delta L)\sqrt{1-\frac{y}{L_G-\Delta L}}}. \tag{3.8}$$

Equation (3.8), as suggested earlier, unrealistically forecasts the electric field at the left-side boundary of $y=L_G-L$ to tend to infinity. In addition, the assumption of *pinch-off* at the right side of this point (i.e., for $L_G-\Delta L<y<L_G$), according to (2.55), predicts,

$$E(y)=-\frac{qN_a}{\epsilon_s}(y-L_G+\Delta L). \tag{3.9}$$

Comparison between (3.8) and (3.9) reveals another problem of this *long-channel model*, which is the discontinuity in the longitudinal electric field. This is shown in Figure 3.4c. As will be seen in Section 3.2.2, these unphysical predictions of the *long-channel model* are somewhat acceptable when ΔL is just a small fraction of L_G.

In a properly designed MOSFET, *channel-length modulation* is minimal. Given this condition, according to (3.6), *gate transconductance* under the *saturation regime* of operation is provided by

$$G_m=\frac{\partial I_D}{\partial V_G}=\mu_n C_{ox}\frac{W}{L_G}(V_G-V_T). \tag{3.10}$$

3.2.1.2 Drain-Induced Barrier Lowering At this point in the discussion, while not yet attempting to offer a list of shortcomings of the *long-channel model*, it is quite educational to briefly visit one of these limitations. This discussion is intended to identify the part of the *channel*, which is in most need of electrostatic dominance of the *gate*.

Although quite innocently, we have so far maintained the potential of the *source* terminal (i.e., reference potential) to be equal to the *channel* potential at its *source* end (i.e., $V(y=0)$), we have not paid much attention to the presence of a *p–n* junction between the *source*-side *carrier* reservoir and the *channel*. In conjunction with this presence, however, the observed increase in ΔL with V_D and the effective increase in the electric field across the *linear* part of the *channel*, particularly in *MOSFETs* of short *gate length*, serve to increase the *channel* population. The increase in the *forward bias* across this junction, according to the explanations of Section 2.4.2.2, can remove the *channel* from the electrostatic control of the *gate*. Electrostatic dominance of the *gate* on this part of the *channel* is of crucial importance in the operation of *MOSFETs*. This effect, which is known as *drain-induced barrier lowering* (*DIBL*), prompts a *gate*-independent rush of *carriers* from the *source* reservoir to the *channel* and causes an increase in the *drain* current even in

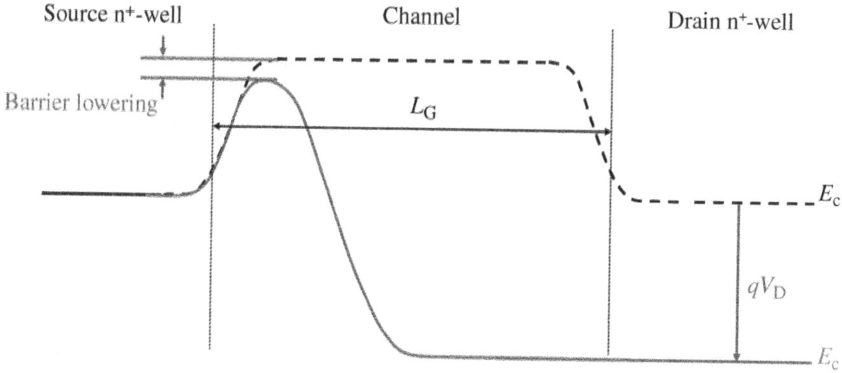

FIGURE 3.5 Schematic illustration of the effect of drain-induced barrier lowering on the conduction band edge diagram of a MOSFET when under thermal equilibrium (the drain-side band edge is shown in dashed curve) and when biased with a large positive V_D, where the band edge is shown by the continuous gray.

excess of the expected increase from *channel-length modulation*. This is only one of the appearances of the so-called *short-channel* operation (which will be studied at length in Section 3.2.2). The *band diagram* representation of Figure 3.5 highlights the appearance of *DIBL*.

In the description of the modulating effect of the *gate* voltage on the *conductivity* of the *channel* of a *field effect* transistor, it is sometimes left out that it is in terms of raising or lowering the potential barrier between the *source* and the *channel* (and not the *carrier* concentration in the *channel*) that this modulation takes effect. While under low *drain* voltages this modulation results in a voltage-controlled resistor, under high *drain* voltages the *field effect* transistor operates as a voltage-controlled current source.

3.2.1.3 Body Effect A careful inspection of (2.172) reveals that in evaluation of the *inversion* charge density, Equation (3.2), by assuming a constant *threshold voltage*, has overlooked the spatial variation of the *channel* potential in the presence of a finite V_D. *Shorting* the *body* and *source* terminals can set the *carrier* reservoir-to-*bulk* potential difference (i.e., V_{CB}) in (2.172) to *zero* at the *source* end of the *channel* (i.e., with the assumption of no potential drop across the *body* and *source MS* and *p–n* junctions). In the presence of a finite *drain bias*, such an assumption is certainly not applicable to the rest of the *channel*. As a result, on the basis of (2.172), (3.2) needs to be modified to

$$Q_n(y) = -C_{ox}[V_G - V(y) + V_B - \phi_{FB} - 2|\phi_b|] + \sqrt{2\epsilon_s q N_a [2|\phi_b| + V(y) - V_B]}. \quad (3.11)$$

This equation, just like (3.2), neglects the role of *drain*-induced electric field in populating the *inversion channel*.

Inserting (3.11), instead of (3.2), into the Equation (3.1) and integrating along the length of the *channel* results in the following equation for the *drain–source* current under the *linear regime* of operation:

$$I_D = \mu_n \frac{W}{L_G}\left\{ C_{ox}\left(V_G - \phi_{FB} - 2|\phi_b| - \frac{1}{2}V_D + V_B\right)V_D \right. \\ \left. - \frac{2}{3}\sqrt{2\epsilon_s q N_a}\left[(2|\phi_b| + V_D - V_B)^{3/2} - (2|\phi_b| - V_B)^{3/2}\right]\right\}. \tag{3.12}$$

This equation, although more accurate than (3.5), is much less convenient to use. In (3.12), instead of assuming the *source* and *body* terminals to be tied together, the potentials of the *drain, gate*, and *body* terminals of a *MOSFET* are compared to that of the *source* as the reference potential.

In this derivation, with the exception of the *body effect* and the leniency in taking any potential at the *source* and *body* terminals, all other aforementioned assumptions of the *long-channel model* are still in place.

Instead of using (3.12), a heuristically modified version of (3.5) has also been proposed. This modified equation, as shown below, takes advantage of a fitting parameter α referred to as *bulk-charge factor*:

$$I_D = \frac{\mu_n W C_{ox}}{L_G}\left[(V_G - V_T)V_D - \frac{\alpha}{2}V_D^2\right]. \tag{3.13}$$

A typical value of this fitting parameter among *long-channel MOSFETs* is 1.5. According to this empirical equation, *drain* current maximizes when $V_D = (V_G - V_T)/\alpha$. This is taken as the modified value of V_{Dsat}. Considering this modified value, under the *saturation regime* of operation,

$$I_D = \frac{1}{2}\mu_n C_{ox}\frac{W}{\alpha(L_G - \Delta L)}(V_G - V_T)^2. \tag{3.14}$$

The evaluation of the *body effect* presented in this section is accurate as long as the assumptions of the *long-channel model* are validated. For *short-channel MOSFETs*, both the *drain-* and *gate-*induced electric fields imposed on the *channel* should be considered in a two-dimensional context to evaluate the *body effect*.

3.2.1.4 Subthreshold Operation Whereas so far in Section 3.2.1, we have focused on modeling the operation of *MOSFETs* when *gate* voltage exceeds the *threshold voltage* (i.e., when the *surface* potential is beyond the *onset of strong inversion*), circuit designers are more and more *biasing* these transistors in the so-called subthreshold regime of operation. As given by its name, in this regime of operation in spite of the *biasing* the *MOS* capacitor in *inversion*, the *surface* potential falls short of the *onset of strong inversion*. The demarcation between the *subthreshold regime* of operation and the operation beyond the *onset of strong inversion* are often expressed in terms of the *gate* voltage at which one sees a transition from the exponential dependence of the *inversion channel carrier* concentration on V_G to a more or less linear dependence.[17] The popularity of this mode of operation has been an outcome of *technology scaling*. As *scaling* improved the packing density of *chips*, managing the heat generated by the large number of transistors[18] of a *chip* and also the battery lifetime became formidable challenges for circuit designers. Among the explored directions for reducing the complexity of heat/battery management and limiting the degrading impact of *self-heating* on *carrier mobility* and device performance is *subthreshold* circuit design. In this design strategy, through *biasing* a *MOS* capacitor short of *strong inversion*, the *drain* current and *self-heating* of a given transistor are reduced. This design strategy also promises longer battery lifetime.

In addition to reduction of *self-heating* in *subthreshold* circuit design, this design strategy was also promoted to reduce battery voltage of *scaled CMOS* circuitry. As will be observed in Section 3.2.2, *scaling*-dictated reduction of the physical dimensions of *MOSFETs* (given the same *bias* conditions) causes an increase in the electric-field strength. As a result, prevention of *breakdown* has imposed a considerable limitation on the maximum battery voltage. This reduction in the battery voltage has shifted the *gate* voltage toward the *subthreshold* conditions.

Understanding all these issues requires an insight into the operation of the so-called short-channel MOSFETs. At this point in our discussion, even with the limited insight of *long-channel models*, it is beneficial to evaluate the situation. Later in Section 3.2.2, based on the insight of the *short-channel models*, the requirements of *scaling* in the *subthreshold regime* of operation will be identified.

The assessment of the *subthreshold* regime of operation is often performed through evaluating the so-called transfer characteristic of a *MOSFET* and its *subthreshold swing* (*SS*). As shown by the example of Figure 3.6, the *transfer characteristic* is the logarithmic representation of the *drain* current versus the linear variation of *gate* voltage at a given *drain–source bias*. In an *NMOS* transistor, as expected from (3.7) or (3.14), for *gate* voltages above the *threshold voltage*, the logarithm of I_D is only slightly variable with V_G. The exponential dependence of the electron concentration at the *Si*/*SiO*$_2$ *interface* on the *surface potential* and on the *gate* voltage, for *gate* voltages less than the *threshold voltage* indicates a faster decline in $\log I_D$ with V_G.[19] As marked in Figure 3.6, the inverse of the slope of the *transfer characteristic* in the *subthreshold regime* is known as *SS*. This value is often reported in terms of the mV reduction in the *gate* voltage (below the *threshold voltage*) needed for reducing the *drain* current by a factor of 10 (i.e., in units of mV/dec). Evaluation of the rate of this decline is at the very core of understanding the *subthreshold* operation of a *MOSFET*. The smaller the *SS*, the more effective is the *gate* in turning the transistor *off*.

Under the *subthreshold* regime of operation, the *MOSFET* should be seen as a front-to-back series combination of two n^+–n^- junctions. This is because under this regime the *gate* voltage of the *NMOS* has not yet been able to produce a sizeable

[17] See Equations (2.157) and (2.164).

[18] That is, currently exceeding the 1 billion mark.

[19] See Equations (2.157) and (2.158).

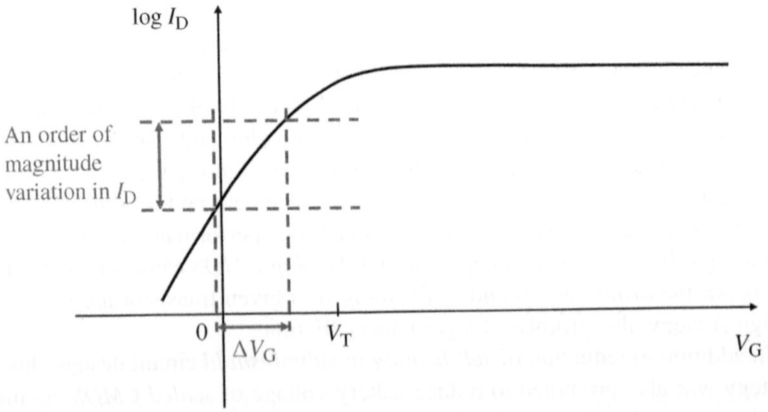

FIGURE 3.6 A typical transfer characteristic of an enhancement-mode n-channel MOSFET. According to this diagram, the value of the subthreshold swing is given by ΔV_G in units of mV/dec. On this diagram, while V_G is variable, V_D is constant.

inversion charge layer at the semiconductor *surface*. It is crucial to point out that under this regime of operation, the n^- surface cannot be purely seen as a large resistivity *channel*. Such a simplification will prevent us from observing the implications of the presence of these two junctions.

According to the above explanation, under the *subthreshold* regime of operation, V_{DS} is primarily dropped across the *depletion* region of the *reverse-biased* n^+–n^- junction at the *drain* side of the *channel*. This leaves the rest of the *channel* with a negligible electric field. As expected, this means that the contribution of *drift transport* to the *subthreshold drain* current is negligible. This picture of close-to-*zero* electric field, at least in part of the *channel*, is especially correct for *long-channel MOSFETs* for which the *depletion* region of the *reverse-biased* n^+–n^- junction of *drain* side cannot reach all the way to that of the *source* side.

Despite the negligible amount of *drift* current, *diffusion* current under this regime of operation becomes sizeable. This is due to the induction of a large gradient of electron-concentration profile along the *channel*. Application of V_{GS} and the resulting increase in the *surface* potential (i.e., ϕ_s) reduces the potential barrier between the *channel* and *source*-side *carrier* reservoir. We can readily see this evolution of the *diffusion* current in analogy with the *base* current of a *BJT* (see the *band diagram* of Fig. 3.7).

As a result of the exponential dependence of current flow across this junction on the height of the *source*-side reservoir to *channel* potential barrier, increase of V_G causes an exponential increase in the *drain* current. This can be intuitively approximated by

$$I_D \approx I_{D_0} \exp\left(\frac{qV_{GS}}{nkT}\right). \tag{3.15}$$

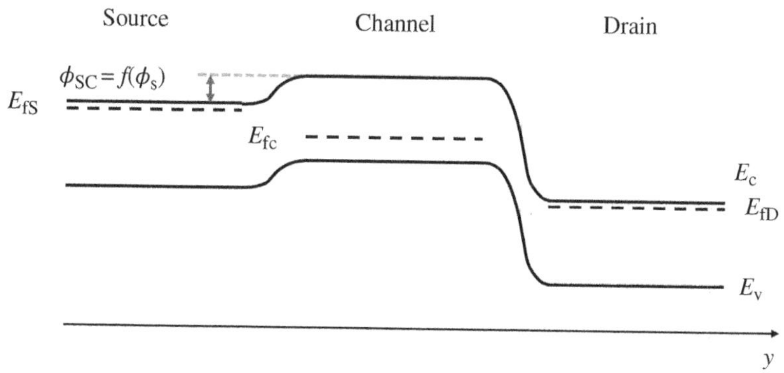

FIGURE 3.7 Band diagram along the channel length of an enhancement-mode MOSFET in subthreshold bias when positive drain and gate biases are applied in common-source configuration. As indicated, the barrier height from source to channel is a function of the surface potential and effectively V_G.

The presence of n in (3.15) represents the fact that the *surface* potential (i.e., ϕ_s) while related to V_{GS} is not equal to it.[20] In terms of the voltage division across the series combination of *oxide-capacitance* C_{ox} and the *depletion-capacitance* C_s, the *surface* potential can be linked in the following form to the parameter n and V_{GS} in (3.15):

$$\frac{1}{n} = \frac{d\phi_s}{dV_{GS}} = \frac{C_{ox}}{C_s + C_{ox}}. \tag{3.16}$$

The *small-signal depletion-capacitance* C_s is defined by ϵ_s/x_d and C_{ox} is equal to ϵ_{ox}/t_{ox}. x_d is the *depletion-region width* and ϵ_s is the dielectric constant of the semiconductor.

The approximate exponential dependence of I_D on V_{GS}, presented in (3.15), indicates a linear *transfer characteristic* in the *subthreshold regime* of operation. According to (3.15) and (3.16), this linear *transfer characteristic* presents an *SS* of

$$\mathrm{SS} = n\frac{kT}{q}Ln10 \cong 60n \text{ mV/dec at room temperature.} \tag{3.17}$$

In order to achieve a minimum value of *subthreshold swing*, n needs to be as close as possible to *one*. As suggested by (3.16), ordinarily this is done by maximizing the *oxide capacitance*. This observation supports our earlier argument that a larger *oxide capacitance* normally prompts a more dominant *gate* action.

In this section, we have assumed an ideal *surface* devoid of *surface states*. The presence of *surface states* modifies (3.17) by making ϕ_s a smaller portion of V_{GS}.

[20] See the discussions of Section 2.4.4.

In addition, in Equation (3.17) the finite thickness of the *inversion layer* and the so-called quantum capacitance have been neglected.[21]

3.2.2 Modeling of the Operation of Short-Channel MOSFET

As we have already mentioned, with *scaling* the *gate length* of *MOSFETs*, due to the enhancement of the *drain*-induced electric field, the electrostatic dominance of the *gate* electrode over the *channel* is becoming less obvious. The large *drain*-induced electric field of the so-called short-channel MOSFETs contradicts a number of major assumptions made in formulating the *long-channel models* of Section 3.2.1. The appearances of these contradictions are known as *short-channel effects* (*SCE*). Among these are *velocity saturation* and the degradation in *differential mobility*.

Many of these appearances, depending on the *bias*, are present to different degrees in all *MOSFETs*. As alluded to earlier, a clear-cut demarcation between the *short-* and the *long-channel MOSFETs* in terms of the *gate length* cannot be made. A common way to demarcate between a *long-channel* and a *short-channel MOSFET* is provided in terms of the roll-off of the *threshold voltage* with the *gate length* in devices of shorter physical *gate length*. In these terms, we can roughly define a *MOSFET* as *short channel* when its *gate length* is comparable to the *depletion-region width* of the *drain* contact. Later in this chapter, in the *subthreshold* regime of operation, a more accurate length scale for the limits of *scaling* before the *SCE* considerably alter the device operation will be presented.

In the so-called short-channel MOSFETs, V_T reduces with the *drain bias* either due to DIBL, *charge sharing* of the *source* and *drain* with the *channel*, or occurrence of sub*surface punch-through*. Due to the concurrent presence of more than just one of these effects at higher *drain* voltage, further roll-off with *gate length* is observed when the *threshold voltage* is experimentally evaluated in the *saturation regime* of operation (i.e., compared to the *linear regime*). These experimental evaluations can be performed by watching the evolution of the *transfer characteristic* from the linear dependence of I_D on V_G to an exponential or via examining the intersection of the linear part of $I_D - V_G$ characteristics with the voltage axis.

The *drain/source charge sharing* refers to the overlap of the *depletion* regions of the *MOS* capacitor and the *p–n*$^+$ junctions of *source* and *drain*. As a result of this overlap, less of the actual *gate* voltage is required to induce *depletion* and instigate the formation of the *inversion channel*. As a result, V_T will become smaller. The closer the two n^+ wells of *source* and *drain* are (i.e., the smaller the L_G), the proportionally smaller is the part of the *depletion* region of the *MOS* capacitor not affected by this issue. As a result, the smaller would be the *threshold voltage*. This makes *charge sharing* a *short-channel* issue alongside the aforementioned mechanism of *DIBL*. As a characterization tool for these types of behaviors, it is worth noting that in devices suffering from *DIBL*, the *transfer characteristic* translates horizontally between the low and the high V_{DS} values. As a result, *DIBL* is quantified in terms of *millivolts*

[21] See the discussions in Section 2.4.2. We will deal with *quantum capacitance* in Section 3.3.

of translation *per volt* of change in the *drain bias*. In a properly designed *MOSFET*, *DIBL* is less than a 100 mV/V.

Subsurface *punch-through*, which is the result of the merger of the two high electric-field *depletion regions* of *source* and *drain* to *bulk p–n*$^+$ junctions, will also become more probable as L_G is reduced. Due to the better electrostatic dominance of the *gate* electrode closer to the insulator/semiconductor *interface* and the reduction of the level of *acceptor* concentration in a *diffusion-doped bulk* from surface down, such a merger is much less probable toward the *surface*. However, below the semiconductor *surface*, especially at higher *drain–source biases*, this merger becomes more plausible. In order to reduce this chance, the junction depths of the *carrier* reservoirs are also among the physical dimensions that are being *scaled* down concurrently with L_G (see as an example Fig. 3.1c). According to *ITRS*, this junction depth (i.e., X_j) is usually designed to be about 30% of the *gate length* in a *planar MOSFET*. To prevent *punch-through*, *ion-implantation doping*, with certain restrictions, is used to increase the *bulk dopant* concentration at the depth where the *punch-through* is most plausible. This *dopant* profile is known as *punch-through implant*. Preventing the leaking of the *punch-through implant* toward the *channel*, and essentially averting the degradation in *channel*'s *carrier mobility* and increase in *threshold voltage*, in *MOSFET*s where L_G is deeply *scaled* requires very small values of not only *projection range* (and essentially ion energies) but also *vertical straggle*. In realizing the shallow profiles needed in implementing the *punch-through* and the *carrier* reservoir *implants*, the so-called *broadening* and *blooming* of the low-energy *ion beam*, however, pose important challenges.

Our goal in this part of the discussion is to provide an approximate representation of the appearances of *SCE* in the operation of *short-channel MOSFETs*. This starts with modifying the governing equations proposed earlier in Section 3.2.1. In this section, according to a chronological order, *short-channel models* with greater versatility and accuracy are presented. The assumptions expressed in Section 3.2.1, unless specifically indicated, remain intact. In terms of the models presented in this section, technological requirements of *scaling* will be formulated and reliability of operation will be assessed.

3.2.2.1 Short-Channel Model: A First-Order Treatment As a first step toward producing a *short-channel model*, the *drain* current of a *short-channel MOSFET*, with incorporation of the *drain*-induced differential-*mobility* degradation, can be represented as

$$I_D = WC_{ox}[V_G - V_T - V(y)]\frac{\mu_{eff}E(y)}{1+\dfrac{|E(y)|}{E_{sat}}}. \tag{3.18}$$

In this equation, which is a modified version of (3.3), variation of *drift* velocity in terms of the *drain*-induced electric field $E(y)$, according to (1.90) with the conveniently made assumption of C_2 equal to *one*, has been incorporated. According to this assumption, knowing the *saturation velocity* v_{sat}, which is an intrinsic property of a given semiconductor, the magnitude of the electric-field strength that induces velocity saturation (i.e., E_{sat}) can be calculated by

$$E_{sat} = \frac{2v_{sat}}{\mu_{eff}}. \tag{3.19}$$

Whereas (3.18) still builds on the foundation of (3.2), which is the dominance of *gate* in forming the *inversion layer*, use of μ_{eff} instead of low-field electron *mobility* (i.e., μ_n) allows for the consideration of the effect of the *gate*-induced electric field in degrading the electron *mobility* along the *channel*. A plausible scenario for this degradation is the enhanced proximity of the centroid of the *inversion layer* to the insulator/semiconductor *interface* with increasing the *gate bias* (which can increase the *interface roughness scattering* and *scattering* due to the presence of *interface states*). Since so far we have assumed the *gate* in charge of the *inversion-layer* induction and the *drain* in charge of *carrier transport*, this is the first instance in the *MOSFET* modeling that we are proposing a correlated involvement of the *gate*- and *drain*-induced electric-field components in one of these two processes. Here, in terms of the *inversion* and *depletion*-layer charge densities (i.e., Q_n and Q_d, respectively),[22] the effective *gate*-induced electric field E_{eff} normal to the semiconductor surface, evaluated approximately at the center of the *inversion* layer, is estimated by

$$E_{eff} = -\frac{Q_n}{2\epsilon_s} - \frac{Q_d}{\epsilon_s}. \tag{3.20}$$

where, according to (2.170) and (2.172) and assumption of *zero* V_{CB},

$$Q_d = -C_{ox}(V_T - 2|\phi_b| - \phi_{FB}). \tag{3.21}$$

Since in the absence of the *drain bias* (3.2) produces

$$Q_n = -C_{ox}(V_G - V_T) \tag{3.22}$$

inserting (3.21) into (3.20) grants the following definition for the effective *gate*-induced electric field:

$$E_{eff} = \frac{\epsilon_{ox}}{\epsilon_s t_{ox}}\left(\frac{1}{2}V_G + \frac{1}{2}V_T - 2|\phi_b| - \phi_{FB}\right). \tag{3.23}$$

In terms of this *effective gate*-induced electric field, μ_{eff} is presented by

$$\mu_{eff} = \frac{\mu_0}{1 + (E_{eff}/E_0)^n}. \tag{3.24}$$

[22] See Section 2.4.

TABLE 3.1 Values of the Parameters of the Empirical Expression (3.24).

Type of the Channel	Parameter		
	μ_0 (cm^2/V s)	E_0 (MV/cm)	n
Electron channel formed at the SiO_2/Si interface	670	0.67	1.6
Hole channel formed at the SiO_2/Si interface	160	0.7	1

Reprinted with permission from Liang et al. (1986). © 1986 IEEE.

The values of the constants of this empirical model (i.e., μ_0, E_0, and n) for *NMOS* and *PMOS silicon inversion channels*, evaluated at room temperature, are presented in Table 3.1.

Through replacing $E(y)$ by $-\partial V(y)/\partial y$ in (3.18), and integrating over the length of the *channel*, we will have

$$I_D = \frac{\mu_{eff} W C_{ox}}{L_G}\left[V_G - V_T - \frac{V_D}{2}\right] V_D \frac{1}{1 + (V_D / E_{sat} L_G)}. \tag{3.25}$$

This equation is the counterpart of (3.5) for evaluating I_D of *MOSFETs* under the *linear regime* of operation.

Due to the magnifying role of *gate-length scaling* on the longitudinal electric field (i.e., $E(y)$), in *short-channel MOSFETs* the *saturation* of *drain* current happens at *drain biases* short of the *pinch-off* mark (i.e., $V_G - V_T$). This is the outcome of *drift-velocity saturation*. Identifying the *drain* voltage at which the *drift velocity* at the *drain* end of the *gate* of a *MOSFET* reaches *saturation velocity* by V_{Dsat}, we can use (3.1) to write

$$I_{Dsat} = W C_{ox} (V_G - V_T - V_{Dsat}) v_{sat}. \tag{3.26}$$

This is of course obvious because at the *drain* end of the *gate*, the density of *inversion*-layer *carriers* flying at the *saturation velocity* is equal to

$$Q_{nD} = -C_{ox}(V_G - V_T - V_{Dsat}). \tag{3.27}$$

The value of V_{Dsat}, also known as the *knee voltage*, can then be calculated by converging (3.25) and (3.26). This convergence results in[23]

$$V_{Dsat} = \frac{E_{sat} L_G (V_G - V_T)}{E_{sat} L_G + (V_G - V_T)}. \tag{3.28}$$

The *drain* voltage V_{Dsat} indicates the onset of I_D saturation (i.e., a transition from the so-called *linear* to the *saturation regime* of operation).

[23] Sometimes, in analogy with parallel resistors, we will write this as $E_{sat} L_G || (V_G - V_T)$.

Equation (3.28) provides an interesting approach to defining a *short-channel MOSFET*. As already observed in Section 3.2.1, while in a *long-channel MOSFET* $V_{Dsat} = V_G - V_T$, the above equation will converge to this definition only if $E_{sat}L_G \gg V_G - V_T$. As a result, we call a *MOSFET long channel* only if $E_{sat}L_G \gg V_G - V_T$. According to this equation, the definition of a *short-channel MOSFET* is not only provided in terms of L_G but also its product with E_{sat} and in comparison to $V_G - V_T$. As E_{sat} is *bias* dependent (see Eqs. (3.19), (3.23), and (3.24)), a physically *short-channel MOSFET* can be *biased* into a *long-channel* mode and vice versa. Now if we insert (3.28) into (3.26), we observe that while for a *long-channel MOSFET* according to (3.6) the *drain* current in the *saturation regime* is quadratically dependent on $V_G - V_T$, for a deeply *short-channel MOSFET* (i.e., when $E_{sat}L_G \ll V_G - V_T$), this dependence is merely linear. For a *drain bias* exceeding V_{Dsat}, since we expect the longitudinal electric field to be stronger and *velocity saturation* to happen sooner in the *channel* (i.e., before its *drain* end), *channel-length modulation* will result. This yields a finite output resistance in the *saturation regime* of operation.[24] Since the presented *short-channel model* under appropriate circumstances is capable of producing the same results as the *long-channel model*, it is also referred to as the *universal model*.

By incorporating the electric-field dependence of *differential mobility*, and especially *drift-velocity saturation*, the present model yields the $I_D - V_D$ characteristics without imposing unrealistic assumptions of infinite electric field and electric-field discontinuity. As we saw in Section 3.2.1, these were associated with the evolution of the *pinched-off channel* in the *long-channel model* (earlier indicated in (3.8) and (3.9)). While citing this merit, we should point out that in preserving (3.2) along the *channel*, the present model still lacks the required insight into *charge sharing*. This issue will be dealt with in Section 3.2.2.2. We will postpone the inclusion of *channel-length modulation* also to that part of the discussion. Figure 3.8 illustrates schematic representations of the variation of the longitudinal electric field and *inversion layer* electron concentration along the *channel* according to the two already presented models next to their predictions of $I_D - V_D$ characteristics. Contrary to the unrealistic assumptions of the evolution of infinite electric field in the latter part of a *MOSFET channel biased* under the *saturation regime* of operation (and *zero carrier* concentration in this part of the *channel*), the presented *short-channel model* merely suggests the evolution of a region of high electric field (and lowered *carrier* concentration) in this latter part of the *channel*.

3.2.2.1.1 MOSFET Scaling: Divergence of Predictions of Short- and Long-Channel Models Equation (3.6) projects an unbound increase of the *drain* current with reducing the L_G. However, at the limit of very small values of *gate length*, (3.26) forecasts only a finite maximum *drain* current given by

$$I_{Dmax} = WC_{ox}(V_G - V_T)v_{sat}. \tag{3.29}$$

[24] We will come back to the description of this *channel-length modulation* after formulating the variation of the longitudinal electric field along the *channel* of a *short-channel MOSFET* in Section 3.2.2.2.

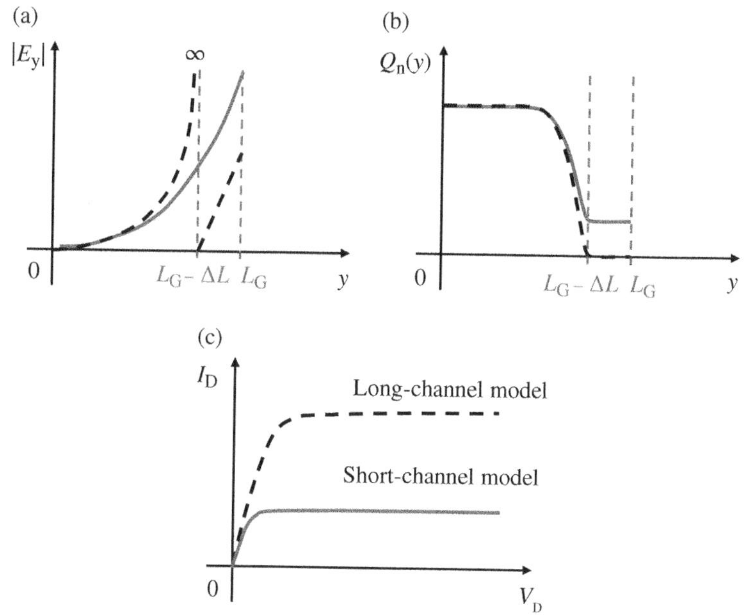

FIGURE 3.8 Profiles of (a) longitudinal electric field and (b) inversion-layer carrier concentration along the channel of a MOSFET biased in saturation according to the so far presented long- and short-channel models. Predictions of the long-channel model are given in black dashed curves, while the predictions of short-channel model are given in continuous gray. (c) Predictions of the drain current–voltage characteristics for one given value of V_G according to these two models for a short-channel transistor.

Even this value of *drain* current is not attainable, since it suggests that the *gate length* has been reduced to *zero*. Unlimited reduction of the *gate length*, aside from physical feasibility, is at odds with the assumption of *drift–diffusion* mode of *transport* used in the so far presented models.

As illustrated in Figure 3.9, while the forecast of *gate-length scaling* according to (3.6) seems acceptable for larger values of L_G, in deeply *scaled MOSFETs* the forecast for the *drain* current based on the *long-channel model* is very much unacceptable.

The hypothetical definition of maximum *drain* current, presented by (3.29), is used in defining a *figure of merit* for *gate-length scaling* in *short-channel MOSFETs*. This *figure of merit*, which is known as *drain* current *ideality factor* K_I, is given by

$$K_I = \frac{I_{Dsat}}{I_{Dmax}} = \frac{V_G - V_T - V_{Dsat}}{V_G - V_T}. \tag{3.30}$$

Based on the definition of V_{Dsat}, the *ideality factor* K_I is not only dependent on the physical dimensions of the transistor but also on the *bias* (because, beside $V_G - V_T$, E_{sat} is also *bias* dependent). The closer K_I is to unity, the better the transistor takes advantage of the *carrier-transport* properties intrinsic to the semiconductor.

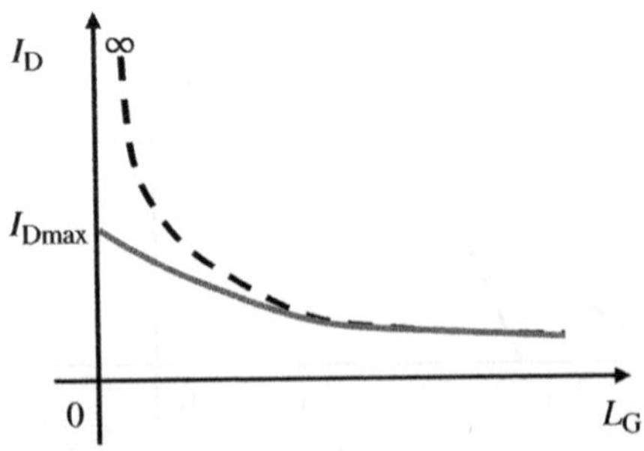

FIGURE 3.9 Prediction of expression (3.6) for scaling of drain current of a MOSFET with gate length (shown in dashed black curve) and real scaling scenario shown in gray continuous curve.

In addition to the *drain* current *ideality factor*, other *figures of merit* have also been used in evaluating the performance of *MOSFETs* and their *scalability* with *gate length*. One such *figure*, from the high-frequency operation point of view, is the *unity gain cutoff frequency* f_T, which is the highest frequency at which a current gain is expected from the transistor. This *figure of merit* is evaluated in terms of the *gate transconductance*. As suggested earlier, the *gate transconductance* is a key indicator in determining how swiftly the *gate* of a *MOSFET* can modulate the *carrier* population of the *inversion channel*. According to the *small-signal* circuit model of a *MOSFET*, the maximum frequency at which a *MOSFET* presents *current gain* is evaluated by

$$f_T = \frac{g_{m\text{-sat}}}{2\pi C_{in}}. \tag{3.31}$$

Here $g_{m\text{-sat}}$ is the *small-signal gate transconductance* of the *MOSFET biased* under the *saturation regime* of operation and C_{in} refers to the total *capacitance* seen at the *gate* terminal.

As expressed in (3.10), in a *long-channel MOSFET*,

$$g_{m\text{-sat}} = \mu_n C_{ox} \frac{W}{L_G}(V_G - V_T) = \frac{2I_{Dsat}}{V_G - V_T}. \tag{3.32}$$

For *long-channel MOSFETs*, in the absence of *charge sharing* with *source* and *drain*, the *oxide capacitance* is directly in control of the input capacitance. As a result, C_{in} is approximated by *two-thirds* of the value of the *oxide capacitance*. Based on this approximation, in these devices the *unity gain cutoff frequency* can be evaluated as

$$f_T = \frac{\mu_n C_{ox}(W/L_G)(V_G - V_T)}{2\pi(2/3)WL_G C_{ox}} = \frac{3\mu_n(V_G - V_T)}{4\pi L_G^2}. \tag{3.33}$$

However, assuming E_{sat} to be approximately V_G independent, the *short-channel model* presented in Section 3.2.2.1 according to (3.26) gives the *gate transconductance* as

$$g_{m\text{-}sat} = W v_{sat} C_{ox} \frac{(V_G - V_T)(V_G - V_T + 2E_{sat}L_G)}{(V_G - V_T + E_{sat}L_G)^2}. \tag{3.34}$$

For this *gate transconductance*, if we make the same assumption for the input capacitance, the *unity gain cutoff frequency* instead of (3.33) will be evaluated by

$$f_T = 3v_{sat} \frac{(V_G - V_T)(V_G - V_T + 2E_{sat}L_G)}{4\pi L_G (V_G - V_T + E_{sat}L_G)^2}. \tag{3.35}$$

The approximation of C_{in} with $2/3WL_GC_{ox}$ is, however, very rough in this case.

Equation (3.35) in *short-channel MOSFETs* simplifies as $E_{sat}L_G$ becomes negligible (i.e., compared to $V_G - V_T$). For very small values of *gate length* (or to be speaking more generically, for very small values of $E_{sat}L_G$), according to (3.35), f_T tends toward

$$f_T = \frac{3v_{sat}}{4\pi L_G}. \tag{3.36}$$

As observed between (3.33) and (3.36), the L_G^{-2} dependence of f_T in the case of *long-channel MOSFETs* is approximately reduced to a L_G^{-1} dependence in *short-channel MOSFETs*. This is clearly an indication of a weaker gain in terms of improving the frequency response with *gate-length scaling* of *MOSFETs* than predicted by the *long-channel models*.

Corresponding to the definition of K_I, in terms of (3.34), another *ideality figure of merit* defined in terms of the *gate transconductance* under the *saturation regime* of operation and its hypothetical maximum value is also quite often spoken of, where

$$K_{g_m} = \frac{g_{m\text{-}sat}}{g_{m\text{-}max}} = \frac{(V_G - V_T)(V_G - V_T + 2E_{sat}L_G)}{(V_G - V_T + E_{sat}L_G)^2}. \tag{3.37}$$

Example

For a *silicon NMOSFET* with the following specifications:

$L_G = 0.1$ μm, $W = 0.1$ μm, $t_{ox} = 5$ nm, $X_j = 0.03$ μm, and $N_a = 10^{17}$ cm^{-3}

calculate the *drain* current at the onset of saturation when $V_{GS} = 0.3$ V and $V_{SB} = 0$. For this *bias* condition also calculate the *drain current ideality factor*.

Assume *aluminum gate* and *SiO_2 gate* dielectric. The saturation velocity of electrons is equal to 10^7 cm/s.

We first use (2.172) to calculate the *threshold voltage*, where

$$C_{ox} = \frac{\epsilon_{ox}}{t_{ox}} = \frac{3.9 \times 8.85 \times 10^{-14}}{5 \times 10^{-7}} \cong 0.69 \times 10^{-6} \mathrm{F/cm^2} \quad \text{and}$$

$$q\phi_{FB} = q\phi_M - q\phi_S = q\phi_M - \left(q\chi + \frac{E_g}{2} + kTLn\frac{N_a}{n_i}\right)$$

$$= 4.1 - \left(4.05 + 0.56 + 0.0258Ln\frac{10^{17}}{1.45 \times 10^{10}}\right) \cong -0.9 \text{ eV} \quad \text{and} \quad \phi_b = -0.4 \text{ V}.$$

Accordingly,

$$V_T = -0.9 + 0.8 + \frac{1}{0.69 \times 10^{-6}}\sqrt{2 \times 11.7 \times 8.85 \times 10^{-14} \times 1.6 \times 10^{-19} \times 10^{17} \times 0.8}$$

$$\cong 0.14 \text{ V}.$$

Now based on (3.23),

$$E_{eff} = \frac{3.9}{11.7 \times 5 \times 10^{-7}}\left(\frac{0.3}{2} + \frac{0.14}{2} - 0.8 + 0.9\right) \cong 2.1 \times 10^5 \text{ V/cm}.$$

For which, based on the model presented in Table 3.1,

$$\mu_{eff} = \frac{670}{1 + \left(\frac{2.1 \times 10^5}{0.67 \times 10^6}\right)^{1.6}} \cong 577 \text{ cm}^2/\text{V s}.$$

Based on this value, (3.19) yields

$$E_{sat} = \frac{2 \times 10^7}{577} \cong 3.4 \times 10^4 \text{ V/cm}.$$

Based on the definition of V_{Dsat} given in (3.28),

$$V_{Dsat} = \frac{E_{sat}L_G \cdot (V_G - V_T)}{E_{sat}L_G + (V_G - V_T)}$$

where $V_G - V_T \cong 0.16$ V and $E_{sat}L_G \cong 0.34$ V yield $V_{Dsat} \cong 0.11$ V.
For this value of the *drain bias*, (3.25) allows calculation of $I_D \cong 3.47 \times 10^{-6}$ A.
For this case according to (3.30),

$$K_I = \frac{0.16 - 0.11}{0.16} \cong 0.3.$$

3.2.2.1.2 MOSFET Scaling: Overview of Technological Needs So far, in terms of the mentioned *figures of merit* in Section 3.2.2.1.1 we have observed that moving from a *long-channel* to a *short-channel* paradigm has reduced the expected gain of *scaling* in terms of the boost in the *drain current drive* and the *switching speed* per *technology node*.[25] However, even with this smaller anticipated gain, *gate-length scaling* has still managed to maintain its status as the workhorse of the semiconductor technology. While we have not yet done justice to the implications of the competition between the *drain-* and *gate*-induced electric fields, the thus far presented discussions have correctly identified the oxide thickness (i.e., t_{ox}) as one of the physical dimensions in need of shrinking (i.e., to keep up with L_G *scaling*).

Relatively recently, the *scaling* of t_{ox} has turned into one of the major setbacks of *CMOS* technology. Reduction of t_{ox} to a few *nanometers* has imposed many orders of magnitude increase on the *gate*-leakage current. This increase has been induced by the *field-emission* process.[26] As observed in (1.93), the probability of this quantum phenomenon is exponentially dependent on the thickness of the *gate* insulator. The resulting diminishing difference between the *on-* and *off-states' drain* currents of *a MOSFET* initiated major research to find a solution. One direction of this research has been formulated based on replacing SiO_2 as the *gate* insulator by other dielectrics, which have higher *permittivities* (often referred to as *high-K dielectrics*). Since higher *permittivity* dielectrics offer a given value of *gate-insulator capacitance* at larger dielectric thicknesses, replacing SiO_2 with *high-K dielectrics* can exponentially reduce the gate's *field-emission* current.

Among the insulators traditionally available in *silicon* technology (such as SiO_2 and often nonstoichiometric Si_xN_y), the dielectric nature, which is defined through the distortions of the *valence shell*, dictates the increase in the *permittivity* to be correlated to the reduction in the size of the *bandgap*. As a result, for those of higher *permittivity*, the reduced *bandgap* in turn serves to increase the *gate* leakage through *thermionic emission*. This was one of the reasons for the limited success of the so-called oxynitrides and the groundbreaking introduction of the *lattice-polarizable* dielectrics into the mainstream *silicon* technology. In this latter group of dielectrics, such as HfO_2, $HfTiO_4$, $SrHfO_3$, relative dielectric constants in the range of 20 (compared to SiO_2's 3.9 and Si_3N_4's 7.5) are realized without suffering from a reduced *bandgap*. In evaluation of the performance of these *high-K dielectrics*, often the so-called notion of *equivalent oxide thickness* (or *EOT*) is invoked. *EOT* is defined as the thickness of a SiO_2 layer, which can offer the same *gate-insulator capacitance* achieved using another dielectric. Evidently for a *high-K dielectric, EOT* is smaller than the physical thickness of the dielectric layer. Despite its advantages, *polycrystalline* HfO_2 is plagued by the large diffusion at its *grain boundaries*, a problem that can be resolved through amorphization via mixing with SiO_2 to form $Hf_xSi_{1-x}O_2$.

[25] Which is often indicated by the reduction of the minimum feature size (i.e., L_G) by a factor of *two*, while the actual size of a typical *MOSFET* is 10–15 times larger than its *gate length*.

[26] See Section 2.5.2.

Despite the appealing features of *high-K dielectrics*, the road to replacing the *dry thermally grown* SiO_2 has been very challenging. This is because the *interface* between SiO_2, as an *amorphous* insulator, and the *crystalline silicon* presents one of the most awe-inspiring surprises of the nature for semiconductor technologies. None of the other so far explored insulators present as clean and smooth of an interface with *silicon*. This is especially more amazing when one considers how little goes into the process of *thermal oxidation*. The increased *scattering*, even for the best of the alternative insulator/*silicon interfaces*, has so far resulted in *carrier mobilities* less than 90% of those realized at the SiO_2/Si *interfaces*. Aside from the *interface roughness*, the *polar* nature of these *high-K* dielectrics is the other contributing factor to this degradation in electron *mobility* (i.e., due to additional *remote polar phonon scattering*).

In addition to the push for adopting *high-K dielectrics*, the need for enhancing the electrostatic dominance of the *gate* over the process of populating the *inversion channel* of *MOSFETs* has brought forward other major structural changes to the topology of these devices. Solutions as broadly ranging as the realization of *MOSFETs* on the ultrathin *silicon* layer of *silicon-on-insulator wafers* (i.e., *SOI*)[27] and use of a buried second *gate*, to forming *MOSFET channels* in *nanowires* and *fins* etched into the *silicon wafers*, have been studied over the past almost two decades. The most popular of these alternatives are known as *FinFETs*.[28] In *planar MOSFETs*, which we have studied thus far in this chapter (such as the one depicted in Fig. 3.3), the placement of the *gate* on the top surface allows only the definition of a *gate*-induced electric field normal to this surface. However, realizing the *silicon channel* in the form of a *fin* etched into the *silicon wafer* extends the possibility of imposing *gate* action also from the sidewalls of the *fin*. A simple *FinFET* is depicted in Figure 3.10. The triple-sided coverage of the *gate* around the *channel* of a *FinFET* increases the electrostatic dominance of the *gate* over the *channel* (hence improves the *gate transconductance* and *subthreshold swing*).[29] These so-called 3-D MOSFETs, as early as 2011, have been in the process of replacing *planar MOSFETs* in the major semiconductor foundries. While we will maintain a focus on *planar silicon MOSFETs* in the rest of this chapter, in Chapter 4, among other novel technologies, *FinFETs* and *UTBSOI MOSFETs* will be dealt with.

The required boost in the per unit area *oxide capacitance*, or generally speaking the *insulator capacitance* of a *MOSFET*, in addition to enhancing the electrostatic dominance of the *gate* over the *channel*, dictates important modification to the *bias* conditions of a *MOSFET*. According to (2.172) a boost in C_{ox} in the absence of any change in the other involved parameters forces a reduction in the *threshold voltage*. With the smaller t_{ox}, the acceptable values of *gate bias* are also forced to be smaller.

[27] Named as *UTBSOI MOSFETs*.

[28] *FinFET* and *UTBSOI* are the results of the pioneering work of the research team of Professor *Chenming Hu* at the University of California–Berkeley. They were demonstrated for the first time in 1999 and 2000, respectively.

[29] An in-depth quantitative evaluation of this multi*gate* effect is presented in Chapter 4.

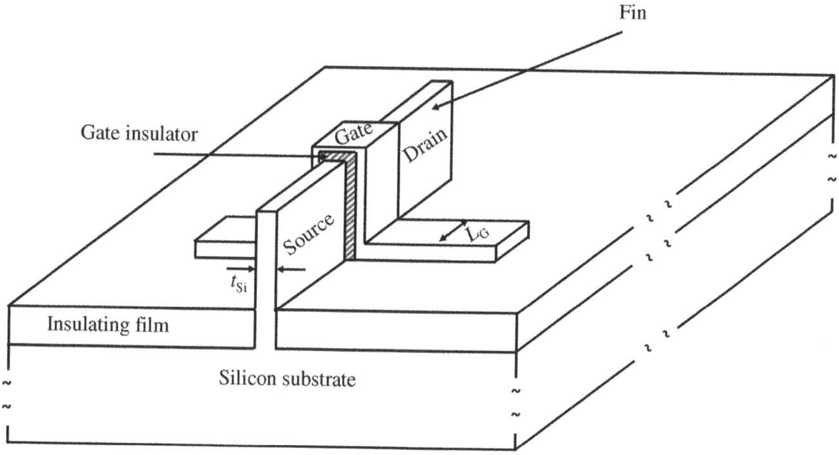

FIGURE 3.10 3-D schematic illustration of a FinFET structure.

This latter requirement is imposed by the need to watch for the *critical* electric field of the insulator as t_{ox} is being reduced. The combination of smaller V_G and V_T, and the resulting smaller difference between the *on* and *off* states of a *MOSFET* switch, are among the other factors that have contributed to the diminishing gain of *technological scaling*. As mentioned in Section 3.2.1.4, this has been one of the contributing factors in the growing importance of the *subthreshold* regime of operation. In Equation (2.172) terms such as ϕ_{FB} and ϕ_b are only logarithmically dependent on the *dopant* concentration. Increase in N_a, although limited by the resulting degradation of *carrier mobility*, is among the factors that are being used in *scaling* strategies to relatively counterbalance the role of C_{ox} increase in reducing V_T (see Fig. 3.1c). While the increase in N_a, theoretically speaking, can be quite safely realized through *ion implantation* with a proper *projection range* and *vertical straggle* (i.e., to minimize the increase in the *dopant* concentration at the insulator/*silicon interface* and as a result the degradation in *channel*'s *carrier mobility*), with the reduction in X_j, satisfying these requirements has become more of a challenge. This technological constraint has been already visited in the description of the *punch-through implant* in Section 3.2.2.

In light of the reduction in V_T and V_G, another important challenge faced by the *scaling* strategies has been the *depletion* of the *polysilicon gate* and the proportional increase in the potential drop across this *depletion region* (i.e., proportional in scale of V_G). So far we have not paid attention to the potential drop across the *polysilicon* layer and the resulting *depletion capacitance* formed in series with the *insulator capacitance*. In *deeply scaled MOSFETs* this issue cannot be neglected. However, since the need for improving the *conductivity* of the *gate* electrode (which is aggravated by the reduced contact area of the *gate* and the resulting increase in the series resistance of this electrode in *scaled MOSFETs*) has already triggered the replacement of *polysilicon* with highly conductive metals such as *copper*[30] and *metal silicides*, the issue of *gate depletion* is less relevant.[31] In terms of these relatively new *gate* materials, the *work function* of the metal (or alloys) should mimic that of the *polysilicon* (i.e., in the case of *n-channel MOSFETs* the *work function* should match that of n^+*-polysilicon*, and for *p-channel MOSFETs* it should match that of p^+*-polysilicon*). This is necessary to guarantee a proper matching of *threshold voltages* in a *CMOS pair*.[32] As a result of this requirement, not just one metal but two different metals in place of *polysilicon* should be adopted in a *CMOS* technology. While at first glance the idea of using one metal with *work function* between these values might seem interesting to entertain, as expected from (2.172), this strategy will result in increase in the *threshold voltages* of both *MOSFETs* in a *CMOS pair*, which is certainly not affordable. Metals such as *tungsten*, *tantalum*, and *silicides* such as *NiSi* have been explored in *CMOS* technology. The use of a metal *gate* not only eliminates the *polysilicon depletion* problem and reduces the *gate*'s series parasitic resistance, it also contributes to increasing the *carrier mobility* through reducing the *transverse* electric field at a given *gate* overdrive.

Reduction of the lateral dimensions of *MOSFETs* (as an example in the form shown in Figure 3.1c) has also promoted the use of high-*conductivity silicides* for reducing the series *Ohmic* contact resistance of the *source* and *drain* terminals. Stoichiometry and the phase of the *silicide* are very important to achieving the required resistivity. Common metals used in forming these *silicides* are *Ti*, *Pt*, *W*, *Ta*, *Mo*, and *Co*. $TiSi_2$, with a resistivity of 13 μΩ cm, is one of the most desirable *silicides* used in *silicon* technology.

While the innovative modifications enumerated in this section kept the *scaling* trend alive, at the same time they contributed to a very definitive addition to the complexity of the fabrication technology. Thanks to these innovations, after many decades of *technological scaling* and sour predictions for the end of the *scaling* roadmap,[33] the

[30] From the technological standpoint it is worth noting that *atomic layer deposition* (for short, *ALD*) and *electroplating* are the technologies used in depositing *copper*. One of the major challenges in *copper* metallization is that *copper* quite easily diffuses through SiO_2 and *Si*. Diffusion of *copper* through SiO_2 can prompt the formation of conductive spikes through the insulator, especially under high electric field and/or high-temperature conditions. This is often referred to as *electromigration*. Since additionally, as indicated in Section 2.3.2.1.4, *copper* can introduce a very deep *recombination center* to the *silicon*'s *bandgap*, such a diffusion needs to be prevented through using a refractory metal barrier layer (or its *nitride*) such as *W*, *Ta*, *WN*, and *TaN*. Adhesion of *copper* is also quite poor, which requires incorporation of an *adhesion-promoter* layer. Due to the dependence of the surface roughness and *grain* sizes on the concentration in the *electrolyte* bath used in *electroplating* (e.g., *cupric sulfate* mixture), the resistivity of the *electroplated copper* is quite dependent on this factor.

[31] *Metal silicides* were introduced into the *silicon CMOS* technology in the 1980s.

[32] That is equal in magnitude but opposite in polarity.

[33] Which is often referred to as *Moore's law*. This is in honor and according to the early observations of *Gordon Moore*, one of the pioneers of semiconductor technology, on the progress of *gate-length scaling* and packing density on the scales of time. These early observations over the past four to five decades have served as the pacemaker of the semiconductor technology. According to the so-called Moore' law, the number of transistors in *logic IC*'s doubles every 18–24 months, while the *gate length* is reduced to *half* its value every 4 years. Since as we speak the *gate length* has approached the 10 nm mark, real worries in terms of clashing with the very foundations of quantum mechanics, such as *Heisenberg's uncertainty principles*, have started to reorient the discussions on the trends to follow beyond the limits of *Moore's law* (i.e., "more than *Moore*" instead of "more *Moore*").

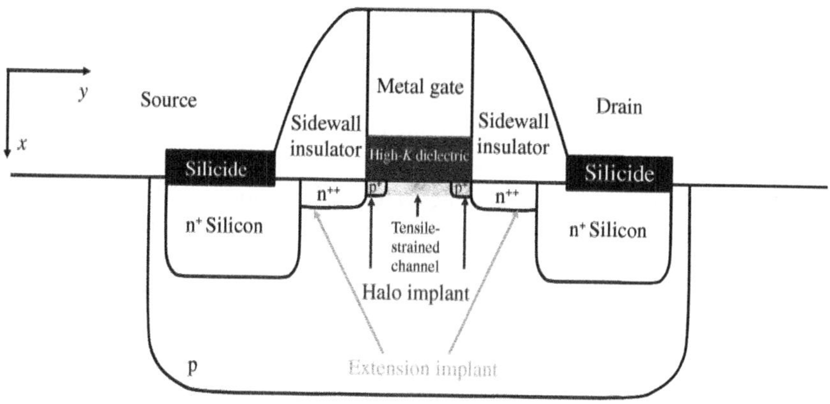

FIGURE 3.11 Cross-sectional view of a modern planar n-channel MOSFET.

scaling trend, although less agile, is still pushing for smaller *gate lengths*. In terms of the aforementioned technological needs of *scaling*, and a few more that will be identified in Section 3.2.2.2, Figure 3.11 depicts a typical *deeply scaled planar MOSFET* structure. A comparison between the cross-sectional views of *MOSFETs* depicted in Figures 3.3 and 3.11, is quite revealing of the drastic increase in the technological complexity of modern *MOS* processes.

As expressed in Figure 3.11, in the modern *planar MOSFET*, the low-temperature requirement of many of the considered *high-K dielectrics* prevents the implementation of the *self-aligned* technology.[34] This is due to the fact that, as indicated earlier, in this *gate-first* process the *gate* stack is expected to withstand the high temperatures of the *annealing* process that follows the *ion-implantation doping* of the *drain* and *source carrier* reservoirs. This technological constraint explains the need for the sidewall dielectric and the n^{++} *doped* shallow *implants* in this structure. In Section 3.2.2.2 the reason for the incorporation of the p^+ *doped* section of this profile (which is known as *halo implant*) is identified. The *halo implant* profiles are implemented through an *angled ion-implantation* process.

Whereas in the *planar MOSFET* technologies *technological nodes* are presented in terms of the *gate length*, the shift of paradigm from *planar* to *3-D MOSFETs* has made the definition of *nodes* not as clear as before. As the later discussions of this chapter and Chapter 4 suggest, the performance boost of *scaling* in terms of *figures of merit* such as f_T has quite substantially plummeted. Realization of *3-D MOSFETs*, however, has been able to make important strides in improving packing density and as a result *chip* functionality.

[34] This is due to the often low *crystallization temperature* of these dielectrics. As noted earlier, crystallization imposes a challenge, in terms of leakage through the *polycrystalline grain boundaries*, on the *gate* dielectric. The crystallization temperature is just 400°C in case of HfO_2. In SiO_2, in the *fused silica* form, this temperature is in excess of 1700°C, which is way beyond the needed *annealing* temperature. Research on finding *high-K dielectrics* of high-enough *crystallization temperature* is ongoing.

3.2.2.2 *Short-Channel Model: Two-Dimensional Treatment of Velocity Saturation*

As suggested earlier, by increasing the *drain* voltage beyond V_{Dsat} electrons not only at the *drain* edge of the *gate* of the *short-channel silicon MOSFET* but even in its expanding vicinity will obtain their maximum *drift velocity* (i.e., *saturation velocity*). For higher values of V_D, the length of this latter part of the *channel*, through which electrons travel at the *saturation velocity*, becomes larger. This part of the *channel* is referred to as the *velocity-saturation region* (or *VSR* for short).[35] Whereas in the rest of the *channel* the so-called universal model of Section 3.2.2.1 is applicable, in the *VSR* correlated impact of both the *gate*- and the *drain*-induced electric-field components needs to be considered. The proportionally longer the *VSR*, the proportionally less relevant is the role of *low-field electron mobility* in determining the average electron velocity and the *drain current drive*. This is due to the fact that under this condition, throughout most of the length of the *channel*, electrons are traveling at their *saturation velocity*. This is certainly more prone to happen in *MOSFETs* of small *gate length* operating beyond the *onset of strong inversion*. As expressed before, under the *subthreshold regime* of operation, the low-field *diffusion* mode of *carrier transport* is dominant; hence, the *low-field transport* properties remain quite relevant even in somewhat *deeply scaled MOSFETs*.

Recently, a renewed interest has been evolving regarding the realization of *GaAs CMOS* technology. While *GaAs channels* enjoy much higher *low-field carrier mobility* than *silicon*, the *saturation velocity* is almost the same among all semiconductors. As a result, with the exception of the *subthreshold* operation, as long as operating under the constraints of the *drift–diffusion transport* formalism, a substantial gain from this switch in semiconductor technology should not be expected. However, under the *ballistic* regime of *transport*, owing to the *temporal velocity overshoot* of both types of *charge carriers* (and even slightly before that, due to the steady-state *overshoot* in the *drift velocity* of electrons), *GaAs CMOS* technology can muster interest (i.e., in terms of improving the *drain* current and frequency response beyond those of *silicon CMOS*). An added value in this switch from *Si* to *GaAs* technology is the presence of a wider *bandgap semi-insulating GaAs* substrate that does not leak the high-frequency *signal*. In the case of the *semi-insulating GaAs* substrates, the substrate isolation is even superior to that of the nontraditional high-resistivity *silicon* substrates. This switch in technology, however, is plagued by the higher cost of production and smaller diameters of *GaAs wafers*.[36] As a solution to these problems, *epitaxial growth* of *GaAs* on *silicon wafers* is being pursued. Such a solution is especially interesting from the point of view of integration of electronic and optoelectronic circuits, which are, respectively, built in *Si* and *GaAs* technologies, on the same *chip*.

As suggested earlier in the explanation of *charge sharing* and *DIBL*, the proper description of the operation of a *short-channel MOSFET* cannot be complete without

[35] Sometimes called the *saturation region*.

[36] Whereas *silicon wafers* of 12 in. diameter have been commercialized and even larger diameters (i.e., 15 in. and 18 in.) are in the queue, currently available commercial *GaAs wafers* have a maximum diameter of 6 in. Evidently, considering the production *yield*, the larger-size *wafers* can support the realization of larger *chips* of higher functionality.

considering both the *gate*- and the *drain*-induced electric-field components in evaluating the *inversion-layer carrier* concentration of the *channel.* As a result, in the presence of these two orthogonal electric-field components, at least in the high electric-field *velocity-saturation region* of a *short-channel MOSFET, Poisson's equation* should be solved two dimensionally.[37] While in the case of a *long-channel MOSFET* we have already addressed the presence of the *drain*-induced longitudinal electric field, this electric-field component has been merely considered in the context of the *carrier transport* along the *channel* and not the formation of *inversion* and *depletion* charge profiles. As observed in Section 3.2.1.1, even in the case of the *long-channel MOSFET,* the assumption of division of roles of the two electric-field components becomes contradictory at the *drain* edge of the *gate* (where the longitudinal electric field becomes the strongest and the approximation becomes least valid). However, in the case of the *long-channel MOSFET* since the length of the part of the *channel* for which the assumptions of the *model* are contradicted is proportionally much smaller than the *gate length,* this approximation remained valid.

Considering the charge concentration profiles composed of the immobile *depletion-region* charge concentration of $-qN_a(x,y)$ and the mobile *inversion-layer carrier* concentration given by $-qn(x,y)$, the 2-D *Poisson's equation* in terms of the *channel* potential $V(x,y)$ is written as[38]

$$\frac{\partial^2 V(x,y)}{\partial x^2} + \frac{\partial^2 V(x,y)}{\partial y^2} = \frac{q[N_a(x,y) + n(x,y)]}{\epsilon_s}. \tag{3.38}$$

In here, according to Figure 3.12, x is the axis normal to the *SiO_2/Si interface*, and y indicates the longitudinal axis along the *length* of the *channel.* In solving (3.38), the presence of a *quantum well* at the *conduction band* edge of the *SiO_2/Si interface* demands the presentation of $n(x,y)$ in terms of energy *subbands* formed in this triangular *quantum well.* Such an evaluation requires a *self-consistent* solution of the *Schrödinger equation* with *Poisson's equation* of (3.38). This *self-consistent* solution is typically produced through the application of numerical techniques. Ease of interpretation of analytical solutions, however, has driven researchers to develop an approximate analytical solution (i.e., without the need for explicit incorporation of the calculation of energy *subbands*). Whereas the presentation of this section is hinged on this premise, later on through describing the assessment of the quantum phenomena, we will talk about the limitations of this framework.[39]

The solutions provided in this section on the basis of the so-called analytical pseudo two-dimensional model are formulated for a purely saturating v_d–E characteristic and within the context of a *self-aligned MOSFET.* Following the foundations to

[37] The reason for focusing on the *VSR* is that in the earlier part of the *channel* the *drain*-induced electric field is much weaker than the *gate*-induced field.

[38] Since along the *width* of the *MOSFET* (i.e., the z-axis) there is no electric field to be considered, the 3-D assessment will not be necessary. All of the indicated quantities are also z independent.

[39] In Chapter 4, it will be indicated that the *quantum phenomena* impacts this approximation to quite different degrees among the *silicon* and *III–V channels.*

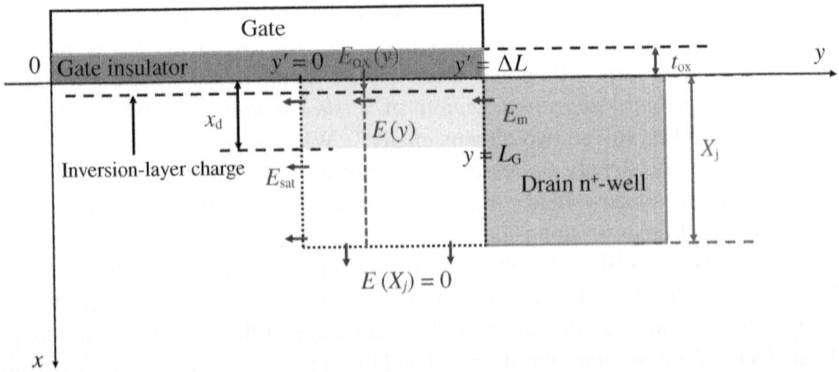

FIGURE 3.12 Cross-sectional presentation of the foundation of Poisson's equation expressed in (3.38). The dotted box identifies the boundaries of the Gaussian box encompassing the velocity-saturation region of the channel. The out-of-plane components of electric field are zero.

be presented in Chapter 5, the same model can be extended to *noninversion field effect transistors*[40] and even realized in semiconductor systems, which demonstrate a hump in the *drift-transport* characteristic.[41]

In the application of the 2-D *Poisson's equation* of (3.38) to the *velocity-saturation region*, a *Gaussian box* (as shown in Fig. 3.12) is considered. This *Gaussian box* encloses not only the *inversion layer* but also the *depletion-region* charge profile.

Gauss's law states that the integral of electric flux emanating normal to all surfaces of a closed volume is proportional to the amount of the charge present within the volume enclosed by these surfaces. This integration will be made easier if we take a *Gaussian box* for the closed volume whose lower surface along the x-direction is defined well below the *SiO_2/Si interface*, so that the *gate*-induced electric field normal to this surface would be negligible. For this reason, this lower surface of the box is defined at the *junction depth X_j* of the *drain* and *source carrier* reservoirs. The top surface of the *Gaussian box* is defined at the *SiO_2/Si* interface. This plane stretches in the y-z plane from the first

[40] These can be either *depletion-mode MOSFETs* or generally speaking *FETs* in which the *channel's* population is not *inversion* based. As a result, in the operation of these transistors a flush boundary between the *carrier* reservoirs of the *source* and *drain* and the *gate* edges is not demanded. As will be observed among the equations derived in this section, while distancing the edge of the *drain* from the *gate* reduces the longitudinal *drain*-induced electric field, and hence increases the *breakdown voltage* of the transistor, the presence of such an opportunity in these so-called noninversion FETs is invaluable to realization of power transistors.

[41] While, as will be seen in this section, for a *channel* with a purely saturating v_d–E considering the *drift transport* will be sufficient, in presence of wide and high peaks in the steady-state v_d–E characteristic of semiconductors such as *GaN*, *diffusion current* in the high electric field region of the *channel* can also become substantial. This is due to the substantial presence of a region of *negative differential mobility* in these *channels*, which in light of *current continuity* can induce an increase in electron population (and hence a finite spatial gradient in the *carrier* concentration), where the *drift velocity* is dropping with the increase of the electric field.

point at which *velocity saturation* happens throughout the length of the *VSR*. According to such a definition, the depth of both the *inversion* and the *depletion* charge profiles of the *VSR* fall within the boundaries of the *Gaussian* box.

As shown in Figure 3.12, there is no electric-field component normal to the front and back surfaces of the *Gaussian box*. As a result, in addition to the lower surface, on the front and back surfaces of the box integration yields *zero*.

Based on the topology of the *Gaussian box*, only on the top and the two side surfaces, defined in *x-z* plane, will integration result in a finite value. According to the definitions presented in Section 3.2.2.1, the electric field normal to the left-side *x-z* surface of the *Gaussian* box is equal to E_{sat}.[42] In order to evaluate the electric field at different points along the *channel*, the right-side boundary of the box can be defined at different values of y along the *VSR*.

Where in Figure 3.12, $y' = 0$ is defined at the point y where the *velocity saturation* initiates (i.e., $V(y) = V_{Dsat}$) and the *gate*-induced electric field within the *gate* oxide normal to the top surface of the *Gaussian* box is denoted as $E_{ox}(y')$, *Gauss's law* yields

$$E_{sat}X_j W - E(y')X_j W - W\frac{\epsilon_{ox}}{\epsilon_s}\int_0^{y'} E_{ox}(y)dy = -\frac{q}{\epsilon_s}W\int_0^{X_j}\int_0^{y'}[n(x,y)+N_a(x,y)]dydx. \tag{3.39}$$

In Equation (3.39), y' can take any value from *zero* to ΔL (which is the length of the *VSR*). Constrained by *current continuity*, since velocity along the length of the *VSR* is constant (i.e., v_{sat}), $n(x,y)$ is also independent of y. In this equation, the longitudinal component of the electric field (i.e., $E(y')$) is assumed to be independent of x. This is a good approximation for large *gate* voltages but not so much for small values of V_G.

Assuming the *bulk dopant* concentration (i.e., N_a) to be constant, by differentiating both sides of Equation (3.39) versus y', we will have

$$X_j\frac{dE(y')}{dy'} + \frac{\epsilon_{ox}}{\epsilon_s}E_{ox}(y') = \frac{q}{\epsilon_s}X_j[n(x,y)+N_a]. \tag{3.40}$$

In (3.40), according to (2.159), the *gate*-induced electric field in the *gate* oxide beyond the *onset of strong inversion* is given by

$$E_{ox}(y') = \frac{V_G - V(y') - \phi_{FB} - 2|\phi_b|}{t_{ox}}. \tag{3.41}$$

Combining (3.40) and (3.41) yields

$$X_j\frac{dE(y')}{dy'} + \frac{\epsilon_{ox}}{\epsilon_s}\frac{V_G - V(y') - \phi_{FB} - 2|\phi_b|}{t_{ox}} = \frac{q}{\epsilon_s}X_j[n(x,y)+N_a]. \tag{3.42}$$

[42] This is based on the assumption of an equipotential *x–z* surface for the n^+-*well* of the *drain* (i.e., with potential V_D). The high *dopant* concentration of this n^+-*well*, and its low resistivity, warrants this assumption.

However since according to (3.41), and the definition of the *VSR*, we know that[43]

$$E_{\text{ox}}(y'=0) = \frac{V_G - V_{\text{Dsat}} - \phi_{\text{FB}} - 2|\phi_b|}{t_{\text{ox}}} = \frac{q}{\epsilon_s} X_j [n(x,y) + N_a] \tag{3.43}$$

Equation (3.42) can be simplified to

$$\epsilon_s X_j \frac{dE(y')}{dy'} = C_{\text{ox}} [V(y') - V_{\text{Dsat}}] \tag{3.44}$$

or equivalently

$$\frac{dE(y')}{dy'} = \frac{V(y') - V_{\text{Dsat}}}{l^2} \tag{3.45}$$

where the length scale l is defined as $l = \left((\epsilon_s/\epsilon_{\text{ox}}) X_j t_{\text{ox}}\right)^{0.5}$.

Using the definition of the electric-field strength, (3.45) converts to

$$\frac{d^2V(y')}{dy'^2} = \frac{V(y') - V_{\text{Dsat}}}{l^2} \tag{3.46}$$

Considering the boundary conditions of $E(y'=0) = E_{\text{sat}}$ and $V(y'=0) = V_{\text{Dsat}}$, Equation (3.46) produces closed-form expressions for the variation of the *channel* potential and electric field along the length of the *VSR*:

$$V(y') = V_{\text{Dsat}} + E_{\text{sat}} l \cdot \sinh\left(\frac{y'}{l}\right) \tag{3.47}$$

and

$$E(y') = E_{\text{sat}} \cdot \cosh\left(\frac{y'}{l}\right) \tag{3.48}$$

where, on the basis of (3.39), we have assumed this electric field to be in the $-y$ direction.

According to (3.47) and (3.48), the characteristic length scale l is an indicator for the rate of the exponential increase of the potential and the electric-field strength along the length of the *velocity-saturation region*. Shrinking l with scaling of t_{ox} and X_j indicates a larger spatial gradient in both *channel* potential and *drain*-induced electric field of the *scaled MOSFETs*.

Equations (3.47) and (3.48) highlight the reasons for the aforementioned need to evaluate the process of *inversion-layer* charge induction within the *VSR* in a

[43] That is on the basis of the 1-D *Poisson's equation* prior to *VSR* in the *channel*.

two-dimensional context. While before $y'=0$ the *channel* potential and the longitudinal electric field are only gradually increasing, beyond this point the demonstrated exponential increase in both of these quantities signals the growing role of the *drain*-induced electric field in populating the *inversion channel*.[44] Even though according to (3.47) and (3.2) along the length of the *VSR* the role of the *gate*-induced electric field in inducing the *inversion layer* is exponentially less important, it is only through the charge induction of the *drain*-induced electric field that the needed constant *carrier* concentration of the *inversion layer* along the length of this region is guaranteed.

According to (3.47) and (3.48), considering the maximum value of the *drain*-induced electric-field strength at the end of the *VSR* (i.e., $y'=\Delta L$, for which $V(y')=V_{\mathrm{D}}$) as E_{m}, where

$$E_{\mathrm{m}}=\left[\frac{(V_{\mathrm{D}}-V_{\mathrm{Dsat}})^2}{l^2}+E_{\mathrm{sat}}^2\right]^{1/2} \tag{3.49}$$

ΔL can be evaluated as

$$\Delta L=lLn\left\{\frac{[(V_{\mathrm{D}}-V_{\mathrm{Dsat}})/l]+E_{\mathrm{m}}}{E_{\mathrm{sat}}}\right\}. \tag{3.50}$$

Instead of defining the characteristic length scale in the spatial variations of potential and electric field within the *VSR* as $\left((\epsilon_{\mathrm{s}}/\epsilon_{\mathrm{ox}})X_{\mathrm{j}}t_{\mathrm{ox}}\right)^{0.5}$, a more accurate two-dimensional model of *MOSFETs* gives this factor in terms of the following equation:

$$l=0.22X_{\mathrm{j}}^{1/2}t_{\mathrm{ox}}^{1/3} \tag{3.51}$$

where all of the length variables are in cm. Equation (3.51) is developed on the basis of the assumption of *silicon channel* and *SiO_2 gate* insulator.

3.2.2.2.1 Channel-Length Modulation in Short-Channel MOSFETs According to the requirement of current continuity along the length of the *channel*, the *drain* current under the *saturation regime* of operation (i.e., when $V_{\mathrm{D}}>V_{\mathrm{Dsat}}$ and $\Delta L\neq 0$) can be calculated by substituting the length of the *linear channel* (i.e., $L_{\mathrm{G}}-\Delta L$) for the *channel*'s physical length (i.e., L_{G}) in (3.26) and (3.28). This modification results in

$$I_{\mathrm{D}}=I_{\mathrm{Dsat}}\frac{V_{\mathrm{G}}-V_{\mathrm{T}}+E_{\mathrm{sat}}L_{\mathrm{G}}}{V_{\mathrm{G}}-V_{\mathrm{T}}+E_{\mathrm{sat}}(L_{\mathrm{G}}-\Delta L)}. \tag{3.52}$$

[44] Later in this section, the inherited assumption of the steady-state *carrier transport* (i.e., from the *long-channel model*), while the electric field is undergoing an exponential spatial variation, will be revisited.

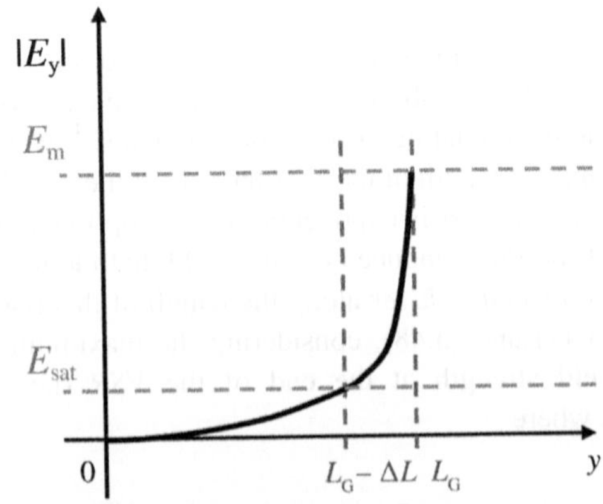

FIGURE 3.13 Prediction of the pseudo 2-D model for the variation of electric field along the channel length of a MOSFET biased in saturation.

Care should be taken that V_{Dsat} itself, as indicated in (3.28), is L_G dependent. Following this substitution, since according to (3.28) and (3.50) ΔL and V_{Dsat} are interdependent, solutions can be only produced recursively.

In replacing L_G by $L_G - \Delta L$, we have followed a mindset similar to the evaluation of the *channel-length modulation* of the *long-channel MOSFETs* (which was expressed in (3.7)). However, evaluation of ΔL according to (3.50) bears very little resemblance to the awkward assumption of *channel pinch-off.* As shown in Figure 3.13, the presented two-dimensional model, without any unrealistic assumption of infinite electric field or discontinuity in the electric field, merely predicts an exponentially increasing electric field along the *VSR*. In compliance with our previous indications, electric field prior to $y' = 0$ is relatively constant (i.e., in the so-called linear channel).

Similar to a *long-channel MOSFET*, increase in ΔL with the *drain* voltage indicates a finite output resistance. This resistance is based on the slope of the I_D–V_D characteristics at the onset of *saturation* and can be evaluated as

$$R_{out} = \left(\frac{\partial I_D}{\partial V_D}\right)^{-1} = \frac{1}{I_{Dsat}}\left(\frac{V_D - V_{Dsat}}{l}\right)\left(L_G + \frac{V_G - V_T}{E_{sat}}\right). \tag{3.53}$$

Example

For a *silicon NMOSFET* with the following specifications:

$$L_G = 0.1\ \mu m,\ W = 0.1\ \mu m,\ t_{ox} = 5\ nm,\ X_j = 0.03\ \mu m,\ \text{and}\ N_a = 10^{17}\ cm^{-3}$$

calculate the *drain* current when $V_{DS} = V_{GS} = 0.3$ V and $V_{SB} = 0$.

Assume *aluminum gate* and SiO_2 *gate* dielectric.

In an earlier example in Section 3.2.2.1.1 we have already evaluated the *drain* current of this transistor at the onset of saturation when $V_{GS} = 0.3$ V and $V_{SB} = 0$. Based on those calculations, we know that $V_T = 0.14$ V and $V_{Dsat} = 0.11$ V. However, in those calculations the *channel-length modulation* was not considered (since performed at the onset of saturation). Hence for the higher *drain* voltage of this case, according to the 2-D model presented in this section, V_{Dsat} requires reevaluation.

According to the specifications of this transistor, the length scale l is given by

$$l=\left(\frac{\epsilon_s}{\epsilon_{ox}}X_j t_{ox}\right)^{0.5}=\left(\frac{11.7}{3.9}\times 0.03\times 10^{-4}\times 5\times 10^{-7}\right)^{0.5}\cong 0.2\times 10^{-5}\ \text{cm}.$$

According to this value and first taking the initial value of V_{Dsat} equal to 0.11 V, the value of V_{Dsat} is evaluated recursively through employing (3.28), (3.49), and (3.50).

At the first iteration, based on (3.49),

$$E_m=\left[\frac{(V_D-V_{Dsat})^2}{l^2}+E_{sat}^2\right]^{1/2}=\left[\frac{(0.3-0.11)^2}{(0.2\times 10^{-5})^2}+\left(3.4\times 10^4\right)^2\right]^{1/2}\cong 10^5\ \text{V/cm}.$$

Using this value, Equation (3.50) leads to

$$\Delta L=0.2\times 10^{-5}\times Ln\left\{\frac{\left[\frac{0.3-0.11}{0.2\times 10^{-5}}\right]+10^5}{3.4\times 10^4}\right\}\cong 0.35\times 10^{-5}\ \text{cm}.$$

Now this value of ΔL needs to be subtracted from L_G to allow reevaluating V_{Dsat} for the second step of iteration: $L_G-\Delta L=0.65\times 10^{-5}$ cm. Accordingly, the new value of V_{Dsat} is given by

$$V_{Dsat}=\frac{3.4\times 10^4\times 0.65\times 10^{-5}\times 0.16}{3.4\times 10^4\times 0.65\times 10^{-5}+0.16}\cong 0.09\ V.$$

Employing this new value of V_{Dsat}, E_m is reevaluated as 1.1×10^5 V/cm, which gives

$$\Delta L=0.2\times 10^{-5}\times Ln\left\{\frac{\left[\frac{0.3-0.09}{0.2\times 10^{-5}}\right]+1.1\times 10^5}{3.4\times 10^4}\right\}\cong 0.37\times 10^{-5}\ \text{cm},$$

yielding the new value of $L_G-\Delta L=0.63\times 10^{-5}$ cm.

Accordingly, $V_{Dsat}=\dfrac{3.4\times 10^4\times 0.63\times 10^{-5}\times 0.16}{3.4\times 10^4\times 0.63\times 10^{-5}+0.16}\cong 0.09$ V, which indicates convergence.

Now (3.52) can be used to calculate the *drain* current:

$$I_D=I_{Dsat}\frac{V_G-V_T+E_{sat}L_G}{V_G-V_T+E_{sat}(L_G-\Delta L)}=3.47\times 10^{-6}\times\frac{0.16+3.4\times 10^4\times 10^{-5}}{0.16+3.4\times 10^4\times 0.63\times 10^{-5}}$$
$$\cong 4.64\times 10^{-6}\ \text{A}.$$

3.2.2.2.2 Hot Carriers in Short-Channel MOSFETs The *critical* electric field and the intensified exponential increase of $E(y')$ with *scaling* of the *MOSFET* technology impose a hard limit on the maximum affordable value of V_D. Beyond this maximum value of *drain* voltage, *avalanche breakdown* will remove I_D from the control of the *gate* and *drain* terminals. However, the situation with *scaling* of the *drain* voltage is not as grim as the evaluation of the *breakdown voltage* on the basis of comparing E_m to the *critical* electric field E_C might suggest. This is since, as observed in Section 2.3.2.1.5, in the presence of a rapidly changing electric field, such as $E(y')$ in (3.48), the *carrier* energy will not be able to instantaneously follow the lead of the electric field. The lag of the *carrier* energy with respect to the electric field only allows the *carrier temperature* at $y' = \Delta L$ to be elevated to levels predicted by exponentially smaller values of electric field, which are present earlier in the *VSR*. Proper evaluation of the *breakdown voltage* of *deeply scaled MOSFETs* requires assessment of the *energy relaxation time*, and the lag between energy and the electric field. Whereas such an evaluation might not be very straightforward, despite its inaccuracy, as a worst case scenario the evaluation of the *channel breakdown voltage* of *MOSFETs* on the basis of E_m is often considered. This assessment yields an even more pessimistic scenario as *scaling* continues.

Even short of the limit of *avalanche breakdown*, *scaling* is faced with another challenge: *hot carrier generation*. According to the definitions presented in Section 1.4.4.1, corresponding to the notion of *carrier temperature*, the term *hot carrier* refers to *carriers* that manage to gain substantial amount of kinetic energy (i.e., much in excess of the average thermal energy) from the electric field in between the *inelastic scattering* events. As will be observed in this section, the *parasitic positive feedback loops* built into the structure of a traditional *planar MOSFET*, even short of the conditions fit for *avalanche breakdown*, yield a relatively small number of *hot carriers* generated within the *VSR* a chance to impose reliability concerns and eventually *breakdown* on the operation of these transistors. Since the more pronounced presence of the *VSR* in *deeply scaled MOSFETs* has strengthened the hold of these secondary effects on the operation of transistors, they are often referred to as *SCE*. As mentioned earlier, the term *SCE* is an appropriate umbrella term for not only the *breakdown* mechanisms and reliability issues that we will speak of in this section, but also the other *short-channel* manifestations that we have so far considered (such as *DIBL* and *charge sharing*). We will show in this section that the *SCE*-related manifestations are further aggravated in presence of pairs of *NMOS* and *PMOS* transistors in the traditional *planar CMOS* technology.

Figure 3.14 identifies the important parasitic components of current induced by *hot carriers* in an *NMOS* transistor. Short of the *avalanche breakdown* conditions, the *hot* electrons generated in the *VSR* are relatively small in number and possibly energy. They can, however, still take part in enhancing the *impact-ionization* rate and producing extra *electron–hole pairs*. As schematically illustrated in Figure 3.14, the strong electric field of the *VSR* quite swiftly separates the electrons and *holes* in these *pairs*. In the presence of a positive V_G and V_D, electrons tend to

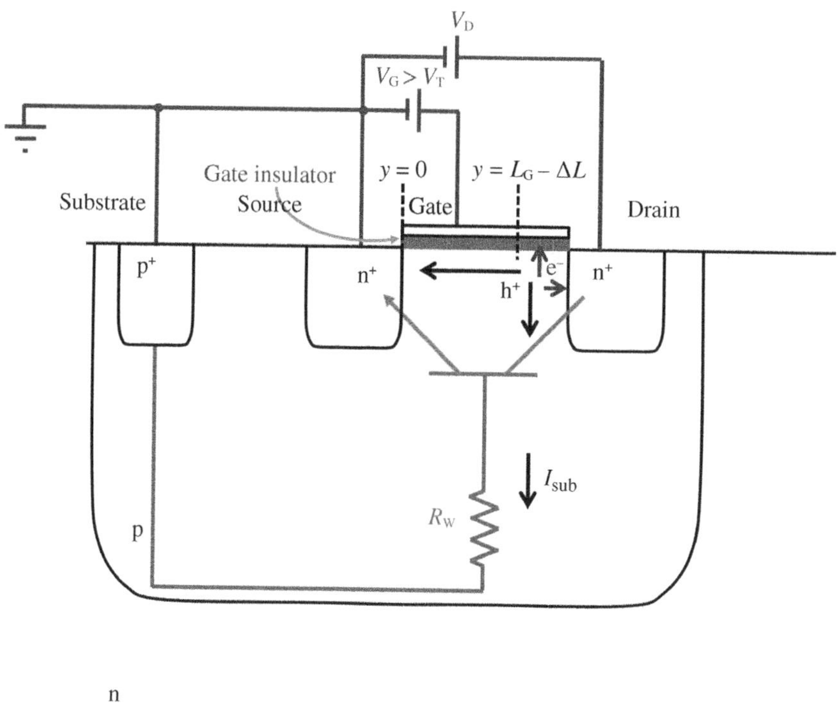

FIGURE 3.14 Trajectories of electrons and holes generated within the VSR of an n-channel MOSFET and the circuit diagram of the positive feedback network resulting from migration of holes toward the substrate. R_W denotes the resistance of the p-well within which the NMOS is realized. Passage of the current I_{sub} induced by migration of holes through this large resistance can provide the turn-on bias for the BE emitter junction of the depicted parasitic BJT and hence establishing the positive feedback loop. This feedback results in the so-called snapback breakdown.

either take the trajectory toward the *drain* or *gate* electrodes, and *holes* will be collected by the *substrate* and *source* contacts. Although ideally the presence of the large *conduction band* discontinuity prevents the migration of electrons toward the *gate* electrode, as will be shown in further details later in this section, *hot* electrons stand a certain chance to make it through the *gate* insulator (i.e., electron migration via the *field-assisted thermionic emission* process). While such a migration can contribute to the *DC gate* current, at the same time, through interaction between the *hot* electrons and the insulator, the migration can impose the possibility of forming *trap sites* within the insulator or at the insulator/semiconductor *interface* and electron trapping in these new (or the already existing) *trap sites*. This latter manifestation, which can be easily tracked via studying the *C–V* characteristic

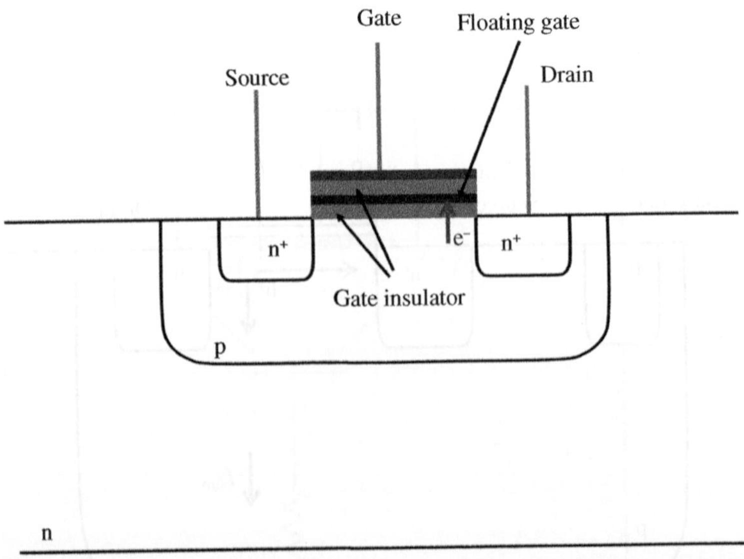

FIGURE 3.15 Illustration of the concept of programming a floating gate through migration of hot electrons generated within the VSR.

of the *MOS* system, results in a change in the *threshold voltage* of the *stressed MOSFET*.[45]

As suggested in Figure 3.15, *hot carrier* injection through the oxide is used in programming *nonvolatile memories*. In these cases, a *floating gate* is sandwiched within the *gate*-insulator layer, and the transfer of *hot* electrons to this *floating gate* is essentially used to change the status of the *threshold voltage* as a *bit* of memory. Ultraviolet illumination can be used for removing the electrons transferred to this *floating gate* and resetting the *state* of the *memory cell*.

The contribution of electrons separated from the *hot carrier*-generated *EHPs* to the *drain* current, which is shown in Figure 3.14, for conditions short of *avalanche breakdown* is not very substantial. Migration of *holes* toward the *source* and *substrate* contacts can, however, make a substantial contribution to I_D.

Migration of *holes* toward the *source* contact bears a considerable resemblance to *DIBL*. The injection of these *holes* across the *source*-side *p–n* junction *forward biases* this junction and lowers the *source*-side *potential barrier* identical to *DIBL*. Lowering this *potential barrier* triggers the removal of the *channel* from the electrostatic control of the *gate* and provokes an I_D increase (and possibly *channel breakdown*). For *channel breakdown* to happen, this mechanism needs to be sustained via the *positive*

[45] Whereas we often speak of *hot* electrons, due to the larger *valence-band* discontinuity of *SiO_2/Si interface* (i.e., compared to ΔE_c), such a manifestation will be much less probable in the case of *hot holes* in a *PMOS* transistor (at least where the *gate* insulator to *channel* contact poses a larger discontinuity on the *valence band* compared to the *conduction band*). More will be said on this topic later in this section.

feedback loop. Such a *feedback loop* is present through the link between the increase in the *drain* current and the growth in terms of the number of *hot carriers* promoting *impact ionization*.

Migration of *holes* toward the *substrate* contact can also promote the *channel breakdown* in conjunction with the presence of a *positive feedback loop*. This *loop* is realized through the connection of the *MOSFET* and the parasitic *BJT* highlighted in Figure 3.14. As illustrated in this figure, the parasitic *BJT* is formed between the *carrier* reservoirs of the *drain* and *source* and the *p-type bulk*. Since *bulk* is very often a *lightly doped* high-resistivity region, current carried by *holes* that are migrating toward the *substrate* contact can produce sufficient voltage drop to *forward bias* the *base–emitter junction* of the parasitic *BJT* and essentially *bias* it in the *forward-active mode* (i.e., to produce *current gain*). Not unlike the previous scenario, this circuit has the possibility of forming a self-sustaining *positive feedback loop*. The mechanism of *channel breakdown* promoted by the presence of this parasitic *BJT* is known as *snap-back breakdown*. Although highly *doping* the *bulk* reduces the chance of *forward biasing* the *base–emitter junction* of the parasitic *BJT*, its adversary roles in reducing the *channel's carrier mobility* and promoting *Zener breakdown* on the *drain*-side's *p–n junction* prevents it from being a very suitable remedy. As will be shown in Chapter 4, *SOI MOSFETs* can provide a better solution to this problem.

The *hot carrier*-generated *substrate* current takes on an even less predictable limiting nature in the operation of *CMOS pairs* (i.e., compared to *NMOS* and *PMOS* technologies). Figure 3.16a presents a schematic cross section of a planar *CMOS pair*. As highlighted in this figure and shown in the circuit schematic of Figure 3.16b, the presence of parasitic *BJT*s and resistors among the *NMOS* and *PMOS* transistors of this pair and the possibility of forming a *positive feedback loop* can impose an important mechanism of *breakdown* related to *hot carrier* generation on the operation of *MOSFETs*. This mechanism of *breakdown*, which is known as *latch-up*, especially at the input–output stages of *CMOS* circuitries that are prone to *signal* overshoot and ringing and essentially imposition of higher than expected electric fields, levies serious challenges on the operation of these circuits. In addition to forming a *positive feedback loop* through the initiation of *hot carrier*-generated *substrate* current I_{gen}, in the presence of radiation-induced *impact ionization* (e.g., in extraterrestrial applications), such a *feedback loop* can substantially harm the reliability of operation of *CMOS* circuitries. In the *feedback loop* depicted in Figure 3.16b, the closed-loop current gain is given by

$$A_I = \frac{R_W \cdot R_{sub} \cdot \beta_n \cdot \beta_p}{(R_W + r_{\pi n}) \cdot (R_{sub} + r_{\pi p})} \tag{3.54}$$

where R_W and R_{sub} are the parasitic resistance of the *well* and the *substrate*, respectively. β_n and β_p are the *common-emitter* current gain of *npn* and *pnp* parasitic *BJTs*, and $r_{\pi n}$ and $r_{\pi p}$ are the series input resistance of the two *BJTs* under this *bias* configuration, respectively. The *positive feedback loop* takes hold of the operation of the *CMOS pair* and causes *breakdown* only when $A_I > 1$.

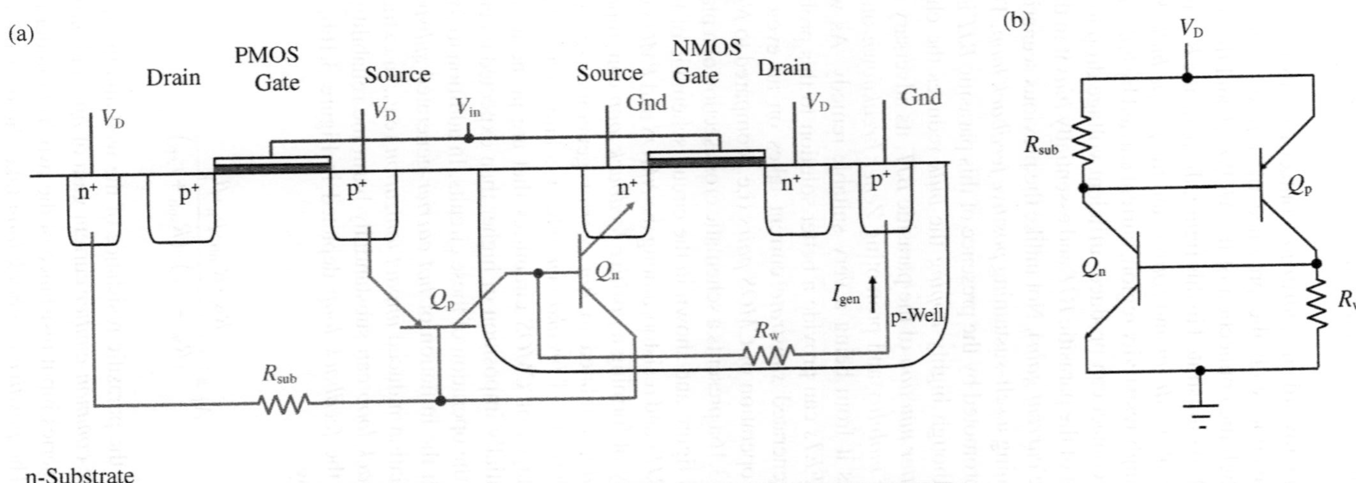

FIGURE 3.16 (a) Illustration of the mechanism of latch-up among CMOS pairs on a cross-sectional view of a n-well CMOS pair. Parasitic elements Q_n, Q_p, R_{sub}, and R_w are indicated in gray. (b) Circuit diagram of the latch-up feedback network.

For (3.54) to apply, it is required that the conditions for bringing the two parasitic *BJTs* into the *forward-active* mode of operation are concurrently sustainable. Toward this end a certain *hold time* is attributed to the operation of the *loop*. Within this context, the sufficient potential drop across R_W should sustain long enough to turn *on* not only the *npn BJT* but also the *pnp BJT* (which is turned *on* after turning *on* the *npn BJT*, and establishing sufficient potential drop across R_{sub}). It is only then that the required potential drop across R_W (for keeping the *npn BJT on*) can be sustained even in the absence of the mechanism that produced I_{gen} at the beginning of this cycle (i.e., *radiation* or *hot carrier* generation within the *VSR*). If these conditions for generation of I_{gen} are swiftly removed (i.e., before the two *BJTs* of the *positive feedback loop* are *on*) the threat of *latch-up* will not be present.

On the basis of (3.54), the essential substitution of a *silicon* substrate by an insulator in the *SOI CMOS* technology can effectively eliminate *latch-up*. As will be observed in Chapter 4, through dissociating the two *BJTs*, the *positive feedback loop* of Figure 3.16b will be eliminated. Among other solutions for the *latch-up* problem are reducing the *BJT* current gain and reduction of R_W and R_{sub}. Reduction of the parasitic resistances can be done through using multiple contacts to the *substrate* and the *well* and by defining *guard rings* for imposing definitive values of potential independent of I_{gen} (which comes at the cost of reduced packing density) or processing the *CMOS pair* on a *lightly doped epitaxial* layer grown on a very *highly doped* substrate (which comes with a price tag associated with the *epitaxial growth*). For reducing the current gain of the parasitic *BJTs*, a variety of solutions including implementing *retrograde doping*, reducing the *excess minority carrier lifetime* of the *base* via introducing *Au-doping*-induced *recombination centers*, and increasing the thickness of *quasi-neutral region* of the *base* through implementing *deep-trench* isolation have been successfully explored.

Hot carrier generation within the *VSR* has also been the cause for another important modification to the structure of *MOSFETs*: the *halo implant*. As we took note on Figure 3.11, this *dopant* profile is realized through *shallow angled ion implantation* of *acceptors* under the *drain* and *source* edges of the *gate* in an *NMOS* transistor (*donors* are *implanted* for *PMOS* transistors). Such an *implant* profile, while making the *inversion* of this part of the *channel* more difficult, creates a more resistive region in the proximity of the *VSR*. Potential drop across this region limits the peak of the electric field developed in the *VSR* (which is formed prior to this *halo-implanted* profile). As a result of the reduction in E_m, the aforementioned negative attributes of the *SCE* will be also reduced. Figure 3.17 illustrates an earlier variant of this technology, which is known as *lightly doped drain* (*LDD*).

Whereas introduction of a high-resistivity extension on the *drain* side of the *channel* has the merit of spreading the potential drop[46] across a longer portion of the *channel length*, and as a result reducing E_m, incorporation of these features on the low electric-field *source* side of the *channel* merely results in degrading the device performance (i.e., reduction of the *gate transconductance*). This is one of the prices

[46] That is, V_D–V_{Dsat}.

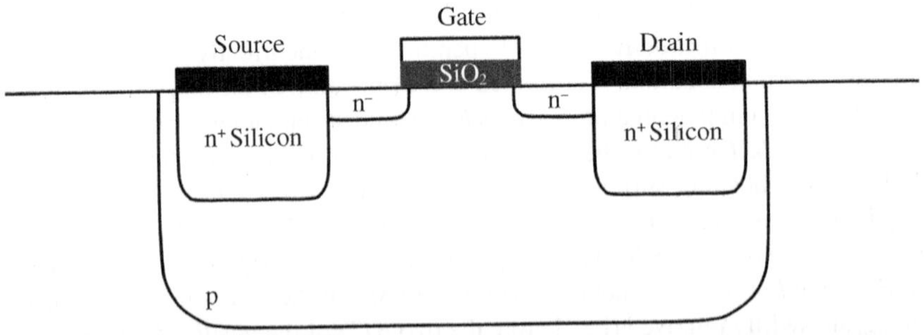

FIGURE 3.17 Cross-sectional view of a NMOS transistor using the LDD technology.

that *CMOS* technology pays for further integration. While these profiles are created after forming the *gate* stack using a *large-angle tilted ion-implantation technology*,[47] the lack of alignment of all the transistors on a *chip* requires a two-step *implantation* at opposite angles, resulting in forming the *halo implant* not only on the *drain* side but also on the *source* side of the *channel*.

In the rest of this section, based on the findings of the *short-channel model* presented in Section 3.2.2.2, the roles played by *hot carriers* in inducing *gate* and *substrate* currents are further explored.

SUBSTRATE CURRENT As we have seen earlier in (2.120), in terms of a first-order approximation of equal *effective mass* for electrons and *holes*, the *generation* of *electron–hole pairs* in a given semiconductor requires the acquisition by the *hot carriers* of an amount of kinetic energy 50% in excess of the *bandgap* energy. While such an acquisition from the electric field is quite possible within the high electric-field *VSR* region, its implications on the device characteristics such as *gate* and *substrate* current must be studied. As noted earlier, in an *NMOS* transistor the electrons generated in the *VSR* are attracted to regions of higher potential (i.e., *drain* and *gate*), while *holes* are attracted by the lower potential contacts of *source* and *substrate*. As the electron and *holes* are quickly removed from the high electric-field region of the *VSR*, normally *avalanche breakdown* will not be instigated instantaneously. However, we already know that a *positive feedback loop* in the forms contributing to *latch-up* or *snapback* can quite possibly contribute to the *channel breakdown*. As a result of these *feedback* systems, each of the possible trajectories for the *excess* electrons and *holes* (which are *generated* in the *VSR*) has the possibility to levy important reliability implications on the operation of the *MOSFET*. Among these trajectories, due to the presence of the *gate*-insulator and *band-edge* discontinuity, the path to the *gate* is the least possible. However, especially at higher *drain biases*, electron migration across this path has been observed to contribute to excessive *gate*-leakage current. Although this *carrier* injection is essential to the operation of *nonvolatile memories*, from the point of view

[47] Known as *large-angle-tilt implanted drain* (or *LATID* for short).

of the operation of a *MOSFET*, this nonideal *gate* current is a nuisance that requires evaluation. The second nonideal component of current is the *substrate* current, which, as indicated earlier, is the culprit in the *latch-up* and *snapback* processes. In this part of our discussion, on the basis of the electric field evaluated in the *VSR*, these two non-ideal current components will be assessed.

We start the evaluation of the *substrate* current by determining the number of *holes* generated through the *impact-ionization* process in the *VSR*. As seen earlier in (2.131), the number of *impact-ionization* events instigated by one *carrier* per unit length is determined in terms of *ionization rate*,[48]

$$\alpha = A_{\mathrm{i}} \exp\left(-\frac{B_{\mathrm{i}}}{E}\right). \tag{3.55}$$

In this equation, B_{i} in the case of *holes* is almost twice that of electrons, making *impact ionization* by *holes* exponentially less likely. As a result, oftentimes we talk about *hot* electrons and not so much about *hot holes*.

Based on (3.55) and assuming *current continuity* along the length of the *channel*, if we assume that none of the *holes* will take the *source*-bound trajectory, the *substrate* current can be written as

$$I_{\mathrm{sub}} = \int_0^{L_{\mathrm{G}}} I_{\mathrm{D}} A_{\mathrm{i}} \exp\left(-\frac{B_{\mathrm{i}}}{E}\right) dy. \tag{3.56}$$

Using the mathematical manipulation of dy with $(dy/dE)dE$ or $-E^2(dy/dE)d(1/E)$, *substrate* current can be written as

$$I_{\mathrm{sub}} = \int_{E_{\mathrm{s}}^{-1}}^{E_{\mathrm{m}}^{-1}} -I_{\mathrm{D}} A_{\mathrm{i}} \exp\left(-\frac{B_{\mathrm{i}}}{E(y)}\right) E^2(y) \frac{dy}{dE} d\left(\frac{1}{E}\right) \tag{3.57}$$

where E_{s} is the electric field at the *source* end of the *channel*. According to (3.49) E_{m} is the maximum longitudinal electric field at the *channel*'s *drain* end.

Based on the profile of the variation of the electric field along the length of the *channel*, since only later in the *VSR* the longitudinal electric field is strong enough to yield a substantial *ionization rate*, (3.57) can be safely evaluated only within this latter part of the *channel* identified by the length scale l. Since in this region, according to (3.48), electric field can be approximated by $E_{\mathrm{sat}}(\exp(y'/l)/2)$, we have

$$\frac{dE}{dy} \approx E_{\mathrm{sat}} \frac{\exp(y'/l)}{2l} = \frac{E}{l}. \tag{3.58}$$

[48] In this simpler form of (2.131), due to the dominance of the exponential dependence of the *ionization rate* on the electric field, A_i has taken the place of A_iE.

According to (3.58),

$$E^2(y)\frac{dy}{dE} \approx lE. \tag{3.59}$$

In inserting (3.59) into (3.57), due to the presence of a dominant exponential dependence on the electric field in $\exp(-B_i/E)$, electric field in (3.59) can be taken as E_m. This approximation yields,

$$I_{sub} = -A_i l E_m I_D \int_{E_s^{-1}}^{E_m^{-1}} \exp\left[-\frac{B_i}{E(y)}\right] d\left(\frac{1}{E}\right) = \frac{A_i}{B_i} l E_m I_D \exp\left(-\frac{B_i}{E}\right)\Bigg|_{E_s}^{E_m}. \tag{3.60}$$

Since E_m is much larger than E_s, (3.60) can be simplified to

$$I_{sub} \approx \frac{A_i}{B_i} l E_m I_D \exp\left(-\frac{B_i}{E_m}\right). \tag{3.61}$$

Based on (3.49), (3.61) can be approximately presented by

$$I_{sub} \approx \frac{A_i}{B_i}(V_D - V_{Dsat}) I_D \exp\left(-\frac{lB_i}{V_D - V_{Dsat}}\right). \tag{3.62}$$

According to (3.62), clearly the most important device-related parameters in determining the *substrate* current are l, V_D, and V_{Dsat}, while B_i and A_i are material-related parameters. In a *silicon NMOS*, the value of B_i is experimentally evaluated to be approximately equal to 1.7×10^6 V/cm while A_i/B_i is about 1.2 V^{-1}.

Despite the ease of use of this compact equation, we should remind ourselves that in a rapidly changing electric field, the above framework produces a very pessimistic estimation. This is caused by overlooking the *energy relaxation time* when evaluating the *ionization rate*.

As indicated earlier, one of the implications of the flow of current through the *substrate* is the instigation of *snapback breakdown* (and in case of *CMOS*, *latch-up*). However, even in the absence of the necessary conditions for any of the *breakdown* processes, as seen in Section 3.2.1.3, the induction of a potential difference across the *substrate* by the passage of the *substrate* current causes a change in the *threshold voltage* and the *drain* current. Since the *substrate* current is dependent on the *drain* voltage, the aforementioned variation of the *drain* current is another appearance of *drain* current modulation by the *drain* voltage.

GATE CURRENT Establishing a current through the *gate* insulator is linked to *carrier* hopping through existing (or *hot carrier* generated) defects in the insulator and/or a quantum-related *tunneling* processes (i.e., *field-emission* or *field-assisted thermionic emission*).

Based on the discussions of Section 2.5.2, in the absence of existing or stress-generated defects within the *gate* insulator, the so-called *lucky-electron tunneling* model[49] presents a simple yet versatile framework for evaluating the *gate* current.[50] In this model, the two aforementioned processes of *hot carrier generation* and *quantum tunneling* are linked to the *gate* current via the additional processes of redirectional *scattering* and *transport* through the insulator. The first of these two additional processes is required to redirect the trajectory of the *hot carriers* traveling along the length of the *channel* to one normal to the insulator/semiconductor *interface*. This redirected *carrier* then needs to be transported all the way across the insulator to contribute to the *gate* current. In the *lucky-electron tunneling* model, the probability of these four processes (which are treated as four independent processes), are used in evaluation of the *gate* current. According to the results of this model, in spite of the common cause (i.e., *hot carriers*), due to the presence of a substantial difference between the *band bending* at the *inversion-channel/buffer* interface and the insulator/semiconductor interface, the *bias* dependence of the *substrate* and the *gate* current are very different from one another. While at least for *gate* dielectrics thicker than a few *nanometers* the *gate* current is much smaller than the *substrate* current, it is also observed that its magnitude maximizes when the *drain* and *gate* voltages are comparable.

In order to formulate the *lucky-electron tunneling* model, it is worth reiterating that both electric-field components (i.e., the *drain*-induced and the *gate*-induced electric field) are involved in inducing the *gate* current. While the first contributes to energizing the *thermal-equilibrium* electrons in the *channel* to rise over the insulator/semiconductor potential barrier, the latter field is responsible for moving (and, delivering) the redirected *hot carriers* toward the *gate* electrode. For a *hot carrier* with an amount of energy corresponding to the barrier height to be contributing to the *gate* current, an almost elastic redirectional collision is required to change the trajectory of the *carrier* at no loss of energy. An energy loss during this *scattering* event can keep the originally *hot* electron behind in the *potential well*. The statistical independence of the processes involved in the *lucky-electron tunneling* makes it possible to calculate the probability of *lucky-electron tunneling* in terms of the product of the probabilities of the four individual processes.

In case of the SiO_2/Si interface, due to the larger size of the *band discontinuity* of the *valence band* compared to the *conduction band*, *lucky-electron tunneling* is more relevant than *lucky-hole tunneling*. This situation is further exacerbated due to the often smaller *mean free path* of *holes* in comparison to electrons. As noted in Section 2.4.1, the barrier height seen by electrons at the SiO_2/Si interface is about 3.1 eV. As a result, neglecting the involvement of quantum *tunneling* only those *hot* electrons of kinetic energy larger than this value will be able to contribute to the *gate* current. This value for the case of *hot holes* in this system is about 3.15 eV.

[49] See S. Tam, P. K. Ko, and C. Hu, "Lucky-electron model of channel hot electron injection in MOSFET's," *IEEE Trans Electron Devices*, vol. **31**, no. 9, pp. 1116–1125, 1984.
[50] While such assumptions are quite sufficient for many of the SiO_2/Si interfaces, especially when *SCE* is not so dominant, the models developed without these assumptions are much more complicated.

In the absence of a finite *energy relaxation time* and assuming a constant energizing longitudinal electric field along the later part of the *VSR* (i.e., instead of the exponentially variable electric field in this region), an electron needs to travel through the *channel* for about a length of $\Delta E_c/qE_m$ before it encounters a *scatterer* to be able to contribute to the *gate* current.[51] In this definition, ΔE_c refers to the *conduction band* discontinuity at the insulator/semiconductor interface.[52] This length should be smaller than the electron *mean free path* (i.e., λ). Based on this explanation and Equation (2.218), the probability of an electron becoming sufficiently *hot* to contribute to *gate* current is expressed by $\exp[-\Delta E_c/(qE_m\lambda)]$.

The probabilities of redirectional collision, injection, and collection according to this model are often evaluated as a function of the *gate*-induced electric field (i.e., $p(E_{ox})$). As a result, in proportionality to the *inversion-layer* electron concentration (i.e., implied through I_D), *gate* current is evaluated as

$$I_G \propto I_D \int_0^l \exp\left(-\frac{\Delta E_c}{qE_m\lambda}\right) p(E_{ox})dy'. \tag{3.63}$$

Assuming small variation of $p(E_{ox})$ along l, Equation (3.63) can also be presented as

$$I_G \propto I_D \exp\left(-\frac{\Delta E_c}{qE_m\lambda}\right). \tag{3.64}$$

3.3 QUANTUM CONFINEMENT AND ELECTROSTATICS OF MOSFET

Whereas the framework presented in Section 3.2 provides us with an acceptable foundation for the analysis of *MOSFETs* operating under the *drift–diffusion transport* regime, this foundation lacks the appreciation of the quantum phenomena. Such a deficiency gains a more pronounced stature as the device dimensions shrink further (i.e., even within the limits of applicability of *drift–diffusion* formalism). In this regard, what is missing from our description under the regime of *strong inversion* is an assessment of the quantum phenomena prompted by the spatial confinement of *carriers* at the insulator/semiconductor interface.[53] While in the presence of the *inversion layer* formed in this *quantum well* the approximations of the *Maxwell–Boltzmann* statistics become inadequate, in light of the discussions of Section 1.3.5 appreciation of the evolution of energy *subbands* in this *degenerate channel*

[51] This is clearly with the assumption of predominance of *thermionic emission* and not the *field emission*.

[52] Evidently, quantum phenomena and the difference between the first *subband* and E_c is overlooked.

[53] In view of the discussion of *quantum capacitance* in Section 2.4.2, see Equations (3.2), (3.11), (3.22), (3.27), and (3.44), in which ignoring that quantum effect the *gate capacitance* under *strong inversion* (i.e., C_G) is taken equal to C_{ox}.

gains significance. With further *scaling* of *MOSFETs*, due to the enhanced spatial confinement of electrons in this *inversion channel*, these quantum effects gain even more significance. Analogous to the description of the *subthreshold* regime of operation in terms of a series combination of the *oxide* (or generally speaking *insulator*) and the *depletion capacitance* (i.e., C_{ox} and C_s, respectively) to yield the *gate capacitance* in (3.16), the existence of energy *subbands* in an *inversion channel biased* in *strong inversion* is often expressed through the incorporation of a so-called inversion capacitance in series with the *oxide capacitance* to define the *gate capacitance*:

$$C_G = \frac{C_{ox} C_{inv}}{C_{ox} + C_{inv}}. \tag{3.65}$$

This approach sufficiently addresses the quantum phenomena under the *drift–diffusion transport* regime.[54]

The smaller the *inversion capacitance*, the smaller is the *gate capacitance* compared to the *oxide capacitance*. This means that the assumption of $C_G = C_{ox}$ built into the model of Section 3.2 would be further from reality.

Still not directly addressing the quantum phenomena, phenomenologically in terms of the distance between the position of the peak of the *carriers' probability density function* (or the centroid of *inversion-layer*) and the insulator/semiconductor interface (i.e., Δ), C_{inv} can be defined as

$$C_{inv} = \frac{\epsilon_s}{\Delta}. \tag{3.66}$$

The value of Δ, also known as the *inversion-layer* thickness, is typically about 5 nm in *MOSFETs*.

In order to better understand the meaning of the *inversion capacitance*, it is worth looking into definition of *quantum capacitance*. These two concepts are closely related and are employed interchangeably in our discussion. *Quantum capacitance* is defined in terms of development of quantized *energy levels* in the presence of spatial confinement of *charge carriers*. To assess this quantum phenomenon, as a simpler example a fully *degenerate* case of a *quantum well* with only one occupied *subband* can be studied at $T = 0$ K. According to Figure 3.18, *gate* potential raises or lowers the *subband* in this *single-level channel*.

For this situation, in terms of the *surface* potential ϕ_s and the *sheet carrier* concentration n_s, the *quantum capacitance* is given by

$$C_q = q\frac{\partial n_s}{\partial \phi_s} \tag{3.67}$$

[54] Beyond this regime, there is much that the model of Section 3.2 is not doing justice to. We will briefly talk about these issues in Chapter 6.

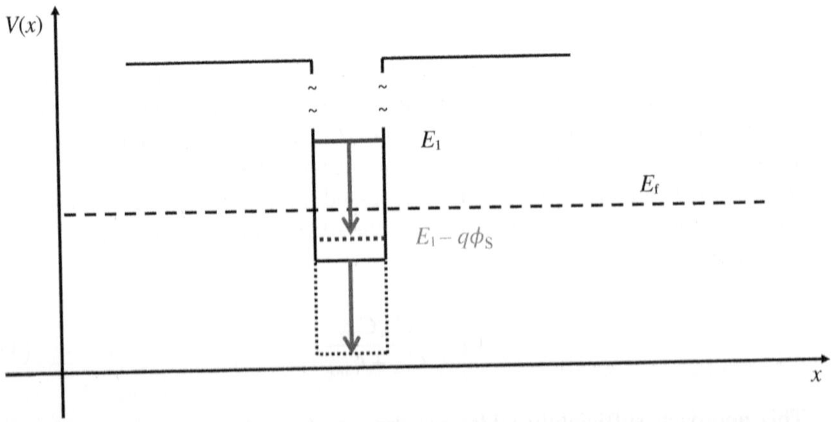

FIGURE 3.18 Variation of the position of the single energy level of a quantum well with the application of the surface potential ϕ_s versus the Fermi level. Potential of the well is only indicated along the *x*-direction. The repositioned energy level and the potential are indicated with the dotted lines.

where according to (1.38),

$$n_s = N_{2D} \times \left(\frac{E_f - E_1'}{kT} \right). \tag{3.68}$$

In (3.68), $N_{2D} = \left(m^* kT / \pi\hbar^2\right)$ and $E_1' = E_1 - q\phi_s$. In this definition, E_1 denotes the first *subband energy level* and use of E_1' indicates the repositioning of this *energy level* with regard to the *Fermi level* (defined by the potential of *drain* and *source* when tied together, while the *gate* is modifying the *surface* potential ϕ_s).

Using Equations (3.67) and (3.68), for the *lattice temperature* of *zero Kelvin*,

$$C_q = q^2 \frac{m^*}{\pi\hbar^2}. \tag{3.69}$$

Equation (3.69) shows that the *quantum capacitance* is proportional to an important material-related property of the *channel*: the *density of states' effective carrier mass*. In terms of this observation, in materials of higher *effective carrier mass*, the *quantum capacitance* will also be larger. In these cases, based on (3.65) the *insulator capacitance* plays a bigger role in defining the *gate capacitance*. Since in order to enhance the *charge-carrier* induction in the *channel* a larger *gate capacitance* is required, semiconductor *channels* of lower *effective mass*, while presenting a better *carrier transport*, impose a larger burden on increasing the *insulator capacitance* to compensate for the loss in *inversion capacitance*. That translates into a greater need for *high-K dielectrics* and thinner insulators, which as indicated previously are not without challenge. This issue, in terms of the competition between the *transport* and *charge-carrier* induction in the *channel*, poses another challenge to the *scaling* of *MOSFETs*.[55]

In order to assess the value of *quantum capacitance* at *lattice temperatures* above *zero Kelvin*, we can evaluate the *sheet carrier concentration* of the *inversion channel* in terms of *Fermi–Dirac statistics* governing the aforementioned *single-subband channel*:

$$n_s = \int_{E_1'}^{\infty} N_{2D} \cdot f_D(E - E_f)dE = \int_0^{\infty} N_{2D} \cdot f_D(E' - E_f + E_1 - q\phi_s)dE'. \quad (3.70)$$

For this situation, the *quantum capacitance* is given as

$$C_q = q\frac{\partial n_s}{\partial \phi_s} = q\int_0^{\infty} N_{2D}\frac{\partial f}{\partial \phi_s}dE = q^2\int_0^{\infty} N_{2D}\left(-\frac{\partial f}{\partial E}\right)dE = q^2\langle N_{2D}(E_f)\rangle \quad (3.71)$$

where the factor $-\partial f/\partial E$ almost acts as a *Dirac delta* function with a width of about kT defined at the *Fermi level*. As a result, the *quantum capacitance* is proportional to the average *density of states* at the *Fermi level*.

3.4 SUBTHRESHOLD OPERATION OF SHORT-CHANNEL MOSFET

In view of the major interest in the operation of *MOSFETs* in the *subthreshold* regime, a substantial body of work has focused on devising models for forecasting the *scalability* of *MOSFETs* in terms of the *subthreshold* operation.[56] The analytical attempts among these models are built on the foundation of solving *Poisson's equation*. According to the assumption of *subthreshold* operation, these models only consider the immobile charge concentration.

Contrary to the earlier expectations of linear dependence of *threshold voltage* and *surface* potential roll-off with L_G in terms of a certain so-called length scale l, the models proposed over the past two decades all agree on the exponential dependence of the roll-off in the form $\exp(-L_G/l)$. Nevertheless, the evaluation of the length scale up until recently had remained a controversial matter. These models, considering a tolerable *threshold voltage* roll-off of about 100 mV, identify the minimum *gate length* of *three* times the length scale. As a result, accurate assessment of the length scale is quite important in forecasting the *scaling* strategies.

In spite of the lingering ambiguity in evaluation of l, the recently published review of Xie et al. compares the assumptions and results of these models versus the trends

[55] From this point of view, a comparison between *silicon* and *GaAs MOSFETs* is interesting. We will talk about this in greater details in Chapter 4. Within this context, the difference between the *transport* and the *DOS effective carrier masses* should be appreciated. These quantities although related, as indicated in Chapter 1, are not the same for all semiconductors.

[56] For a full treatment of these models, seeQ. Xie, J. Xu, and Y. Taur, "Review and critique of analytic models of MOSFET short-channel effects in subthreshold," *IEEE Trans Electron Dev*, vol. **59**, no. 6, pp. 1569–1579, 2012.

obtained from 2-D numerical simulations. This tends to propose the approach of Frank et al. as the most accurate representation of the *scalability* of *subthreshold* operation, especially in connection with the use of *high-K dielectrics*.[57] In this part of our discussion, the major results of this model with the indication of the background assumptions are presented.

The approach of Frank et al. in evaluation of the length scale, which is referred to as the *2-region* (and as expressed in Chapter 4, in case of *FinFETs*: *3-region*) *length-scale model*, builds on the foundation of solving *Poisson's equation* both in the *gate* dielectric and the *depleted bulk*. In an ideal dielectric in the absence of charge, *Poisson's equation* written in the dielectric turns into a *Laplace equation*.[58] Xie et al. point out that since the use of *high-K dielectrics* allows the physical thickness (and not the *EOT*) of the dielectric to grow, if in need of reduction of *field emission* a thick dielectric is chosen, the appreciation of the differences between the *permittivities* of the semiconductor and the insulator is considerably important. Whereas through solving *Poisson's equation* by combining the insulator and the semiconductor in one region with a combined thickness of $x_{dm} + (\epsilon_{Si}/\epsilon_i)t_i$ (where x_{dm}, t_i, ϵ_{Si}, and ϵ_i are the maximum width of the *depletion region*, insulator thickness, and the *permittivities* of silicon and insulator, respectively),[59] the effect of the vertical electric field in the insulator and its continuity is acknowledged, such a single-region model falters with the lateral electric field in the insulator.[60]

The *2-region model* of Frank et al., which is especially suited for thick dielectrics, builds on the foundation of the boundary conditions expressed in terms of the *gate*,

[57] Source: D. Frank, Y. Taur, and H. Wong, "Generalized scale length for two-dimensional effects in MOSFETs," *IEEE Electron Dev Lett*, vol. **19**, no. 10, 1998.

[58] Named after *Pierre-Simon Laplace*.

[59] That is, through invoking the notion of *equivalent thickness*.

[60] The error is evidently more massive if one evaluates the *Poisson's equation* only in the semiconductor (i.e., as we did in Section 3.2.2.2 in evaluation of the *velocity saturation region* by assuming a constant vertical electric field E_{ox} in the oxide). This can be readily noticed in terms of the 2-D *Laplace equation* written for potential $\phi(x,y)$ in insulator,

$$\frac{\partial^2\phi(x,y)}{\partial x^2} + \frac{\partial^2\phi(x,y)}{\partial y^2} = 0$$

where in terms of the implicit assumption of constant vertical electric field in the insulator (i.e., *x*-direction), as the first term is taken equal to *zero*, the second terms also become *zero*. This implies a constant *zero* lateral electric field in the insulator, since in this layer only a vertical electric field has been assumed. However, this causes a conflict at the interface since on the semiconductor side of the interface the lateral electric field takes on a finite value (hence resulting in electric-field discontinuity). Although such an assumption has been part of our evaluation of the top boundary condition for assessing the *velocity-saturation region* (i.e., in evaluation of E_{ox} in equation (3.41)), since for large electric fields developed in the *VSR* lack of appreciation of *energy relaxation time*, even as it is, builds in a certain level of overestimation into the model, often not much of an attention is given to these effects that are counted secondary. This is especially applicable when we acknowledge the fact that even with *high-K dielectrics*, due to the relatively limited maximum achievable dielectric constant, the physical thickness of the insulator is often very small. We will reveal more on these limits as we go through this section. Due to the relatively simple form of equations developed in Section 3.2.2.2, although nowadays more in relative terms, those models are still in use when it comes to speculating on the reliability in operation of *MOSFETs* correlated to *hot-carrier* phenomena.

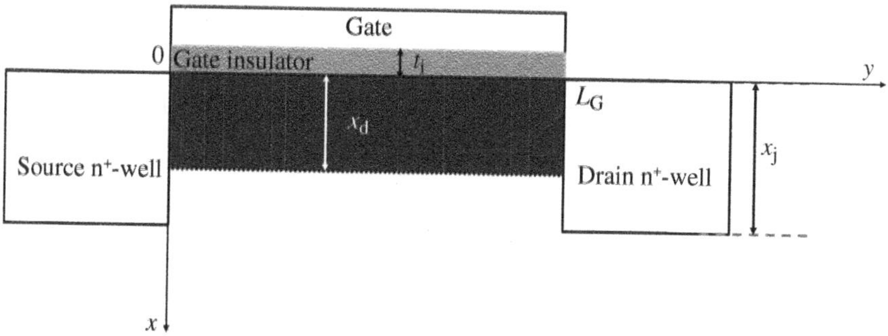

FIGURE 3.19 Illustrative depiction of the foundation of the 2-region model of Frank et al. The two mentioned rectangles of this model are identified in two shades of gray.

source, and *drain* potentials (at the top, left and right-side boundaries), continuity of electric field at the interface of the rectangles identified in the insulator and the semiconductor, and *zero* potential at the bottom interface of the semiconductor's rectangle. These rectangles are identified in Figure 3.19.

As expressed in this list of boundary conditions, *Frank*'s model instead of assuming a *zero* electric field and potential at the bottom boundary of the rectangle defined on the semiconductor side, only assumes a *zero* potential. Although Frank et al. through evaluation of their model versus the 2-D numerical model suggest this as the correct assumption, they are bound to call for a *Dirac delta doping* profile (i.e., in their implementation: a *retrograde junction*) at this bottom boundary to support the discontinuity of the electric field. Although limiting, such an assumption has succeeded in yielding a relatively manageable foundation for evaluating the length scale.

This model results in the following transcendental equation for evaluating the eigenvalues k_n of the potential,

$$\frac{1}{\epsilon_i}\tan(k_n t_i) + \frac{1}{\epsilon_{Si}}\tan(k_n x_{dm}) = 0 \tag{3.72}$$

where x_{dm}, t_i, ϵ_{Si}, and ϵ_i are the maximum width of the *depletion region*, insulator thickness, and the *permittivities* of silicon and insulator, respectively. The potential in terms of the dimensions expressed in Figure 3.19 and these eigenvalues is calculated as

$$\phi(x,y) = \phi_s\left(1 - \frac{x}{x_d}\right)^2 + u_1(x,y) + u_2(x,y) \tag{3.73}$$

where

$$u_1(x,y) = \sum_{n=1}^{\infty} \frac{b_{n1}\sinh[k_n(L_G - y)] + c_{n1}\sinh[k_n y]}{\sin(k_n L_G)}\sin[k_n(x + t_i)] \quad \text{for} -t_i \le x \le 0 \tag{3.74}$$

and

$$u_2(x,y)=\sum_{n=1}^{\infty}\frac{b_{n2}\sinh[k_n(L_G-y)]+c_{n2}\sinh[k_n y]}{\sin(k_n L_G)}\sin[k_n(x-x_{dm})]\quad \text{for}\, 0\leq x\leq x_{dm}. \tag{3.75}$$

In (3.73), the first term is quite evidently the *long-channel* term.

In terms of this potential, each *SCE* term presented in (3.74) and (3.75) takes on a characteristic proportional to $\exp(-k_n(y'/2))$, where clearly according to the previously provided definition of the length scale,[61] the length scale is given as

$$l=\frac{2}{k_n}. \tag{3.76}$$

According to this notion, the transcendental equation of (3.72) is often expressed as

$$\frac{1}{\epsilon_i}\tan\left(\frac{2}{l}t_i\right)+\frac{1}{\epsilon_{Si}}\tan\left(\frac{2}{l}x_{dm}\right)=0. \tag{3.77}$$

Whereas this transcendental equation provides an infinite number of solutions, due to exponential dependence of the *SCE* on k_n, the smallest solution would be dominant. As an example, for *silicon* as the semiconductor and the *relative dielectric constant* of the insulator taken as 7.8, t_i of 5 nm, and x_{dm} of 10 nm, Equation (3.72) yields k_1, k_2, and k_3 of 0.20, 0.43, and 0.63 nm^{-1}, respectively, which marks the dominance of the first eigenvalue.

A number of important observations are made based on this model:

- The length scale l is greater than $(2/\pi)x_{dm}$ and $(2/\pi)t_i$, meaning that both of these factors are in need of *scaling* (one in terms of reduction of the physical thickness of the insulator, the other through increasing the *bulk dopant* concentration, since x_{dm} is the maximum *depletion-region* width).
- If $t_i=x_{dm}$, independent of the dielectric constants $l=(4/\pi)t_i$.
- If $\epsilon_i=\epsilon_{Si}$, $l=\frac{2}{\pi}(t_i+x_{dm})$.
- If $t_i\ll x_{dm}$, $l\approx\frac{2}{\pi}\left(\frac{\epsilon_{Si}}{\epsilon_i}t_i+x_{dm}\right)$.
- In general $t_i<\frac{\pi}{4}l$, or $\frac{\pi}{12}L_{G\text{-min}}$, even if $\epsilon_i\gg\epsilon_{Si}$.

Whereas larger t_i and x_{dm} in terms of control of the *gate* leakage and reduction of the mobility degrading effect of the increased *bulk dopant* concentration are sought after, increasing the *permittivity* of the *gate* insulator provides only a saturating amount of gain toward improving these factors. This is because while the vertical

[61] For example, in (3.47) and (3.48).

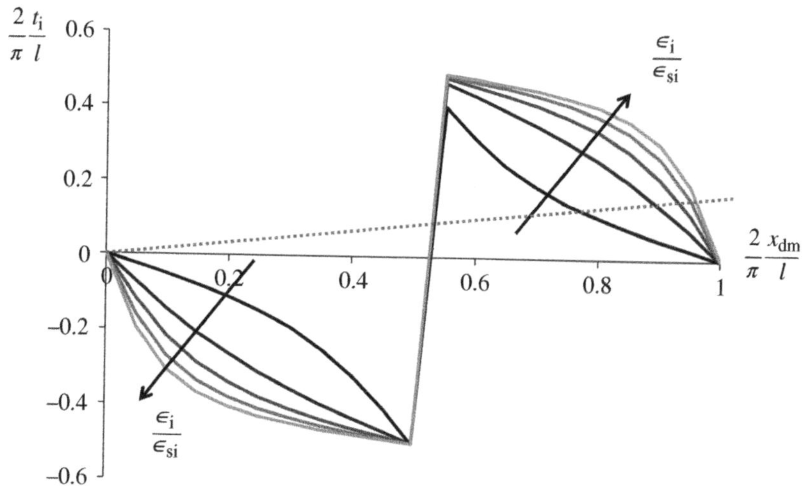

FIGURE 3.20 Graphical tool developed on the basis of the 2-region model referred to in the text. This graphical depiction is capable of illustrating the limit of effectiveness of increasing the gate dielectric constant of a MOSFET in improving the device performance. For a given value of gate length, hence $l = L_G/3$, as indicated by the arrow in progressive lighting of the shades of gray, increasing the ratio of the dielectric constant of the gate insulator to silicon from 0.5 to 4.5 originally heralds a considerable increase in x_{dm} with increasing t_i; however, increasing ϵ_i/ϵ_{Si} to higher values presents curves that are closely packed and as a result little improvement in x_{dm} and t_i. As indicated in the text, an increase in x_{dm} is in favor of mobility improvement of the channel, while an increase of t_i lowers the gate leakage. For illustrative purposes, the superimposed dotted line presents an arbitrary ratio of t_i/x_{dm} taken as a design parameter.

electric field is altered by switching to a *high-K dielectric*, the insulator's lateral fringing electric field is unaffected. As a result, the improvement with the increased dielectric constant saturates. According to this, for the 12 nm *gate length* no matter how high the *permittivity* of the *gate* dielectric, the thickness of the *gate* dielectric needs to be smaller than 3.14 nm. The future of course might have more to say about *scaling* as the use of dielectrics with *anisotropic permittivities* becomes more prevalent and as the quantum effects take over the operation of a *MOSFET*.

Often a graphical solution is presented in terms of evaluating the limits of *scaling* according to the model expressed by (3.72). An example is provided in Figure 3.20. The interested reader is referred to Xie's review paper.

Instead of the *2-regions* solution, a *3-region* solution is extended to the new technological variants of *UTBSOI* and *FinFET*. In case of *double-gate MOSFETs* such as these, the transcendental equation takes the form of

$$\tan\left(\frac{2}{l}t_i\right)\cdot\tan\left(\frac{1}{l}t_{Si}\right) = \frac{\epsilon_i}{\epsilon_{Si}} \tag{3.78}$$

for which graphical solutions are also presented to evaluate the length scale. According to the results of this model, it has been substantiated that *double-gate MOSFETs scale* better than *bulk MOSFETs*. This is because the physical thickness of the *channel* can be made much smaller than x_{dm} even in the absence of *doping*. In such a case, regardless of the ratio of the *permittivities* of the *gate* dielectric and the semiconductor, $l \geq (2/\pi)t_{Si}$ and $l \geq (4/\pi)t_i$. As a result, the minimum *gate length* is given by

$$L_{G\text{-min}} = 3 \max\left(\frac{2}{\pi}t_{Si}, \frac{4}{\pi}t_i\right) \tag{3.79}$$

This is certainly not just because of the presence of two *gates* but because the thickness of the semiconductor t_{Si} can be made much smaller than x_{dm}.

3.5 LIMITS OF SCALING: A RECAP

As suggested in Section 1.4, in spite of the *drift–diffusion transport* formalism assumed in our presented assessment of the operation of *MOSFETs*, starting from the 100 nm *technology node carrier transport* signaled a paradigm change in the operation of these devices. In this transition, it is the reduction in frequency of *scattering* events in *short-channel* devices that has resulted in *ballistic charge-carrier transport* between the *source* and *drain* terminals. This has resulted in incompatibility of the operation of these transistors with the assumption of *drift–diffusion transport*. While not a sharp transformation, with the ongoing *scaling* of *MOSFETs*, such a change is gaining a more prominent status. The chance of *ballistic transport* is also improved when the *channel* is realized in an *intrinsic* semiconductor and when the *lattice temperature* is maintained low.

Under the assumption of *ballistic transport*, the *drain* current is expressed in terms of elementary parameters and not in terms of the *carrier mobility* and the *diffusion constant*. Unlike a *drift–diffusion channel*, the *drain* current in a *ballistic channel* is independent of the *channel length* (i.e., *gate length*) and is only proportional to the *gate width*. Despite this change, as will be corroborated further in Chapter 6, *linear* and *saturation* regimes of operation are still present in the *drain* current–voltage characteristics of *ballistic MOSFETs* (even in extremely small *molecular-scale channels* composed of one *energy level*).

The mechanism of current *saturation* under the *ballistic transport* condition, even without invoking too many details, can be easily understood. In terms of quantum mechanics, the *drain* current consists of two components of opposite signs. The first component is due to *carrier* propagation toward the *drain* with the energy distribution characterized by E_{fS} (where the *source* and *drain* are assumed to be ideal reservoirs with the *Fermi energies* E_{fS} and E_{fD}, respectively). The second component is due to *carrier* propagation toward the *source* characterized by the *drain Fermi* energy (i.e., $E_{fD} = E_{fS} - qV_D$). When *drain* and *source* are under *thermal equilibrium* (i.e., $V_D = 0$), these two components cancel each other. However, with application of positive values of *drain* voltage, the former component gradually dominates. For sufficiently large values of V_D, the *drain Fermi level* drops below the lowest *energy level* of the propagating *state* at the *source*-side potential barrier. As a result, the latter term becomes negligibly small while the former, which is characterized only by the *source* electrode, dominates. Beyond this point, the resulting independence of the *drain* current from V_D gives rise to *saturation* of the *drain* current.

In a *ballistic channel*, the value of the *saturation drain* current is calculated in terms of the product of the occupied *density of states* at the *source* side of the *channel* and the velocity of the injected *carriers* into the *channel* at this point. In this regard, *Natori* has suggested a convenient *figure of merit* for gauging the extent of *ballistic* operation of a *MOSFET channel*.[62]

This *figure of merit* is given by the notion of *injection velocity*, which in terms of operation of a *ballistic MOSFET* (i.e., free of *scattering*) supplying a *saturation drain* current I_{sat} is defined as

$$v_{inj} = \frac{I_{sat}}{WC_G(V_G - V_T)} \tag{3.80}$$

where C_G is the effective *gate* capacitance.

The calculated value of v_{inj} in a fully *ballistic channel* is about, or slightly above, the electron *saturation velocity* (i.e., 10^7 cm/s). The closer v_{inj} is to this number, the stronger the hold of *ballistic transport* on the operation of the *MOSFET*.

Natori's calculations show that the *drain* current under *saturation* is proportional to $(V_G - V_T)^{3/2}$. This is because the mean *carrier* velocity as suggested by (1.57) is proportional to $\sqrt{qn_s = C_G(V_G - V_T)}$. This owes to the fact that *carriers* are *degenerate Fermi* particles flying at *Fermi velocity*. This equation is in contrast to that of the classical *drift–diffusion MOSFETs* presented in Section 3.2 where the *saturation drain* current had a proportionality to $(V_{GS} - V_T)^2$ or $(V_{GS} - V_T)$ in *long-* and *short-channel MOSFETs*, respectively.

REFERENCE

M. Liang, J. Choi, P. Ko, and C. Hu, Inversion-layer capacitance and mobility of very thin gate-oxide MOSFETs, *IEEE Trans Electron Devices*, vol. **33**, no. 3, pp. 409–413, 1986.

FURTHER READING

The following references have been used in preparation of the materials presented in this chapter. They are also suggested as sources for further reading.

[62] See K. Natori, "Ballistic metal-oxide semiconductor field effect transistor," *J Appl Phys*, vol. **78**, no. 8, pp. 4879–4890, 1994.

PHYSICS OF SEMICONDUCTOR DEVICES

C. C. Hu, *Modern Semiconductor Devices for Integrated Circuits*, Pearson Higher Education, Upper Saddle River, NJ, 2010.

K. Lee, M. Shur, T. A. Fjeldly, and T. Ytterdal, *Semiconductor Device Modeling for VLSI*, Simon & Schuster Company, Englewood Cliffs, NJ, 1993.

M. Lundstrum and J. Guo, *Nanoscale Transistors*, Springer, New York, 2006.

R. S. Muller, T. I. Kamins, and M. Chan, *Device Electronics for Integrated Circuits*, Wiley, Hoboken, NJ, 2003.

R. F. Pierret, *Semiconductor Device Fundamentals*, Addison Wesley, Reading, MA, 1996.

B. G. Streetman and S. K. Banerjee, *Solid State Electronic Devices*, Prentice Hall, Englewood Cliffs, NJ, 2006.

S. M. Sze, *Semiconductor Device Physics and Technology*, Wiley, Hoboken, NJ, 2002.

S. M. Sze and K. K. Ng, *Physics of Semiconductor Devices*, Wiley, Hoboken, NJ, 2007.

Y. Taur and T. H. Ning, *Fundamentals of Modern VLSI Devices*, Cambridge University Press, Cambridge, UK, 2009.

Y. Tsividis, *Operation and Modeling of the MOS Transistor*, Oxford University Press, Oxford, UK, 1999.

T. Ytterdal, Y. Cheng, T. A. Fjeldly, *Device Modeling for Analog and RF CMOS Circuit Design*, Wiley, Hoboken, NJ, 2003.

MICROFABRICATION TECHNOLOGY AND MATERIAL CHARACTERIZATION

S. A. Campbell, *Fabrication Engineering at Micro- and Nanoscale*, Oxford University Press, Oxford, UK, 2008.

M. J. Madou, *Fundamentals of Microfabrication and Nanotechnology*, CRC Press, Boca Raton, FL, 2012.

J. D. Plummer, M. D. Deal, and P. B. Griffin, *Silicon VLSI Technology: Fundamentals, Practice and Modeling*, Prentice Hall, Upper Saddle River, NJ, 2000.

D. K. Schroder, *Semiconductor Material and Device Characterization*, Wiley, Hoboken, NJ, 2006.

S. Wolf and R. N. Tauber, *Silicon Processing for the VLSI Era*, Lattice Press, Sunset Beach, CA, 2000.

PROBLEMS

3.1 Prove Equations (3.8) and (3.9).

3.2 Prove Equation (3.12).

3.3 Prove Equation (3.20).

3.4 For a silicon PMOSFET with the following specifications:

$L_G = 0.1\ \mu m$, $W = 0.1\ \mu m$, $t_{ox} = 5$ nm, $X_j = 0.03\ \mu m$, and $N_d = 10^{17}\ cm^{-3}$

calculate the drain current at the onset of saturation when $V_{GS} = -0.3$ V and $V_{SB} = 0$. Assume a p$^+$-polysilicon gate and SiO_2 gate dielectric. Is this MOSFET short or long channel? Is there a condition of operation for which you expect a transition between short- and long-channel modes of operation?

The saturation velocity of holes in silicon is taken as 7×10^6 cm/s.

3.5 Consider a silicon NMOSFET with the following specifications:

$L_G = 0.03\ \mu m$, $W = 0.1\ \mu m$, $t_{ox} = 3$ nm, $X_j = 0.01\ \mu m$, and $N_a = 5 \times 10^{17}\ cm^{-3}$

calculate the drain current at the onset of saturation when $V_{GS} = V_T + 0.2$ V and $V_{SB} = 0$. Also perform this calculation when V_{DS} is 0.2 V beyond V_{Dsat}. Assume an n$^+$-polysilicon gate and treat the SiO_2 gate dielectric ideally.

3.6 For the NMOSTFET described in Problem 3.5, replace the SiO_2 dielectric with a film of HfO_2 of the same thickness and redo the calculations.

3.7 For the NMOSFET described in Problem 3.5, if the n$^+$-polysilicon gate were to be replaced by a p$^+$-polysilicon gate, what threshold voltage adjustment recipe would you suggest for maintaining the value of the threshold voltage? Describe the impact that this change can have on the subthreshold swing and the punch-through current.

3.8 For the NMOSTFET described in Problem 3.5, if X_j were to be increased by a factor of two, carefully describe the impact on the following parameters:

a. Threshold voltage

b. Subthreshold current given a fixed drain bias and $V_{GS} = 0$

c. Punch-through current given a fixed drain bias and $V_{GS} = 0$

3.9 Prove that according to the model presented in Section 3.2.2 for a short-channel MOSFET the drain current at the onset of saturation scales linearly with $V_{GS}-V_T$ and discuss the bases according to which this should change to $(V_{GS}-V_T)^{1.5}$.

3.10 Consider a CMOS pair fabricated in the double-well CMOS technology with the use of n$^+$- and p$^+$-polysilicon gates for the NMOSFET and PMOSFET, respectively. Having known that

for NMOSFET, $L_G = 0.1\ \mu m$, $W = 0.1\ \mu m$, $t_{ox} = 5$ nm, $X_j = 0.03\ \mu m$, and $N_a = 10^{17}\ cm^{-3}$

and

for PMOSFET, $L_G = 0.1\ \mu m$, $t_{ox} = 5$ nm, and $X_j = 0.03\ \mu m$

calculate the values of N_d and W for the PMOSFET to guarantee matching between the transistors.

If we were to blindly replace the polysilicon gates with a metal of work function matched to intrinsic silicon, calculate the threshold voltage of each device and produce the thermal equilibrium band diagram of each device under the gate.

3.11 Retrograde doping is one of the techniques used to avoid latch-up in CMOS technology. Perform a literature search on this technique and provide full description of this process. Explain how it can help with preventing latch-up.

3.12 On the basis of the discussions presented in Section 3.4, write a MATLAB code to reproduce Figure 3.20.

4

FROM SCALING-DRIVEN TECHNOLOGICAL VARIATIONS TO NOVEL DIMENSIONS IN MISFETs

Although the absolute majority of *field effect transistors* today are *inversion-channel MOSFETs* operating according to the principles outlined in Chapter 3, implementation of the *gate*-controlled *channel* resistance in a variety of other *FETs* is based on other distinct sets of principles. As mentioned in Section 2.4, *junction FETs* (i.e., *JFETs*) and *metal–semiconductor FETs* (i.e., *MESFETs*) are the more traditional examples of these alternative device structures. While according to the discussions of Chapter 2, both *p–n* and *MS* junctions offer the possibility of creating *depletion regions* within the semiconductor and *pinching off* the *drain–source channel* (i.e., when *biased* in *reverse*), the use of these junctions in place of the *MOS gate* mechanism in *JFETs* and *MESFETs*, respectively, allows for operating a *FET* between the *depletion* and the *accumulation* regimes of operation (but not the *inversion*). In light of the discussions of Chapter 2, the operation of these transistors can be well understood with a little tweaking of the principles expressed in Chapter 3. *Inversion*-independent operation of these *FETs* reduces the burden of device *microfabrication* by eliminating the need for creating a *gate* precisely lined up with the boundaries of the *highly doped minority-carrier* reservoirs that are needed for forming the *inversion channel* in *MOSFETs*. Such structures can in turn offer lower values of parasitic *gate capacitance* and maximum longitudinal electric field. In addition, in the case of *MESFETs*, the often smaller *gate capacitance* of an *MS junction* in comparison to the *gate capacitance* of a *MOSFET* offers these devices a higher frequency of operation. As a result, as traditional *FETs*, *MESFETs*, and *JFETs* have found their market niches in microwave- and power-electronic circuits.

Field Effect Transistors, A Comprehensive Overview: From Basic Concepts to Novel Technologies, First Edition. Pouya Valizadeh.

In this chapter, some of the other more recently pursued alternatives to *MOSFET* structures are presented. Some of these alternatives are based on older but obscure proposals of the past that in light of the recent technological breakthroughs are gaining a more prominent status. From this list of alternatives, due to their more prominent stature, *heterojunction FETs* (i.e., *HFETs*) have been excluded. These devices will be presented in Chapter 5. While each section of this chapter presents a different alternative, the common thread among these sections is the comparison to *MOSFETs* and what the alternative can offer beyond traditional *MOSFETs*.

4.1 *FinFET, UTBSOI*, AND OTHER MULTIPLE-GATE *FETs*

As indicated in Section 3.2.2.1.2, *scaling* of *MOSFET* technology has resulted in a dire need for major structural changes to the *planar MOSFET* if *scaling* is to continue. At the forefront of these changes is the realization of the so-called *multiple-gate MOSFETs* (especially *FinFETs*) and *ultrathin body silicon-on-insulator MOSFETs* (i.e., *UTBSOI*).[1] Figure 4.1 provides schematic illustrations of a number of these alternatives.

Among the major advantages of *FinFETs* and *UTBSOI MOSFETs* is the absence of *channel doping*. *Random dopant fluctuation* has become one of the major challenges of *deeply scaled MOSFET* technologies. This is because the random fluctuation in the *dopant* concentration of the *channel* results in random fluctuation in transistor characteristics such as the *threshold voltage*. The more prominent nature of this fluctuation among these devices is a manifestation of the reduction of the number of *dopant* atoms per transistor (i.e., within the small volume of a *deeply scaled MOSFET*). Realization of *FinFETs* and *UTBSOI MOSFETs* on *undoped silicon channels* eliminates this problem. In implementing *CMOS pairs* in these technologies, the required symmetry of the *threshold voltages* of the n- and the p-channel *MOSFETs* is satisfied by using a *gate* metal with a *midgap work function*.[2] As pointed out in Section 3.2.2.1.2, problems such as *gate depletion* have turned the adoption of metal instead of *polysilicon gates* to be a necessity for *deeply scaled bulk planar MOSFETs*. The same requirement is certainly present in the case of *multiple-gate MOSFETs*. In the case of these devices, among the explored options for the *gate* electrode are *metal nitrides* such as *TiN*, *TiSiN*, *TaN*, and *TaCN*, which are stable at the high temperatures of *CMOS* processing (i.e., in excess of 1000°C). These *metal nitrides* have *work functions* in the range of 4.4 to 4.7 eV after high-temperature *anneal*. In these cases, the value of the *work function* depends on a number of factors such as layer thickness, *alloy* composition, *crystal orientation*, and the *gate*-dielectric material. These *metal nitrides*,

[1] *IBM* calls *UTBSOI*, *ETSOI* (or extremely thin *SOI*). Since 2001, *FinFETs* and *UTBSOIs* have been highlighted by the *ITRS*.

[2] As can be expected, in the absence of *channel doping*, this situation is unlike what was indicated in Section 3.2.2.1.2 (i.e., with respect to the requirement for two metals of different *work functions* for *PMOS* and *NMOS transistors*, when *polysilicon gate* was replaced in *scaled planar CMOS* technologies with low-resistivity *silicides*).

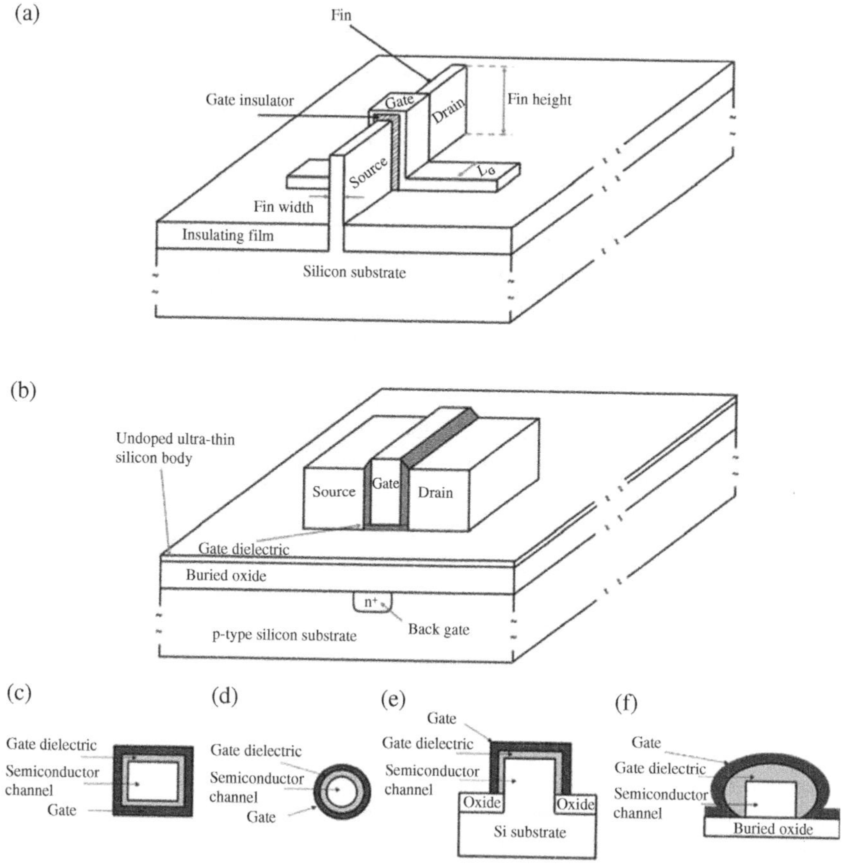

FIGURE 4.1 Schematic 3-D depictions of (a) a *FinFET* and (b) a *UTBSOI MOSFET*. Cross-sectional views of gate coverage of (c) a rectangular gate-all-around *MOSFET*, (d) a cylindrical gate-all-around *MOSFET*, (e) a Π-gate, and (f) an Ω-gate *MOSFET* structure.

while not suitable for application to *planar CMOS*, have the right *work function* for low-power, *multiple-gate CMOS chips*.

In addition to these choices, pertaining to the relevance of the value of the *gate work function* to the proper operation of these *chips*, a novel avenue explored has been the realization of this electrode using *tunable work-function* materials. This is done either through employing *fully silicided* (also known as *FUSI*) metal *gates*, metal layer *interdiffusion* of multiple metal layers of different thicknesses, or implementation of *nitrogen-implanted molybdenum gates*. Among these solutions, *silicides* such as *NiSi* and $CoSi_2$, owing to their *midgap work functions*, are the most explored *silicides* for realizing *gate* electrodes in *silicon multiple-gate MOSFETs*. For adjusting the *work function* of a *silicide*, a high-concentration *dopant* profile is *implanted* into the *poly-silicon* layer prior to the *gate-silicide* formation. The effect of this *dopant* profile on the *effective work function* of the *FUSI gate* depends on the type of the *gate* dielectric.

For example, on an *oxynitride gate* dielectric, *boron* and *aluminum doping* of *FUSI* have been observed to increase the *work function* up to 4.9 eV. *Phosphorus, arsenic,* and *antimony doping* have been shown to reduce the *work function* down to 4.2 eV. *Dopant segregation*[3] throughout the *silicide* formation process has been known as the cause of this behavior. However, in the presence of *high-K dielectrics*, such a *segregation* effect is less observable. It has been also reported that *boron doping* due to *Fermi-level pinning* has less impact on the *work function*. However, this *Fermi-level pinning* has been reported to reduce in *nickel-rich silicides*. For Ni_2Si a *work function* of 4.85 eV has been reported. Also, extremely high concentrations of *ytterbium* in *NiSi* have been reported to reduce the *work function* to 4.22 eV.

As mentioned earlier, the other metal explored as the *gate* material is *molybdenum*. This metal, in addition to its low resistivity, enjoys a high melting temperature of 2623°C. *Molybdenum* has a *work function* of 4.9 eV, which can be reduced to 4.5 eV through *nitrogen implant*. The amount of shift in the *work function* is variable with the *implanted nitrogen dose* and the conditions of thermal *anneal* succeeding the *implant*.

In addition to the elimination of the concerns related to *random dopant fluctuation*, a number of other concerns are also diminished by the effective elimination of the *bulk* in *multiple-gate* and *UTBSOI MOSFETs*. This list includes *punch-through breakdown, body effect, bulk leakage*, and also degradation in the *subthreshold swing* due to the presence of a series *depletion capacitance* with the *gate insulator capacitance*. However, by the same token, in these device alternatives, due to the more pronounced *carrier* confinement, *quantum capacitance* plays a more important role.

In spite of the lure of the device alternatives depicted in Figure 4.1, a careful look into their topography, dimensions, and the need for new materials reveals that implementation of these devices imposes significant burdens on technologists. These burdens are in addition to the challenges faced by *lithography* in realizing the shrinking *gate length*. So far, this extra cost has been deemed affordable for prolonging the *scaling* trend of *CMOS* technology. However, even with the advent of these technologies, aside from problems associated with technological realization of small features, other problems sustain. As will be expressed in Chapter 6, physics-based limitations imposed by quantum effects (and eventually *uncertainty principles*) and failure of proper explanation of *carrier* population through statistical means are the hard limits that the *scaling* strategies are increasingly faced with.

While, as shown in Figure 4.1a, *fin channels* of a *FinFET* are realized through the dense implementation of *shallow-trench isolation* technology, the *UTBSOI MOSFET* shown in Figure 4.1b needs to produce a very uniform *silicon* film, with a thickness of a few *nanometers* across the *wafer*. As a result, the major additional burden of realizing *FinFETs* (i.e., additional to that of the *planar MOSFET* technologies) rests with the process engineers. However, *UTBSOI*'s main burden resides with the *wafer-production* engineers. Despite the original pessimism, on the *UTBSOI* front

[3] *Segregation* is defined in terms of the natural tendency of impurities to move across the interface of two materials of different "maximum acceptable concentration of impurities at *thermal equilibrium*" (also known as *solid solubility*).

surprisingly high uniformities have been reported by *wafer* producers over the past few years.[4] This *wafer*-production process can take advantage of long periods of *oxidation* of the approximately 100 nm thick *silicon* film of typical *SOI wafers* with subsequent *chemical etching* followed by *chemical/mechanical polishing* (i.e., *CMP*). Neglecting the challenges of producing the required film uniformity, the burden of realizing the 3-D structure of *FinFETs* was originally expected to shift the attention of the industry in favor of *UTBSOI*. However, it seems that the major foundries have seen the improved packing density of the 3-D implementation, which is offered through the reduced footprint of *FinFETs*, as a sufficiently attractive incentive for keeping their focus on resolving the challenges of this technology and on commercializing it instead of the *UTBSOI*. The reduced footprint of *FinFETs*, as suggested in Figure 4.1a, comes from the fact that on these devices the effective *gate width* is approximately given by the sum of *two times* the *height* and the *width* of the *fin*. Since the *height* is out of plane, its role in determining the *gate width* contributes to an improved packing density. In addition to this, the reported *fin pitch* of 60 nm, or even smaller, plays a major part in boosting the packing density.

As shown in Figure 4.1, an advantage of *UTBSOI*, which is at least not normally offered by *FinFETs*, is the availability of a *second gate* (i.e., *back gate*). This, however, can be implemented with a slight modification of the *FinFET* structure through *CMP'ing* and removing the top gate of a *fin* structure to produce two independent *side gates* (to be *biased* separately).

While examining the advantages of these novel technologies, it should also be indicated that since *FinFETs* and *UTBSOI MOSFETs* have a reduced requirement of *scaling* of the *gate-insulator thickness*,[5] the strength of E_{eff} in these devices can be reduced compared to that of *scaled planar MOSFETs*.[6] Consequently, due to the lowering of *interface scattering*, an improved *carrier mobility* can be expected.

First *FinFET* reported by *Hu*'s team at the University of California-Berkeley was realized on 20 nm wide *fins* with *gate lengths* as small as 10 nm. No *threshold-voltage* roll-off was observed on *fins* smaller than the *gate length* (i.e., in this case, *gate length* as small as 20 nm did not show *threshold-voltage* roll-off).

As mentioned earlier in this section, due to the absence of a *bulk*, *FinFET* and *UTBSOI MOSFET* have a different mechanism of operation than that of *planar MOSFETs*. In these cases, since *bulk* is virtually absent, no *depletion region* will be formed (i.e., as mentioned in Chapter 3, no *depletion capacitance*). As a result, in the presence of *carrier* reservoirs, without the need for *depletion*, the *intrinsic channel* (i.e., *fin* in *FinFET* or the thin *body* in *UTBSOI*) will be populated by electrons or *holes* corresponding to the polarity of the *gate* voltage. Based on this observation, speaking of *strong inversion* as done in *planar MOSFETs* is not possible in these technologies. Unless the *channel* is quite highly populated, the *gate*-to-*channel* capacitance of these devices remains small, which, from the point of view of lowering the *switching time* of the transistor, is quite interesting. Corresponding to the previously provided descriptions

[4] Based on pioneering report of *SoiTEC* of France.
[5] See the discussion in Section 3.4.
[6] For definition of E_{eff} in a *planar MOSFET*, see Equation (3.23).

of the incentives to operate under the so-called *subthreshold* regime of operation, this is the mode of operation that these devices are often operated at.

Since the definition of the *threshold voltage* in terms of the *onset of strong inversion* is not an option, *threshold voltage* in these novel *FET* varieties is expressed by the demarcation observed on the *transfer characteristic* between the linear and the exponential dependence of *drain* current on the *gate* voltage. Regarding the evaluation of the *surface* potential in the aforementioned 3-*region*-based solutions to the 2-D *Poisson equation*, which was spoken of in Section 3.4, models have been proposed to evaluate this quantity in *FinFET* and *UTBSOI* and to assess the device *scalability* and *short-channel effects*. The foundations of these models and their proposition of length scale in the evaluation of *short-channel effects* closely follow the foundations of the 2-*region* solution of the 2-D *Poisson equation* provided for the *subthreshold* operation of *short-channel planar MOSFETs* in Section 3.4. According to these models, the *body* thickness in a *FinFET* is evaluated to be about L_G. However, in the case of *UTB-SOI*, it should be a *third* of this value for acceptable V_T roll-off. Based on this, *FinFET* is expected to be more *scalable*.

Due to the aforementioned structural/operational differences between traditional *planar MOSFETs* and either the 2-D or the 3-D thin-*body* transistors (i.e., *UTBSOI MOSFET* and *FinFET*, respectively), the *subthreshold swing* and the *threshold voltage* of these novel transistors are much less sensitive to variations in L_G and V_D. In addition to alleviating the problem of *random dopant fluctuation*, these devices do not suffer from *ionized-impurity scattering*, which is quite important in *low-field* operation under the *subthreshold* regime. In both of these designs (i.e., *FinFET* and *UTB-SOI*), by eliminating the *bulk dopant* concentration as a *scaling* parameter, *body* thickness has been added as a new *scaling* dimension.

All of the aforementioned properties give these transistors higher *on* current, lower leakage, lower required *drain* voltage, and as a result lower power consumption and better *scalability*. While at the moment these devices are performing to the satisfaction of the standards, further *scaling* of the *gate length* and the increased chance of *drain–source* quantum *tunneling* (culminating in increasing the *off-state drain* current) are expected to put a halt on further *scaling* over the next few years. This is unless a new physical phenomenon is brought into the equation.

A recent contribution in this area has been made through adoption of *ferroelectric* layers in the *gate* stack of a *MOSFET*. Whereas based on equations (3.16) and (3.17) in Section 3.2.1.4 we indicated the minimum value of the *subthreshold swing* at room temperature to be about 60 mV/dec, implementation of a *negative oxide capacitance* can lift this lower limit from the value of the *SS*. Using a *bilayer* of a *ferroelectric* and a *dielectric* (e.g., $Pb(Zr_{0.2}Ti_{0.8})O_3$ and $SrTiO_3$, respectively), a composite *series capacitance* larger than the two constituent capacitances has been recorded by Khan et al., which indicates the realization of a *negative capacitance* across the *ferroelectric* layer.[7] Such a factor can reduce *n* in equation (3.16) below 1, hence allowing *SS*

[7] See A. Khan, D. Bhowmik, P. Yu, S. Kim, X. Pan, R. Ramesh, and S. Salahuddin, "Experimental evidence of ferroelectric negative capacitance in nanoscale heterostructures," *Appl Phys Lett*, vol. **99**, pp. 113501–113503, 2011.

to become smaller than the 60 mV/dec value. It has been observed that considering the low *Curie temperature*[8] of the *ferroelectric*, temperature can serve as an effective tuning parameter on the value of this *negative capacitance*. However, it should be mentioned that the use of *ferroelectric* materials brings worry of *ferroelectric hysteresis* and its possible contributions to a *memory effect* on the value of the *threshold voltage*. Nonetheless, recent publications in this area suggest that under low *drain biases* this issue can remain under control.[9]

While theoretically such a *bilayer* implementation of the *gate* stack can allow for overcoming the *Boltzmann limit*, expressed in (3.16) and (3.17), from a technological point of view, it still faces many challenges. The gain in overcoming these challenges could, however, be much more than the *Boltzmann-limited* gain of the adoption of the *high-K* dielectrics (see Section 3.2.2.1.2). This is because going beyond the *Boltzmann limit* can tremendously reduce the *off-state* current and the *gate bias* (hence the power dissipation burden of the modern *CMOS ULSI chips*). Crystalline $SrTiO_3$ has been shown to be able to serve as a *perovskite*[10] *template* for the growth of the crystalline *ferroelectric* layer on *silicon*.

Applying the *ferroelectric effect* to a *high-K* dielectric such as *hafnium oxide* in the form of *ZrHfO*, while having fabrication compatibility with the modern *CMOS* technology, has been shown to offer very steep *transfer characteristics* with stable room-temperature values of *SS* in the range of 5–60 mV/dec.[11]

4.1.1 Quantitative Assessment of the Advantages of *SOI* and Multiple-Gate *MOSFETs*

In addition to *FinFET* and *UTBSOI*, presently other alternative device structures are also being explored to improve the electrostatic dominance of the *gate* over the *channel* (which is more commonly known as *electrostatic integrity* of the *gate*) and to extend the life of the *scaling* strategies. Earlier in this section, a number of these alternatives were schematically depicted in Figure 4.1. Deliberation on the advantages and the disadvantages of these alternatives needs a quantitative basis. To satisfy this requirement, in this section, a group of tools that are currently being used in performing quantitative assessment among these more recent technologies and a few older ones are presented. As will be observed among these discussions, in the quest for improving the *electrostatic integrity* of the *gate* via realization of *multiple-gate FETs*, these transistors should ultimately take the form of a wirelike *channel*, which is connected to *source* and *drain* terminals at its two ends, while the *gate* is wrapped around its middle section.

[8] Named after *Pierre Curie*, is the temperature at which material's *permanent magnetism* transitions to *induced magnetism*.

[9] C. Cheng and A. Chin, "Low-voltage steep turn-on pMOSFET using ferroelectric high-*K* gate dielectric," *IEEE Electron Device Lett*, vol. **35**, no. 2, pp. 274–276, 2014.

[10] The name *perovskite* refers to those materials sharing the crystal structure of $CaTiO_3$ (with the *oxygen* taking the *face-center* positions). *Perovskite*is named after *Lev Perovski*.

[11] C. Cheng and A. Chin, "Low-voltage steep turn-on pMOSFET using ferroelectric high-*K* gate dielectric," *IEEE Electron Device Lett*, vol. **35**, no. 2, pp. 274–276, 2014.

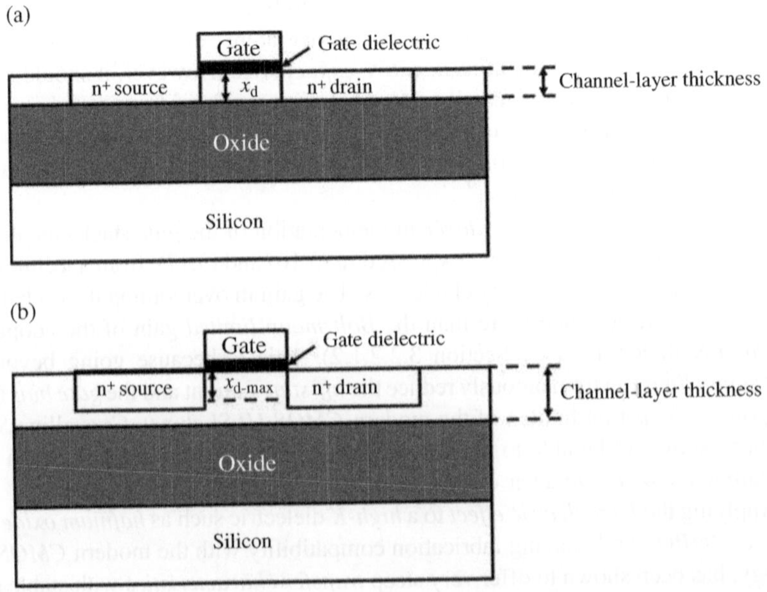

FIGURE 4.2 Simplified cross-sectional views of (a) a fully depleted *SOI* n-channel *MOSFET* and (b) a partially depleted *SOI* n-channel *MOSFET*.

As ancestors to some of these new technologies and short of the technological means to realize the ultrathin *intrinsic body* of the *UTBSOI*, *partially depleted* and *fully depleted SOI* technologies were developed in the 1990s.[12] Figure 4.2 provides schematic illustrations of transistors in these technologies. In these alternatives, the *planar MOSFET* structure, instead of realization on a semiconducting substrate, is realized on an *SOI wafer*. The thickness of the *silicon* layer of the *SOI* in the *fully depleted* alternative is smaller among the two and is within the range of the junction depth of the *source* and *drain carrier* reservoirs. These alternatives, in addition to alleviating a number of *SCE*-related reliability concerns such as *latch-up* (although more so in the *fully depleted* version),[13] provided the opportunity to improve the *electrostatic integrity* of the *gate* and to reduce the burdens imposed by *SCE* on further *scaling* of *MOSFETs*. However, these advantages fall short of those already enumerated for *FinFET* and *UTBSOI*.

These technologies have certainly captured the attention of the industry as not only viable but even more so inevitable replacements for *bulk planar MOSFETs*. In this section, while maintaining a special interest on *FinFETs* and other *multiple-gate FETs*,[14] the quantitative assessment will also provide coverage of the *SOI MOSFETs*.

[12] Also known as *PDSOI* and *FDSOI*, respectively.

[13] See the discussion in Section 3.2.2.2.2.

[14] Depending on the technology, a few names have so far been used, which refer to the members of this broad family: *multiple-gate transistor*, *wrapped-gate transistor*, *double-gate transistors*, *tri(ple)-gate transistor*, and *gate-all-around transistor*.

In the present quantitative assessment, in order to evaluate the *scaling* implications, tools generally known as the *voltage-doping transformation model* (or *VDT*), developed by Skotnicki et al., are presented. In spite of their simplicity, these tools provide valuable insights into the operation and *scalability* of modern *MOSFETs*.[15] These evaluation tools are used in assessing the effect of *scaling* of device parameters (i.e., physical dimensions, *doping*-related quantities, and *bias* conditions) on *DIBL* and *SCE's* contributions to *threshold-voltage* roll-off (i.e., $\Delta V_{\text{T-DIBL}}$ and $\Delta V_{\text{T-SCE}}$, respectively). The following equations defining $\Delta V_{\text{T-SCE}}$ and $\Delta V_{\text{T-DIBL}}$ in planar *MOSFETs* are derived from *VDT*[16]:

$$\Delta V_{\text{T-SCE}} = 0.64 \frac{\epsilon_{\text{Si}}}{\epsilon_{\text{ox}}} \left[1 + \frac{X_{\text{j}}^2}{L_{\text{eff-G}}^2}\right] \frac{t_{\text{ox}}}{L_{\text{eff-G}}} \frac{x_{\text{d-max}}}{L_{\text{eff-G}}} \phi_{\text{bi}} \equiv 0.64 \frac{\epsilon_{\text{Si}}}{\epsilon_{\text{ox}}} EI \cdot \phi_{\text{bi}} \tag{4.1}$$

and

$$\Delta V_{\text{T-DIBL}} = 0.80 \frac{\epsilon_{\text{Si}}}{\epsilon_{\text{ox}}} \left[1 + \frac{X_{\text{j}}^2}{L_{\text{eff-G}}^2}\right] \frac{t_{\text{ox}}}{L_{\text{eff-G}}} \frac{x_{\text{d-max}}}{L_{\text{eff-G}}} V_{\text{D}} \equiv 0.80 \frac{\epsilon_{\text{Si}}}{\epsilon_{\text{ox}}} EI \cdot V_{\text{D}}. \tag{4.2}$$

In (4.1) and (4.2), $L_{\text{eff-G}}$ is the *effective gate length* (in contrast to the physical length of the *gate*, which also includes the extension of the *depletion regions* at the *source* and *drain* junctions), ϕ_{bi} is the built-in potential of *drain* and *source* contacts, t_{ox} is the thickness of the *gate* oxide, X_{j} is the junction depth of the *Ohmic* contacts, and $x_{\text{d-max}}$ is the maximum penetration depth of the *gate*-induced electric field in the *channel* region (defined as the depth of the maximum *depletion region* under the *gate* in a *bulk MOSFET*). The parameter *EI*, defined in these equations,[17] is referred to as the *electrostatic integrity factor*. This parameter is representative of the way the electric flux from the *drain*, in competition with the electric flux emanating from the *gate*, influences the *channel* potential. The smaller the *EI*, the greater is the electrostatic dominance of the *gate* over the *channel*. In terms of (4.1) and (4.2), the *threshold voltage* of a *bulk MOSFET* with effective *channel length* $L_{\text{eff-G}}$ can be calculated as

$$V_{\text{T}} = V_{\text{T}\infty} - \Delta V_{\text{T-SCE}} - \Delta V_{\text{T-DIBL}}. \tag{4.3}$$

In (4.3), $V_{\text{T}\infty}$ represents the *threshold voltage* of a *long-channel NMOSFET* as calculated in (2.172). While it has been already known that *DIBL* and *SCE* result in *threshold-voltage* roll-off, Equations (4.1)–(4.3) quantify the amount of this roll-off.

[15] For a more thorough description, see T. Skotnicki, G. Merckel, and T. Pedron, "The voltage-doping transformation: a new approach to the modeling of MOSFET short-channel effects," *IEEE Electron Device Lett*, vol. **9**, no. 3, pp. 109–112, 1988.

[16] See T. Skotnicki, "Heading for decananometer CMOS—is navigation among icebergs still a viable strategy?" *Proceedings of the 30th European Solid-State Device Research Conference*, September 11–13, 2000, Cork, Ireland. IEEE, Piscataway, NJ, pp. 19–33.

[17] That is, $EI = \left[1 + \frac{X_{\text{j}}^2}{L_{\text{eff-G}}^2}\right] \frac{t_{\text{ox}}}{L_{\text{eff-G}}} \frac{x_{\text{d-max}}}{L_{\text{eff-G}}}$.

Instead of the implicit description of the interplay between different parameters involved in determining the extent of the undesired *short-channel effects*, *VDT* presents an explicit means of evaluation. For example, in terms of (4.1)–(4.3), *VDT* quantifies how much the *short-channel effects* can be reduced by reducing the *junction depth* and *gate*-oxide thickness and by increasing the *dopant* concentration (essentially reducing $x_{d\text{-max}}$). Without these tools, designers have been forced to adopt implicit design rules for minimization of *short-channel effects*.[18]

As discussed in Chapter 3, *short-channel effects* are caused by reduction of the electrostatic control of the *gate* over the *channel*. We took note in that part of the discussion that this is due to a competition between the *gate*-induced electric field and the electric flux emanating from the *drain* and source *Ohmic* contacts. Since in a *bulk MOSFET* the electric field of the *source* and *drain* contacts propagates throughout the *depletion regions* associated with these contacts, the higher the *doping* level of the *channel*, the lower the chance of propagation of these field lines throughout the *channel length* (and essentially less *short-channel effects*). However, this remedy reaches its limits when demanding *channel doping* levels exceeding $10^{19}\ \text{cm}^{-3}$ (needed for *very short-channel* devices). This limit is clearly set by the negative impacts on the device operation (e.g., in terms of reduced *carrier mobility* and increased *threshold voltage*).

As expressed in Chapter 3 and also earlier in this chapter, the limitations imposed by *SCE*, however, can be overcome by introducing new degrees of freedom to the equation. One such opening is provided through switching from a semiconducting substrate to an *SOI* substrate. As observed in Figure 4.2, depending on the thickness of the semiconducting *channel* layer, we can either have a *FDSOI* or a *PDSOI MOSFET*. In a *FDSOI MOSFET*, the larger part of the electric flux emanating from *drain* propagates through the *buried oxide* before reaching the *channel*. In the case of a thin *buried-oxide* layer sandwiched between the *channel* layer and a heavily *doped* underlying layer serving as the ground plane, these field lines, instead of terminating in the *channel* and competing with the *gate*-induced electric field, will terminate on this ground plane. In this fashion, although *FDSOI* technology manages to very effectively control the *SCE*, it adds to the *junction-capacitance* and *body-effect* problems. As a result, like any other practical solution, a compromise needs to be made between these competing factors.

In comparison to *FDSOI*, *double-gate MOSFETs* offer a more efficient solution to the *SCE*-related roll-off of the *threshold voltage* due to their small thickness of the *channel* layer. In the first generation of these devices, which was proposed by Sekigawa et al.,[19] the two *gates* of the proposed *FET* were connected together. In this structure, as depicted in Figure 4.3, because of the small thickness of the *channel* layer, most of the electric flux propagating from the *source* and *drain* contacts terminates on the

[18] Such as using $(X_j/L_{\text{eff-G}})^2 = 1/3$, $t_{ox}/L_{\text{eff-G}} = 1/30$, and $x_{d\text{-max}}/L_{\text{eff-G}} = 1/3$ to obtain ΔV_{DIBL} of 29 mV at $V_D = 1$ V.

[19] See T. Sekigawa and Y. Hayashi, "Calculated threshold-voltage characteristics of an XMOS transistor having an additional bottom gate," *Solid State Electron*, vol. **27**, no. 8, pp. 827–828, 1984.

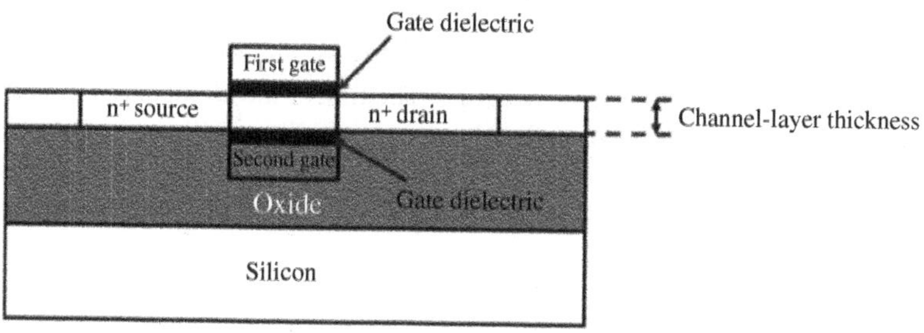

FIGURE 4.3 Simplified cross-sectional view of a dual-gate *MOSFET.*

bottom *gate*. As a result of this termination, only a limited amount of the *drain*-induced electric flux can reach the *channel* without imposing a *body-effect* problem.

In order to offer quantitative means of assessment, the aforementioned *VDT* tools have been proven to be readily applicable to the aforementioned *FDSOI* and *double-gate MOSFET* technologies.[20] Following the definition of *EI* for a *bulk MOSFET* in (4.1) and (4.2), the *EI* factor of the *FDSOI MOSFETs* is defined as

$$\mathrm{EI} = \left[1 + \frac{t_{\mathrm{Si}}^2}{L_{\mathrm{eff\text{-}G}}^2}\right] \frac{t_{\mathrm{ox}}}{L_{\mathrm{eff\text{-}G}}} \frac{t_{\mathrm{Si}} + \alpha t_{\mathrm{b\text{-}ox}}}{L_{\mathrm{eff\text{-}G}}}. \tag{4.4}$$

As observed in (4.4) and anticipated from our earlier qualitative discussions on the *FDSOI* devices, in this definition X_j is replaced by the thickness of the *channel* layer t_{si}. Also, in (4.4), knowing that the *gate* electric field can propagate throughout the entire thickness of the *channel* layer and a portion of the *buried oxide*, the role of $x_{d\text{-max}}$ in (4.1) and (4.2) is swapped for $t_{si} + \alpha t_{b\text{-ox}}$. $t_{b\text{-ox}}$ is the thickness of the *buried-oxide* layer, and α indicates the extent of penetration of the electric field in the *buried oxide*. According to (4.4), as suggested earlier, reducing $t_{b\text{-ox}}$ is anticipated to improve the *SCE*.

In developing the form of the *EI* factor applicable to the *double-gate MOSFET* depicted in Figure 4.3, we take note that the effective value substituting $x_{d\text{-max}}$ in (4.1) and (4.2), and also X_j, for each *gate* is equal to $t_{Si}/2$. As a result,

$$\mathrm{EI} = \frac{1}{2}\left[1 + \frac{t_{\mathrm{Si}}^2/4}{L_{\mathrm{eff\text{-}G}}^2}\right] \frac{t_{\mathrm{ox}}}{L_{\mathrm{eff\text{-}G}}} \frac{t_{\mathrm{Si}}/2}{L_{\mathrm{eff\text{-}G}}}. \tag{4.5}$$

The other noticeable difference between (4.4) and (4.5) is the presence of the factor 1/2 behind the brackets in (4.5), which is due to the presence of two, and not just one, *gate* electrodes.

Expressed in terms of the *EI* factors presented in (4.4) and (4.5), the *double-gate MOSFET* has the advantage of appearing twice as thin as the equivalent *FDSOI* transistor. As a result, the *double-gate MOSFET* offers an extended lifeline to the

[20] See T. Skotnicki, "Ultimate scaling of SOI MOSFETs," *MIGAS short course,* Villard de Lans, France, 2004.

gate-length scaling of *CMOS* technology with less technological constraint. Following the same logic, the so-called *surrounding-gate* (or *quadruple-gate*) *SOI MOSFET* is rightly expected to offer an even better *EI*. The first of the members of the *surrounding-gate MOSFET family* were realized on vertical *silicon pillars*. In terms of these device architectures, for *gate length* as small as 5 nm (which surrounds the perimeter of the *pillar* like a ribbon) and *pillar* diameter of 3 nm, full functionality has been reported as early as 2006.[21]

4.1.2 Multiple-Gate *MOSFETs*: A Complementary Perspective on the Implementation and Physics of Operation

As indicated in Section 3.2.2.2, the potential profile across the *channel* of a *MOSFET* can be obtained by solving *Poisson's equation*. While in *bulk planar MOSFETs* solving the two-dimensional *Poisson's equation* was sufficient, in the case of *multiple-gate MOSFETs* 3-D solutions become necessary. Neglecting the quantum phenomena, in a *multiple-gate NMOSFET* the following *Poisson's equation* should be solved using *Schottky's depletion approximation*[22]

$$\frac{\partial^2 V(x,y,z)}{\partial x^2} + \frac{\partial^2 V(x,y,z)}{\partial y^2} + \frac{\partial^2 V(x,y,z)}{\partial z^2} = \frac{q[N_a(x,y,z) + n(x,y,z)]}{\epsilon_s}. \tag{4.6}$$

Equivalently, in terms of the electric-field components along the three axes of the *Cartesian* coordinate system (i.e., E_x, E_y, and E_z), (4.6) can be presented as

$$\frac{\partial E_x(x,y,z)}{\partial x} + \frac{\partial E_y(x,y,z)}{\partial y} + \frac{\partial E_z(x,y,z)}{\partial z} = -\frac{q[N_a(x,y,z) + n(x,y,z)]}{\epsilon_s}. \tag{4.7}$$

This latter form clearly indicates that the sum of the spatial variations of the three components of electric field at a given point in the *channel* is constant. An increase in one is approximately an indication of a reduction in either one or both of the other two electric-field components. It is in terms of this behavior that earlier in our discussion a *zero-sum* competition between the *drain* and *gate* electrodes in imposing electrostatic dominance over the *channel* was implicitly indicated.

[21] See H. Lee, L. Yu, S. Ryu, J. Han, K. Jeon, D. Jang, K. Kim, J. Lee, J. Kim, J. Cheol, G. Lee, O. Sub, P. Chang, W. Bae, L. Mok, J. Yang, J. Yoo, S. Kim, and C. Yang-Kyu, "Sub 5 nm all-around gate FinFET for ultimate scaling," *Proceedings of Symposium on VLSI Technology,* Honolulu, HI, June 13–15, 2006. Digest of Technical Papers. IEEE, Piscataway, NJ, pp. 58–59; andN. Singh, F. Lim, W. Fang, S. Rustagi, L. Bera, A. Agarwal, C. Tung, K. Hoe, S. Omampuliyur, D. Tripathi, A. Adeyeye, G. Lo, N. Balasubramanian, and D. Kwong, "Ultra narrow silicon nanowire gate-all-around CMOS devices: impact of diameter, channel orientation and low temperature on device performance," *Proceedings of IEEE Electron Devices Meeting*, San Francisco, CA, December 11–13, 2006. IEEE, Piscataway, NJ, pp. 1–4.

[22] This equation, if treated within the context of a 1-*region model*, neglects the variation of the electric field within the *gate* insulator. As a result, as it was indicated in Section 3.4, not unlike the model of Section 3.2.2.2, it has a number of limitations. This happens even prior to the dominance of quantum phenomena.

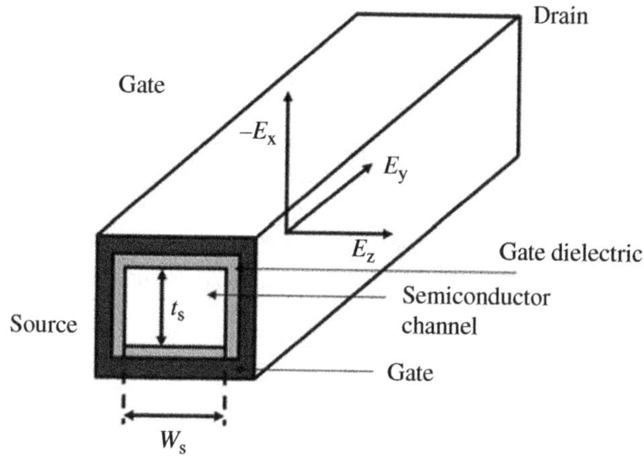

FIGURE 4.4 Components of electric field governing the charge status of the channel of a gate-all-around *MOSFET*.

According to Figure 4.4, in Equation (4.7) the y-component of the electric field (i.e., E_y) represents the penetration of the *drain*-induced electric field in the *channel*. This is the component inducing *short-channel effects*. The impact of this component on the device behavior can be controlled by increasing the *gate length* (and as a result reducing E_y) or by increasing the control of the *gate* over the *channel*. The increase in *gate* control can be achieved by enhancing $(\partial E_x(x,y,z))/\partial x$ in the *top* and *bottom gates* and through enhancing $(\partial E_z(x,y,z))/\partial z$ in the *lateral gates*. According to the structure of the *multiple-gate MOSFET* shown in Figure 4.4 and our earlier discussion, this can be done by reducing the thickness of the semiconductor *channel* layer (i.e., t_s) and/or the width of the *fin*-looking *channel* (i.e., W_s).

According to numerical simulations, in a properly designed *short-channel MOSFET*, the *effective gate length* of the *MOSFET* must be larger than 5–10 times the length scale l (also known as the *natural length*).[23]

For a *double-gate MOSFET*, solving (4.6) through assuming boundary conditions conceptually matching those of Section 3.2.2.2 prompts the *drain*-induced electric field to take on an exponential characteristic like the one expressed in (3.48). In this case, however, the length scale l is given by

$$l=\sqrt{\frac{\epsilon_s t_{ox} t_s}{2\epsilon_{ox}}}. \tag{4.8}$$

Comparing this length scale with the one developed in Section 3.2.2.2 again shows that the role of X_j is now played by $t_s/2$. Since the reduction of the length scale signals

[23] In case of the *bulk planar MOSFET*, the length scale l was defined in (3.51) and in the pertinent discussions to (3.48).

less *drain* influence on the *channel* potential, the indicated faster down*scaling* of l with t_s (i.e., compared to dependence of the length scale of a *bulk planar MOSFET* on X_j) is very promising. Such an effect can be beneficial to relieving some of the technological hassles in trimming the oxide (or generally speaking *gate*-dielectric) thickness and as a result in adopting *high-K dielectrics*. According to this observation, in the case of a *double-gate MOSFET*, this device is half as thick as a *single-gate* transistor.

Application of the aforementioned *1-region model* to a *quadruple-gate SOI MOSFET* of rectangular cross section by assuming $\partial^2 V/\partial x^2 = \partial^2 V/\partial z^2$ at the center of the device yields the length scale of

$$l = \sqrt{\frac{\epsilon_s t_{ox} t_s}{4\epsilon_{ox}}}. \tag{4.9}$$

for the exponential variation of the *drain*-induced electric field along the length of the *channel*.

In contrast to the rectangular cross section of the *channel*, in the case of the *gate-all-around cylindrical MOSFETs*, the length scale is given by[24]

$$l = \sqrt{\frac{2\epsilon_s {t_s}^2 Ln(1 + (2t_{ox}/t_s)) + \epsilon_{ox} {t_s}^2}{16\epsilon_{ox}}}, \tag{4.10}$$

where t_s stands for the diameter of the semiconductor cylinder.

A comparison between Equations (4.8) and (4.9)–(4.10) reveals that with a keen interest in limiting the *short-channel effects*, *quadruple-gate MOSFETs* present more *scalable* device architectures. It is according to the merits enumerated in this section that as early as 2011 many of the major semiconductor foundries have moved toward replacing *bulk planar MOSFETs* by these 3-D alternative structures. In view of this drastic change, in order to keep track of the *scaling* trend and technological improvements, in addition to properly expressing the *modes* of operation of the transistor, appropriate language needs to be introduced into our discussion to replace that of the older technologies.

As indicated in Section 3.2.2.1, in the age of *FinFETs* and other *multiple-gate FETs*, the definition of *technological nodes* according to the *gate length* is becoming quite obsolete. A major difference between the language often used in expressing *bulk planar MOSFET* and *multiple-gate FinFET* technologies is that while in the former the *minimum feature size* is defined by the *gate length*, in the latter it stands for the *width* of the *fin*.

In the case of *double-gate FinFETs*, as the *channels* are formed on the sidewalls of the narrow *fins*, the effective *channel width* is almost equal to $2n$ times the *fin height* where n is an integer number equal to the number of *fins* in the device structure. In

[24] See C. P. Auth and J. D. Plummer, "Scaling theory for cylindrical, fully depleted, surrounding-gate MOSFET's," *IEEE Elec Device Lett*, vol. **18**, no. 2, pp. 74–76, 1997.

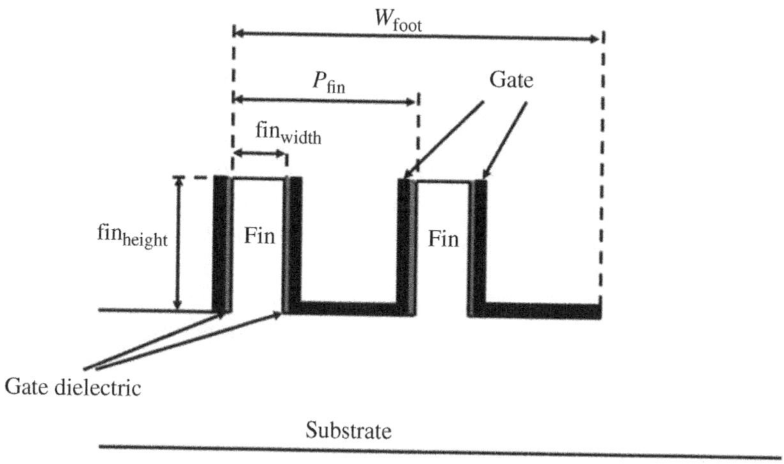

FIGURE 4.5 Cross-sectional view of a double-gate 2-fin *FinFET*. As illustrated in this figure, gate is only covering the sides of each fin.

these device architectures, the sum of the spacing between every two *fins* and the *fin width* is referred to as *fin pitch* (here represented by P_{fin}). P_{fin} is a parameter limited by *lithography*. These quantities are expressed in Figure 4.5.

The information presented in this figure comes in handy in assessing the increase in packing density of *FinFET chips* compared to that of *bulk planar MOSFETs*. According to Figure 4.5, while in *planar MOSFET* the device's *gate width W* is defined by the footprint of the *gate* on the substrate, W_{foot}, for *FinFETs* the *effective gate width* (i.e., W_{eff}) is related to W_{foot} by

$$W_{eff} = 2 \times fin_{height} \times \frac{W_{foot}}{P_{fin}}. \tag{4.11}$$

In terms of packing density, in order to make the *FinFET* layout competitive, it is required that $W_{eff} \geq W_{foot}$ (or equivalently $2 \times fin_{height} \geq P_{fin}$).

***4.1.2.1 Implications of Crystal Orientation on the Operation of* FinFETs** The type of *channel*-formation process used in the realization of *multiple-gate FinFETs* demands closer attention to *crystal orientation*. As it was indicated in Chapter 1, *carrier mobility* among other things is dependent on the surface orientation of the *channel*. This dependence is caused by the anisotropy built into the *effective carrier mass* and the *surface-scattering rate*. For electrons, the *silicon* surface with the highest *mobility* is (*100*), whereas the lowest-*mobility* surface is (*110*). For *holes*, the opposite of this order is valid. For both *charge carriers*, the value of *mobility* for the (*111*) *silicon* surface is somewhere between the values observed on these other two surfaces.

Whereas in *planar MOSFETs* surface orientation of the *channel* is defined by the surface of the *wafer* (which, as indicated in Section 3.1, is predominantly the (*100*) *silicon* surface), in the case of *FinFETs* the *orientation* of the *fin* sidewalls and as a result that of the *channels* formed in their vicinity are defined by the *orientation* of the *fins* with respect to the *flats*[25] of the *wafer*. As a result, *carrier-transport* properties at these surfaces are defined by *lithography*. The *(110) fin* sidewall surfaces are realized by patterning *fins* at 0 or 90° angles with respect to the *<110> notch* on *silicon wafers* of *(001)* surface orientation. This follows the same layout style as that of the conventional *planar* devices. The *(100) fin* sidewall surfaces are the result of rotation of the *fin* layout by a 45° angle with respect to the *<110> notch*.

The ability to define *fins* with both (*110*) and (*100*) sidewall surface orientations on the same *chip* gives an alternative path to optimizing electron and *hole mobilities* of a *CMOS pair* simultaneously. This is while in *planar CMOS* technology the imbalance between the *current drives* of *NMOS* and *PMOS* transistors (which is caused by the lower *mobility* of *holes*) can be only mended by adopting a larger *gate width* for the *PMOS* transistor. However, this new ability comes with a price tag of the increased layout area and higher process-complexity *lithography*.

4.1.2.2 Technological Implications of* FIN *Structure on the Drain and Source Contacts *Scaling* of *CMOS* technology has not only strained the implementation of the *gate* electrode, but at the same time it has added to process complexity in realizing *drain* and *source Ohmic* contacts. In this regard, a second look into the structures of the *multiple-gate MOSFETs* depicted in Figure 4.1 is quite informative. In the particular case of a *FinFET*, as we took note earlier in this section, the need for reduction of *short-channel effects* has been pushing for the *fin width* to take on values even smaller than the *gate length*. As a result, even at the current *technology node*, the *fin width* has contracted to values on the order of 10 nm. Such a small dimension creates a huge hassle in realization of *Ohmic* contacts. This is because making *source* and *drain Ohmic* contacts to these small *fins* imposes a high parasitic *contact resistance* on the device operation.

A popular remedy for alleviating this parasitic resistance is to widen the *etched* profile of a *fin* outside the *gated* area. This is done through the use of *selective epitaxial growth* (or *SEG*). The structure realized by this method is often referred to as a *raised source* and *drain* (or *RSD*) *fin*. Figure 4.6 presents a schematic depiction of this structure next to its circuit-level model. While such an implementation manages to reduce the *source* and *drain* components of parasitic contact resistance, it does not result in improving the parasitic resistance in the extended regions of the *fin* (which is indicated by R_{ext} in Fig. 4.6b). This is the resistance in the *source-* and *drain-extension* regions under the *gate spacer*. This resistance remains high due to the small thickness of the *fin*. While bringing the raised *source* and *drain* regions closer to the *gate* can reduce

[25] Which are the standard cuts made to the periphery of *wafers*, intended to encode the *crystal orientation* of the surface of a *wafer*, in addition to its *doping* type. As examples, in *silicon* technology *p-type wafers* of (*111*) surface have only one *flat* perpendicular to the *<110>* direction; however, on an *n-type* (*111*) *wafer* in addition to this larger so-called primary flat, there is a smaller *secondary-flat* cut at 45° *radial angle* with this one. For the (*100*) surface, for p-type and n-type *wafers* the angles between the *primary flat* made perpendicular to the *<110> direction* and the *secondary flat* are 90° and 135°, respectively. These *cuts* are present in *silicon wafers* smaller than 6″ diameter. For larger-size *wafers*, instead of *flats*, *notches* are formed at the periphery. This is to limit the surface area cut out of the *wafer*.

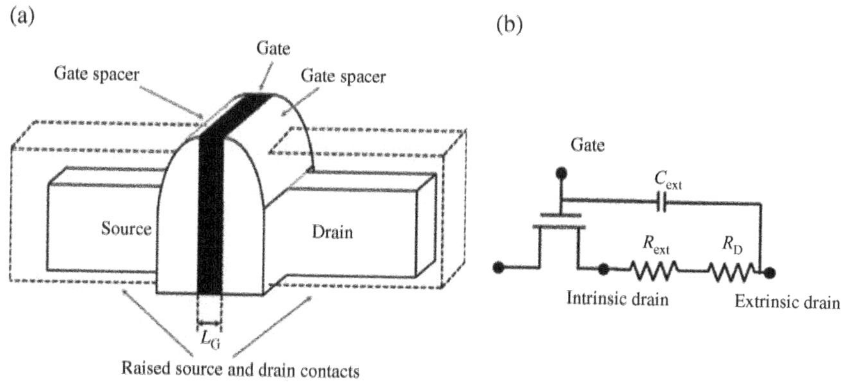

FIGURE 4.6 (a) Schematic depiction of the structure of a *FinFET* with raised source and drain contacts. (b) Circuit diagram of the drain-side major parasitic resistive and capacitive elements associated with the raised structure in (a). Employing the raised source and drain contacts reduces R_D. Extending these raised profiles toward the gate (i.e., by reducing the width of the spacer), while results in improving R_{ext} worsens C_{ext}. Adapted with permission from Colinge (2008, p. 88). Copyright 2008, Springer Science + Business Media LLC.

this resistance, it causes an increase in the parasitic capacitance C_{ext} (indicated in Fig. 4.6b). As a result, a compromise needs to be made between these two competing factors regarding the frequency response.

4.1.2.3 Implications of Charge Sharing and Quantum Phenomena in Multiple-Gate* FET *Architectures: A First Glance The presence of corners in the *nonplanar* insulator/semiconductor interfaces of the *triple-*, *quadruple-*, Π-, or Ω-*gate* structures depicted in Figure 4.1 has been observed to induce a premature *inversion channel* at these corners. This is due to the *charge-sharing effect* of the adjacent *gates*. The formation of this premature *channel* has been indicated by the observation of multiple *threshold-voltage* values (i.e., for the corner *channels* and the *channels* formed at the *top* or *sidewall* insulator/semiconductor *interfaces*). Inspecting the profile of $\partial g_m/\partial V_G$ versus *gate* voltage is capable of clearly demarcating these multiple values of *threshold voltage*. The hump in such a characteristic indicates the formation of an *inversion channel*. In devices of large *dopant* concentration and small radii, at the corners often a double-hump $\partial g_m/\partial V_G$ profile representing a double *threshold-voltage* situation is observed. Clearly, the formation of the premature *inversion channel* caused by the so-called corner effect has a degrading impact on the value of the *subthreshold swing*. Such a *corner effect* can be prevented by switching from the *fin-structured channel* to a *cylindrical channel*.

The other important implication of the topology of *multiple-gate MOSFETs* that we have so far overlooked is a more pronounced presence of quantum phenomena. We have already paid a visit to the quantum effect in *bulk planar MOSFETs* of shrinking *gate length* in Section 3.3. As indicated earlier, in addition to the *gate-length scaling*, in *multiple-gate MOSFETs scaling* has also resulted in contraction of the *width* of

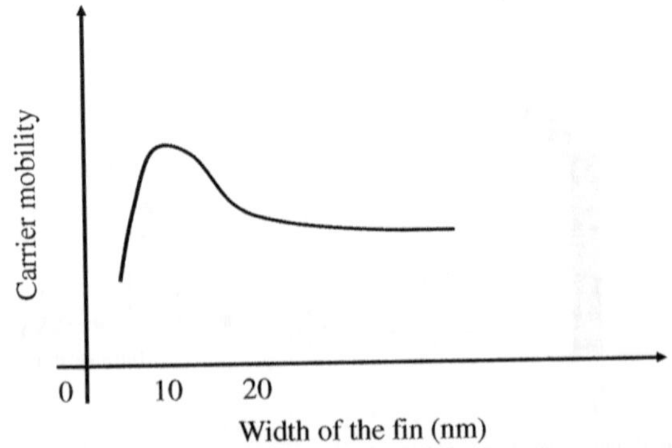

FIGURE 4.7 Approximate trend of variation of carrier mobility with the width of the fin.

the semiconductor *channel* to values less than 10 nm. Considering the architecture of these devices, under these conditions of spatial confinement, electrons tend to reveal more of their quantum nature (i.e., to extents exceeding that of *deeply scaled planar MOSFETs*). In such small dimensions, either a *two-dimensional electron gas* (in a *double-gate MOSFET*) or a *one-dimensional electron gas* (in a *triple-* or a *quadruple-gate MOSFET*) is formed. Evaluation of the operation of these devices without paying close attention to the formation of energy *subbands* and their attributed *density of state functions* will not be possible. One interesting aspect of this quantum nature in *multiple-gate MOSFETs*, such as a *FinFET*, is that the *inversion channel* need not be formed in the close vicinity of the insulator/semiconductor *interface*. Instead, such a *channel* can be formed as a *volumetric inversion channel* in the center of the *fin*.[26] Such a difference in the induction mode of the *inversion channel* has important consequences on the *carrier mobility* and also on the *threshold voltage* of these devices.

The aforementioned removal of the center of the *inversion channel* from the vicinity of the *interface* has been observed to result in an early improvement in *carrier mobility* of thin-film devices. This is due to the reduction of *interface-roughness scattering*. However, with further reduction of the size of the *fin*, a roll-off of *carrier mobility* is observed. This roll-off is caused by additional *intersubband scattering* and increased *interface-roughness scattering*. In this case, even the center of *fin* is not far enough from the insulator/semiconductor *interface*. Figure 4.7 provides a schematic depiction of the dependence of *carrier mobility* on the *width* of the *fin*.

In small-size devices facing quantum effects, formation of energy *subbands* and the resulting increase in the energy at the effective edge of the *conduction-band* (and the *valence band*) cause the *threshold voltage* to increase. Considering the energy *subbands*, both *longitudinal-* and *transverse-mass subbands* should be included.

As indicated earlier, the thickness of the semiconductor *channel* layer is an important parameter in the operation of *fully depleted double-gate MOSFETs*. In addition to

[26] Discussion of *quantum capacitance* and assumption of $\Delta \neq 0$ in a *bulk planar MOSFET* in Section 3.3, although rooted in the same phenomenon, due to the larger volume of *semiconductor* in those devices, does not really result in *volumetric inversion* to this extent.

what was indicated earlier, this parameter influences the definition of the energy *subbands*. The electron *transport* properties in these *channels*, even excluding the impact of film thickness on the *scattering* rates, are closely influenced by the way electrons occupy the energy *subbands* developed by the spatial confinement. Depending on the energy of the lowest *transverse* and *longitudinal subbands*, these *subbands* can sway the *effective transport mass* of the *inversion* electron population to take a greater notice of one or the other *subband. Transport* properties of the *channel* are heavily determined by these *subband* energy values (which, as we know, are partially determined by the thickness of the *channel* layer).

Careful description of the aforementioned semiconductor thickness effect reveals that while in *silicon* films of 5–10 nm thickness the major contribution to surface electron density is in the *transverse mass level*, for film thicknesses below 5 nm the lowest *longitudinal mass level* becomes the main contributor to electron population. At the same time, due to the stronger rise among all *energy levels*, which is caused by further spatial confinement, the total electron surface density also reduces.

In addition to the issues already enumerated in this chapter, it should be indicated that probably the most important concern in making of *multiple-gate MOSFETs* is the technological difficulty in realization of the semiconductor *channel* in these structures. In those *multiple-gate MOSFETs* that are realized using *dry etching* of *SOI wafers* (i.e., according to realization without the use of *epitaxial growth* techniques), after this *fin-isolation etching* step, *sidewalls* are rough and require a subsequent step of *oxidation* and H_2 *anneal* for smoothening. The complexity of *etch* and the need for additional steps to reduce the *surface roughness* and *surface state density* are factors that can be limiting to the fabrication *yield.*

4.1.3 Strain Engineering: From Bulk to Multiple-Gate *MOSFETs*

In spite of the fundamental nature of strain in materials, the implications of strain on device operation are controllable through appropriate choice of technology and device dimensions. Strain is not only present in *pseudomorphically* grown *epitaxial heterostructures* but also in *homogeneous* films. While some films possess intrinsic stress, in others *phonon-induced lattice vibrations* can induce film stress at higher temperatures.

As indicated in Section 1.7.2.2, *carrier mobility* in terms of *Drude*'s model is calculated on the basis of *mean scattering time* and *carriers' effective mass* (i.e., $|\mu| = (q\tau)/m^*$). Both of these are quantities that can be modified through application of strain on the crystal, a technique that is commonly referred to as *strain engineering. Strain engineering* of *carrier mobility* in novel *MOSFETs* has been the basis of an important line of research. *Strain engineering* has an anisotropic influence on *carrier mobility*. As a result, its outcome is highly dependent on the surface orientation and the direction of the *channel.*

4.1.3.1 Uniaxial Strain in* CMOS *Technology Implementation of *biaxial* strain in *silicon channels* through forming the *channel* at *lattice-mismatched heterojunctions*

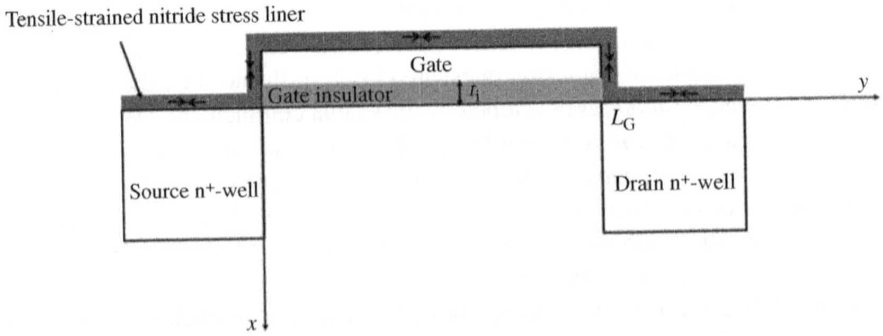

FIGURE 4.8 Schematic depiction of induction of stress in different directions along the channel of a *MOSFET* covered with a tensile-strained nitride stress liner.

(such as *Si/SiGe*) has been explored for about three decades.[27] More recently, the application of *uniaxial* strain through implementation of *stressors* has come to the spotlight for boosting the performance of *planar CMOS pairs*. Using these techniques, in *PMOS* and *NMOS* transistors, about 50 and 30% improvement in the *drain* current drive have been reported, respectively.[28]

In addition to *Si/SiGe heterojunctions, nitride stress liners* are often used in modern *silicon NMOS* transistors to improve performance. These *nitride* films, depending on their *deposition* condition, can be either under *tensile* or *compressive* strain. Through *deposition* of a *nitride* layer under *tensile* strain on the *gate* and over the *source* and *drain* regions, this film produces *tensile* strain along the length of the *channel*. This is since the *nitride* film covering the *source* and *drain* regions tends to shrink. However, by the same token, as shown in Figure 4.8, the *nitride* film covering the sidewalls of the *gate* imposes a *compressive* strain in the direction perpendicular to the *channel*, on the *gate*, and eventually on the *channel*. In *channels* realized on *(100) silicon* surface at 0° angle with the <*110*> direction, the <*110*> electron *mobility* is observed to improve by applying *tensile* strain along the *channel* direction and *compressive* strain normal to the surface. Hence, following the layout style of conventional *planar* devices, a net improvement in electron *mobility* is achieved with the use of this kind of *stressor*. *Hole-mobility* improvement can be achieved by exerting the opposite type of *strain* along the *channel* direction, while it does not show much sensitivity to the strain perpendicular to the *silicon channel*. In this case, a *compressive*

[27] We will start this section with the more novel approaches, while more will be presented on these traditional modes of forming the *strained channel* toward the end of the section.

[28] M. Horstmann, A. Wei, T. Kammler, J. Hontschel, H. Bierstedt, T. Feudel, K. Frohberg, M. Gerhardt, A. Hellmich, K. Hempel, J. Hohage, P. Javorka, J. Klais, G. Koerner, M. Lenski, A. Neu, R. Otterbach, P. Press, C. Reichel, M. Trentsch, B. Trui, H. Salz, M. Schaller, H. Engelmann, O. Herzog, H. Ruelke, P. Hubler, R. Stephan, D. Greenlaw, M. Raab, N. Kepler, H. Chen, D. Chidambarrao, D. Fried, J. Holt, W. Lee, H. Nii, S. Panda, T. Sato, A. Waite, S. Liming, K. Rim, D. Schepis, M. Khare, S. Huang, J. Pellerin, and L. Su, "Integration and optimization of embedded-SiGe, compressive and tensile stressed liner films and stress memorization in advanced SOI technologies," *Proceedings of International Electron Devices Meeting*, Washington, DC, December 5–7, 2005. IEDM Technical Digest. IEEE, Piscataway, NJ, pp. 233–236.

stress liner, instead of a *tensile stress liner*, is employed. Since in devices with longer *gate length* the *stressors* implemented in this fashion are further removed from the *channel*, among these devices, they are not as effective in improving the *carrier mobility*. Aside from the *(100)/<110>* example taken in this paragraph for the surface and the *carrier-transport* direction, other alternative surface and *channel* directions present different behaviors with the application of these types of *stress liners*. As a result, in *FinFETs* in addition to what was expressed in Section 4.1.2.1, different orientations of the *channels* realized on a given *chip* can take advantage of the strain dependence of the *carrier mobility* as another dimension in improving the *chip* performance.

Another novel *strain-engineering* approach explored in modern *CMOS* technology is the incorporation of the embedded *SiGe* and *SiC source* and *drain* contacts. While originally this endeavor was aimed at improving the contact resistance of *PMOS* devices (i.e., due to the lower *Schottky* barrier height for *holes* in *SiGe* compared to that of *Si* and the improved *boron activation* in *SiGe*), use of *SiGe source* and *drain* structures experimentally resulted in a much larger than expected improvement in *drain* current drive. Later on, it was found out that the excess improvement in the *current drive* is due to the introduction of *compressive* strain along the *channel* and a resulting improvement in the value of the *hole mobility*.[29] *SiGe*'s larger *lattice constant* is the cause of this *compressive* strain on the *silicon channel*. This situation is schematically illustrated in Figure 4.9. Using this technique, from 25 to 40% improvement has been reported on the current drive of *multiple-gate MOSFETs*. This amount of improvement is less than the improvement observed in *bulk planar MOSFETs* of *(001)/<110>* orientation. This has been explained by the reduced volume of *SiGe* in *multiple-gate MOSFETs*.

Implementation of *SiC source* and *drain* contacts, due to the smaller *lattice constant* of *SiC* compared to that of *Si*, is bound to induce a *tensile* strain along the *channel*. This has been also explored for improving the current drive of *silicon NMOS* transistors. In the case of *multiple-gate NMOS*, it has been shown that 1% *carbon* incorporation in $Si_{1-x}C_x$ can result in about 20% improvement in the current drive.

Local strain can also be induced from the *gate* electrode. This is especially important in *FinFETs* and other *multiple-gate* transistors. In these devices, since the *gate* wraps around the *fin*, the stress directly applied from the *gate* will be more capable of improving the *channel's carrier mobility* (i.e., in comparison to the stress applied through *source* and *drain* contacts). Some of the *metal-deposition* processes can

[29] This conclusion was reached after much internal debate in *Intel*. For early publication on this matter, see S. Thompson, N. Anand, M. Armstrong, C. Auth, B. Arcot, M. Alavi, P. Bai, J. Bielefeld, R. Bigwood, J. Brandenburg, M. Buehler, S. Cea, V. Chikarmane, C. Choi, R. Frankovic, T. Ghani, G. Glass, W. Han, T. Hoffmann, M. Hussein, P. Jacob, A. Jain, C. Jan, S. Joshi, C. Kenyon, J. Klaus, S. Klopcic, J. Luce, Z. Ma, B. Mcintyre, K. Mistry, A. Murthy, P. Nguyen, H. Pearson, T. Sandford, R. Schweinfurth, R. Shaheed, S. Sivakumar, M. Taylor, B. Tufts, C. Wallace, P. Wang, C. Weber, and M. Bohr, "A 90 nm logic technology featuring 50 nm strained silicon channel transistors, 7 layers of Cu interconnects, low K ILD, and 1 μm^2 SRAM cell," *Proceedings of International Electron Devices Meeting*, San Francisco, CA, December 8–11, 2002. IEEE Piscataway, NJ, pp. 61–64.

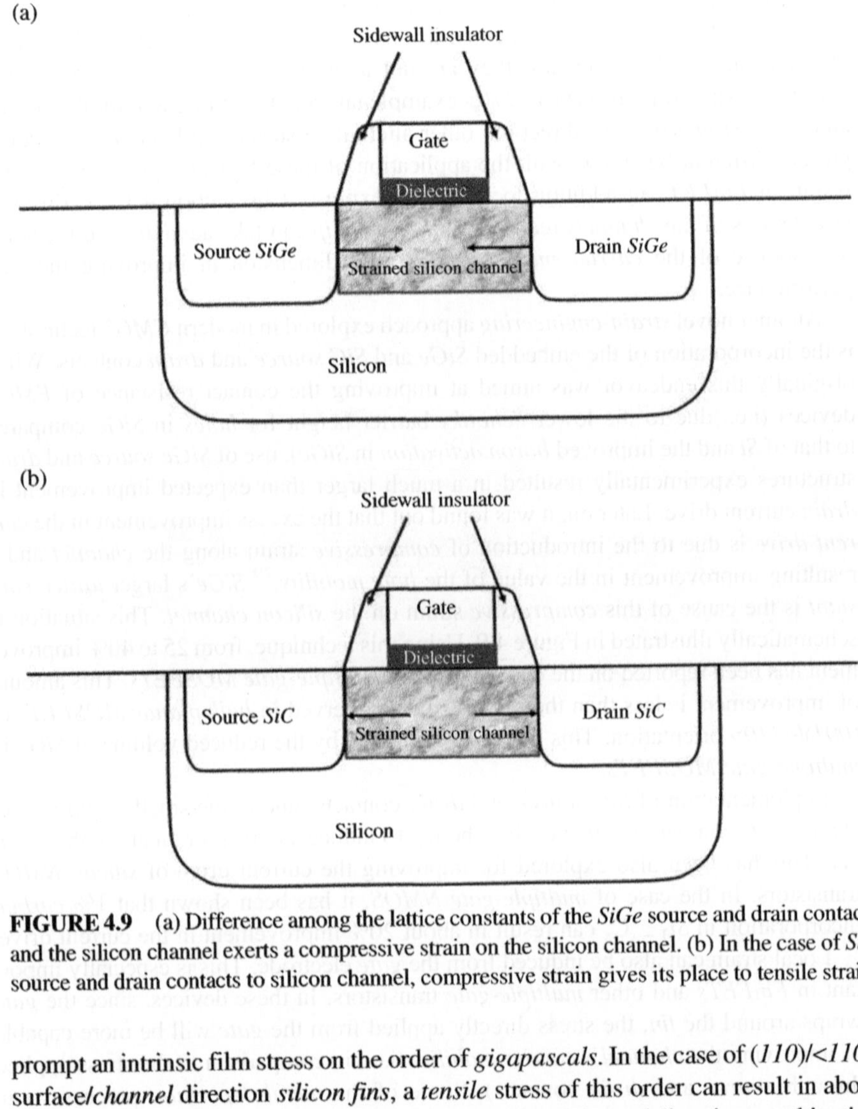

FIGURE 4.9 (a) Difference among the lattice constants of the *SiGe* source and drain contacts and the silicon channel exerts a compressive strain on the silicon channel. (b) In the case of *SiC* source and drain contacts to silicon channel, compressive strain gives its place to tensile strain.

prompt an intrinsic film stress on the order of *gigapascals*. In the case of *(110)*/<*110*> surface/*channel* direction *silicon fins*, a *tensile* stress of this order can result in about 100% improvement in electron *mobility*. This surface/*channel* direction combination is a particular case of superb sensitivity. Such a behavior is not prevalent in the case of other surface/direction combinations (and neither is it in the case of *hole mobility*). As the *gate* is always close to the *channel*, in contrast to the stress induced by the *source* and *drain*, the *gate*-induced stress, where present, is not layout dependent and is almost equally implementable in both *short-* and *long-channel* devices.

4.1.3.2 Biaxial Strain in CMOS *Technology*

In addition to the aforementioned techniques, the application of stress on the *channel* via the substrate has been also

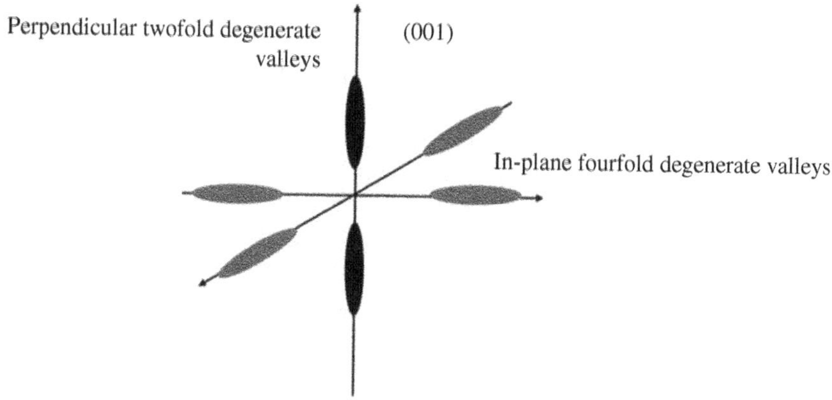

FIGURE 4.10 *E–k* representation of the lift of the sixfold degeneracy of the conduction band of silicon into an in-plane fourfold and an out-of-plane twofold degeneracy.

explored in the form of *strained SOI* substrates. This is done by growing *silicon* on *SiGe*. As mentioned earlier in this section, besides the use of *SiGe* in the *source* and *drain* contacts of *PMOSFETs* for implementing *stressors*, a more mainstream use of *SiGe* has been based on employing a *two-dimensional epitaxial growth* technique for making *biaxially strained Si/SiGe channels*. Theoretical and experimental studies have shown considerable electron *mobility* enhancement in *pseudomorphically grown silicon channels* realized on *relaxed* $Si_{1-x}Ge_x$ films.

As indicated in Section 1.6.2.2.2, the *biaxial* strain in these *channels* causes the sixfold degeneracy of the *conduction-band valley* of *silicon* to be lifted into a twofold degeneracy (in the lowered *valleys* with longitudinal *effective-mass* axis normal to the *heterointerface*) and a fourfold degeneracy (in the raised *valleys* with longitudinal *effective-mass* axis parallel to the *heterointerface*). This is shown in Figure 4.10.

This lift of *degeneracy* results in the redistribution of *carriers* between the aforementioned twofold and fourfold *degenerate valleys*. Among these two groups, the lowered twofold *degenerate valleys* (also referred to as *unprimed valleys*) are more densely populated. In these *valleys* electrons demonstrate a smaller conduction *effective mass* (i.e., the *transverse mass*) in *carrier transport* parallel to the *heterointerface*. Caused by the large-enough splitting between the two groups of *valleys*, the *intervalley phonon scattering* rate is also effectively reduced in these *strained silicon channels*. Both of these factors contribute to improvement in electron *mobility*. At the same time, the reduced *intervalley scattering* rates cause the *energy relaxation time constant* to become longer. This extension of the *energy relaxation time constant* contributes to a spectacular non-steady-state electron velocity *overshoot* in these *strained channels*. In spite of these improvements, *deep scaling* still results in familiar problems of *DIBL* and *SCE* in the *bulk strained* $Si/Si_{1-x}Ge_x$CMOS transistors. This requires attention by improving the *EI*.

One of the important observations made on these *epitaxial MOSFETs* is that the *volumetric inversion*, indicated in Section 4.1.2.3, tends to reduce with *strain*. This is because the electron concentration in the center of the *channel* reduces with the increase in strain.

When extending the possibility of *strained epitaxy* to *multiple-gate MOSFETs*, it should be pointed out that with the reduction of the dimensions of the *fin* to *nanometer* scales, *bulk* crystal symmetry is altered and the information gathered on its basis such as the *density of states* functions and *band* structure undergo severe modifications. This is due to the spatial confinement of electrons in the directions perpendicular to the *carrier-transport* direction. As a result, especially in small *FinFETs*, such information cannot be inherited from the *bulk* semiconductor. While in *silicon nanowires* of *diameter* smaller than 5 nm such concerns must be taken into account, for nanowires of larger dimensions the approximations based on the less mathematically intense parabolic *effective-mass* approach are still favored.

4.1.3.3 Uniaxial versus Biaxial Strain in* CMOS *Technology As expressed in the previous two sections, in spite of their enticing nature, incorporation of *biaxial* and *uniaxial* stress to *CMOS logic* technologies has faced a few difficulties. These difficulties include the requirement of two different types of strain for improvement of *mobility* in n- and p-type *uniaxially strained channels* (i.e., *tensile* and *compressive*, respectively). However, in the case of *biaxial* strain, the most important impeding issues include the presence of defects in the *relaxed SiGe* film (and its propagation into the *silicon channel* layer) and loss in performance at high electric fields.

As mentioned in Section 1.6.2.2.2, the lift in *degeneracy* between the *heavy-* and the *light-hole bands* of the *valence band* is also expected to boost the *hole mobility* of the *biaxially strained channel* of a *PMOS* transistor. However, Fischetti et al. have shown that the *hole-mobility* improvement realized by strain-caused separation of *light-* and *heavy-hole bands* in *biaxially strained silicon channels* grown on *relaxed SiGe* is reversed through application of large *gate*-induced electric fields normal to the *channel* of a *planar MOSFET*.[30] This is due to the impact of higher *gate* voltage in further confining the centroid of the *inversion layer* toward the insulator/*silicon interface*. This itself causes a shift of the *heavy-* and *light-hole band degeneracy*. Since the shifts caused by these phenomena cancel one another, at high vertical electric fields the *hole-mobility* improvement of the *biaxially strained channel* vanishes.

Such a field-induced degradation, however, is not present in *uniaxially strained channels*. In *uniaxially strained channels*, the substantial improvement in *mobility* can be sustained at high vertical electric fields. It should also be indicated that it can be materialized at much smaller levels of stress. The large amount of stress required in *biaxially strained* films used in *planar CMOS* to realize the required *carrier-mobility* improvement is a cause for reliability concerns. This is because,

[30] M. V. Fischetti, Z. Ren, P. M. Solomon, M. Yang, and K. Rim, "Six-band k.p calculation of the hole mobility in silicon inversion layers: Dependence on surface orientation, strain, and silicon thickness," *J Appl Phys*, vol. **94**, pp. 1079–1095, 2003.

as indicated in Section 1.6.2.2.1, large *compressive* strain (especially for thicker films) leads to film *relaxation*.

Because of the increased vertical electric field in *nanoscale MOSFETs*, the preservation of *mobility* improvement of *uniaxially strained channels* makes this type of strain induction especially useful in *nanoscale MOSFETs*. This is one of the contributing factors to the acceptance of 3-D *FinFETs* to mainstream *CMOS* manufacturing smaller than the 22 nm *technology node* (i.e., as substitutes for the *bulk planar MOSFETs*).[31]

In *uniaxially strained channels*, most of the improvement in *current drive* is materialized by the improvement in *conductivity effective mass*. In *biaxially strained channels* the improvement is considerably made through the reduction in *scattering*. As a result, in *nanoscale* devices in which *scattering* is not too frequent, the *uniaxial strain* results in much higher *current-drive* improvements.

Another determining factor on the choice of *uniaxial* strain versus *biaxial* has been the effect of strain on the *threshold voltage*. Process-induced *uniaxial* strain has been observed to offer about a five *times* smaller shift in the *threshold voltage* in comparison to the shift inherited from growth-induced *biaxial* strain. This smaller shift in *threshold voltage* translates into a lesser need for adjusting the *channel doping* level to maintain the value of the *threshold voltage*. The large value of *threshold-voltage adjustment doping* needed in *biaxially tensile*-strained *channels* results in approximately 50% loss of expected enhancement in *carrier mobility*.[32]

As a result of these observations, *uniaxial* stress as a form of feature enhancement for *CMOS* technology has been employed in all high-performance logic families after the 90 nm *technology node*.

Through a variety of means including the use of *stressors*, among the *channels* of a *multiple-gate MOSFET* an asymmetry in the *carrier-transport* behavior can be established if the thickness of the semiconductor is large enough. As an example, this asymmetry can be established through varying the *compound* semiconductor composition of the *channel*.[33] This idea is similar to that employed in the *velocity-modulation transistor* (*VMT*), which is to be presented in Section 4.2.

[31] For further reading on the early explorations on this matter, see S. Thompson, G. Sun, K. Wu, J. Lim, and T. Nishida, "Key Differences For Process-induced Uniaxial vs. Substrate-induced Biaxial-Stressed Si and Ge Channel MOSFETs," *Proceedings of International Electron Devices Meeting*, San Francisco, CA, December 13–15, 2004. IEEE, Piscataway, NJ, pp. 221–224.

[32] S. Thompson, N. Anand, M. Armstrong, C. Auth, B. Arcot, M. Alavi, P. Bai, J. Bielefeld, R. Bigwood, J. Brandenburg, M. Buehler, S. Cea, V. Chikarmane, C. Choi, R. Frankovic, T. Ghani, G. Glass, W. Han, T. Hoffmann, M. Hussein, P. Jacob, A. Jain, C. Jan, S. Joshi, C. Kenyon, J. Klaus, S. Klopcic, J. Luce, Z. Ma, B. Mcintyre, K. Mistry, A. Murthy, P. Nguyen, H. Pearson, T. Sandford, R. Schweinfurth, R. Shaheed, S. Sivakumar, M. Taylor, B. Tufts, C. Wallace, P. Wang, C. Weber, and M. Bohr, "A 90 nm logic technology featuring 50 nm strained silicon channel transistors, 7 layers of Cu interconnects, low K ILD, and 1 μm^2 SRAM cell," *Proceedings of International Electron Devices Meeting*, San Francisco, CA, December 8–11, 2002. IEEE, Piscataway, NJ, pp. 61–64.

[33] We will explore this in greater depth within the context of *HFETs* in Chapter 5.

4.1.4 Limitations of the Introduction of III–V Channels to Multiple-Gate and Other Modern *CMOS* Technologies

As mentioned in Section 3.2.2.2, the advance of *scaling* in *CMOS* technology and the quest for achieving higher switching speeds have renewed the zeal for improving the *channel*'s *carrier mobility* through the adoption of semiconductors of higher *carrier mobility* than *silicon*. For that reason, high-*mobility compound III–V* semiconductors are yet again on the agenda. However, along with the expected improvement in *low-field carrier mobility*, there are reservations with regard to factors such as increased *DIBL* and *subthreshold swing* (due to the larger *dielectric constant* among some, and the smaller *density of states* among many, compared to *silicon*), which might upset this trend.

As a result of these adverse effects, the performance gain expected from switching to *III–V channels* has turned out to be much smaller than the expected 100% improvement. On many occasions, a negative impact has even been observed. Results of an analysis performed by *Skotnicki* and *Boeuf*, which are presented in this section, provide a framework for understanding the applications and conditions appropriate for inclusion of *III–V channels* in modern *CMOS* technology.

It is important to highlight that there is a relationship between the distance from the insulator/semiconductor *interface* to the centroid of the *inversion layer* formed in a *MOS* structure (i.e., Δ)[34] and the *DOS* of the *channel*. In materials with smaller *DOS*, this space is larger. The definition of the *quantum capacitance*, which was presented in Section 2.4.2, is also given in terms of the same quantity. As a result, the lower *DOS* of *III–V compound* semiconductors results in a larger value of equivalent *gate*-dielectric thickness in the *inversion-mode* operation (defined as $t_{\text{inv}} = t_{\text{ox}} + (\epsilon_{\text{ox}}/\epsilon_{\text{s}})\Delta$).[35] The increase of t_{inv} can have a multifold impact on the device performance. Among these impacts are:

- The *effective oxide capacitance* defined as $C_{ox} = \epsilon_{\text{ox}}/t_{\text{inv}}$ will be reduced, which results in reduced population of the *inversion channel* for identical *bias* conditions.
- *Subthreshold swing* will be degraded due to the reduction of the *effective oxide capacitance.*
- *DIBL* will be worsened due to the same reduction in the value of this effective capacitance.

In addition to these observations, as we have mentioned in Section 3.2.2.2, if we fail to pay attention to the relevance of the nonlinearity of *drift-transport* characteristics of the *channels* in *deeply scaled MOSFETs* to the device operation, the difference in electron *mobility* of *III–V* and *silicon channels* might give us the wrong impression. This is because *deep scaling* of *MOSFETs* has shifted the *transport* paradigm (along most of the *channel length*) from *low-field* to the *high-field transport.*

[34] Sometimes called the *Dark Space*.

[35] ϵ_s is the dielectric constant of the semiconductor *channel*.

As a result, in such *channels* the average electron velocity along the *channel* is defined predominantly by the *saturation velocity* and not the *low-field mobility*. Hence, in spite of the tremendous gain expected from the improved *low-field* electron *mobility* of *III–V* semiconductors, a much less gain in terms of *current drive* is obtainable. Even an eight times *mobility* improvement (often presented by a *mobility improvement factor*, k_{μ}) has been observed to offer not more than 40% gain in the *current drive* of a 22 nm *channel MOSFET*. The improvement in *saturation velocity* is often presented in terms of a factor k_{v_s}.

However, in even physically shorter *channels*, other contributing factors such as *quasiballistic transport* and non-steady-state *velocity overshoot* (which are much more obvious in *compound* semiconductors such as *InGaAs* and *InGaN*) become dominant. These factors have a stronger presence in the case of the *compound* semiconductors than *silicon*. The *quasiballistic transport* in *short-channel MOSFETs* is often mathematically envisioned by an effective increase in k_{v_s}.[36]

Based on these observations, *Skotnicki* and *Boeuf* propose that for *low-power MOSFET* switches (i.e., with very low *off-state* current density I_{off} in the order of 10 nA/μm) *III–V channels* will produce a performance inferior to that of the *silicon channels*. Additionally, in the case of high-performance *CMOS* (i.e., with I_{off} current in the order of 2 μA/μm), although a performance improvement is expected, the expected improvement is about 10% lower than the improvement offered by the *strained silicon channel* alternative.[37] Concluding these observations, it seems that the proper approach to *carrier-mobility* improvement is via physically *straining* the *silicon channel* and not switching to *III–V* semiconductors.

4.2 VELOCITY-MODULATION TRANSISTOR

The ultimate limit on the *switching speed* of *FETs*, as long as the modulation of *drain* current through the *gate* voltage is envisioned via the V_g dependence of the *channel's carrier* concentration, is a factor determined by the *transit time* of *charge carriers* along the *channel length*. This is the case among all *FETs* that we have spoken of so far in this volume.[38] In these devices the modulation of *carrier*-concentration through the *gate* voltage is inevitably linked to the lateral distribution of *carriers* at different points along the *channel*. As a result, a certain amount of time defined by the *transit time* of *carriers* through the *channel length* would be required for modulation of the *charge-carrier* concentration. This *transit time*, which is related to the *channel length* and the average velocity of *charge carriers* along the *channel*, imposes

[36] D. Fleury, G. Bidal, A. Cros, F. Boeuf, T. Skotnicki, and G. Ghibaudo, "New experimental insight into ballisticity of transport in strained bulk MOSFETs," *Proceedings of Symposium on VLSI Technology*, Honolulu, HI, June 16–18, 2009. IEEE, Piscataway, NJ, pp. 16–17.

[37] T. Skotnicki and F. Boeuf, "How can high mobility channel materials boost or degrade performance in advanced CMOS," *Proceedings of Symposium on VLSI Technology*, Honolulu, HI, June 15–17, 2010. IEEE, Piscataway, NJ, pp. 153–154.

[38] See Equations (3.1) and (3.2).

a hard limit on the *switching speed* of these *FETs*. The so-called velocity-modulation transistors aim at overcoming this barrier through modifying the aforementioned modulation principle.

4.2.1 *VMT*: Basic Principles of Operation

In the *velocity-modulation transistors* (also known as *VMTs*), the aforementioned hard limit set by the *transit time* is alleviated by controlling the *channel's effective carrier mobility* (and as a result average *carrier* velocity) via the *gate* voltage while the *carrier* density per unit area (i.e., Q_n) is kept to a constant value. The modulation of the *drift velocity* rather than the *carrier* concentration is the reason for the name *velocity-modulation transistor*.

In the pioneering proposal of *Sakaki*[39] on *VMTs*, as an example for such a concept, a *dual-channel FET* has been proposed (see Fig. 4.11). The principles of the operation of this device are presented in this section. In Section 4.2.2, by going through the seminal work of *Cohen* and *Webb*[40] on physical realization of this concept, actual implications of this device structure on the operation of the transistor and its relevance to *deeply scaled FET* technologies are explored.

In order to clarify the description of the *velocity modulation*, according to Figure 4.11, consider a *FET* with *drain* current density per-unit *channel width* along the *y*-axis as J_D. As shown in Figure 4.11, the *gate*-defined electric field is applied along the *x*-axis. In this device, the *sheet conductance G* of the *channel* is defined by

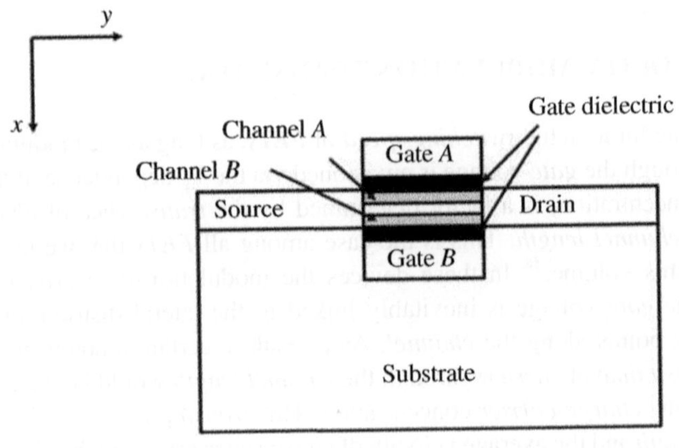

FIGURE 4.11 Cross-sectional view of a *VMT*. As indicated in this figure, two parallel channels are established between the drain and source terminals. Mobilities of the two channels are μ_A and μ_B, respectively.

[39] H. Sakaki, "Velocity Modulation Transistor (VMT)-A new field effect transistor concept," *Jpn J Appl Phys*, vol. **21**, no. 6, pp. L381–L383, June 1982.

[40] E. Cohen, Operation, development and application of the velocity modulated transistor, PhD dissertation, Purdue University, May 1998.

$$G \equiv \frac{J_D}{E_y} = \frac{Q_n v_{eff}}{E_y} = Q_n \mu_{eff}, \quad (4.12)$$

where v_{eff} and μ_{eff} are the *effective* values of *drift velocity* and *carrier mobility* along the *channel length*, respectively. In Equation (4.12), E_y refers to the *drain*-induced electric field.

According to (4.12), by modifying the *gate* voltage V_g by a small amount (i.e., ∂V_g), the *channel's sheet conductance* undergoes a modulation that can be expressed by

$$\partial G = \mu_{eff} \partial Q_n + Q_n \partial \mu_{eff}. \quad (4.13)$$

Equation (4.13) is provided based on the assumption that Q_n and μ_{eff} are independent quantities. Also, this description is that of a transistor operating in the *linear* regime of operation.

Equation (4.13) signifies the main difference between the operation of a *velocity-modulation transistor* and that of all other forms of *FETs* that we have so far spoken of in this volume. Whereas in a *VMT* the second term in (4.13) is dominant, in other *FETs* the first term is responsible for modulating the *channel's sheet conductance*.

As explained earlier, the key goal in the design of a *VMT* is to devise a control structure in which the *gate* voltage is predominantly responsible for modifying the *carrier velocity* (or *carrier mobility*) and not so much the *carrier* concentration. After *Sakaki*, as depicted in Figure 4.11, such a structure is envisioned through realizing two parallel *channels* of different *carrier mobilities* between the *drain* and the *source* (i.e., μ_A and μ_B for *channels* A and B, respectively). Figure 4.12a provides an example of implementation of such a *dual-channel* structure in the *III–V* $Al_xGa_{1-x}As$ system. In this structure, two so-called *modulation-doped*[41] *AlGaAs/GaAs heterojunctions* are grown on top of one another. Here, the *donor* and *acceptor* concentrations are chosen according to the scheme presented in Figure 4.12b, so that when $V_{GA} = V_{GB} = 0$ the two *channels* have equal electron concentrations (i.e., $Q_{n\text{-}A} = Q_{n\text{-}B} = Q_n/2$). V_{GA} and V_{GB} are the potentials applied to the *gates* responsible for *channels A* and *B*, respectively. Also in this scheme, in order to produce a *high-* and a *low-mobility channel*, high levels of impurities $(N_a > N_d \sim 5 \times 10^{16} cm^{-3})$ are intentionally introduced into *channel B*. As a result, due to the enhanced *ionized-impurity scattering*, μ_B is lowered in comparison to μ_A. In this proposal, at low temperatures *Sakaki* expected the electron *mobilities* of the *low-* and *high-mobility channels* to be in the order of 10^3 and 10^5 cm^2/V s, respectively. Although a big difference between the two values is needed, what is more important is to offer a large difference at room temperature (i.e., to offer a *VMT* operation without needing to cool down the transistor).

[41] In Chapter 5, a careful explanation of *modulation doping* will be presented. At this point in our discussion, however, it suffices to say that according to what was said in Section 2.3.2.1, the *bandgap* discontinuity at a *heterojunction*plays a role in moving the *charge carriers* across the *heterointerface*, from the side that *dopants* are introduced into to the opposite side. This is to establish *thermal equilibrium* at the *heterojunction*. This form of introduction of *charge carriers* to certain *undoped*regions of a *heterostructure*is known as *modulation doping*.

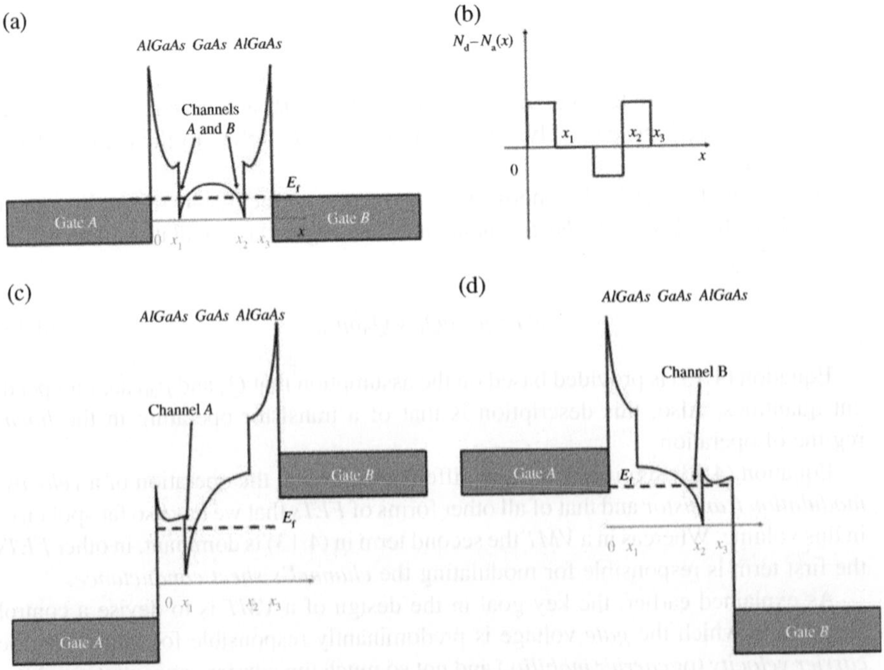

FIGURE 4.12 (a) Approximate conduction-band edge diagram of a dual-gate *VMT* under thermal equilibrium. (b) Distribution profile of the dopants in the heterostructure of the *VMT* presented in (a). (c) Approximate conduction-band edge diagram of the dual-gate *VMT* when a positive gate bias is applied to *gate A*, simultaneous with an equal negative bias applied to *gate B*. (d) Approximate conduction-band edge diagram of the dual-gate *VMT* when a negative gate bias is applied to *gate A*, simultaneous with an equal positive bias applied to *gate B*. In case (c) *channel B* is depopulated, while as illustrated through raising the *Fermi level* with respect to conduction-band edge of the *channel A*, the electron population of this channel is enhanced. The opposite of this scenario has happened in (d). Reprinted (adapted) with permission from Sakaki (1982). Copyright 1982, The Japan Society of Applied Physics.

Stepwise application of a large positive voltage to *gate A* (i.e., $V_{gA} = V_g$) while a negative voltage of similar magnitude is simultaneously applied to *gate B* (i.e., $V_{gB} = -V_g$) results in modification of *channel*-potential distribution in the form shown in Figure 4.12c. As a result of this *channel*-potential variation, those electrons that existed in *channel B* before the application of the negative potential to *gate B* will now be dumped into *channel A*. This charge transfer requires a certain amount of time for reaching steady state. Given a negative V_g, the opposite of this charge transfer will take place between the two *channels*. In such a fashion, depending on whether the *high-* or the *low-mobility channel* is predominantly populated, given a finite *drain–source bias*, the *drain* current can increase or decrease after changing V_g between positive and negative values. These *states* of *drain* current present the opportunity to define the *on-* and *off-states* of a *switch*. For analog applications when the device is operating under the *saturation regime*, because of the smaller difference of *saturation drift velocity* among the two *channels* (i.e., compared to the difference in *carrier mobility* among the two *channels*), a tangible advantage to this structure cannot be foreseen.

In theory, this independence of the *switching time* from the *transit time* along the *channel length* can promote the *switching time* to be reduced to *picoseconds*, even in relatively *long-channel* transistors. This is the time required for charge transfer between the *high-* and the *low-mobility channels*.

In a *VMT*, in spite of the aforementioned *real-space* electron transfer, the combined electron concentration of the two *channels* is still expected to remain intact. For a positive V_g, given a finite *drain–source bias*, the *carrier transport* along the *channel length* will be dominated by the *high-mobility channel A*. This results in a boost in G. Figure 4.12d demonstrates the effect of the sign reversal of the voltages applied to the two *gates* on the *channel* potential. In the case of dumping *channel A* electrons into the *channel B*, reduction of G occurs. The major implicit assumption in this explanation of the operation of a *VMT* is that the two *gate signals* are completely synchronous.

The main merit of *VMT's* operation is in its reliance on the transversal charge transfer across a few tens of nanometers-thick separation layer between *channels A* and *B* to turn *off* the *switch*. As mentioned earlier in this section, this is in contrast to the *charge-carrier transport* along the *channel length* that needs to take place in switching *off* the traditional *FETs*. Such an opportunity fosters a gain with regard to lowering the demand in *gate-length scaling*. Given the same *gate length*, the boost in *switching speed* of a *VMT* compared to that of a *MOSFET* is more tangible in *FETs* with longer *channels* than the already *deeply scaled* transistors.

We also have to highlight that the operation of the *VMT* presented in this section is a low-voltage operation. The *real-space transfer* (i.e., *RST*) of *charge carriers* that takes place in these devices has a different origin than the *hot-carrier RST* process.[42] In the presented *VMT*, the *gate* voltage induces the *interchannel* transfer of *cold* electrons, whereas in the case of the *drain-induced real-space transfer*, the *drain*-induced electric field increases the *electron temperature* and essentially makes it possible for the *hot* electrons to transfer in *real space*.

4.2.2 Real-Space Transfer: Speed and Functionality

The discovery of *RST* in a multilayer *heterostructure* is credited to *Gribnikov*.[43] As presented earlier, the basic idea behind the operation of *VMT* through *RST* is very

[42] While in Section 3.2.2.2.2 we have already spoken of *hot carriers* and their negative implications in inducing *gate* leakage and a number of reliability concerns, based on the discussions of this section and Section 2.5.1.2, one can expect this mode of *real-space transfer* in a multiple-*channel* device to be capable of offering new dimensions in functionality.

[43] Z. Gribnikov, "Negative differential conductivity in a multilayer heterostructure," *Sov Phys Semicond*, vol. **6**, pp. 1204–1205, 1973. While in this section we will maintain a focus on *RST's* role in the operation of *VMTs*, in Chapter 5 more will be said on *RST* and its relevance to realization of *negative differential resistance* (which is the topic explored in the above publication) and high-frequency oscillation.

simple: *interchannel* transfer of *carriers* from the *channel* that they were originally traveling in into another *channel* with different *carrier-transport* properties. Employing *real-space transfer* in the operation of transistors has the possibility of improving their speed and functionality. From the point of view of improving the *switching speed*, as indicated earlier, *VMTs* have an advantage over traditional *FETs*. While in a *MOSFET* the *channel* is being filled and emptied by the longitudinal transfer of *carrier* from *source* to *drain*, in *VMTs* this process is materialized by the transversal transfer of *carriers* between a *high-velocity primary channel* and a *low-velocity secondary channel*. In this respect, the *VMT* has the ability to take advantage of *epitaxially* defined layers not only to materialize the *mobility* improvement in one of the *channels* (i.e., through offering the possibility of spatially separating *charge carriers* from *ionized impurities*[44]) but also to define the separation of the *channels epitaxially* and not *lithographically*.

As already noted in Section 4.2.1, *real-space transfer* can be either materialized through the movement of *hot carriers* out of a *quantum well* or via the potential-controlled transfer of *cool carriers* from one *potential well* into another. In the first method the *carrier* transfer from the *high-mobility* to the *low-mobility quantum-well channel* is often observed in the form of a *negative differential resistance* (i.e., *NDR*) device. In the second method a *VMT* can be realized. While a *VMT* aims at switching the transistor between the *on-* and *off-states* through the *RST* phenomenon, *NDR* devices realized through this concept aim at modulation of *carrier mobility* without turning the *conduction channel* off.

As shown in Figure 4.11, the two *channels* of a *VMT* are *shorted* to one another at the *source* and at the *drain* contacts. As a result, they are subject to the same *drain–source bias*. The *VMT* proposed by *Sakaki*, as indicated in this figure, has four terminals (*drain*, *source*, top gate, and bottom *gate*). In this structure, improvising the bottom *gate* in a *FET* imposes difficulties on the device fabrication process. As explained in Section 4.2.1, in this device the *drain* current is determined by the *bias state* of the two *gates* that operate in tandem to maintain a constant value of the total *channel* electron concentration.

With respect to the *switching time*, two components can be identified in a *VMT*: the *interchannel transit time* (τ_i) and the steady-state *settling time* (τ_s). Based on the benchmarks adopted in the design of *VMTs*, τ_i is usually about 0.2 ps while τ_s has been estimated to be as low as 1.3 ps.[45] Related to the *interchannel transit time*, it should be indicated that:

1. This process is a *conduction* event. In this transversal *charge-carrier* transfer, *carriers* are dumped from one *channel* to the other due to the existence of a large *gate*-induced electric field transverse to the *channel*. As a result, the *interchannel transit time* is not *capacitance limited* by the size of the *gate*.

[44] More will be said on this topic in Chapter 5, when presenting the operating principles of *high electron mobility transistors* (i.e., *HEMTs*).

[45] H. Sakaki, "Velocity Modulation Transistor (VMT)-A new field effect transistor concept," *Jpn J Appl Phys*, vol. **21**, no. 6, pp. L381–L383, June 1982.

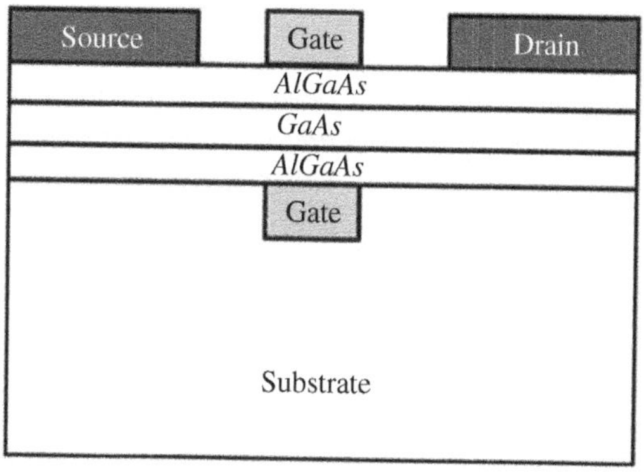

FIGURE 4.13 Cross-sectional view of the heterostructure of the dual-gate *AlGaAs/GaAs/AlGaAs VMT* of Figure 4.12.

2. Considering the short *interchannel* distance that one expects for the successful operation of a *VMT*, the transverse *charge-carrier transport* is either simply *ballistic* or it is defined by the non-steady-state peak of the *drift velocity*. As a result, the choice of materials with high temporal peak of *drift velocity* as *VMT channels* is quite enticing.

By taking 0 and 0.6 V between the two *gates* in a *VMT* realized in an $Al_xGa_{1-x}As$ system (i.e., like the one presented in Fig. 4.12), Cohen[46] estimated an *interchannel* electric-field strength of approximately 15 kV/cm. According to a semiclassical treatment, over a *ballistic* flight duration *t*, electrons move a distance given by

$$d = \frac{qE}{2m^*}t^2, \tag{4.14}$$

where *E* represents the transverse electric-field strength and m^* is the *effective mass* of electrons along the transverse direction to the *channel*. A small 0.2 ps flight time would be required for a travel distance of 79 nm in *GaAs*. This is the distance indicated for the *interchannel* separation in the work of Kizilyalli and Hess.[47] The aforementioned simple evaluation demonstrates an acceptable correlation to the finding of *Kizilyalli* and *Hess* where in their *Monte Carlo*-based assessment of *VMT* operation, they assumed an *AlGaAs/GaAs/AlGaAs double-heterojunction* structure with 100 nm thick *GaAs* layer, *gate length* of 0.4 μm, and *drain–source* separation of 1 μm. A schematic representation of this structure is provided in Figure 4.13. As indicated in Section 1.4, in *charge-carrier* flight over a small distance, such an assumption of *ballistic transport* is reasonable. Under these circumstances, electron–electron

[46] E. Cohen, Operation, development and application of the velocity modulated transistor, PhD dissertation, Purdue University, May 1998.

[47] L. Kizilyalli and K. Hess, "Ensemble Monte Carlo simulation of a velocity modulation field effect transistor VMT," *Jpn J Appl Phys*, vol. **26**, pp. 1519–1524, 1987.

scattering is fast enough to interfere (i.e., with a time constant of about 150 fs). However, these interactions conserve momentum and as a result do not contribute to velocity degradation. Considering other *scattering* mechanisms, it should be pointed out that *polar optical deformation potential scattering*[48] has a momentum *scattering time* of about 1 ps for Γ-*valley* electrons with speeds below about 4.2×10^7 cm/s. At greater speeds the *scattering time* is about 200 fs. However, since the electrons are at lower speeds for most of their flight through a *GaAs VMT channel* of the aforementioned dimensions, this will not pose a problem. Nonetheless, this situation will change if the *gate length* were to be made smaller.

As a result, among the two aforementioned *time constants*, τ_s constitutes the dominant time constant in determining the *switching time*. In the *Monte Carlo* study referred to in the work of *Cohen*,[49] in the evaluation of τ_s the device was rendered stable when the *drain* and *source* currents reached 90% of their steady-state values.

One important factor contributing to the relatively long *settling time* is the excess momentum normal to the *heterointerface*. Due to this excess perpendicular momentum in a *double-heterojunction VMT*, upon reaching the wider *bandgap AlGaAs barrier layer*, electrons will be prone to backward reflection. In this reflection, there is little speed reduction. As a result, these electrons will find themselves moving against the transverse electric field. This causes them to later fall back in *harmonic motion* unless they undergo *scattering* in the meanwhile. This excess momentum eventually decays due to the presence of *scattering* mechanisms such as *polar optical phonon* and *interface scattering*. Since the excess transverse momentum is defined proportional to the square root of the *interchannel* distance, reduction of the distance between *channels* can be also used to reduce this portion of τ_S.

It is worth indicating that τ_s is not a *time constant* defined in terms of a *gate*-defined capacitive event. As a result, the *switching time* of the intrinsic *VMT* (excluding the parasitic elements) is independent of *gate capacitance*. In conclusion to these observations, in the design of *VMTs* the importance of the *interchannel* distance in the operation of the device is not less than that of the *gate length*. The previously referred to *Monte Carlo*-based study calculated the *switching time* of the aforementioned *submicron-gate FET* in the absence of the two *channels* needed for *VMT* operation, while everything else was identical, to be about 3.5 ps. However, in the case of the *VMT*, the same study predicted *switching times* in the range of 1.5–3 ps. In this case, as indicated earlier, it was observed that electrons experience *non-steady-state velocity overshoot*.

Based on these observations, the operation of a *VMT* requires a number of conditions to be met. The following list identifies the four basic requirements for making of a high-speed *VMT*:

1. Existence of two parallel *channels* of widely different *carrier velocities*. It is worth reminding the reader once more that because of the lack of tangible

[48] See Section 1.7.3.7.1.

[49] E. Cohen, Operation, development and application of the velocity modulated transistor, PhD dissertation, Purdue University, May 1998.

difference between the *saturation velocities* of most semiconductors, the *channel* should be *biased* in low *drain–source* electric fields (or probably into the *velocity-overshoot* regime).

2. Existence of a very small *interchannel* distance.
3. Definition of two *gates* (i.e., a top and a bottom *gate*).
4. Biasing the two *gates* so that the sum of the *carrier* concentrations in the two *channels* remains constant.

4.2.2.1 Detecting Real-Space Transfer Verifying *RST* is obviously an important step toward guaranteeing *VMT* operation. The layer structure of the first single-*source*/single-*drain* transistor offering verifiable *real-space transfer* at room temperature is presented in Figure 4.14a.[50] The transistor realized on this layer structure is composed of two *channels*, with an *interchannel* separation of 70 nm. The ratio of the room-temperature electron *mobilities* of the two *channels* was reported as 2.5. This *mobility* ratio has been observed to be sufficiently high to provide current modulation in the form of a *VMT* at room temperature. These two n-type *channels* are defined above a p^+ layer envisioned to function as the bottom *gate* of the *VMT*.

The approximate *conduction band-edge* diagram of this layer structure is presented in Figure 4.14b. As shown in this diagram, a *2DEG* is only formed at the narrow-*bandgap* side of the top *AlGaAs/GaAs heterointerface*, not at the bottom one. In the implementation of this design, lowering the temperature during crystal growth causes adverse effects on *carrier mobility*. This method had been employed for lowering the *mobility* of the lower *channel*.

On the layer structure presented in Figure 4.14, Cohen et al. built different types of devices for detection of *DC real-space transfer*. The schematic structure of one of these devices, with three top *gates*, is provided in Figure 4.15. In this device, the top *channel* takes part in *screening* the bottom *channel* from the effects of the top *gates*. As a result, the top *gates* will *deplete* the *2DEG* of this *channel* before they get to deplete the bottom *channel*. In the *bias* scheme explored for detecting *RST* in this structure, first *gate* 1 (which is the *gate* closest to the *source*) is negatively *biased* with respect to the *source*. As a result, with increasing this negative *bias*, the top *channel* beneath *gate* 1 will be entirely *depleted* and the bottom *channel* will still be populated.

Accordingly, in the absence of *real-space electron transfer* into this *channel*, the *2DEG* that was *heteroepitaxially* built into the top *channel* would cease to deliver current to the *drain* terminal (i.e., even when $V_{DS} \neq 0$). This is due to the fact that under this *bias* condition, the *2DEG channel* is broken under *gate* 1. Obviously, this is correct after a brief transit period, which is important to be considered for a *gate length* longer than a *micron*. As a result of this *zero*-current situation, the *2DEG channel* potential beyond the *depleted gate* 1 would be equal to the *drain* potential. However,

[50] See note 49.

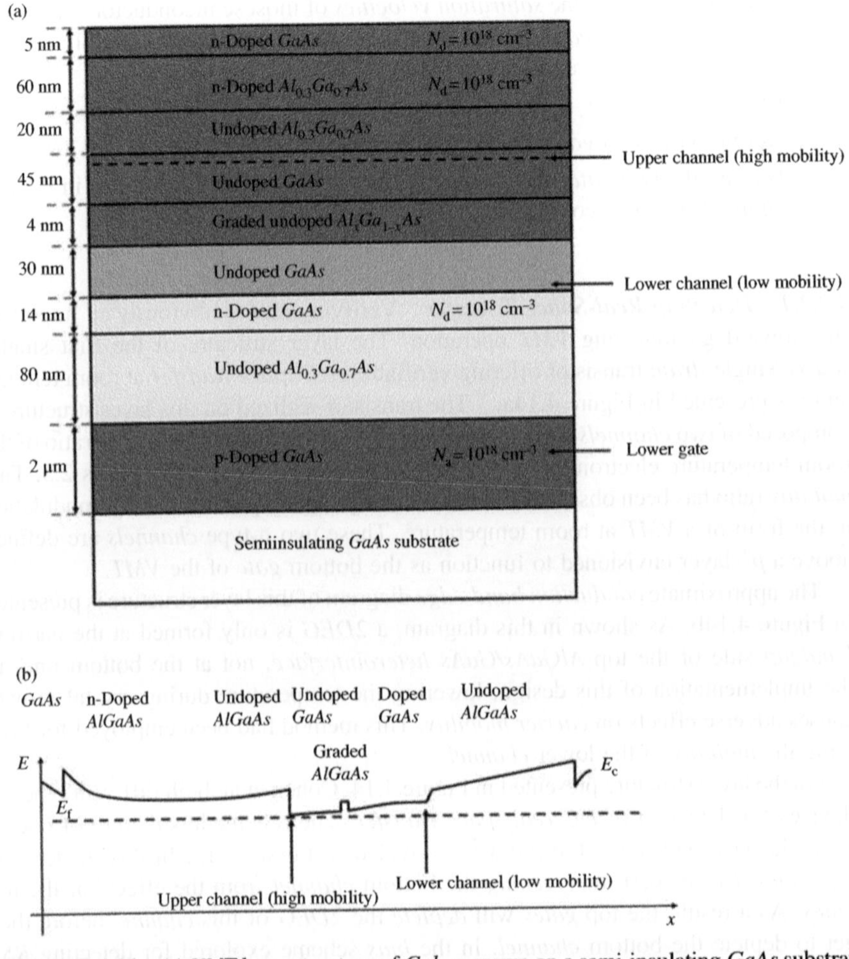

FIGURE 4.14 (a) *VMT* layer structure of *Cohen* grown on a semi-insulating *GaAs* substrate. Layers indicated by a darker shade of gray were grown at growth temperature of 600°C, while white and lighter backgrounds indicate 550°C growth temperature and growth under temperature ramp-up, respectively. Indicated thicknesses of the layers are not in scale. (b) Approximate conduction-band bending of the layer structure depicted in (a). Reproduced with permission from Cohen et al. (1997). Copyright 1997, AIP Publishing LLC.

given a finite *drain–source bias*, along the bottom *channel* the potential changes more monotonically.

As a result of the difference among the spatial profiles of potential variation along the two *channels*, beyond *gate* 1 a tangible potential difference will form between the two *channels*. The presence of this potential difference over the 70 nm *interchannel* distance is the driving force for the *real-space carrier transfer*. To detect the *RST*, the potential of the middle *gate* (i.e., *gate* 2) is tied to the *source*. As a result, for small *drain–source biases*, the *carrier* populations of the two *channels* under *gate* 2 remain

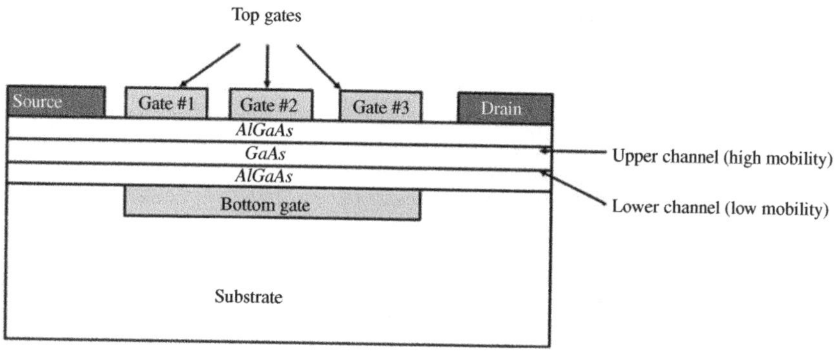

FIGURE 4.15 *RST* detection device of *Cohen.*

near their *thermal-equilibrium* values. In this structure, the presence of the *RST* can be verified by slowly sweeping the potential of *gate* 3 toward small negative values. Due to *screening*, the small V_{g3} *bias* will only modulate the *2DEG* (while the lower *channel* is left relatively unaffected). In the presence of *real-space transfer* of electrons into the *2DEG channel* and hence conduction through the top *channel*, *drain* current modulation by V_{g3} will be observed. However, in its absence a modulation in the *drain* current will be only observed when V_{g3} takes on more negative values (i.e., when affecting the bottom-*channel*'s population).

In the previously presented *biasing* scheme, *RST* is observed only when qV_{DS} is smaller than the potential-barrier height between the two *channels*. At higher V_{DS} values, since the *channel* confinement will be overwhelmed by the presence of more energetic electrons and larger *band bending*, differentiating the signatures of *RST* through this structure will not be possible.

4.2.2.2 Technological Challenges and Limitations of* VMT *Design: A Brief Summary On the layer structure of Figure 4.14 and in the *RST* detection device depicted in Figure 4.15, Cohen et al. implemented the bottom *gate* through the p^+-*doped* layer. In detecting the *DC RST*, the potential of this *gate* was left floating. The same structure, although without multiple top *gates*, is used in the implementation of a *VMT*. The bottom *gate* is formed by *metallization* of the *etched* shoulder of the so-called *mesa* structure[51] defined on the *epilayer* of Figure 4.14. For a *VMT*, this structure is shown in Figure 4.16. Of course, usage of a p^+ layer, although avoiding many of the complications present in realization of the metallic bottom *gate*, complicates the

[51] In *microfabrication* of semiconductor devices and circuits on *heteroepitaxially* formed *channels*, since the *channel* is formed continuously throughout the *wafer*, securing electrical isolation between neighboring devices built on a *wafer* is only possible if the *channel* is made discontinuous in areas surrounding each device. One way to realize this is through *etching* and physical removal of parts of the layer structure (hence the *channel*) from the areas surrounding each device. The *unetched* areas over which the active regions of individual devices are realized, because of their shapes, are known as *mesas*. As will be seen in Chapter 5, this isolation technology, which is known as *mesa isolation*, is the traditional mode of electrical isolation used among *III–V HFET* technologies.

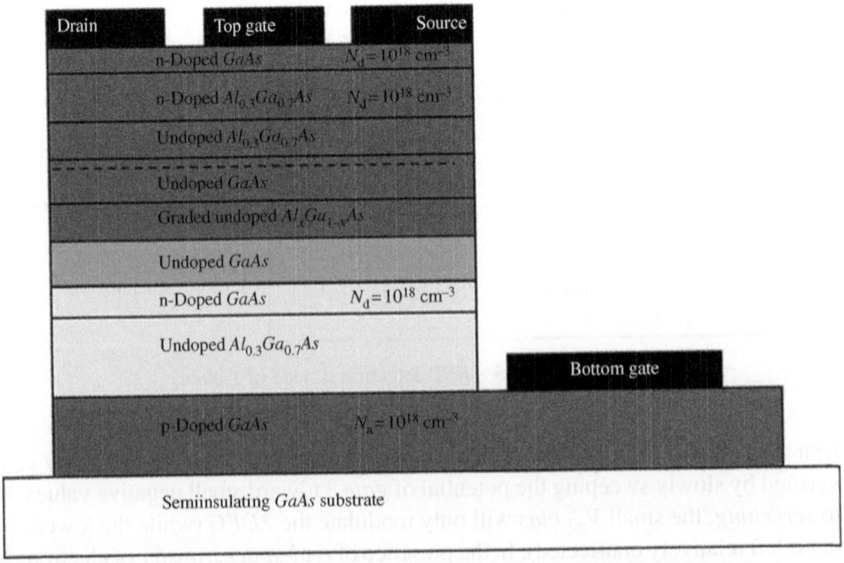

FIGURE 4.16 Mesa structure of *Cohen*'s *VMT* on the layer structure of Figure 4.14.

operation of the back *gate* by increasing the *depletion capacitance* and limiting the frequency response.

As indicated earlier, the proper operation of a *VMT* relies on a *biasing* scheme that guarantees the total number of majority *carriers* among the two *channels* to remain constant at all times. As a result, in order to determine the proper operating conditions of these transistors, the *carrier* concentration versus depth profile, on a device like the one depicted in Figure 4.16, must be calculated under simultaneous *biasing* of the two *gates*. Addressing this need, the in-depth discussion on the formation of *2DEG channels* and their *carrier* population at *III–V heterojunctions* provides the foundation for quantitative assessment of the operation of *VMTs*. This is presented in Chapter 5.

As mentioned in Section 4.2.2.1, the implementation of the lower *low-mobility channel* in the *VMT* structure of Figure 4.16 is done by lowering the crystal growth temperature during the growth of this lower *channel*. The *arsenic precipitate* formation during growth at 200–250°C with 1–2% excess *As* and suspension of the *As precipitates* in the regular *GaAs matrix* are counted responsible for the lower *carrier mobility* at the bottom *heterointerface*. These *precipitates* have been reported to act as buried *Schottky surfaces* (i.e., complete with the associated *surface states*).[52] While this phenomenon can certainly secure a lower electron *mobility* for the bottom *channel*, it can at the same time impact the crystalline perfection of the upper *hetero-interface* and degrade the top *channel*'s electron *mobility*.

[52] See A. C. Warren, J. M. Woodall, P. D. Kirchner, X. Yin, F. Pollak, M. R. Melloch, N. Otsuka, and K. Mahalingam, "Role of excess As in low-temperature grown GaAs," *Phys Rev B*, vol. **46**, pp. 4617–4620, 1992.

Controllability of *mobility* and *charge-carrier* concentration through variation of *dopant* concentration and *anneal* conditions provide low-temperature-grown *GaAs* (i.e., *LT-GaAs*) with enough versatility for use in *RST* design. The *electron affinity* of *LT-GaAs* is the same as that of *GaAs*. As a result, the bidirectional *interchannel* transfer of electrons is possible.

As will be observed in Chapter 5, while in the so-called *polar AlGaN/GaN/AlGaN heterostructures* the top *channel* gets its *polar 2DEG* without the need for *doping* the *barrier AlGaN* layer, the bottom *heterointerface* requires *doping* to form a *channel*. Hence, this *channel* has a higher probability of imposing *Coulomb scattering* on *charge carriers*. Thus, as a way to implement a *VMT* without lowering the temperature during growth of the lower *channel*, this *polar* layer structure might be of interest.

4.3 RESONANT-GATE AND RESONANT-CHANNEL TRANSISTORS

Incorporating an element of mechanical *resonance* in the structure of a *field effect transistor* takes its roots from the *resonant-gate transistor* (i.e., *RGT*) of Nathanson et al.[53] Such a transistor is not only capable of *signal amplification* but also *signal mixing* and *filtering*. Figure 4.17 presents the schematic view of such a transistor. In this *RGT*, the cantilever structure of the *gate* electrode gives it the chance to *resonate* at its *natural frequency*. Of course, *resonance* will take place only when the

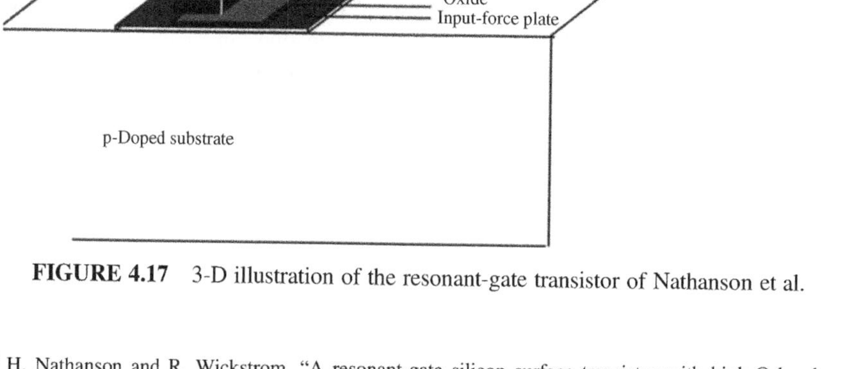

FIGURE 4.17 3-D illustration of the resonant-gate transistor of Nathanson et al.

[53] H. Nathanson and R. Wickstrom, "A resonant-gate silicon surface transistor with high-Q bandpass properties," *Appl Phys Lett*, vol. **7**, pp. 84–86, August 1965; H. Nathanson, W. Newell, R. Wickstrom, and J. Davis, "The resonant gate transistor," *IEEE Trans Electron Devices*, vol. **14**, no. 3, pp. 117–133, March 1967.

signal applied across the air gap shares its frequency with the *natural frequency* of the cantilever *gate*. Nathanson et al. described their *resonant-gate transistor* in terms of an *electrostatically excitable tuning fork with the field effect transistor readout*. In their reports dating back to 1965, they showed experimental results on *RGTs* with *gold vibrating beams* that operated in the frequency range of 1–100 kHz. In their work, a 5 kHz *RGT* was about 0.1 mm long, offering *resonance quality factors* (*Q factor*) as high as 500 and overall input–output voltage gain of about 10 dB.

Based on Figure 4.17, it can be pointed out that the *resonant-gate transistor* of Nathanson et al. is composed of three essential elements:

1. An electrostatic *input actuator* for converting the input electrical *signal* into a mechanical force.
2. A mechanical *resonator* (i.e., a cantilever) for realization of high *quality factor resonance*. This *resonator* should be sufficiently isolated from its surroundings in order to obtain the required high *Q factor*.[54]
3. An *output transducer* in the form of a *field effect transistor* for sensing the motion of the *mechanical resonator* and converting it to an *electrical signal*.

The high *quality factor* achievable through mechanical *resonance* has been pursued as an enticing avenue for *on-chip* incorporation of *resonating elements* needed for realization of *filters*, *mixers*, and *oscillators* of acceptable *phase noise* in *transceiver*[55] circuits. While in these circuits high *quality-factor* frequency-selective elements are traditionally incorporated as *off-chip* elements,[56] their *on-chip* cohabitation with the transistors can pave the way for a tremendous reduction in size (and even the power consumption) of the circuit.

As early as the 1960s, one of the best-known challenges to improving the functionality of telecommunication systems has been the realization of high *quality-factor* tuning components in *integrated circuit* technologies. Solutions such as incorporating *RGTs* hence seem promising. However, a major concern about the *on-chip* realization of high *quality-factor resonators* is their ultrasensitivity to *process variations*. After many decades since *Nathanson's* pioneering invention, this concern still lingers around and requires careful attention. As an example, one can see that given a *Q factor* of 1000 (which means a *passband* equal to onetenth of 1% of the *resonance*

[54] Issues such as *damping* and also *particle loading* on the *resonator* have negative impacts on the operation of these *vibrating elements* and their *Q factor*. As a result, these elements should be kept under low pressure. For longer *resonators*, which are needed for *resonance* at lower frequencies, maintaining the isolation from the mechanical movements in the environment is quite challenging. As will be seen in this section, requiring shorter length in high-frequency *resonators* makes achieving isolation less cumbersome. Whereas at the time of Nathanson et al., technological sophistication of the day only allowed the realization of *resonance* at about tens to hundreds of kilohertz, major strides made over the past few decades in semiconductor manufacturing technology have made the realization of *on-chip resonating elements* serving gigahertz applications of telecommunication systems possible.

[55] That is, *transmitter/receiver*.

[56] Such as *surface acoustic wave* (i.e., *SAW*) *filters* and *crystal oscillator tanks*.

frequency), a slight variation in the order of less than 1% in any of the parameters involved in the definition of the *resonance frequency* (such as the length of the *resonator*) results in shifting the desired *signal* completely out of the *passband*. As will be shown in Section 4.3.1, boosting the *resonance frequency* requires *resonating elements* of even smaller dimensions. With an increase in frequency the sensitivity problem will be further aggravated. It is expected that this *tolerance* problem will be worsened for a case in which an *active* element is also present in the *resonance feedback loop* (such as in the *RGT*). This is due to the sensitivity of the *Q factor* to component variations. In these cases, a limited percentage of variation in a component often results in *Q times* bigger variations in the *Q factor*.[57]

Figure 4.17 depicts a *field effect transistor* with a *singly clamped resonating element* (i.e., a cantilever). A *resonant-gate transistor* with a *doubly clamped gate* electrode can be equally well imagined. The *natural frequency* of the *resonating element* in these structures is predominantly determined by the *length* and the *Young's modulus*[58] of the *resonating element* itself.

As presented in Figure 4.17, Nathanson et al. envisioned the implementation of their *resonant-gate transistor* in the form of an overhanging *gate* electrode above the surface of the semiconductor with air taking the place of the *gate* insulator. The idea of frequency selection in a *FET* architecture can nowadays be also implemented with the use of a semiconducting *nanotubes* (or depending on its construction *nanowire*) as the *channel* (hence realizing a *resonant-channel FET*, instead of a *resonant-gate FET*). As an example, a *nanotube/nanowire channel* can be anchored at raised *drain* and *source* terminals where an *Ω-gate* is surrounding the *nanotube/nanowire channel* with an air gap implemented with the use of a *sacrificial material*.[59] Figure 4.18 depicts a possible device structure for a *nanotube/nanowire resonant-channel FET*.

One of the advantages of such an implementation is in the definition of the size of the *resonating element* (in this case the *channel*), which is nowadays readily realizable in *nano*scales, without directly needing *lithography* or *etching*. As will be pointed out in Section 4.3.1, this size reduction can possibly boost the frequency of operation of the *resonator*. Nonetheless, the formidable challenge of aligning the *nanotube/nanowire* with the rest of the transistor structure lies with this technology. This is not unlike the similar challenge in realization of all other *nanotube/nanowire FETs* (to be discussed in Section 4.4). As indicated in Section 4.1, the *gate-all-around* nature

[57] W. Newell, "Selectivity and sensitivity in functional blocks," *Proceedings of IRE*, vol. **50**, p. 2517, December 1962.

[58] Named after *Thomas Young*.

[59] A *sacrificial layer* is a layer that is usually removed through *wet etching* toward the end steps of the *microfabrication*process. Removal of the *sacrificial layer* provides a chance for creating air gaps, which, for example, are needed for producing the possibility of mechanical movement on a *chip*. This layer is not a structural layer. The removal step of the *sacrificial layer*, due to its role in allowing mechanical movements, is known as *release*.

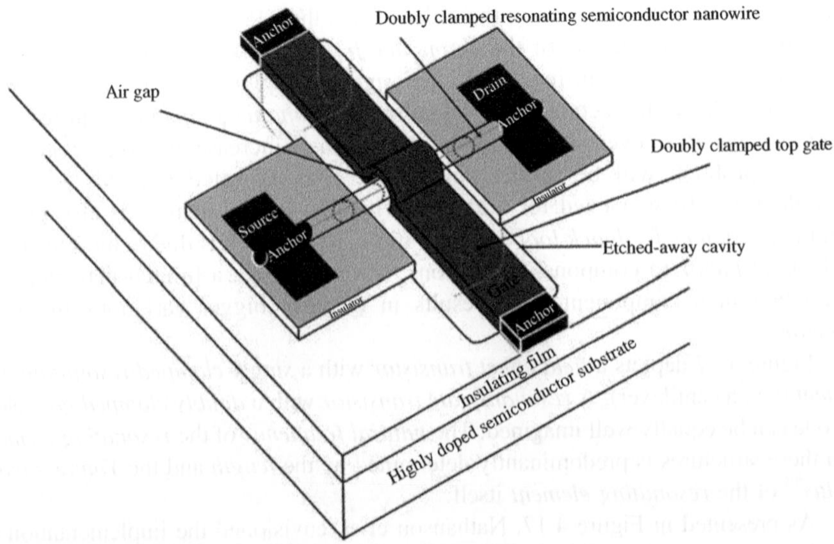

FIGURE 4.18 3-D illustration of a resonating-channel transistor. Employing the highly doped substrate as the second gate can be envisioned to offer the possibility of resonating the doubly clamped all-around top gate.

of a *nanotube* or *nanowire* implementation is another advantage of this form of a *resonating FET* architecture.[60]

4.3.1 Resonant-Gate Transistor: Principles of Operation

The main difference among the principles of operation of traditional *field effect transistors* (like the ones expressed in Chapter 3) and the *resonant-gate transistor* is the presence of a *resonating element* in the structure of the latter. In order to understand the principles of operation of *RGTs*, it is crucial to add to the already presented principles of operation of *FETs* some information on the operation of the *resonating element*. In this section, we will present this information in terms of a simplified *lumped mass-spring* model. A more elaborate *distributed* model for a *doubly clamped resonating beam* is presented by Bannon et al.[61] For the sake of brevity, while briefly presenting the structure of the *resonators* reported in this work, we refrain from fully presenting this model. The interested reader is strongly encouraged to explore this

[60] One can speculate on mixing the two ideas of *resonant-gate* and *resonant-channel FET* as an interesting device concept. This is since such a structure can have applications as a *filter mixer*. Of course, in *mixing* the two ideas the *natural resonance* of the *resonant gate*, depending on its shape, can take different forms (e.g., in the form of a circular *resonating* structure or a *beam resonating* structure). Realizing the form of the circle for predicting the *resonance*, however, might be deemed a *microfabrication* challenge. Also, the small overlap between the *nanotube* and the *gate* might pose another challenge on the operation of such an element.

[61] F. Bannon, J. Clark, and C. Nguyen, "High-Q HF microelectromechanical filters," *IEEE J Solid State Circuits*, vol. **35**, no. 4, pp. 512–526, 2000.

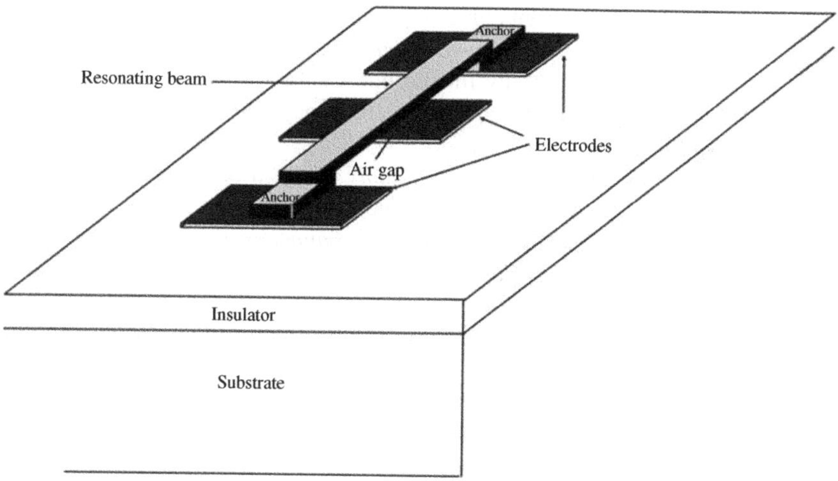

FIGURE 4.19 3-D illustration of a doubly clamped beam resonator.

model in depth. In the case of *doubly clamped beams*, the presentation of Bannon et al. on *resonating beams* provides an interesting perspective. This presentation not only covers the basics of *resonance*, but even beyond that goal it deals with a number of important secondary effects such as *beam softening* and *beam hardening* (which, considering the issue of *tolerance*, require attention in the design of elements such as *RGTs*).

This pioneering work, in the area of *doubly clamped polysilicon MEMS*[62] *resonators* for achieving a high *quality-factor resonating-element on-chip*, builds on principles quite similar to *Nathanson*'s *RGT* proposal. However, instead of using a cantilever metallic-*gate resonator* built into an *active* element (i.e., a *FET*), this work takes advantage of a *resonating* semiconductor *beam* in the form of an essentially two-terminal *passive* element. Figure 4.19 presents the schematic view of a *doubly clamped resonator*. In this structure, since the *beam* is *highly doped*, the electrodes envisioned at its two *anchors* cannot be *biased* independently. Consequently, no *DC* current will pass along the *beam*. In a device like this, the impact of the *DC* voltage difference between the *polarizing* electrode and the *beam* on the *beam*'s *conductivity* (if at all present) is a parasitic effect. This parasitic effect is addressed in the form of the partial *depletion* of the *highly doped polysilicon beam* (i.e., due to the application of a *polarization voltage* between the *polarizing* electrode and the *beam*).[63] On these *resonators*, application of an input *AC signal* to the *polarizing* electrode, instead of charge modulation in the traditional sense of a *FET*, only results in *beam resonance* (if the *signal's* frequency matches the *beam's natural frequency*). This *resonance* induces a time dependence in the *beam*-to-electrode capacitance. As a result of this

[62] Acronym for *microelectromechanical systems*.

[63] We will address the role of *polarization voltage* in the operation of these circuit elements shortly in this section.

vibration, a *small-signal* current at the *resonance frequency* will be detectable in the *AC* current fed by the source of the *polarization voltage* (i.e., V_P) while other frequencies fed to the *gate* are essentially filtered out. Caused by the absence of *transistor action*, these devices are only capable of frequency selection (i.e., *filtering*) and not *signal amplification*.

Whereas the implementation of this *doubly clamped beam resonator* is deeply anchored in the use of a *highly doped* conductor-like *beam*, it is expected that using a *light-to-moderately doped* semiconducting *channel* in the form of a *resonant-channel FET* (as expressed in Fig. 4.18) might not only offer frequency selectivity but also a transistor-like *transconductance gain* at the *resonance frequency*.[64]

As mentioned earlier, the enticingly high *quality factor* of mechanical *resonance* among these *passive* and *active* circuit elements is their most valuable asset. Although the *quality factors* of the *doubly clamped beam resonators* are in the order of *thousands*, detection of their *output signal* is a challenge. This is due to the high *output impedance* of these devices (especially if designed for very high frequencies) and the resulting difficulty in *impedance matching*.[65] As a result of poor *matching*, often the amplitude of the detected *AC signal* is very limited. This makes the immediate *signal amplification* a major requirement for adopting these devices into telecommunication systems. While a *signal* amplifier in the form of a *FET* placed in tandem with the *resonator* (i.e., to *amplify* the frequency-selected *signal* of the *resonator*) can also fulfill the same task, the proposed merger of the tasks of *filtering* and *amplification* in one *depletion-mode resonant-channel FET* is speculated to have the possibility of reducing the aforementioned burden of *signal amplification*. Such a proposed *resonant-channel FET* ceases to *resonate* when the semiconducting *channel* is *depleted*. This is because, as will be observed shortly, two conductors are required for *biasing* the *resonator* (i.e., in the *polarization* process). This device, as depicted in Figure 4.18, can be envisioned in the form of a *depletion-mode nanotube/nanowire FET*.

As illustrated in Figure 4.17, in the *RGT* structure of Nathanson et al., in a place toward the end of the cantilever *resonating beam*, an insulated input-force plate is devised. However, in the case of a *doubly clamped beam resonator*, based on the requirement for efficient mechanical actuation, the input-force plate must be envisioned in the middle of the *beam*.[66] Application of a potential difference between the *resonating element* and this force plate results in exertion of an electrostatic force on this element. As suggested earlier in this section, this element will demonstrate an appreciable vibration only when excited by an electrostatic force at its mechanical

[64] While we will leave deeper discussions of this proposal out of this volume, it should be pointed out that in light of the discussions of this section on the principles of operation of *RGTs*, it is expected that especially for *deeply scaled FETs* such an implement will suffer from severe nonlinearity. This is since a nonuniform *channel* potential seems to play a strong role in determining the *resonance frequency*. In *long-channel FETs*, where the *gradual-channel approximation* is valid, since *channel* potential is only slowly varying, this challenge seems to be much less imposing.

[65] For understanding the roots of this high *output impedance*, see F. Bannon, J. Clark, and C. Nguyen, "High-Q HF microelectromechanical filters," *IEEE J Solid State Circuits*, vol. **35**, no. 4, pp. 512–526, 2000.

[66] That is determined in terms of position dependence of *beam* stiffness.

resonance-frequency. In a *RGT*, as a result of this vibration, a time variation will be imposed on the *channel's* electron concentration. This temporal variation of *charge-carrier* concentration is detectable in terms of the time variation of the *drain* current. In the operation of this device, the so-called *DC polarization voltage* controls both the force *signal* and the induced charge concentration of the *channel*. As will be demonstrated shortly, the voltage *gain* of a *RGT* at *resonance frequency* is proportional to the square of the *polarization voltage* (i.e., V_p^2).[67]

Figure 4.20 presents the aforementioned simplified *electromechanical* model of Nathanson et al. for the operation of the *RGT*. As shown in the figure, this simplified model represents the *resonator* in terms of the *lumped mass* and *spring constant* elements. As a first-order model, attention is not paid to the position dependence of the size of the air gap between the cantilever and the *silicon* surface. Whereas this deflection-caused spatial variation is part of the reality of the operation of *RGT*, this model takes into account only an *effective* value of the distance between the *gate* and the surface-defined *channel*. The model also fails to fully appreciate the impact of deflection on the *spring constant* of the *micromechanical resonator*. In a more accurate model (like the aforementioned distributed model of Bannon et al.), these effects, which are known as *beam softening* and *beam hardening*, are incorporated.[68]

In Figure 4.20b, the *small-signal* equivalent circuit to the *mass-spring* mechanical circuit of Figure 4.20a is presented. In this equivalent circuit, the *FET* is represented by its *small-signal* equivalent. In terms of this model, according to the *biases* applied to the electrodes indicated in Figure 4.20a and the structural parameters of the *RGT*, the electrostatic force exerted on the cantilever is

$$F = -\frac{1}{2}(V_P + v_{in}\sin(\omega t))^2 \frac{\partial C_{fb}}{\partial \delta_e} = \frac{1}{2}\left[V_P^2 + 2V_P v_{in}\sin(\omega t) + \frac{v_{in}^2}{2}(1-\cos(2\omega t))\right]\frac{\epsilon_0 A}{\delta_e^2}, \tag{4.15}$$

where C_{fb} is the capacitance between the input-force plate and the cantilever, A is the overlapping area of this parallel-plate capacitor, and δ_e is the equilibrium spacing between its two plates. V_P is called the *polarization voltage*, and $v_{in}\sin(\omega t)$ is the sinusoidal *input signal* of angular frequency ω, applied across C_{fb}. Since the dielectric of the capacitor is air, $\epsilon_r = 1$.

In the force equation of (4.15), the first term represents the constant component of force prompted by the *polarization voltage*. This component causes the dependence of δ_e on V_P, leading to a recursive nature in (4.15). This dependence is why V_P is referred to as the *polarization voltage*. The second term is the dominant time-varying component of the force *signal*. Under *small-signal* conditions of operation, the third and

[67] See Equation (4.19).

[68] Despite the aforementioned differences in operation of the *passive resonating-beam* element presented in Figure 4.19 and the *resonant-gate* or *resonant-channel FETs* depicted in Figures 4.17 and 4.18, respectively, since the principles of *resonance* are more readily explainable in the former simpler elements, from the perspective of the discussion of this section, they can well serve as educational tools.

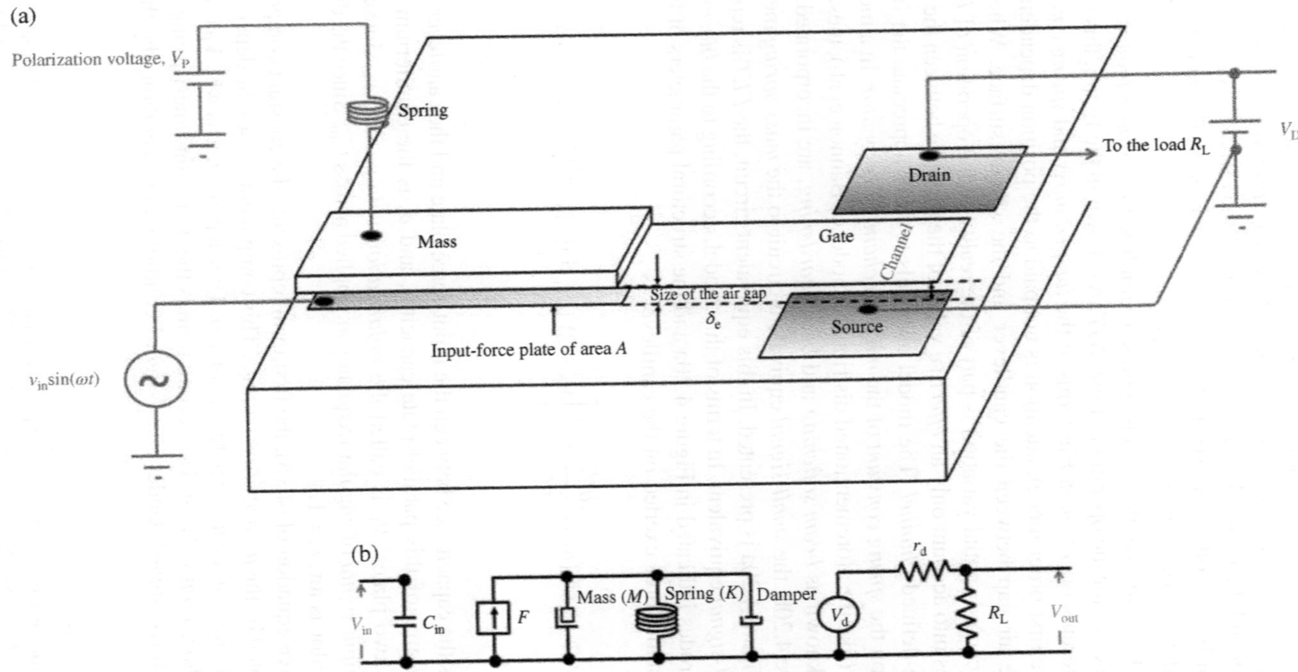

FIGURE 4.20 (a) Proposed electromechanical model of Nathanson et al. for the operation of the *RGT* of Figure 4.17. (b) Equivalent small-signal circuit diagram of (a). v_d is a function of the size of the air gap δ. Reprinted, with permission, from Nathanson et al. (1967). Copyright 1967, IEEE.

fourth terms have small impacts on the overall force. In terms of this approximation, the dominant time-dependent component of force is given by

$$F_\omega = F\sin(\omega t) = V_P v_{in}\frac{\epsilon_0 A}{\delta_e^2}\sin(\omega t). \tag{4.16}$$

This sinusoidal force results in induction of a sinusoidal deflection on the cantilever. The *amplitude* and *phase* of this deflection are given by the standard *resonance equation*

$$\delta(\omega) = \frac{F/K}{1-(\omega^2/\omega_r^2)+j(1/Q)(\omega/\omega_r)}, \tag{4.17}$$

where K is the *dynamic spring constant* of the *resonating element*. ω_r, which is the *angular resonance frequency*, in terms of the *equivalent mass* of the *resonator* (M) is given by $\sqrt{K/M}$. Q is the *quality factor* of the *resonator*, which is calculated in terms of *resonator's damping constant* (D) by $K/\omega D$.[69]

Clearly, the *resonance frequency* increases with increasing the *beam*'s *dynamic spring constant* and reducing its *effective mass*. This observation dictates smaller dimensions for realizing higher-frequency *resonators*.[70]

As discussed earlier, the deflection expressed in (4.17) is then sensed by the surface *channel* of the *FET*. From the perspective of the operation of *FETs*, it can be expected that at a given constant value of *bias* the temporal variation in the *gate*-induced electric field, caused by this vibration of the *gate* in a *RGT*, will have an equivalent effect to that of a sinusoidal *signal* overriding the *gate bias* in a conventional *MOSFET*.

Therefore, according to this analogy, the amplitude of the *AC signal* of the *drain* can be formulated as

$$v_d = -A_v\delta_e\delta\frac{\partial}{\partial\delta_e}\left(\frac{V_P}{\delta_e}\right) = A_v\frac{\delta}{\delta_e}V_P, \tag{4.18}$$

where A_v is the voltage gain of the *clamped-gate* equivalent *FET* (which is determined according to the structural and *bias* conditions of the transistor) and δ is the sinusoidal *resonator* deflection given in the frequency domain by (4.17). In Equation (4.18), V_P/δ_e is the *gate*-induced electric field, transverse to the *channel*, at the equilibrium position.

According to Figure 4.20b, the overall voltage *transfer function* at the *resonance-frequency*, with the assumption of *matched resistive load*, is given by

[69] Also, $j=\sqrt{-1}$.

[70] In a *doubly clamped resonator*, the *resonance frequency* is determined in proportionality to $\sqrt{E/\rho}\times(t/L^2)$, where E is the *Young's modulus*, ρ is the mass density, and t and L are the thickness and the length of the *beam*, respectively.

$$\left(\frac{v_{\text{out}}}{v_{\text{in}}}\right)_{\text{r}} = -jQV_{\text{P}}^{2}\frac{A_{\text{v}}R_{\text{L}}}{r_{\text{d}}+R_{\text{L}}}\frac{\epsilon_0 A}{K\delta_{\text{e}}^{3}} = -j\frac{QV_{\text{P}}^{2}A_{\text{v}}}{2}\frac{\epsilon_0 A}{K\delta_{\text{e}}^{3}}. \tag{4.19}$$

At frequencies away from *resonance*, the *transfer function* decays in a fashion similar to that of a conventional *LC tank*.

Another point that we have to make is that the equilibrium position of the *resonator* can be calculated by equating the constant component of the electrostatic force, calculated in (4.15), and the counterbalancing restoring force of the spring

$$\frac{1}{2}\frac{\epsilon_0 A V_{\text{P}}^{2}}{\delta_{\text{e}}^{2}} = K_0(\delta_0-\delta_{\text{e}}), \tag{4.20}$$

where δ_0 is the size of the air gap at V_{P} of *zero volt* and K_0 is the *mechanical spring constant*[71] of the *resonator*.

Equation (4.20) yields δ_{e} as a function of V_{P}. However, care must be taken that beyond a maximum value of V_{P} the restoring force of the spring fails to maintain the equilibrium. Hence, beyond this limit the two plates of the capacitor will be *shorted* together. This value of V_{P} is referred to as the *pull-in voltage*, represented by V_{PI}. Setting the derivative of V_{P} versus δ_{e} equal to 0 yields $\delta_{\text{e}} = (2/3)\delta_0$. According to which, the *pull-in voltage* is given by

$$V_{\text{PI}} = \sqrt{\frac{8}{27}\frac{K_0\delta_0^{3}}{\epsilon_0 A}}. \tag{4.21}$$

Beyond this *polarization-voltage* size of the air gap is immediately reduced to 0. In terms of this *pull-in voltage*, the normalized value of V_{P} is given by

$$\frac{V_{\text{P}}}{V_{\text{PI}}} = \frac{\delta_{\text{e}}}{\delta_0}\sqrt{\frac{27}{4}\left(1-\frac{\delta_{\text{e}}}{\delta_0}\right)}. \tag{4.22}$$

Under normal operating conditions, the *normalized* V_{P} has a value in the vicinity of 0.5, and $(\delta_{\text{e}}/\delta_0) \cong 1$. Under these conditions, Equation (4.22) is approximated by

$$\frac{V_{\text{P}}}{V_{\text{PI}}} \cong \sqrt{\frac{27}{4}\left(1-\frac{\delta_{\text{e}}}{\delta_0}\right)}. \tag{4.23}$$

As mentioned earlier, an important highlight in the interaction between the electrostatic force and the restoring force of the spring is the dependence of the *dynamic*

[71] As will be seen shortly, this is different from the *dynamic spring constant* (i.e., *K*).

spring constant (K) on V_P.[72] This dependence is clarified through differentiating the net force versus δ_e:

$$K = \frac{\partial}{\partial \delta_e}\left[\frac{1}{2}\frac{\epsilon_0 A V_P^2}{\delta_e^2} - K_0(\delta_0 - \delta_e)\right] = -\frac{\epsilon_0 A V_P^2}{\delta_e^3} + K_0 = K_0\left(3 - 2\frac{\delta_0}{\delta_e}\right). \tag{4.24}$$

The change in the spring constant obviously results in a shift in the *resonance frequency* (ω_r) away from the mechanical *resonance frequency*, ω_0. According to (4.23) and (4.24),

$$\frac{\omega_r}{\omega_0} = \frac{\sqrt{K/M}}{\sqrt{K_0/M}} = \sqrt{3 - 2\frac{\delta_0}{\delta_e}} \cong 1 - \frac{4}{27}\left(\frac{V_P}{V_{PI}}\right)^2. \tag{4.25}$$

4.3.2 Resonant-Channel Transistor: Principles of Operation

In this section, a summary of one of the few existing reports on realization of *resonant-channel FETs* is provided.

Over the recent years, the realization of *carbon nanotubes* has stirred up interest in their utilization in a variety of different applications including the *channel* of a *FET* (used either as a chemical sensor or an electrical switch) or the suspended body of a mechanical *resonator*.[73] The possibility of precise, and yet *etching*-less, definition of these small conducting and semiconducting *channels* and their superb *transport* characteristics have been the major contributors to this tremendous interest. Whereas achieving the frequency extension of *doubly clamped resonators*, such as those mentioned in Section 4.3.1, to the gigahertz range requires precise definition of very small dimensions through *etching*, this frequency extension can be achieved easily via employing the *etching*-independent definition of the small geometry of a *carbon nanotube* (or, for that reason, other lesser-known *nanotubes* and *nanowires*). As indicated in Section 4.3.1, small *tolerance* in the definition of the dimensions of the *high-Q resonators*, especially at higher frequencies, leaves an insupportable strain on the *etching* techniques normally needed for realization of the required geometry. As a result, removing the need for *etching* in realization of these dimensions tremendously reduces the *micro-/nanofabrication* challenge.

In a *resonator* application (i.e., in the form of a *resonant-channel transistor*), a suspended *carbon nanotube* can be used in the form of a *frequency mixer* for detection of mechanical vibration. In the work of Witkamp et al., a single *gate*-dependent *resonance* has been observed whose operation is attributed to the fundamental mode of

[72] That is *spring softening*.

[73] We will deal with *nanotube* and *nanowire FETs* in Section 4.4; however, at this point in our discussion, although slightly ahead of schedule, it is deemed suitable to present an example of their application in the form of the merger of a *FET* and a *resonator*. This discussion will be carried out purely from a structural point of view.

vibration of the suspended *nanotube*.[74] Just like the *MEMS resonators*, spoken of in Section 4.3.1, in these *nanoelectromechanical resonators* operation is based on the impact of mechanical motion on changing the electrical properties of the *nanotube channel*, and vice versa. It should be mentioned that *NEMS*[75] *resonators*, in addition to the aforementioned indication of frequency improvement, also promise an improvement in power consumption. Earlier in the development of *carbon nanotubes*, these *resonators*, instead of being used in the form of an electronic device, were being explored for studying the fundamental limitations of mechanical motion. The low mass, very high *Young's modulus* (i.e., 1.25 TPa), and high *scalability* with length of the *single-walled carbon nanotubes* (*SWCNTs*) have made them an ideal choice for applications in *nanoelectromechanical systems*.[76] Room-temperature indications of *thermally excited bending-mode* vibration of *single* and *doubly clamped SWCNTs* have been also reported.

Witkamp et al. reported on the observation of a single *gate*-dependent *resonance* in a number of *resonant-channel transistors* demonstrating *resonance frequencies* identifiable by the calculated *bending-mode resonance frequencies* of the *nanotubes* in the range of 20–100 MHz. These showed *quality factors* in the range of 6–300. Figure 4.21 presents a schematic depiction of these devices. These suspended *nanotube* devices were realized on a *highly doped p-type silicon* substrate with a 500 nm thick insulating SiO_2 layer covering its surface. The *highly doped silicon* substrate served as the *gate* electrode used in setting the *nanotube* in motion and tuning its *oscillation frequency*. In this work, *carbon nanotubes* were grown in a *chemical vapor deposition* (i.e., *CVD*) setup optimized for growing *SWCNT*.[77] *Atomic force microscopy* (i.e., *AFM*) was then used for identifying the location of the *nanotube*. Subsequently, *Au/Cr drain* and *source* electrodes were *deposited* and *patterned* on the two ends of the *nanotube* using an *electron-beam lithography* system. A *release* process was then followed to remove the *sacrificial* SiO_2 from beneath the *nanotube* while using the *source* and *drain* electrodes as protective *masks* for part of the oxide to which the *nanotube* is supposed to remain anchored.

As in the case of *MEMS resonators*, application of an *AC gate signal* v_g serves to actuate the suspended *nanotube*. The driving force of this actuation is induced through the displacement dependence of the *gate capacitance* C_G. Application of this *AC signal*, if frequency matched to the *natural frequency* of the *doubly clamped nanotube*, results in induction of an oscillatory movement in the *nanotube*. For a case where the displacement amplitude averaged along the *nanotube* is *nonzero*, the *gate capacitance* will vary with time (i.e., $C_G^{mech}(t)$). This results in *gate* modulation of the *channel's* charge concentration (and hence modulation of the conductance of the semiconducting *channel*):

[74] B. Witkamp, M. Poot, and H. Van der Zant, "Bending-mode vibration of a suspended nanotube resonator," *Nano Lett*, vol. **6**, no. 12, pp. 2904–2908, 2006.

[75] Acronym for *nanoelectromechanical systems*.

[76] See note 54.

[77] J. Kong, H. Soh, A. Cassell, C. Quate, and H. Dai, "Synthesis of individual single-walled carbon nanotubes on patterned silicon wafers," *Nature*, vol. **395**, no. 6705, pp. 878–881, 1998.

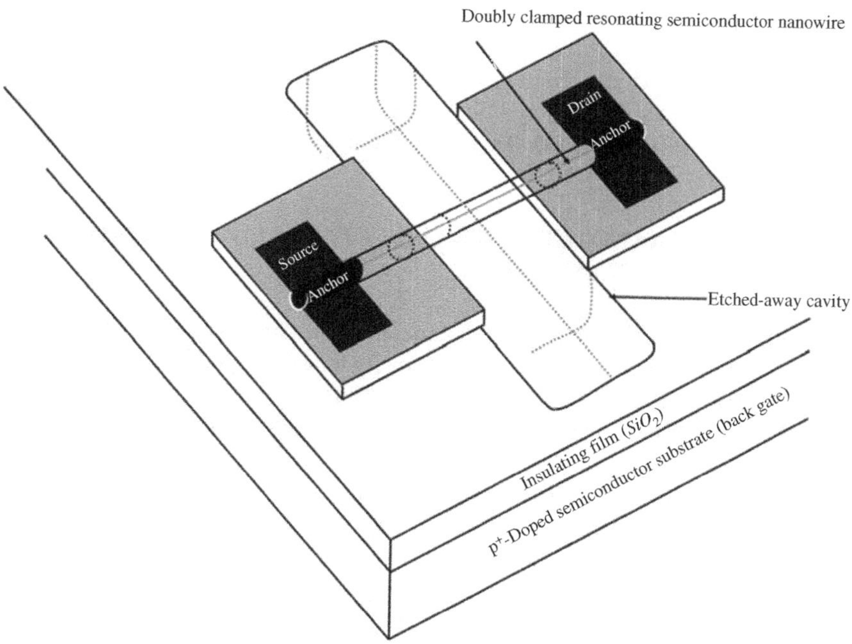

FIGURE 4.21 3-D schematic of the resonating *nanowire* channel transistor.

$$Q_{\mathrm{g}}^{\mathrm{mech}}(t) = C_{\mathrm{G}}^{\mathrm{mech}}(t) \times V_{\mathrm{G}}, \tag{4.26}$$

where V_{G} is the *DC gate bias*.[78]

In these devices, the mechanically contributed charge modulation expressed in (4.26) supplements the typical charge modulation of the *channel's* electron concentration by the *gate* voltage:

$$Q_{\mathrm{g}}^{\mathrm{direct}}(t) = v_{\mathrm{g}}(t) \times C_{\mathrm{G}}. \tag{4.27}$$

The second modulation term, which is presented in (4.27), is even present for the *off-resonance AC signals* applied to the *gate*. The motion of the *nanotube* can be detected by measuring the conductance of this *channel* and studying its frequency spectrum. In the work of Witkamp et al., the suspended length of the reported devices has been within the range of 1.1–1.5 μm with diameters measured by *AFM* to be in the range of 1–3 nm. Witkamp et al. also observed *resonance* at *harmonics* of *eigenfrequency* while applying *AC gate signals* of larger amplitudes (i.e., in a more nonlinear operation mode).

Whereas the structure reported in this work has a back *gate*, as suggested in Figure 4.18, an Ω-*gate* structure with the possibility of oscillation can be also

[78] This is obviously about the early *linear regime* of operation of the transistor.

envisioned in the form of the *resonant-gate/resonant-channel FET* with *mixing* functionality.

To avoid viscous gas *damping* and also *particle loading* on the *resonator*, the transistors presented in Section 4.3 are studied under low pressures (i.e., as low as 10^{-5} Torr for *nanoresonators*). This is an important limitation of these devices.

4.4 CARBON NANOTUBE *FET* AND *FETs* REALIZED ON OTHER NANOTUBE AND NANOWIRES

As will be observed in this section, unique properties of *single-walled carbon nanotubes* (such as *ballistic transport* with relatively long *mean free path*, superb compatibility with high-K dielectrics, and surface stability) have promoted them as excellent candidates for the *channel* of future high-performance *FETs*. It is particularly with regard to the question of *gate-length scaling* of *FETs* that *carbon nanotubes* (i.e., *CNTs*) have been proclaimed as one of the most promising contenders to take over *silicon* as the *channel* material.

CNT-FETs (also known as *CNFETs*) are ultrathin body devices. As a result, among these devices the severe *mobility* degradation typically observed in *deeply scaled silicon MOSFETs* of *nanometer* dimensions is not observed. Due to this property, even in *deeply scaled CNFETs* (i.e., of *channel lengths* in the order of a few *nanometers*), *long-channel* characteristics still prevail. This makes *nanotubes* extremely attractive choices for *channel-length scaling*. Another interesting property of *carbon nanotubes* is that they are not only realizable in semiconducting but also in metallic nature. This possibility is a consequence of the extreme sensitivity of the *band* structure of *carbon nanotubes* to the exact position of *carbon* atoms forming the *tube*. A *carbon nanotube* is formed by rolling up a finite sheet of *graphite*. Figure 4.22 illustrates the results of this roll-up in two of its famous forms, known as *arm chair* and *zigzag* (named based on the resemblance of the lineup of *C–C* covalent bonds to these shapes). The form of the lineup has tremendous impacts on the structure of the *dispersion diagram* of the *nanotube*. Both metallic and semiconducting *CNTs* (which are distinguishable in terms of the angle of the roll-up and the diameter of the *nanotube*) present very striking electrical, mechanical, and thermal properties.[79] As a result, these *nanotubes* have been explored not only as superb candidates for the *channel* of *deeply scaled FETs* but also in realization of low-resistivity *contacts* and *interconnects* on a *chip*.

Appenzeller et al. in their seminal 2002 paper[80] reported on the observation of room-temperature 1-D *transport* properties in *carbon nanotubes* of very small

[79] It is interesting to mention that through selecting the proper diameter, the *bandgap* of semiconducting *CNTs* can be tuned within a wide range of energy.

[80] J. Appenzeller, J. Knoch, R. Martel, V. Derycke, S. Wind, and P. Avouris, "Carbon nanotube electronics," *IEEE Trans Nanotechnol*, vol. **1**, no. 4, pp. 184–189, 2002.

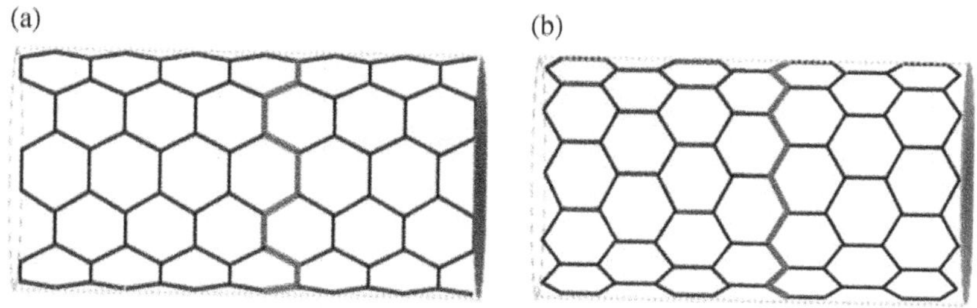

FIGURE 4.22 Side view of the lineup of the covalent bonds in (a) an armchair and (b) a *zigzag* rolled carbon nanotube. In each case a set of bonds are highlighted in gray to demonstrate the resemblance of these names to the shapes of the bond lineup.

diameters (i.e., 1.4 nm). As a result, in a device application context, these *nanotubes* can effectively offer suppression of *scattering* (i.e., due to the limited *phase space* available in the 1-D case), without needing to cool down the *lattice*.[81] In metallic *CNTs*, room-temperature *mean free paths* as long as *several microns* are reported. Consequently, these *nanotubes* can offer *conductivity eight times* higher than that of *copper*. This fact is the motivation for employing metallic *CNTs* in the mesh networks used in *chip interconnects*. The absence of *inelastic scattering* and the good *thermal conductivity* of *CNTs* promote current densities exceeding 10^9 A/cm^2.

Appenzeller et al. have also reported on the observation of *carrier transport* through more than one 1-D *subband* under certain *electric-field* conditions (i.e., for high V_{GS} in *CNFETs*).[82] In this observation, they used *Pt doping* of *nanotube* to make the impact of different *bands* visible.

Certainly with an eye for fulfilling its vast potentials, beyond a single-device operation, the *CNFET* technology needs to satisfy the demand of *chip*-level integration. Although even at the infancy of research on *CNTs* successful demonstrations (i.e., of *CNFET*-based *intermolecular inverters*) have proven these devices viable of *chip*-level integration,[83] this technology still faces an important challenge with *channel* alignment. Section 4.4.1.4, through presenting a simple *microfabrication* process of a *CNFET*, provides a brief coverage of this challenge.

Although not as exciting of a topic as *CNFETs*, other *nanotube* and *nanowire*[84] *FETs* have also managed to gain the attention of research community. In

[81] See the discussion in Section 1.7.3.8.

[82] J. Appenzeller, J. Knoch, M. Radosavljević, and P. Avouris, "Multimode transport Schottky-barrier carbon nanotube field effect transistor," *Phys Rev Lett*, vol. **92**, no. 22, pp. 226802-1–4, 2004.

[83] P. Avouris, "Carbon nanotube electronics," *Chem Phys*, vol. **281**, nos. 2–3, pp. 429–445, 2002; A. Javey, H. Kim, M. Brink, Q. Wang, A. Ural, J. Guo, P. McIntyre, P. McEuen, M. Lundstrom, and H. Dai, "High-kappa dielectrics for advanced carbon-nanotube transistors and logic gates," *Nat Mater*, vol. **1**, no. 4, pp. 241–246, 2002.

[84] The structural difference between a *nanowire* and a *nanotube* is that while the former ideally has a filled-cylindrical form, the structure of the latter, although ideally cylindrical, is tubular.

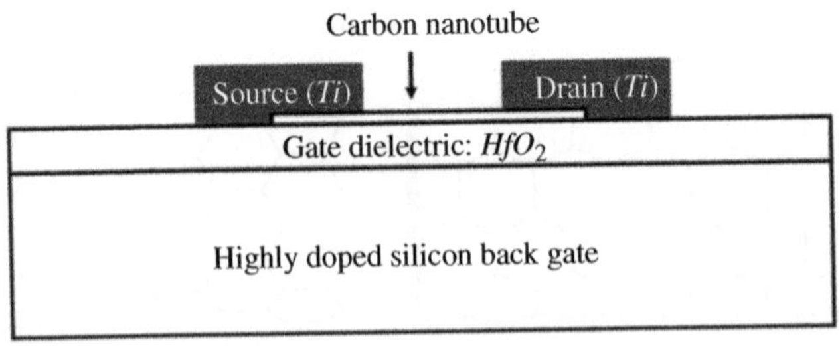

FIGURE 4.23 Cross-sectional view of a back-gated *CNFET*. Gate dielectric is very thin in the order of less than 10 nm of equivalent SiO_2 thickness.

Section 4.4.2, a very brief overview of some of the recent developments in these areas is presented.

4.4.1 *CNFETs* versus *MOSFETs*: Differences in Principles of Operation and Realization

In the early stages of the realization of *CNFETs*, it was expected that the operation of these devices should naturally mimic that of *MOSFETs*. In which case, due to the availability of high-quality *Ohmic* contacts, the transistor performance is dominated by the *transport* properties of the *channel*. However, Appenzeller et al. proposed that in these *CNFETs*, such as the one depicted in Figure 4.23, a new regime of operation is prevalent, in which case the device operation is almost exclusively determined by the *Schottky barriers* formed between the semiconducting *nanotube* and the *drain–source* electrodes.[85] In order to provide an overview of the evolution of this new technology in a more or less real-time fashion, in this section, we first provide a summary of the observations made by these investigators and others who agree with this explanation. Toward the end of the section, however, a seemingly diverging observation made by Javey et al. on *palladium metallization* of *CNTs* and their corresponding explanations are presented.[86] Since this is a technology in rapid evolution, the results presented in this section can only provide a summary of the developments within a limited period of time. Although by the time of the publication of this volume handicapped from the perspective of providing an up-to-date coverage of *CNFETs*, this partial coverage is deemed educational as an extremely interesting example of how experiments lead scientific analysis in developing a new technology.

***4.4.1.1 Schottky Barrier* CNFETs** Among the *CNFETs* available early in the development of this technology, amid other reasons, the lack of the possibility of

[85] J. Appenzeller, J. Knoch, R. Martel, V. Derycke, S. Wind, and P. Avouris, "Carbon nanotube electronics," *IEEE Trans Nanotechnol*, vol. **1**, no. 4, pp. 184–189, 2002.

[86] A Javey and H. Dai, "Carbon nanotube electronics," *Proceeding of 19th International Conference on VLSI Design*, Hyderabad, India, January 3–7, 2006. IEEE, Piscataway, NJ.

heavily *doping* the parts of the *nanotube* overlapping the *source* and *drain* electrodes contributed to failure in producing *tunneling Ohmic* contacts to the *channel* (i.e., essentially via extending the 3-D metal electrodes). These devices were noticed to often exhibit *p-type Schottky barrier FET*[87] characters. Nevertheless, Appenzeller et al. have proposed that the exhibition of this *p-type transport* character[88] is not due to the inherent *p-type* nature of the *channel*. In spite of the conceptual similarity of the operation of these *CNFETs* to that of the conventional *bulk Schottky barrier MOSFET* (i.e., *SB-MOSFET*), these transistors operate differently.[89] According to Appenzeller et al., this is due to the extreme geometrical difference between the metal electrode and the *nanotube channel*.

The presence of an extremely small region of transition between the 3-D metal electrode and the almost 1-D *carbon nanotube* forces the *Schottky barriers* formed at these junctions to be much thinner than in conventional *bulk SB-MOSFETs*. In *SB-CNFETs* under the applied *drain*- and *gate*-induced electric fields, this is the key contributing factor to the observation of *carrier-transport* conditions unlike those common among *bulk SB-MOSFETs*.

The output characteristics of *bulk planar SB-MOSFETs* differ from those of regular *bulk planar MOSFETs*. These differences are particularly apparent in the low-V_{DS} range of *bias*. In this regime, *SB-MOSFETs* demonstrate exponential, rather than linear, *drain* current–voltage characteristics. This behavior indicates the dominant role of *thermionic emission* through the *blocking* contacts formed at the *drain* and *source interfaces* to the *channel*. In *SB-MOSFETs* the assumption of *diffusive carrier transport* ignores the presence of the *Schottky barriers* at the metal–semiconductor interfaces. Hence, part of the reality of the operation of these devices will be lost to this assumption. However, due to the aforementioned difference of geometries in a *CNFET*, adopting a *diffusive* model of *carrier transport* in the description of the electrical characteristics of these devices has been shown to be inadequate even in more ways than in *SB-MOSFETs*. As will be indicated shortly, this is especially highlighted in the *bias* dependence of the *subthreshold* characteristics of *CNFETs*.

As suggested earlier, the physics of the *Schottky blocking* contact is considerably different if the size of the contact region is smaller than the *depletion-region width* on the semiconductor side. Under this condition, the role of the sample boundaries on the definition of *channel* potential becomes considerably more important. In order to

[87] That is a *FET* that has *Schottky barriers*, instead of *Ohmic* contacts, at the *drain* and *source* ends of the *channel*.
[88] Which is simply distinguishable via studying the *drain* current under different polarities of V_{DS} and V_{GS}.
[89] The *Schottky barrier* nature of *CNFET* has been also indicated in the work of others, including: A. Petrov and S. Rotkin, "Transport in nanotubes: Effects of remote impurity scattering," *Phys Rev B*, vol. **70**, pp. 035408-1–10, 2004; M. Hazani, D. Shvarts, S. Peled, V. Sidorov, and R. Noaman, "Self-assembled carbon nanotube-based field effect transistors," *Appl Phys Lett*, vol. **58**, no. 21, pp. 5025–5027, 2004; K. Alam and R. Lake, "Leakage and performance of zero-Schottky-barrier carbon nanotube transistor," *J Appl Phys*, vol. **98**, pp. 06437-1–9, 2005; A. Franklin, R. Sayer, T. Sands, D. Janes, and T. Fisher, "Vertical carbon nanotube devices with nanoscale length controlled without lithography," *IEEE Trans Electron Devices*, vol. **8**, no. 4, pp. 469–476, 2009; S. Salamat, X. Ho, J. A. Rogers, and M. Alam, "Intrinsic performance variability in aligned array CNFETs," *IEEE Trans Nanotechnol*, vol. **10**, no. 3, pp. 439–444, 2011.

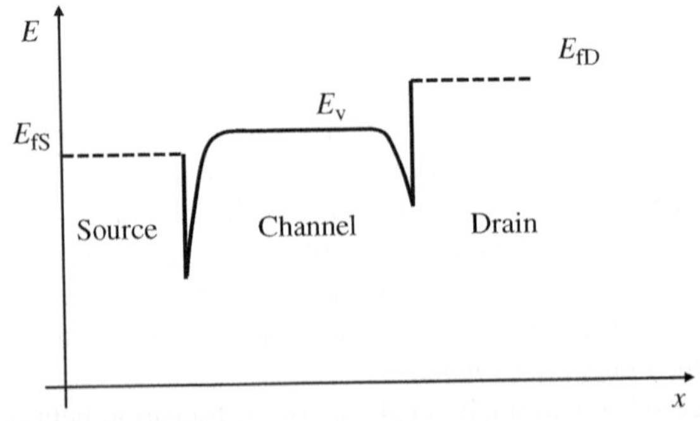

FIGURE 4.24 Approximate valence-band bending along the channel of a biased p-type *SB-CNFET*.

understand this situation, consider an *ultrathin body SB-MOSFET* in which the *gate* potential has started to form an *inversion* layer. Since the *silicon* layer is very thin, it is not just the boundary of the *silicon* layer with the *gate* electrode that is affecting the *inversion* process. Instead, *channel* potential is also a function of the other boundaries of the *active* layer. The impact of the other boundaries to *silicon* translates into an effective reduction of the *Schottky barrier width* in comparison to that of a *bulk SB-MOSFET*. The thinner the body, the larger is this reduction of the *Schottky barrier width*. In *CNFETs* in which the actual body thickness is just about 1.5 nm, the *Schottky barriers* can be extremely thin, which predicts a drastic increase in the *tunneling* process through this *barrier*. This situation is illustrated in Figure 4.24.

In spite of the increased *tunneling* probability through this *thinned barrier*, it is important to point out that the contacts to the ultrathin body semiconductor will not be exactly *Ohmic*. As a result, the distinction between a *SB-CNFET* and a *MOSFET* is not purely academic. Whereas in the case of an *Ohmic* contact, the *Ohmic* behavior is independent of the *drain*- and *gate*-induced electric-field components, in a *SB-CNFET* the *Schottky* barrier width (hence the degree of similarity to an *Ohmic* appearance) depends on these internal electric-field conditions. As a result of this difference, in the case of the *SB-CNFET* the framework of a regular *MOSFET* cannot be used in the analysis of the *drain* current–voltage characteristics nor for extracting the *channel's carrier-mobility* information.

On the assumption of *p-type* characteristic for the *channel* of *CNFET*, attempts have been made to convert the *source* and *drain blocking* contacts into *Ohmic* contacts by employing a higher *work-function* metal such as *Ni* or *Au* instead of the more usual *Ti*. According to this interpretation, the application of these higher *work-function* metals was expected to result in the lineup of metal's *Fermi level* with the *carbon nanotube's valence-band* edge and to prompt the effective removal of the barriers to *hole injection* into the *channel*. However, as reported by Appenzeller et al., in experimentation with these metals such an expectation was not warranted.[90] Instead,

[90] This is although due to the very small presence of *surface states*, *Fermi-level pinning* at the *nanotube* to metal interfaces is almost absent. Later in this section, more will be said on this topic.

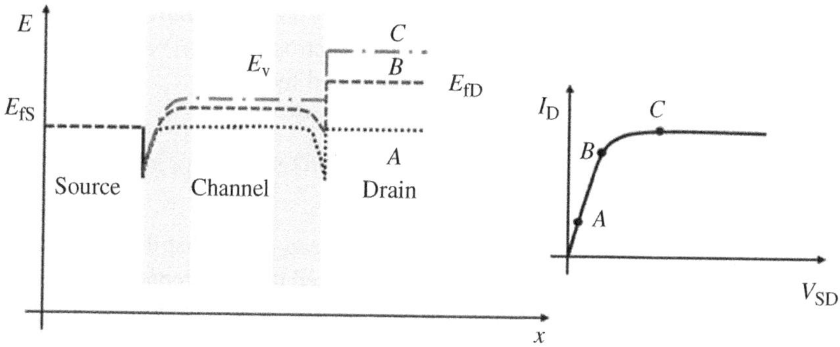

FIGURE 4.25 Approximate valence-band edge diagrams of a p-channel *SB-CNFET* corresponding to three different bias scenarios demonstrated in the *I–V* characteristics of the inset (i.e., points *A*, *B*, and *C*). For all these cases gate–source voltage is above the threshold voltage. Reprinted, with permission, from Appenzeller et al. (2002). Copyright 2002, IEEE.

it was found that independent of whether *Ti*, Ti_xC_y, *Co*, or *Au* were employed as the metal electrode, a *blocking Schottky* behavior was prevalent at the *source* and *drain* terminals. The analysis referred to in the work of Appenzeller et al. was primarily made on the basis of the behavior of *subthreshold swing* (i.e., *SS*) as a function of temperature and dielectric thickness. A particularly important observation was that among the devices such as the one presented in Figure 4.23, *SS* does not decrease with the reduction of temperature. This is unlike the expectation from a conventional *bulk MOSFET*. Surprisingly, they observed a clear trend of *scaling* of *SS* with the *gate-oxide* thickness. These observations were indicative of the fact that *CNFETs* behave as *SB-CNFETs* instead of *MOSFET*-like behavior (and not even exactly as *SB-MOSFETs*).

In addition, it has been shown that the value of the *SS* of a *CNFET* is dependent on the *drain–source bias*. This additional observation also supports the description that the *carrier transport* through a *CNFET channel* is governed by the presence of *Schottky barriers* at the *source* and *drain* contacts.

Figure 4.25, through explaining the behavior of a *SB-CNFET*, underlines the similarities between the output characteristics of a conventional *MOSFET* and a *SB-CNFET*. In this case, an *undoped nanotube* with *p-type* conduction behavior is considered. In this figure, light gray areas indicate the *band lineup* close to the *source* and *drain* contacts while *Fermi levels* of these contacts are indicated to be somewhere within the *bandgap* of the *nanotube*. Above the *threshold voltage*, the *source*-side *valence-band* edge is close to the *Fermi level* of the *source* electrode.

In case A of Figure 4.25, a small *drain* voltage is considered. Under this condition, for *drain–source* conduction to materialize, the two *Schottky barriers* formed between the *source* and *drain* contacts and the *nanotube channel* should be overcome. Assuming *ballistic transport* inside the *nanotube* and neglecting *coherence*,[91] through taking

[91] See the discussions in Section 1.7.

into account the *transmission tunneling* probability from the *source* electrode to the *nanotube* and from the *nanotube* to the *drain* electrode as $T_1(E)$ and $T_2(E)$, respectively, the *drain–source* current can be represented by[92]

$$I_d = \frac{4q}{h}\int \left(f_D^S(E) - f_D^D(E, V_{DS})\right) \times T_1(E) \times T_2(E, V_{DS})dE. \tag{4.28}$$

In this representation, f_D^S and f_D^D stand for the *Fermi–Dirac* distribution functions at *source* and *drain*, respectively. The factor 4 indicated in (4.28) represents the two 1-D modes of *transport* around the *Fermi wave vector* (i.e., $\pm K_F$). In this case, the *channel* is assumed long enough so that the *tunneling* probability of the *source* contact is not affected by the *drain*-induced electric field (the identical condition is also expected on the *source* side). The shape of the *Schottky barrier* in the *source* region is determined by the *gate*-induced electric field.

Considering a fixed value for the *gate* voltage, $T_1(E)$ remains constant while the *drain* voltage is increasing (i.e., case B in Fig. 4.25). However, under this condition both $T_2(E)$ and the number of empty *states* on the *drain* side increase. Assuming dominance of *tunneling* over *thermionic emission*, a linear dependence between the *drain* current and V_{DS} is established.

Beyond a certain value of V_{DS}, the *drain*-side *Schottky barrier* is completely suppressed. As a result, all *carriers* in the *states* that can *tunnel* through the *source*-side *Schottky barrier* contribute to the *drain* current. This change in circumstances results in *drain* current saturation (i.e., case C in Fig. 4.25). As larger *gate* voltages result in further *thinning* of the *barrier* on the *source* side (i.e., improved $T_1(E)$), under these conditions saturation of I_D happens at higher current levels.

Appenzeller et al. claim that their studies, since showing a drastic dependence of the output characteristics on the type of the contact made to the same *nanotube*, illustrate that the internal processes of *nanotubes* are not dramatically involved in determining the output characteristics of *CNFETs*. As presented so far in this section, in their analyses, the *scattering* inside the *nanotube* was not taken into consideration.

Two distinct features are often spotted on the *transfer characteristics* of simple *back-gated SB-CNFETs* such as the one depicted in Figure 4.23. The first feature is the existence of an *ambipolar* mode of conduction, which is observed through the existence of large I_{on} current for both positive and negative *gate–source* voltages at any given V_{DS} (while the current drops when V_{GS} is close to 0). The second feature is the presence of a strong V_{DS}-dependent I_{off}. In this case, it is observed that as the *drain* becomes more negatively *biased*, the minimum current I_{off} goes through an exponential increase, while the *gate* voltage corresponding to this minimum current also shifts to more negative values.[93] These observations are illustrated in Figure 4.26. The two left and right *drain* current branches of Figure 4.26, which are the currents carried by *holes* and electrons, are referred to as *p*- and *n-branches*, respectively.

[92] More will be said on the topic of the *transmission* formalism in Chapter 6.

[93] Y. Lin, J. Appenzeller, J. Knoch, and P. Avouris, "High performance carbon nanotube field-effect transistor with tunable polarities," *IEEE Trans Nanotechnol*, vol. **4**, no. 5, pp. 481–489, 2005.

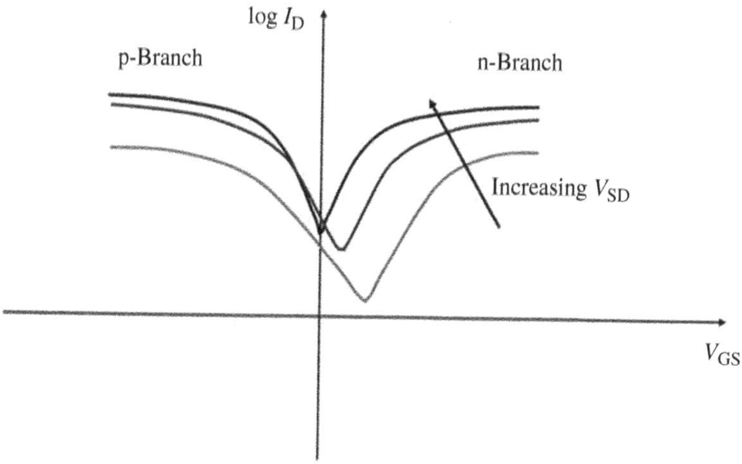

FIGURE 4.26 Typical transfer characteristic of a back-gated SB-CNFET. Reprinted, with permission, from Lin et al. (2005). Copyright 2005, IEEE.

In Figure 4.27, for further explaining the trends illustrated in Figure 4.26, *band diagrams* of a *SB-CNFET biased* both under negative and positive values of V_{GS} are depicted for two different values of V_{DS}. Based on these diagrams, a major identifiable difference between a conventional *MOSFET* and a *CNFET* is that in the latter, unlike the former, switching is dominated by the modulation of *Schottky barriers* formed at the *nanotube*/metal contacts. As illustrated in Figure 4.27a and b, under both cases of sufficiently negative and sufficiently positive *gate–source* voltages, *Schottky barriers* are effectively thinned to enable *carrier* injection from *source* or *drain* into the *channel* (i.e., *hole* injection from *source* under $V_{GS} < 0$ and electron injection from *drain* under $V_{GS} > 0$). The value of this *tunneling*-induced current in *CNFETs* can be much more significant compared to a 3-D *bulk MOSFET*. This is because, as indicated earlier, a *barrier width* as small as a couple of *nanometers* can be obtained for the case of an ultrathin body object when thin *gate* dielectrics are employed.[94]

It is also important to notice that since, in the case of *nanotubes*, the *carrier's mean free path* can be as long as 500 nm, only a slight *band bending* in the *nanotube* body is assumed in the *band-diagram* depiction of Figure 4.27. This is true even at room temperature where a nearly *ballistic transport* through a 300 nm long *nanotube* prevails, hence resulting in no potential drop across the *nanotube*.

According to Figure 4.27b, unless the *nanotube* is very short, application of a more negative V_{DS} does not affect the barrier seen by the *holes* (i.e., for *hole* injection, which is the dominant factor of *transport* in *p-branch*). However, such a V_{DS} will decrease the width of the barrier imposed on the electrons. As a result, in order to

[94] J. Appenzeller, M. Radosavljevic, J. Knoch, and P. Avouris, "Tunneling verses thermionic emission in one dimensional semiconductors," *Phys Rev Lett*, vol. **92**, pp. 048 301-1–048 301-4, 2004.

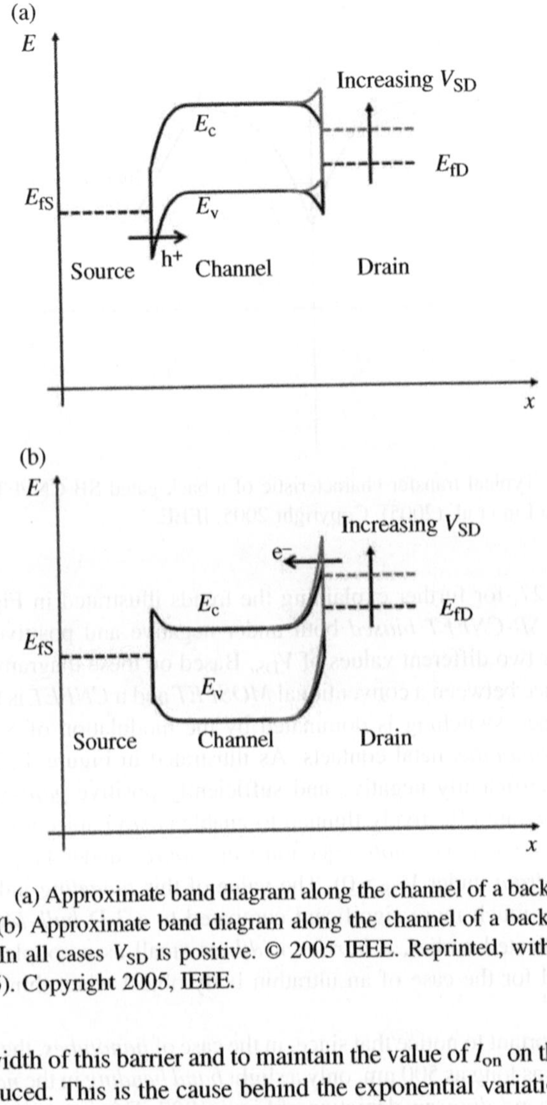

FIGURE 4.27 (a) Approximate band diagram along the channel of a back-gated *SB-CNFET* when $V_{GS} < 0$. (b) Approximate band diagram along the channel of a back-gated *SB-CNFET* when $V_{GS} > 0$. In all cases V_{SD} is positive. © 2005 IEEE. Reprinted, with permission, from Lin et al. (2005). Copyright 2005, IEEE.

increase the width of this barrier and to maintain the value of I_{on} on the *n-branch*, V_{GS} should be reduced. This is the cause behind the exponential variation of I_{off} and the previously indicated negative shift of *n-branch*, with V_{SD}.

As indicated in Chapter 3, it is a well-known fact that the switching performance of *MOSFETs*, measured in terms of the *subthreshold swing* and the *on-state* current, can often be improved by decreasing the *gate*-oxide thickness. However, the different mode of operation of *SB-CNFETs* requires a closer look into this strategy. In a *SB-CNFET*, it is observed that the enhanced control of the *gate* always also results in a higher I_{off}, with a more pronounced V_{DS} dependence of the *subthreshold* characteristic.

The observations made in Figure 4.26 also deserve a more direct address from the application's point of view. It should be emphasized that while in a *CMOS* technology both *n*- and *p-type* transistors with *unipolar* characteristics are required, a thin-oxide *SB-CNFET* is usually *ambipolar*.

These so-called disadvantages of *CNFET* technology, however, far from being an inherent property of *carbon nanotubes* are the result of a prevailing mentality biased toward the *scaling* trends of *MOSFET* technology.

As will be reviewed shortly, over the recent years a number of techniques of realization of *drain* and *source* contacts in *CNFETs* have managed to dethrone the *Schottky barrier* as the only means of contact formation. Among these solutions are the realization of *doped Ohmic* contacts and *Ohmic* contact realized with the use of *palladium*. As reported by Appenzeller et al., implementing a high *dopant*-concentration profile at the *source* and *drain* contact areas of *CNTs* produces *CNFETs* with characteristics resembling those of conventional *MOSFETs*. This resemblance indicates the realization of *tunneling Ohmic* contacts to the *carbon nanotubes* by *thinning* the *barrier* (i.e., under all circumstances of the *gate*- and the *drain*-induced electric fields).[95] Achieving controlled *doping* of *nanotubes* through the traditional mode of *ion-implantation doping*, however, has been proven to be difficult. This is due to the evident incompatibility between the *ion implantation* used in *planar* semiconductor technologies and *carbon nanotube* processing. While an almost benign process in *bulk silicon* processing, in this case the substitution of *carbon* atoms (from which the *carbon nanotube* is composed of) by the *implanted* impurities would quite likely destroy the desired *nanotube* properties. This difficulty has necessitated the exploration of other techniques of *doping*. Examples of which, we will encounter in Section 4.4.1.3.

4.4.1.2* CNFETs *with Ohmic Source and Drain Contacts According to Appenzeller et al., failure in realizing *Ohmic* contacts by increasing the metal *work function* and yet success in achieving such a character through *doping* the semiconductor reveal a mistake in interpreting *carbon nanotubes* as inherently *p-type* semiconductors.[96] As indicated in Section 4.4.1, observations of Javey et al. on realization of *Pd Ohmic* contacts to *undoped CNTs*, however, seem to have presented contradictory evidence on this matter.[97] This team has demonstrated that the *MS* junction realized with the use of the large *work-function Pd* forms an *Ohmic* contact to the *undoped CNT*, hence alleviating the challenges imposed by the presence of *Schottky* contacts in *SB-CNFETs* (which were discussed in Section 4.4.1.1). By employing this technique, it has been shown that *Ohmic* contacts in *CNFETs* can be also realized in the absence of *doping* and diverting the drawbacks associated with it. Studies by Javey et al. have also presented an interesting perspective on the issue of the absence of

[95] J. Appenzeller, Y. Lin, J. Knoch, Z. Chen, and P. Avouris, "Comparing carbon nanotube transistor- The ideal choice: A novel tunneling device design," *IEEE Trans Electron Devices*, vol. **52**, no. 12, pp. 2568–2576, 2005.

[96] J. Appenzeller, J. Knoch, R. Martel, V. Derycke, S. Wind, and P. Avouris, "Carbon nanotube electronics," *IEEE Trans Nanotechnol*, vol. **1**, no. 4, pp. 184–189, 2002.

[97] A Javey and H. Dai, "Carbon nanotube electronics," *Proceeding of 19th International Conference on VLSI Design*, Hyderabad, India, January 3–7, 2006. IEEE, Piscataway, NJ.

Fermi-level pinning[98] in *MS* contacts made to *CNTs* (while, on the contrary, present in *thin-body MS* contacts made to *silicon*).

The study of the formation of *Ohmic* contacts to *CNTs* started with the desire to understand their intrinsic electronic-*transport* properties. A *FET* structure is often used for this purpose. Only in the absence of a significant energy loss in the *drain* and *source* terminals of a *CNFET* is such a study meaningful. As mentioned in Section 4.4.1, many of the original studies, however, remained entangled by the assumptions of *Ohmic* contact formation in *MOSFETs* (i.e., via realization of *tunneling* contacts through heavy *doping* of the *drain* and *source* areas of the *channel*). Whereas this has been the remedy for solving the challenge imposed by *Fermi-level pinning* on the formation of *Ohmic* contacts to *intrinsic planar Si channels*, the dramatically different picture of *surface states* (and also *doping*) in *carbon nanotubes* has been demonstrated to both allow and require a different approach for the definition of *Ohmic* contacts.

Metal-induced gap states[99] (or *surface states*) formed at the metal–semiconductor interfaces are widely believed to result in the formation of *planar dipoles* causing the *pinning* of the *Schottky* barrier height. As discussed earlier, *Fermi-level pinning* is observed to result in a value of barrier height independent of the metal *work function*. This *dipole* layer (also, as discussed in Section 1.1.6, known as a *surface layer*) in the case of a 1-D *channel* such as a *carbon nanotube* is of molecular scales. As a result of the 0-D scale of the *dipoles*, electrostatic circumstances drastically different from those of the *planar dipoles* observed at the surface of *planar MOSFETs* will be established. Whereas in the *planar dipoles* the potential is constant outside of the *dipole layer*, in *molecular*-scale *dipoles* the potential is observed to decay to 0 in regions far away from the *source* and *drain* interfaces. This decay causes the *surface states* to be unable to strongly *pin* the *Fermi level*. As a result, the *Schottky* barrier height to *carbon nanotubes*, unlike *planar silicon MOSFETs*, is deemed quite tunable through adoption of *metals* of different *work functions*.

Accordingly, using *palladium* with a *work function* of about 5.12 eV, a barrier height of *zero* to the *valence band* of *chemically intrinsic nanotubes* with *diameters* greater than 1.6 nm (i.e., *bandgaps* smaller or equal to 0.6 eV) has been reported. The use of this large *work-function* metal in contact with a *p-type*-behaving *nanotube* has granted good *source* and *drain Ohmic* contacts and high current densities even at relatively low *bias* voltages.[100] The *p-type CNFETs* realized through using *Pd drain* and *source* contacts have been shown by Javey et al. to offer a *linear* behavior in their *on state*. Electrical conductance of the *channels* of these *CNFETs* has been observed to improve by *freezing* the *acoustic phonons* at low temperatures. At *cryogenic* temperatures, this conductance was shown to approach the *quantum conductance limit* of $G = 4q^2/h$, which is a good indication of the dominance of *ballistic transport* (i.e., without any significant *scattering* event of the *carriers*, not only in the *nanotube*

[98] See Section 2.2.2.1.4.

[99] Also known as *MIGS*.

[100] A. Javey, J. Guo, Q. Wang, M. Lundstrom, and H. Dai, "Ballistic carbon nanotube field-effect transistors," *Nature*, vol. **424**, no. 6949, pp. 654–657, 2003.

channel but also at the *source-* and *drain*-contact interfaces). This performance is in clear contrast to the operation of *SB-CNFETs*, realized, for example, by using *Ti*, *Ni*, and *Cr* at the *drain* and *source* contacts. Due to elimination of *thermionic emission* and *thermally assisted tunneling*, *SB-CNFETs* exhibit a nearly insulating characteristic at low temperatures. Realization of *zero* height for the *Schottky* barriers at the *drain* and *source* contacts has been proven to allow *CNFETs* to operate near the performance limit of an ideal 1-D system.

Considering *nickel's work function of* 5.15 eV, which is even in excess of that of *palladium*, observation of this contrast presents itself as an interesting phenomenon, which demands a more extensive look into the chemistry of the *Ohmic Pd* contacts to the *CNTs*.

It is also of interest to know that exposure to *molecular hydrogen* offers the possibility of reversible modification of the *surface work function* of *Pd*. Dissociation of *hydrogen molecules* at the surface of *Pd* results in *atomic hydrogen* that can dissolve in *Pd*, causing an effective lowering of its *work function*. This metal *work-function* modulation has been used to study the variation of the electrical conductance and to show that the two are correlated, hence showing that there is no dominant *pinning* involved.

4.4.1.3* CNFETs *of n-Type Characteristics Although originally *p-type* characteristics were observed among the *CNFETs*, more recently *n-type CNFETs* have also been reported. Javey et al. have reported on physically *short-channel* (*gate length* ~80 nm) *n-type CNFETs*.[101] These devices were realized using *chemically doped* contacts in the form of *potassium doping*[102] of the *source* and *drain* regions and *atomic-layer deposition* (i.e., *ALD*) of HfO_2 high-K dielectric. According to this fabrication technology, *n-type* characteristics were observed among *CNFETs* realized on *nanotubes* of diameters about 1.6 nm (i.e., *energy bandgap* of about 0.6 eV). *Chemical doping* was reported to have resulted in suppression of *Schottky barriers* at the *drain* and *source* contacts. Reported values of the *subthreshold swing* of these high I_{on} devices are about 70 mV/dec. These devices exhibited negligible *ambipolar* conduction and high I_{on}/I_{off} current ratios up to 10^6 at a *drain bias* of 0.5 V.[103] In a normalized sense, these numbers are compared favorably with those of the state-of-the-art *silicon*

[101] A. Javey, R. Tu, D. Farmer, J. Guo, R. Gordon, and H. Dai, "High performance n-type carbon nanotube field effect transistors with chemically doped contacts," *Nanoletters*, vol. **5**, no. 2, pp. 345–348, 2005.

[102] This is done in *potassium* vapor. In this technique, *n-doping* of the *nanotube* is performed using *K*-vapor generated by applying a current across an *alkaline*-metal dispenser for a few minutes (i.e., 1–10 min depending on the desired *doping* level).

[103] While in Section 4.1 we mentioned that the use of *ferroelectrics* in the *gate* stack of a *MOSFET* can remove the *Boltzmann limit* from the value of the *subthreshold swing*, it is worthwhile indicating that such a lower limit can be also lifted by employing the *tunneling* mechanism for *carrier* injection into the *channel* at its *source* side (i.e., in a so-called *tunneling FET*, also known as *TFET*). This approach to reduction of *SS*, however, causes a reduction of the ratio between the *on-state* and the *off-state* currents of the *switch*. Read-out of an *SS* larger than the *Boltzmann's* lower limit of 60 mV/dec at room temperature and observation of a specifically *n-type* (and not a *p-type* or an *ambipolar*) character, in addition to the large I_{on}/I_{off} ratio, can be interpreted as signs that an *Ohmic* behavior is established at the *chemically doped source* and *drain* contacts to these physically very short *channels*.

NMOSFETs (i.e., considering the diameter of the *nanotube* as the normalization parameter used in place of the *gate width* of a *MOSFET*). This is a very important evidence for the suitability of studying these devices as replacements for *silicon MOSFETs* in *CMOS* technology. An important property of *carbon nanotubes* is that they have an almost symmetric *dispersion diagram*. As a result, in *CNTs* the *effective transport masses* of electrons and *holes* are quite identical. This is another indicator that *CNT channels* are suitable for adoption in *CMOS* technology.

Through *K-doping* of *carbon nanotubes* similar in principles to conventional *NMOSFETs*, Javey et al. have realized n^+-i-n^+ profiles along the *SWCNT channels*. Schematic representations of this *doping* profile and a device realized according to this process are shown in Figure 4.28. In this implementation, they have reported on a technology based on patterning the top *gate* stack (with the *gate length* of about 80 nm) on individual *SWCNTs* placed on a SiO_2-covered p^+-*silicon* substrate. In this *dual-gate* structure, the p^+-*doped* substrate can be employed as the back *gate*. The top *gate* stack, realized by *ALD* and *lift-off* techniques,[104] was composed of an approximately 8 nm thick HfO_2 dielectric layer and 0.5/15 nm *Ti/Pd gate* metals.

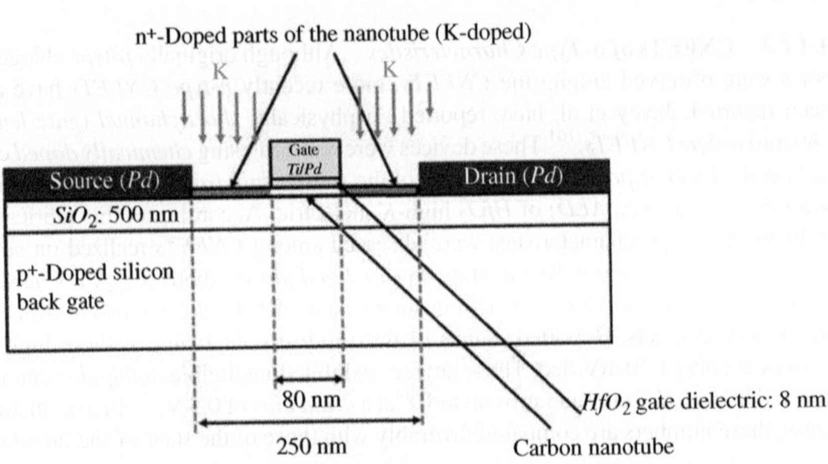

FIGURE 4.28 Cross-sectional depiction of the dual-gate n^+-i-n^+ doped *CNFET* of Javey et al. The middle part of the nanotube is kept intrinsic. Dimensions are not in scale. Reprinted (adapted) with permission from Javey et al. (2005). Copyright 2005, American Chemical Society.

[104] *Lift-off* is a metal-patterning technique used in *microfabrication* processes. In this technique, *metal* is deposited on an already patterned layer of *photoresist* (i.e., *PR*). As a result, in the openings made through this *sacrificial photoresist* layer, metal film will adhere to the surface of the material residing beneath this layer. However, elsewhere metal will sit on the *PR* film. Through dipping such a *wafer* in a *solvent bath* suitable for removal of the *PR*, depending on the thickness of the metal film and the conformity of its coverage over the *PR*, metal film in those parts that are originally sitting on the *PR* tends to break away from the rest and float on the surface of the *bath*. This process can be encouraged by applying *ultrasonic* vibration to the *solvent bath*.

Drain and *source* electrodes were realized by the 15 nm thick *Pd* film with electrode distance of about 250 nm. The top *gate* stack was not overlapping these *Pd* electrodes. In the process of *doping*, parts of the *nanotubes* outside the *gate* stack and uncovered by the *source* and *drain* electrodes were fully exposed to *K*-vapor to form n^+ regions. The reason for the selective *doping* of the *nanotube* (i.e., using the *gate* stack as a *mask*) is that the *heavy n-type doping* along the entire length of the *nanotube* will result in a *metallic-type transport* throughout the *channel*, hence no *transistor action*.

This study indicated that due to the reactivity of *potassium* atoms, upon exposure to ambient this type of *chemical doping* is reversible. For that reason, the reported measurements were conducted under vacuum (about 10^{-5} Torr). Javey et al. have proposed the use of *surface passivation* or adoption of *air-stable dopants* for alleviation of this problem.

In the design presented in Figure 4.28, implementation of an exceedingly large contact *doping* not only resulted in the positive outcome of higher *on-state* current (i.e., due to the improved metal–semiconductor *transmission probability* and the reduced *series resistance*) but also adversely caused higher leakage current and more *ambipolar* conduction. This observation, as illustrated in Figure 4.29, is due to *band-to-band tunneling*. Based on this observation, moderation in *doping* is a major element in the success of this technology.

In *silicon MOSFETs* where *band-to-band tunneling* is not significant, the leakage current is mostly independent of the *source* and *drain doping* profiles. This is especially true for length scales larger than 100 nm. The observed enhanced *band-to-band tunneling* in *CNFETs* with increased *doping* level of the *source* and *drain*, according to Javey et al., is attributed to *CNTs*' smaller energy *bandgaps*, smaller

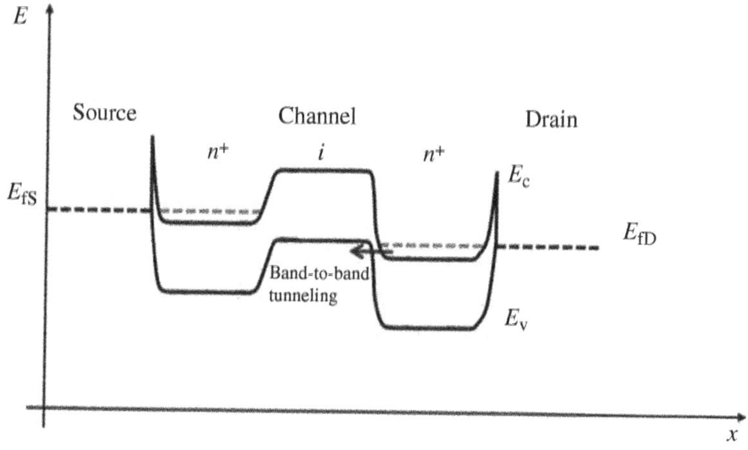

FIGURE 4.29 Approximate off-state band diagram of the *CNFET* of Figure 4.28 under positive V_{DS}. While the absence of energy states between the *Fermi levels* of source and drain forecasts zero current across the intrinsic channel, the indicated increased probability of band-to-band tunneling between the n^+-doped parts of the nanotube and its intrinsic middle increases the off-state current in an *ambipolar* fashion. Reprinted (adapted) with permission from Javey et al. (2005). Copyright 2005, American Chemical Society.

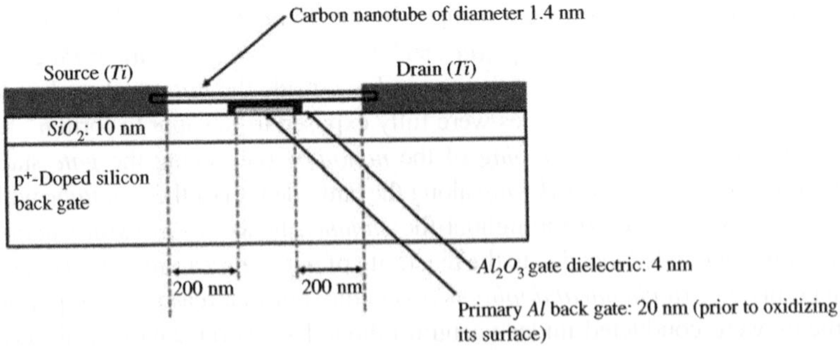

FIGURE 4.30 Cross-sectional view of the electrostatically doped dual back-gated *CNFET* of Lin et al. Dimensions are not in scale.

effective carrier masses, and symmetry of their *conduction* and *valence-band states*.

In addition to *chemical doping* (for which, in this section, we have only provided an example for realization of the *n-type CNFET*), *n-i-n* and *p-i-p doping* profiles of the *nanotubes* are also achievable through the so-called *electrostatic doping*. Implementation of the *dual-gate CNFETs* instead of the simple back-*gated CNFETs* (such as the one depicted in Fig. 4.23) has been successfully demonstrated to be able to eliminate the *ambipolar* characteristics of *transport* and to create pure *n*- or *p-type FETs* with excellent *off-state* performance and *SS*. In *n*- and *p-type CNFETs* realized on *doped n-i-n* and *p-i-p nanotubes*, respectively, instead of *contact switching* by modulating the *source* and *drain Schottky barriers*, operation takes place through varying the *band* potential in the *undoped* middle portion of the *nanotube*.

Before closing this discussion, on the basis of the structures reported by Lin et al., it is worthwhile to pay a brief visit to the realization of so-called *electrostatic doping* in dual back-*gated CNFETs*.[105] In these back-*gated* structures, in the middle of the *nanotube* in addition to the regular back *gate*, there exists a primary *aluminum* back *gate* residing underneath the *nanotube*. This is separated from it by a 4 nm thick Al_2O_3 insulator (see Fig. 4.30). According to the dimensions employed by Lin et al., the fringing electric field of this primary *gate* has been shown to be incapable of affecting the *drain* and *source* contacts. This metal layer has been used to *screen* the *nanotube* in its middle section from the reach of the electric field emanating from the *silicon* back *gate*. The *silicon* back *gate*, however, is in charge of controlling the behavior of the *source* and *drain Schottky barriers* in the *access regions* of the *nanotube* (which are the parts of the *channel* located between the edges of the *gate* and the *drain–source* electrodes).

The *band diagrams* illustrated in Figure 4.31 clearly demonstrate how even in the absence of *chemical doping* the *unipolar* behavior prevails in the operation of the dual-*gate CNFET* of Figure 4.30.

[105] Y. Lin, J. Appenzeller, J. Knoch, and P. Avouris, "High performance carbon nanotube field-effect transistor with tunable polarities," *IEEE Trans Nanotechnol*, vol. **4**, no. 5, pp. 481–489, 2005.

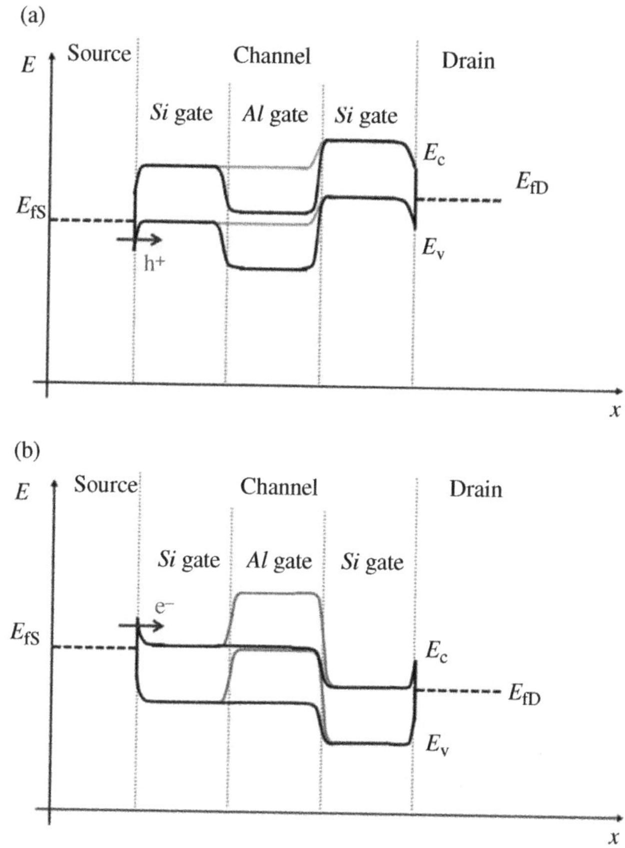

FIGURE 4.31 (a) Approximate band diagram along the channel of the dual back-gated *CNFET* of Figure 4.30 when V_{GS} of the silicon back gate and V_{DS} are both negative. For the part of the channel overlapping the *Al* gate when the V_{GS} of the *Al* gate is negative band edges are illustrated in gray, while when positive the band edges are shown in black. (b) Approximate band diagram along the channel of the dual back-gated *CNFET* of Figure 4.30 when V_{GS} of the silicon back gate and V_{DS} are both positive. For the part of the channel overlapping the *Al* gate when the V_{GS} of the *Al* gate is negative band edges are illustrated in gray, while when positive the band edges are shown in black. Reprinted, with permission, from Lin et al. (2005). Copyright 2005, IEEE.

As shown in Figure 4.31, in the structure of Figure 4.30 in order to turn the switch *on*, the *bias* of the primary *gate* $V_{GS\text{-}Al}$ should be made positive if a positive $V_{GS\text{-}Si}$ is already applied to the back *silicon gate*. However, a negative *bias* applied to this primary *gate* will establish a barrier in the middle of the *nanotube*, which prevents current conduction in the *n-branch*. The opposite argument applies to the *p-branch* when the $V_{GS\text{-}Si}$ is smaller than 0. Such a dual-*gate CNFET* depending on the polarity of $V_{GS\text{-}Si}$ can behave as a *unipolar n-* or *p-type FET* (i.e., instead of an *ambipolar CNFET*). This mode of realization of *unipolar CNFETs* is often referred to as *electrostatic doping*.

In these devices, under a sufficiently negative (or positive) $V_{GS\text{-}Si}$, the *source* and *drain Schottky barriers* are made thin enough to allow for *holes* (or electrons) to *tunnel* into the *nanotube channel*. As a result of this *carrier* injection, regions in their vicinity take *p-type* (or *n-type*) characters. As illustrated in the *band diagrams* of Figure 4.31, in this fashion, as in the case of *chemical doping*, *n-i-n* and *p-i*-p profiles are created along the *nanotube*. This is the origin of the name *electrostatic doping*.

4.4.1.4* CNFETs*: Alignment Challenge and Other Fabrication Feats Over the past few years, a number of fabrication technologies for realization of *CNFETs* have been developed. Many of these technologies rely on *dispersion spinning* of solutions containing *nanotubes*. As indicated in Section 4.4.1, one of the earliest fabrication technologies of *CNFETs* relies on the use of a single back *gate* for the transistor. In this technology, a desired *gate* dielectric is either grown or deposited on top of a *degenerately doped silicon wafer*, which serves as the back *gate*. Then, *spinning* a 1,2-*dichloroethane* solution containing *nanotubes* (e.g., which are produced through the *laser-ablation process*) is used for dispersing the *nanotubes* on the surface of the dielectric. Due to the randomness of the *dispersion* process of *CNTs*, however, in this technique the surface density and the orientation of the *nanotubes* cannot be controlled. As a result, this technique does not allow for parallel processing of a number of devices, which is needed in realization of circuits. Following the *spinning* step and after evaporation of the solution, *source* and *drain* contacts (e.g., made out of *titanium*) are evaporated above a *single-walled carbon nanotube*. This can be accomplished by printing a predefined pattern of contacts. In this technique, depending on the size of the *wafer* and the density of the *nanotubes*, one can only rely on a finite chance for finding a properly oriented semiconducting *nanotube channel* with the metallic *drain–source* electrodes. *Nanotubes* that are not anchored down by the evaporated *titanium pads* can be removed by *stripping* the *photoresist* film used in *patterning* the metal contacts.

While unsuitable for realization of circuits, this relatively inexpensive process avoids the need for *mask registration* (i.e., through adoption of a *one-step masking*) and can be used for studying the behavior of individual *CNFETs*.

To achieve a proper lineup between the *nanotubes*, a few technological alternatives such as *dispersion spinning* of *nanotubes* on a *grooved* surface in combination with moving and aligning the *nanotubes* with an *AFM* tip to the grooves have been explored. Although *dispersion* of the *nanotubes* from a solution on a *grooved* surface is clearly expected to provide an improved chance for placing the *nanotubes* inside the *grooves*, the probability of success of such a process is not completely on par with the high expectations of *channel* lineup in a *ULSI* circuit. While the aforementioned process of moving the misplaced *nanotubes* into the empty *grooves* has been explored, such a process imposes a considerable strain on the throughput of the fabrication technology.

In spite of this difficulty, *CNT* technology has been demonstrated to be considerably resourceful in overcoming this challenge, especially through the proper choice of the crystal growth process. Realization of *single-* and *multiwalled carbon nanotubes* through a variety of growth techniques such as *arc discharge*, *laser ablation*, *pyrolysis*, and *CVD* has been proven by a number of researchers. Among these growth techniques, *CVD* growth on prepatterned *catalyst* thin films deposited on a substrate has shown the highest degree of versatility and controllability. In contrast to the aforementioned *dispersion-spinning* technique, this growth mode offers the possibility of selective growth of well-aligned *nanotubes* on a substrate that can be used for realization of vertically oriented *CNFETs*.[106] It has been proven that using the *CVD* technique, *nanotubes* can be aligned by employing the flow of gas and application of electric fields as control mechanisms.[107]

Adoption of high-*K* dielectrics to *CNFETs* is also an important area of interest in the evolution of these devices. As indicated in Section 3.2.2.1.2, adoption of high-*K* dielectric materials (such as ZrO_2 and HfO_2, with *relative dielectric constants* between 15 and 25) to *CMOS* technology has been one of the most challenging endeavors prompted by the *scaling* trends of this technology over the past two decades. However, adoption of these materials, since enabling higher *on-state* current densities (and as a result *switching-speed* improvement), is highly desirable. Although, as mentioned before, such a transition from SiO_2 to a high-*K* dielectric has resulted in a loss of *carrier mobility* in *silicon MOSFETs*, in the case of *carbon nanotubes* the absence of *dangling bonds* at the *nanotube/dielectric* interface and the weak noncovalent bonding interactions between the two materials have been shown to result in no loss in *carrier mobility*. Since in this case there is only weak *van der Waals bonding* between the high-*K* dielectric and the *nanotube*, the *nanotube's* physical nature and its intrinsic chemistry will remain intact when placed in contact with the dielectric.

In the previously mentioned work of Appenzeller et al., a 20 nm thick film of HfO_2 was used as the *gate* dielectric and the *drain–source* separation was about 300 nm.[108] According to the previously presented technique of fabrication, Appenzeller et al. reported on *CNFETs* with performance characteristics resembling those of conventional *p-type MOSFETs* (i.e., a *subthreshold swing* in the order of 200 mV/dec, *current drive* in the order of 250 μA/μm, and maximum *gate transconductance* of 400 μS/μm).[109] As mentioned in Section 4.4.1.1, the *p-type* behavior, however, was suggested not to be a manifestation of a certain *doping* level in the *nanotube* but instead an evidence of dominant role of the *Schottky barriers* present at the *source* and *drain* interfaces to the *nanotube*.

4.4.2 Other Nanotube and Nanowire Transistors

While *carbon nanotubes* have managed to attract much interest as the *channel* of *field effect transistors* used not only in switching but also in sensory applications, an ongoing research is also being pursued toward realization and application of *nanowires* in a

[106] A. Franklin, R. Sayer, T. Sands, D. Janes, and T. Fisher, "Vertical carbon nanotube devices with nanoscale length controlled without lithography," *IEEE Trans Electron Devices*, vol. **8**, no. 4, pp. 469–476, 2009.

[107] G. Amaratunga, A. Teh, S. Cha, G. Ho, J. Jang, Y. Yang, Y. Choi, K. Teo, S. Dalal, D. Kang, N. Rupesinghe, W. Milne, D. Hasko, M. Welland, and J. Kim, "Nanotube and nanowire transistors," *Proceedings of International Semiconductor Conference*, Sinaia, Romania, October 3–5, 2005. IEEE, Piscataway, NJ, pp. 3–8.

[108] J. Appenzeller, J. Knoch, R. Martel, V. Derycke, S. Wind, and P. Avouris, "Carbon nanotube electronics," *IEEE Trans Nanotechnol*, vol. **1**, no. 4, pp. 184–189, 2002.

[109] As mentioned earlier, in reporting on the performance of *CNFETs*, we should pay attention that in the normalized representation of the data such as I_D–V_D, G_m–V_{GS}, or the *transfer characteristic*, instead of presenting these data per unit width of the *gate*, normalization is done with respect to the diameter of the *nanotube*.

large number of other semiconducting families. Among these *nanowires*, in addition to the *III–V* semiconductors, are the semiconductor *oxides* such as *ZnO* and TiO_2. Many of the *CNT* device concepts discussed in Section 4.4.1 potentially have the possibility of being applied to these other *nanowire* structures.

ZnO is just one of the examples of semiconducting *nanowires* recently made available to electronic and sensory explorations. These *nanowires* demonstrate *n-type* electrical conductance. Considering their *direct* wide *bandgap* of 3.37 eV, they have managed to generate quite a bit of interest in electronic applications.[110] *ZnO* crystallizes in the *wurtzite* crystalline form. So far, a variety of gas-phase and solution-phase approaches have been employed in synthesizing *ZnO nanowires*, *tetrapods*, *combs*, *ribbons*, and *rings*. These *nanostructures* have been studied in a variety of applications such as room-temperature *ultraviolet* (i.e., *UV*) *lasing cavities*, *UV/visible photodetectors*, and *active* components in *solar cells* and gas sensors. One of the major challenges to their application, however, has been the ability to reproducibly form high-quality *Ohmic* contacts to these *ZnO nanostructures*. In addition to imposing problems on forming devices and circuits with these *nanostructures*, this challenge has created difficulty in assessing the intrinsic electronic properties (such as *carrier* type/concentration and *mobility*) of these *nanostructures*.

The observed trends among *ZnO nanostructures* have so far indicated that they are intrinsically *n-type* semiconductors. This has been shown to primarily owe to the presence of *oxygen vacancies* and/or *zinc interstitials*.

Many of the earlier studies on *ZnO nanowires* and *nanoribbons* were focused on their application as *UV photodetectors*. Among these studies, the use of *Au* contacts (with a *Schottky barrier* height of about 0.7 eV) resulted in a large contact resistance at the *Au–ZnO* interface. *UV excitation* has been observed to increase the *carrier* concentration in the *ZnO channel* through not only *electron–hole pair generation* but also *photodesorption* of electron-trapping species from the *nanowire surface*. These two processes encourage both *tunneling* and *thermionic emission* over the *blocking Schottky barriers*, resulting in a higher detectable current level. These studies crucially highlighted the importance of the *surface* condition of these *nanostructures*, which were later explored as gas sensors.

While in many sensory applications the correlation between the *surface* potential and the current drive of a *gateless FET* structure has been seen as an avenue for making sensors, such applications seem more promising in the case of *nanotube* and *nanowire FETs*. Considering the substantial difference between the surface-to-volume ratios of *bulk* versus *nanoscale* transistors, the modulation of the *surface* potential

[110] See S. Ju, K. Lee, D. Janes, M. Yoon, A. Facchetti, and T. Marks, "Low operating voltage single ZnO nanowire field-effect transistors enabled by self-assembled organic gate nanodielectrics," *Nano Lett*, vol. **5**, no. 11, pp. 2281–2286, 2005; J. Goldberger, D. Sirbuly, M. Law, and P. Young, "ZnO nanowire transistors," *Phys Chem B*, vol. **109**, pp. 9–14, 2005; S. Cha, J. Jang, Y. Choi, G. Ho, D. Kang, D. Hasko, M. Welland, and G. Amaratunga, "High performance ZnO nanowire field effect transistors," *Proceedings of European Solid State Device Research Conference*, Grenoble, France, September 12–16, 2005. IEEE, Piscataway, NJ, pp. 217–220; S. Ju, D. Janes, G. Lu, A. Fachetti, and T. Marks, "Effects of bias stress on ZnO nanowire field effect transistors fabricated with organic gate nanodielectrics," *Appl Phys Lett*, vol. **89**, pp. 193506-1–193506-3, 2006.

through exposure to liquid or gaseous substances can have a larger impact on the current drive of the *nanostructured* transistors. As noted in Section 4.4.1.4, since in *CNFETs Fermi-level pinning* is virtually absent, sensory use of the modulating effect of the *surface* potential can be much more efficient.

Aside from *nanostructures*, *ZnO* thin films prepared by different *sputtering* techniques have been also used in realization of thin-film transistors. In these transistors, low electron *mobilities* of 0.03–3 $cm^2/V\ s$ have been reported. These values are smaller than the range of *Hall effect*[111] mobilities measured in single-crystal *ZnO epilayers* (i.e., 100–205 $cm^2/V\ s$).[112]

4.5 *spinFET*

The combined roles of the *magnetized* contacts in preferential *injection* and *detection* of *charge carriers* of certain *spin orientation* and current modulation through flipping the *spin* of the *charge carriers* flowing through a narrow-*bandgap* semiconducting *channel* have been proposed by Datta et al. to be capable of materializing a new type of *field effect transistor*.[113]

Datta et al. proposed that through the so-called *spin–orbit coupling* phenomenon, the *gate* electrode of *FETs* is capable of modulating the *spin* of *charge carriers* traveling in the narrow-*bandgap* semiconductor *channels*. The name often used for the *field effect transistors* built on these principles is *spinFET*.

4.5.1 *spinFET*: Principles of Operation

The operation principles of a *spinFET* have been visualized based on the reported observations of *quantum-interference effects* in low-temperature manipulation of electron *transport* in *nanostructures*. These *quantum-interference effects* have been envisioned in Datta et al.'s proposal as counterparts of a number of well-known principles of operation of microwave or optical networks.

The basic operation of the *spinFET* can be understood in analogy with the *electro-optical light modulator* shown in Figure 4.32. An *electro-optical modulator* is composed of a *polarizer* at the input that *polarizes* the light at 45° angle to the *y-axis* (in the *y–z plane*), an *electro-optical channel*, and an *analyzer* placed at the output.

Among these three components, the operation of the *polarizer* can be represented in terms of the following linear combination of the *z*- and the *y-polarized* light:

$$\begin{bmatrix}1\\1\end{bmatrix}_{45^\circ\ \text{polarized}} = \begin{bmatrix}1\\0\end{bmatrix}_{z-\text{polarized}} + \begin{bmatrix}0\\1\end{bmatrix}_{y-\text{polarized}}. \tag{4.29}$$

[111] See Section 1.4.3.

[112] J. Goldberger, D. Sirbuly, M. Law, and P. Young, "ZnO nanowire transistors," *Phys Chem B*, vol. **109**, pp. 9–14, 2005.

[113] S. Datta and B. Das, "Electronic analog of the electro-optic modulator," *Appl Phys Lett*, vol. **56**, no. 7, pp. 665–667, 1990.

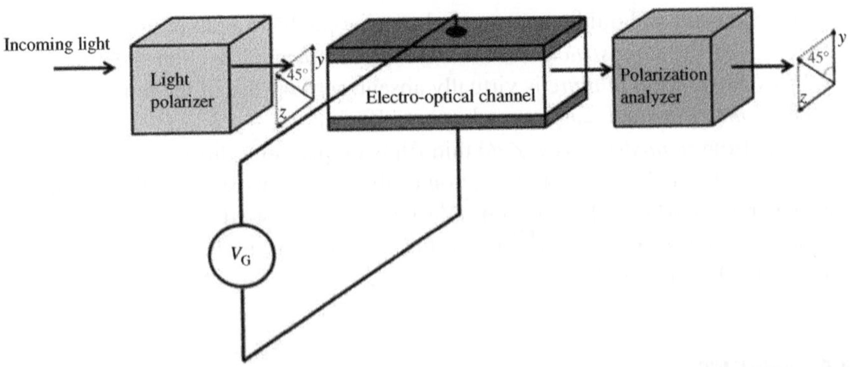

FIGURE 4.32 Illustrative sketch of an electro-optical modulator. Reproduced with permission from Datta and Das (1990). Copyright 1990, AIP Publishing LLC.

In passing the light through the *electro-optical channel*, the two *polarizations* of light encounter different *phase shifts* (i.e., K_1L and K_2L), where K_1, K_2, and L are the *wave vectors* of the two *polarizations* and the length of the *channel*, respectively. This is due to the fact that the *electro-optical effect* forces the dielectric constant ϵ_{zz} to be slightly different from ϵ_{yy}. As a result of this difference in *phase shift*, the light coming out of the *electro-optical channel* is represented by $\begin{bmatrix}\exp(jK_1L)\\ \exp(jK_2L)\end{bmatrix}$. Since the *analyzer*, which is placed at the output, allows the component along $\begin{bmatrix}1\\ 1\end{bmatrix}$ to pass through, the output power P_0 will be given in terms of the following proportionality:

$$P_0 \propto \left| [1 \;\; 1] \times \begin{bmatrix}\exp(jK_1L)\\ \exp(jK_2L)\end{bmatrix} \right|^2 = 4\cos^2\left[\frac{(K_1-K_2)L}{2}\right]. \tag{4.30}$$

As expressed in (4.30), since the *electro-optical effect* modulates the *phase shift* (i.e., $\Delta\Theta = (K_1 - K_2)L$), it can be employed in modulating the output power.

Figure 4.33 illustrates Datta et al.'s proposed *spinFET*, defined in analogy with this *electro-optical modulator*. In this implementation, the *polarizer* and the *analyzer* have been replaced by *ferromagnetic* contacts made to a semiconducting *channel*. In the *ferromagnetic* material, at the *Fermi level* the *density of states* for electrons of one *spin* is considerably larger than that of the other. This fact results in preferential *injection* and also *detection* on the *analyzer* side of electrons of a particular *spin*.[114] The *electro-optical effect* allows for tunability of *spin–orbit coupling* through the *gate*-induced electric field. This property is expected from narrow-*bandgap* semiconductors such as *InGaAs*.

[114] According to this proposal, using *permalloy* contacts *spin*-current *polarization* up to about 50% has been experimentally reported.

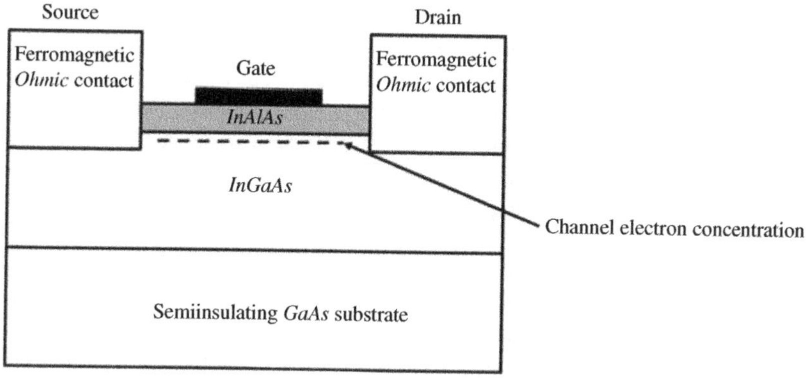

FIGURE 4.33 Proposed *SpinFET* of Datta et al. Reproduced with permission from Datta and Das (1990). Copyright 1990, AIP Publishing LLC.

The *2DEG channels* established at the interface of *heterojunctions* made to narrow-*bandgap* semiconductors,[115] even in the absence of magnetic fields, have been shown to exhibit energy splitting between the *up-spin* and the *down-spin* electrons. This *zero* magnetic field *spin splitting* is believed to be caused by the *Rashba* term in the *effective-mass Hamiltonian*.[116] The origin of this term is in the *gate*-induced electric field normal to the *heterointerface*. Based on this, *polarization* is tunable through changing the *gate*-induced electric field. For a successful *spinFET*, the important condition is that this effect should be large enough to produce a *phase-shift* difference of π within a *mean free paths* (i.e., between the *up-spin* and the *down-spin* electrons). This explains the attention to narrow-*bandgap* semiconductors and the low-temperature operation of these devices. *Graphene*, owing to its long *mean free path* (i.e., even at room temperature), has been shown to have tremendous potentials in realization of room-temperature *spinFETs*. In the case of *InGaAs/InAlAs heterostructures*, the experimentally observed *zero* magnetic field *spin splitting* has been thought to be able to grant the required *phase shift* over a travel path of 0.67 μm (which is shorter than the *mean free path* of greater than 1 μm, commonly observed in high-*mobility* semiconductors at *cryogenic* temperatures).

As expressed earlier, the *spin–orbit coupling constant* η is known to be dependent on the *gate*-induced electric field normal to the *heterointerface*. As a result, it is controllable by the application of a *gate* voltage. According to Datta et al., this is expected to extend the opportunity of implementing a *gate*-voltage-controlled mechanism of *differential phase shifting*.

Without presenting the details, it is very important to point out that this *differential phase shift* is expected to be identical among all of the *subbands* formed at the *heterointerface*. Accordingly, the *2DEG* electrons occupying different *subbands* are

[115] See the discussion in Section 2.5.1.2, while more will be said on the *2DEG* formation in *Type-I* lined-up *heterostructures* in Chapter 5.
[116] Named after *Emmanuel Rashba*.

expected to be equally affected by this phenomenon. This provides an important advantage in a device application context. The operation of *quantum-interference* devices, on the basis of fear of *multimode* washout, needs to be restricted to low temperatures and low voltages to secure the *single-mode* operation. Due to this property, *spinFETs* are expected to be able to operate well even under temperature and *bias* conditions that cause *intersubband scattering*. However, as will be observed in Section 4.5.2, other circumstances have still managed to restrict the operation of these transistors mostly to *cryogenic* temperatures.

From an application point of view, *spinFETs* have been hailed for the possibility of offering another degree of freedom in controlling the status of a *FET* switch (i.e., in addition to turning the *channel on* and *pinching* it *off*).

Over the years a number of confirmations for the predictions of Datta et al. have come to light. The earliest of these was published by Nitta et al., who have reported on *spin–orbit coupling* in an *inverted* $In_{0.53}Ga_{0.47}As/In_{0.52}Al_{0.48}As$ *quantum well* that can be controlled by applying a *gate* voltage.[117] In this study, the *heterostructure* was grown *lattice-matched* to a *Fe-doped* semi-insulating *InP* substrate. Also a 6 nm thick *undoped* $In_{0.52}Al_{0.48}As$ *spacer* layer[118] was used to reduce the *scattering*. As theoretically speculated, they demonstrated that even without any external magnetic field, application of an electric field normal to the *heterointerface* yields an *effective* magnetic field capable of exciting electrons and *lifting* the *spin degeneracy*. They argued in favor of the presence of two alternative mechanisms involved in this *spin splitting*, which are rooted in the host crystal electric field and the *heterointerface* electric field (i.e., the so-called *Rashba* mechanism[119]). Between these two origins, Lommer et al. have pointed out that the second mechanism is dominant among the narrow-*bandgap* semiconductors.[120]

4.5.2 *spinFET*: Challenges in Realization

As it was pointed out in Section 4.5.1, on the basis of the *spinFET* proposal, one can observe that a semiconductor-based *spintronics* technology is in need of at least four essential components:

1. An efficient *spin-polarized carrier injector*
2. An efficient *spin-transport* medium of sufficiently long *spin lifetime* within the hot semiconductor medium

[117] J. Nitta, T. Kazaki, and H. Takayanagi, "Gate control of spin-orbit interaction in an inverted $In_{0.53}Ga_{0.47}As/In_{0.52}Al_{0.48}As$ heterostructure," *Phys Rev Lett*, vol. **78**, no. 7, pp. 1335–1338, 1997.

[118] *Spacer* is a thin *undoped* layer incorporated between the narrow-*bandgap channel* layer and the wide-*bandgap barrier* layer of a Type-I lined-up *heterostructure*. Incorporating this layer is meant to physically increase the distance between the *2DEG* electrons and the *dopants* residing in the often-*doped barrier* layer (i.e., to improve the electron *mobility* of the *2DEG*). More will be said on this topic in Chapter 5.

[119] E. Rashba, "Properties of semiconductors with an extremum loop-cyclotron and combinational resonance in a magnetic field perpendicular to the plane of the loop," *Sov Phys Solid State*, vol. **2**, p. 1109, 1960.

[120] G. Lommer, F. Malcher, and U. Rössler, "Spin splitting in semiconductor heterostructures for B → 0," *Phys Rev Lett*, vol. **60**, pp. 728–731, 1998.

3. An effective *spin-manipulator*
4. An effective *spin detector* of *charge carriers*

Among these four components, the first one has been the most elusive. This is because often an *injection efficiency* smaller than 1% is reported from *MS* contacts made with *ferromagnetic* metals. As will be indicated in the rest of this section, in spite of this challenge, using contacts with *ferromagnetic materials* as *spin injectors* has a number of advantages seemingly unsurmountable by many of the other solutions. Among these are availability of a source of *spin-polarized* electrons and not *holes*, high *Curie temperature*, low *coercive fields*, fast switching times, and a well-developed material technology (since used for many decades in the record industry).

In terms of the second component, long *spin diffusion* with characteristic length of many *microns* (and *spin lifetime* longer than 100 ns) has been reported from the *III-As channels*.

As suggested earlier, since in *2DEG* systems the *Rashba effect* can be used for controlling the *spin precession* via a *gate* voltage, such systems can provide the third component of this list.

In terms of providing the fourth component of such a *spintronic* system, a number of detection mechanisms such as *giant magnetoresistance*-like behavior (i.e., *GMR*) at *Schottky* contacts have been explored by the investigators.

Considering the aforementioned challenge with the realization of the first component of such a system, a lively research is being pursued in this area. The issue of *spin transport* through the interface of two dissimilar materials (i.e., a metal and a semiconductor) is still not fully understood. Although based on the experimental data it has been shown that the interface of a *ferromagnetic* metal and a *heterostructure* is capable of injecting *net spin-polarized* population of electrons into the semiconductor, the efficiency of this technique is very low.

Recently proposed models provided by several groups offer some insight into the issue of *spin transport* across the metal–semiconductor *interface*. Among these, Schmidt et al. have indicated that for successful *spin injection* one of the following two conditions should be met:

1. The conductivities of the *ferromagnetic* metal and the semiconductor should be closely matched.
2. The *ferromagnetic* metal should be 100% *polarized*.[121]

They have shown that without satisfaction of either of these conditions, a very low *spin polarization* (i.e., less than 1%) should be expected. According to Schmidt et al., with the possible exception of CrO_2, no *ferromagnet* has been found to satisfy these conditions.

[121]G. Schmidt, D. Ferrand, L. Molenkamp, A. Filip, and B. van Wees, "Fundamental obstacle for electrical spin injection from a ferromagnetic metal into a diffusive semiconductor," *Phys Rev B*, vol. **62**, p. R4790, 2000.

As another form of *spin injection* from a *ferromagnetic* metal to the semiconductor, a *tunneling* metal–insulator–semiconductor barrier has been proposed by *Rashba*.[122] According to this proposal, it is expected that inserting a *tunneling* barrier between the metal and the semiconductor will provide the essential means for controlling the interface resistance and for supporting the difference in *Fermi levels* between the *spin-up* and the *spin-down bands*. This approach is speculated to possibly circumvent the aforementioned issues revealed by Schmidt et al. It is expected that this solution will enable the use of common *ferromagnetic* metals with robust magnetization. A typical *spin polarization* of about 50% has been speculated on the basis of these contacts.

Aside from these proposals, another reasonable approach to *spin injection* has been based on the use of the so-called *magnetic semiconductors* as *spin injectors*. The alleviation of dissimilarity of the materials in contact (i.e., *magnetic semiconductor* and *heterostructure*) paves the way for a more efficient *spin-injection* process. The design of these *spin-injection* interfaces is in principle based on the familiar *bandgap* engineering techniques used in all semiconductor–semiconductor interfaces.

Ferromagnetic semiconductors form a broad class of semiconducting *compounds* that incorporate a magnetic element. Incorporation of this magnetic element has been shown to lead to pronounced magnetic properties such as *spin-polarized carrier* population. *Ferromagnetic semiconductors* possess a *long-range ferromagnetic* order below their characteristic *Curie temperature*. Classic examples of these semiconductors are *europium chalcogenides* and the *chalcogenide spinels*. Luckily, both of these groups of materials have been extensively studied several decades before the inception of the idea of *spintronics*. According to the experimental results from the operation of the junctions formed with these semiconductors, the *lift* in *spin degeneracy* at the *band* edges has the capacity to yield a *spin-filter* effect. This effect on *EuS*- and *EuSe*-based single-barrier *heterostructures* was originally investigated by Esaki et al. in 1967.[123] According to their experiments, the difference among the potential barriers observed by *carriers* of opposite *spins*, which results from *lifting* the *spin degeneracy*, can be used in realizing a *net spin polarization* of the induced current in the semiconductor.

Although interesting, from an application point of view the low *Curie temperature* of these semiconductors makes them quite inacceptable for incorporation into the mainstream device *microfabrication* technology.

After many years of disinterest, interest in *ferromagnetic semiconductors* was revived with the discovery of *spontaneous ferromagnetic order* in $In_{1-x}Mn_xAs$ in 1989[124] and $Ga_{1-x}Mn_xAs$ in 1996.[125] Unlike the *europium chalcogenides*, these

[122] E. Rashba, "Theory of electrical spin injection: Tunnel contacts as a solution of the conductivity mismatch problem," *Phys Rev B.*, vol. **62**, P. R16267, 2000.

[123] L. Esaki, P. Stiles, and S. von Molnar, "Magnetointernal field emission in junctions of magnetic insulators," *Phys Rev Lett*, vol. **19**, p. 852, 1967.

[124] H. Munekata, H. Ohno, S. von Molnar, A. Segmüller, L. Chang, and L. Esaki, "Diluted magnetic III-V semiconductors," *Phys Rev Lett*, vol. **63**, p. 1849, 1989.

[125] J. De Boeck, R. Oesterholt, A. Van Esch, H. Bender, C. Bruynseraede, C. Van Hoof, and G. Borghs, "Nanometer-scale magnetic MnAs particles in GaAs grown by molecular beam epitaxy," *Appl Phys Lett*, vol. **68**, p. 2744, 1996;H. Ohno, A. Shen, F. Matsukura, A. Oiwa, A. Endo, S. Katsumoto, and Y. Iye, "(Ga,Mn)As: A new diluted magnetic semiconductor based on GaAs," *Appl Phys Lett*, vol. **69**, p. 363, 1996; H. Ohno, "Making nonmagnetic semiconductors ferromagnetic," *Science*, vol. **281**, pp. 951–956, 1998.

magnetic semiconductors are considerably receptive to the idea of employment in *microfabrication* technologies. This is because in these cases *ferromagnetic* properties are realized in semiconductor hosts that are already widely employed in realization of high-performance mainstream electronic devices.

Another important group of *magnetic semiconductors* are semiconductors diluted (not exactly alloyed) with about 5% *manganese*. This level of *Mn* incorporation not only offers a *ferromagnetic* manifestation but also a high level of *acceptor doping*. Theory predicts that the *ferromagnetic* manifestations of this group of *p-type* semiconductors should remain stable in a wide variety of semiconductor hosts. In this regard, $Ga_{1-x}Mn_xAs$, $In_{1-x}Mn_xAs$, and Mn_xGe_{1-x} have been studied extensively. Whereas these *diluted semiconductors* offer a source of *spin-polarized holes*, among them the high *hole masses* and the short *spin lifetimes* (caused by the strong *spin–orbit coupling*) have been observed to impose serious drawbacks on the device applications. In terms of *spin-scattering lifetime*, electrons are much more promising than *holes*. Due to this, an ongoing research on realization of *spin-polarized* electron profiles is underway. Success in this endeavor will be pivotal to realization of *spinFETs* operating at room temperature, instead of at *cryogenic* temperatures. High *Curie temperature ferromagnetic semiconductors* such as *MnGaN* and especially *MnInN* are deemed to have the capacity to obtain an application niche in the area of *spintronics*.

One of the most studied semiconductors in realization of *spin-polarized* electron populations is $Zn_{1-x}Mn_xSe$. Granting the possibility of realizing *spin-polarized* electron concentrations is due to the fact that this semiconductor can be *doped n-type*. *ZnMnSe* is referred to as a *Brillouin paramagnet*. As a result of being a *paramagnet*, this material acquires a net magnetization only in the presence of a magnetic field. Therefore, although an interesting material, *ZnMnSe* is not exactly suited for application in devices such as the *spinFET* of Figure 4.33 (which is supposed to operate without the presence of a magnetic field). An interesting property of this semiconductor is that since both *ZnSe* and *GaAs* have *zinc blende* crystalline structures with a *lattice mismatch* of only 0.25%, this *paramagnetic semiconductor* offers a very suitable match to integration with the superb *GaAs channel*. The *lattice-mismatch* and *conduction-band* discontinuity at the $Zn_{1-x}Mn_xSe/Al_yGa_{1-y}As$ *heterointerface* is controllable through the proper selection of *Al* and *Mn mole fractions* in these compounds. These *diluted magnetic semiconductors* offer a convenient source of *spin-polarized carriers* (with the *spin injection efficiency* into another semiconductor exceeding 80%). However, this is achieved at low temperatures (i.e., below 10 K) and high applied magnetic fields (i.e., larger than 1 T).

Regarding the application of *magnetic semiconductors* as *spin injectors*, it is worthwhile mentioning that an inverse correlation exists between the *diffusive spin-injection efficiency* and the density of *interface* defects.[126] Due to the presence of *interface* defects associated with *stacking faults* in all *cubic heteroepitaxial* systems, these defects are deemed as a potential source of loss in *spin polarization* among semiconductor *spintronic* devices realized in such structures.

[126] R. Stroud, A. Hanbicki, Y. Park, G. Kioseoglou, A. Petukhov, B. Jonker, G. Itskos, and A. Petrou, "Reduction of spin injection efficiency by interface defect spin scattering in ZnMnSe/AlGaAs-GaAs spin-polarized light-emitting diodes," *Phys Rev Lett*, vol. **89**, p. 166602, 2002.

Besides *spin injection* from the *ferromagnetic MS* contacts and the use of *ferromagnetic semiconductors*, *optical pumping* of *spin-polarized carrier* populations into semiconductor *heterostructures* is the other approach to realization of the first component in a *spintronic* setup.

In quantifying the success of *spin injection*, an interesting approach is based on one of the distinguishing characteristics of semiconductor's *radiative recombination* of *charge carriers* and the consequent emission of light.[127] In a *spin-LED*, a *net-polarized* population of *charge carriers* is electrically injected from the metal contact into the semiconductor. Upon *radiative recombination*, those *carriers* that have managed to retain their *spin polarization* (i.e., before going through *radiative recombination*) emit *circularly polarized* light of *positive* or *negative helicity*. In terms of this technique of characterization, the *spin polarization* of the emitted light can be treated as an accurate measure of the *carrier* population's *spin polarization* in the semiconductor.[128]

REFERENCES

J. Appenzeller, J. Knoch, R. Martel, V. Derycke, S. J. Wind, and P. Avouris, "Carbon nanotube electronics," *IEEE Trans Nanotechnol*, vol. **1**, no. 4, pp. 184–189, 2002.

E. B. Cohen, K. J. Webb, D. B. Janes, and M. R. Melloch, "Real space transfer in a velocity modulated transistor structure," *Appl Phys Lett*, vol. **70**, no. 21, pp. 2864–2866, 1997.

J. Colinge (Ed), *FinFETs and Other Multi-Gate Transistors*, Springer, New York, 2008.

S. Datta and B. Das, "Electronic analog of the electro-optical modulator," *J Appl Phys*, vol. **56**, no. 7, pp. 665–667, 1990.

A. Javey, R. Tu, D. B. Farmer, J. Guo, R. G. Gordon, and H. Dai, "High-performance n-type carbon nanotube field-effect transistors with chemically-doped contacts," *Nanoletters*, vol. **5**, no. 2, pp. 345–348, 2005.

Y. Lin, J. Appenzeller, J. Knoch, and P. Avouris, "High performance carbon nanotube field-effect transistor with tunable polarities," *IEEE Trans Nanotechnol*, vol. **4**, no. 5, pp. 481–489, 2005.

H. Nathanson, W. E. Newell, R. A. Wickstrom, and J. R. Davis, Jr., The resonant-gate transistor, *IEEE Trans Electron Devices*, vol. **14**, no. 3, pp. 117–133, 1967.

H. Sakaki, "Velocity modulation transistor (VMT): A new field-effect transistor concept," *Jpn J Appl Phys*, vol. **21**, no. 6, pp. L381–L383, 1982.

FURTHER READING

In addition to those references mentioned throughout this chapter, the following references have been used in the preparation of the materials presented in this chapter. They are also suggested as sources for further reading.

[127] This principle is used in the so-called *spin-polarized light-emitting diodes* (i.e., *spin-LED*).

[128] B. Jonker, "Progress toward electrical injection of spin-polarized electrons into semiconductors," *Proc IEEE*, vol. **91**, no. 5, pp. 727–740, 2003.

G. Bir and G. Pikus, *Symmetry and Strain-Induced Effects in Semiconductors,* Halsted Press, Hoboken, NJ, 1974.

J. Burghartz, *Guide to State-of-the-Art Electron Devices,* IEEE Press, Hoboken, NJ, 2013.

J. Colinge (Ed), *FinFETs and Other Multi-Gate Transistors,* Springer, New York, 2008.

P. Harris, *Carbon Nanotubes and Related Structures, New Materials for the Twenty-First Century,* Cambridge University Press, Cambridge, UK, 1999.

C. Hu, *Modern Semiconductor Devices for Integrated Circuits,* Prentice Hall, Upper Saddle River, NJ, 2010.

A. Javey and J. Kong (Eds), *Carbon Nanotube Electronics,* Springer, New York, 2009.

M. Lundstrom and J. Guo, *Nanoscale Transistors, Device Physics, Modeling and Simulation,* Springer, New York, 2006.

E. Nagaev, *Physics of Magnetic Semiconductors,* Mir Publisher, Moscow, 1983.

C. Nguyen, *Micromechanical Signal Processors,* PhD Dissertation, University of California-Berkeley, Berkeley, CA, 1994.

R. Saito, G. Dresselhaus, and M. S. Dresselhaus, *Physical Properties of Carbon Nanotubes,* Imperial College Press, London, 1998.

Y. Sun, S. Thompson, and T. Nishida, *Strain Effects in Semiconductors, Theory and Device Applications,* Springer, New York, 2010.

Y. Taur and T. Ning, *Fundamentals of Modern VLSI Devices,* Cambridge University Press, Cambridge, UK, 2009.

PROBLEMS

4.1 Through drawing band diagrams demonstrate why the 4.4–4.7 eV work function of some metal nitrides makes them suitable for the gate of the low-power multiple-gate CMOS transistors and not so for the planar CMOS.

4.2 Analyze the following seminal article and summarize the foundation of the voltage-doping transformation model. Assess the applicability of these assumptions to modern CMOS transistors:

T. Skotnicki, G. Merckel, and T. Pedron, "The voltage-doping transformation: a new approach to modeling of MOSFET short-channel effects," *IEEE Electron Device Lett*, vol. **9**, no. 3, pp. 109–112, 1988.

While starting your research with this article, perform a survey for more recent publications on this topic and use them in your analysis.

4.3 Using the layer structure presented in Figure 4.14a, try to duplicate the conduction-band edge diagram presented in Figure 4.14b. Indicate the sources employed for acquiring the constant required in these calculations.

4.4 Read the following seminal article and on its basis carefully formulate the calculation framework for the resonance of a doubly clamped beam resonator:

F. Bannon, J. Clark, and C. Nguyen, "High-Q HF microelectromechanical filters," *IEEE J Solid State Circ*, vol. **35**, no. 4, pp. 512–526, 2000.

4.5 Perform a literature survey on the topic of SB-MOSFETs and summarize the foundations of the operation of these transistors.

4.6 Perform a literature survey on the topic of tunneling FETs and summarize the foundations of the operation of these transistors.

4.7 Perform a literature survey on the dependence of the size of the bandgap of CNTs on their diameter.

4.8 Perform a literature survey on the application of CNFETs as chemical sensors and summarize your findings on the topic of suitability of CNT medium for these sorts of applications.

4.9 As mentioned in Section 4.5, spin injection from magnetic metals to semiconductors through a tunneling MIS barrier has been deemed suitable for realization of SpinFETs. Perform a literature survey and summarize your findings on the latest developments on this topic.

5

HETEROJUNCTION FETs

The inception of the idea of realizing potential barriers in the form of semiconductor *heterojunctions* goes back to the early 1950s. However, in this case, it took the technology more than two decades to catch up with the theory. Not too far after the demonstration of *transistor action* in *germanium* technology with the use of *point-contact* metal–semiconductor junctions, a variety of techniques for the realization of *metallurgical p–n homojunctions* started to appear. Concurrent with these developments, the idea of implementing *p–n* junctions between two different semiconductors (i.e., *heterojunctions*) was also theoretically considered first by Gubanov[1] and a few years later by Kromer.[2] Early attempts for the realization of *Ge/Si* and *Ge/GaAs heterojunctions* (through the *alloying* and the *vapor deposition* processes, respectively) resulted in very high dislocation densities at the *heterointerfaces*. As a result of the disappointing quality of these interfaces, early research on *heterostructures* waned.

The realization of crystal growth techniques, such as *liquid-phase epitaxy* (*LPE*), *molecular beam epitaxy* (*MBE*), and *metalorganic chemical vapor deposition*

[1] A. Gubanov, "On the theory of contact of two semiconductors with mixed conductivity," *Zh Eksp Teor Fiz*, vol. **21**, p. 79, 1951 (In Russian); A. Gubanov, "ТЕОРИЯ КОНТАКТА ДВУХ ПОЛУПРОВОДНИКОВ С РАЗЛИЧНЫМ ТИПОМ ПРОВОДИМОСТИ [Theory of contact between two semiconductors with different types of conductivity]," ЖУРНАЛ ТЕХНИЧЕСКОИ ФИСИКИ *(USSR J Tech Phys)*, vol. **20**, pp. 1287–1301, 1950 (In Russian); A. Gubanov, "ТЕОРИЯ КОНТАКТНЫХ ЯВЛеНИЙ В ПОЛУПРОВОДНИАХ [Theory of contact phenomena in semiconductors]," ЖУРНАЛ ТЕХНИЧЕСКОИ ФИСИКИ *(USSR J Tech Phys)*, vol. **22**, pp. 729–735, 1952 (In Russian).
[2] H. Kromer, "Theory of a wide-gap emitter for transistors," *Proc IRE*, vol. **45**, no. 11, pp. 1535–1537, 1957.

Field Effect Transistors, A Comprehensive Overview: From Basic Concepts to Novel Technologies, First Edition. Pouya Valizadeh.

(*MOCVD*) in the 1970s, paved the way for the inception of important groups of both electronic and optoelectronic devices. The revival of interest in *heteroepitaxy* is owed to the successful fabrication and demonstration of *continuous-wave* room-temperature *double-heterostructure AlGaAs/GaAs/AlGaAs lasers* in 1971.[3] The success of *heteroepitaxy* in the $Al_xGa_{1-x}As/GaAs$ system is especially due to unique properties of this material system pertaining to the *lattice matching* of *GaAs* and *AlGaAs* within a wide range of *Al mole fractions* (i.e., *x*).

Regarding the topic of this volume, the most important contribution of this technological revolution has been the realization of *heterojunction field effect transistors* (*HFETs*). These are *FETs* for which the *channel* is formed in the vicinity of *hetero-interface(s)* of two (or three) semiconductors of different values of both *bandgap* and *electron affinity*. Whereas the *lateral lattice constants* of these semiconductors can be either matched or mismatched, in order to avoid the presence of *dangling bonds* and *trap levels* in the vicinity of the *channel*, crystal growth is often performed *pseudomorphically*.[4] This mode of crystal growth, when the constituent semiconductors of the *heterostructure* do not share a common *freestanding lateral lattice constant*, results in the introduction of mechanical strain at the *heterointerfaces*. We open the discussion of this chapter with a section on challenges and rewards of *heteroepitaxy* among a few well-known semiconductor families.

Although *HFETs* share many of the principles of operation mentioned for *FETs* in the previous two chapters, owing to their different forms of charge induction in the *channel*, they have extended an easier path to the realization of very high-speed switches and very high-gain/low-noise transistors suitable for microwave and millimeter-wave applications. The role of quantum phenomena in the operation of these *heterostructure*-based devices is even more pivotal. As a result, part of this chapter's presentation of the *HFETs* deals with a brief review of quantum phenomena in semiconductor *heterostructures*. As a continuation of the discussion presented earlier in Chapter 1, this review will look into the evolution of *energy subbands* within the *heterostructure quantum wells*.

The realization of *HFETs* on *lattice-mismatch heterostructures* formed in the *III-Nitride* material system in the mid-1990s,[5] bearing in mind the so-called *polar* properties of these *heterostructures*, demands a more careful look into the induction of two components of *polarization* (i.e., *spontaneous* and *piezoelectric*) at these *pseudomorphically* grown *heterointerfaces*. Later in this chapter, after the presentation of the operation principles of the so-called *nonpolar modulation-doped HFETs*, *polarization doping* for the specific example of *Wurtzite III-Nitride heterostructure* will be dealt with.

[3] I. Hayashi, M. Panish, and F. Reinhart, "GaAs–$Al_xGa_{1-x}As$ double heterostructure injection lasers," *J Appl Phys*, vol. **42**, p. 1929, 1971.

[4] See the discussion in Section 1.6.2.2.

[5] See M. Khan, M. Shur, Q. Chen, and J. Kuznia, "Current/voltage characteristic collapse in AlGaN/GaN heterostructure insulated gate field effect transistors at high drain bias," *Electron Lett*, vol. **30**, no. 25, pp. 2175–2176, 1994.

5.1 CHALLENGES AND REWARDS OF HETEROEPITAXY

In the early days of *heteroepitaxy*, the growth of clean *heterointerfaces* between *lattice-mismatch* semiconductors was a major challenge. This restricted the implementation of sharp *heterointerfaces* only to a limited number of *lattice-match* semiconductors. The larger the difference between the *lateral lattice constants* of the two semiconductors forming the *heterointerface*, the bigger of an obstacle this issue was. As a result, the incorporation of the so-called buffer layers with values of *lattice constant* between those of the layers intended for forming the *heterostructure* was studied (this is sometimes referred to as *interface grading*). As examples, in order to grow *GaInAs* (or *GaAsSb*) on *GaAs* substrates, *buffer layers* were originally grown on *GaAs* to make a gradual transition from the *lattice* of *GaAs* to that of $Ga_{1-x}In_xAs$ (or $GaAs_{1-y}Sb_y$). This can be done in the form of increasing the *In* (or *Sb*) composition either gradually or in very small steps (i.e., constant across very thin layers) from its value at the interface of *GaAs* substrate (i.e., $x = 0$ and $y = 0$) to the maximum value required in the design of the overlayer. The small difference in *lattice constant* between the two immediate *monolayers* has been shown to allow a very clean interface. This technological development was not suitable for forming a sharp *heterointerface* between *GaAs* and $Ga_{1-x}In_xAs$ (or $GaAs_{1-y}Sb_y$), with value of x (or y) selected for the overlayer.

In overcoming these challenges, through precise control of temperature, pressure, and growth rate, *strained heteroepitaxy* of thin *epilayers* has come a long way since the early days. A very interesting example of this success is the realization of sharp *pseudomorphic heterojunctions* between the two widely *lattice-mismatch* semiconductors of *GaN* and *AlN* in the early 2000s.[6] While often in the evaluation of *lattice matching* we refer to the values of *lattice constants* evaluated at room temperature, it is necessary to appreciate that the growth temperature and operation temperature (i.e., due to *self-heating* of the electronic devices under *bias* and the temperature of the environment) are widely different from 300 K. As a result, any consideration of the reliability of operation of *heterojunction*-based devices requires the appreciation of thermal-expansion behavior of the constituting semiconductors of the *heterostructure*.

In Figure 5.1, the approximate values of *bandgap* energy and *lattice constant* of a number of *elemental* and *compound* semiconductors are presented. For the *alloyed* semiconductors shown in this figure, variations of the size of the *bandgap* and *lattice constant* with the compositional proportion of the parent semiconductors are presented by the curves connecting the dots representing these parents. An important property of *compound* semiconductors is their miscibility over the entire (or at least

[6] For the *Wurtzite GaN*, the *lattice constant* a at room temperature is 3.189 Å (while c is 5.186 Å). The value of a for *AlN* is 3.112 Å (while c is 4.982 Å). Considering these values of a, according to Equation (1.105), a high strain of 2.47% is developed at the *Wurtzite tensile-strained AlN/GaN heterointerface* grown along the *c-axis*. Due to this large *tensile strain*, in order to avoid crack formation, the thickness of the *AlN epilayers* is restricted to only a few *nanometers*.

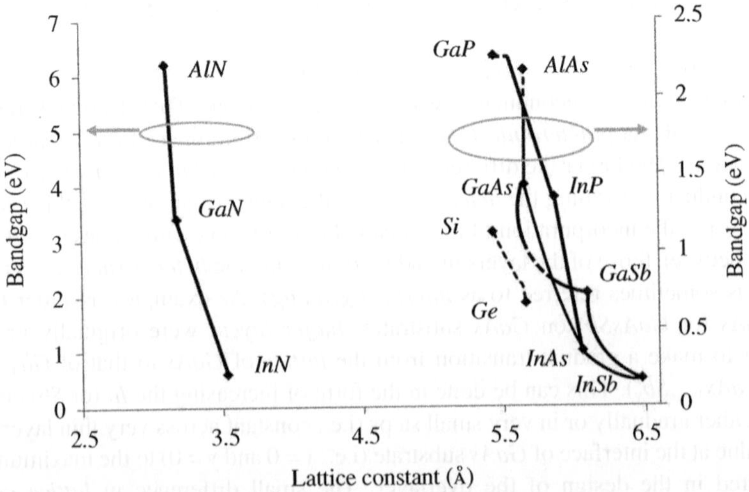

FIGURE 5.1 Approximate variation of the room-temperature bandgap versus lattice constant among a number of semiconductors. Diamond marks represent these information for both elemental and compound binary semiconductors. The curves connecting these symbols represent these information for the alloys made out of the semiconductors represented by the diamond marks at the two ends of each curve. The dashed parts of these curves represent an indirect bandgap characteristic.

part of the) composition range. In this figure, the dashed portions of the curves indicate the range of compositions resulting in an *indirect bandgap* semiconductor. For example, in the case of *AlGaAs* while on the basis of the *band structures* of *GaAs* and $Al_xGa_{1-x}As$ for *x* smaller than 0.45 a *direct bandgap* is observed,[7] for larger values of *Al mole fraction*, the *bandgap* of *AlGaAs* takes on an *indirect* character.[8] For the *indirect cubic* semiconductors, the *indirect bandgap* might be defined between either the X_6 or L_6 and the Γ_8 points (i.e., identified at two different values of $\vec{k}$ in the *Brillouin zone*). The symbols Γ_n, X_n, and L_n refer to *group* notions used in identifying the symmetry of the *Brillouin zone* at a so-called reciprocal lattice point.[9]

As shown in Figure 5.1, the *indirect bandgaps* of some *alloys* such as *InGaP* (which are defined by the *X valley* of the *conduction band*) change very little with the proportional composition of the parents in the semiconductor *alloy*. This causes a sharp transition in the slope of the solid curve (i.e., *direct bandgap*) and the dashed curve (i.e., *indirect bandgap*).

[7] This *bandgap* is defined by the difference in energy between the *conduction-band* minima (i.e., Γ_6) and the *valence-band* maxima (i.e., Γ_8), which are all located at $\vec{k} = [000]$ in the *Brillouin zone*.

[8] This is defined by the energy-separation between X_6 and Γ_8 points in the *Brillouin zone*.

[9] We have had earlier encounters with these concepts in Chapter 1, including the mention of the *k-space* as a *Fourier-space* in Section 1.7.3. The notion of a *reciprocal lattice* refers to the result of this transformation.

5.1.1 Lattice Matching and the Substrate Challenge

Exploring alternative *III–V* and *II–VI* semiconductor technologies has required much investment toward creating proper substrates with acceptable *lattice matching* to each of these technologies. The realization of commercial *GaAs* and *InP* substrates is the major result of these investments. As indicated earlier in Section 1.6.2.2.1, the presence of *lattice mismatch* even in the case of *strained pseudomorphic epitaxy* poses difficulties on the realization of thick *epilayers* (especially so for the *tensile-strained* among these).

Among the major *III–V* semiconductors, the creation of a *lattice-match* substrate to *GaN* has been proven to be the most challenging of all. This is because the traditional technique used in the realization of semiconductor substrates (i.e., *Czochralski wafer-fabrication process*[10]) relies on melting the *electronic-grade*[11] *polycrystalline charge* in a crucible. Nevertheless, due to the high melting temperature of *GaN*,[12] in this semiconductor technology, finding an appropriate crucible has not been possible. So far, a few research teams have been successful in devising an alternative technique for realizing *free-standing GaN* substrates. However, the size of the resulting substrate has remained limited.[13] This is due to the high-pressure requirement of this growth technique. As a result, thus far, much of the work on this semiconductor family has been based on the adoption of *lattice-mismatch* substrates such as *SiC*, *sapphire*, and *silicon* and the incorporation of intermediate *buffer layers* between the substrate and the thin *epilayers* intended for forming a sharp *heterointerface*. In these endeavors, the differences in the crystalline structure and the value of the *lattice constant* between the substrate and the *epilayers* have caused the *III-Nitride epilayers* grown on these substrates to suffer from high dislocation densities (i.e., $>10^9\ cm^{-2}$). As expected, these issues have so far contributed to performance and reliability concerns among the devices fabricated on these *epilayers*.

The availability of relatively large-size *GaAs* and *InP* substrates has been instrumental in the success of a number of *III–V* compound semiconductor technologies. Among these *III–V compounds*, the *ternary AlGaAs compound* (which is *lattice-matched* to the *GaAs* substrate) and the *quaternary GaInAsP compound* (which for certain compositions is *lattice-matched* to the *InP* substrate) especially stand tall. The *lattice-mismatch* $\Delta a/a$ (defined in terms of the *free-standing lattice constant* a and the difference of its value between the two layers of the *heterostructure*) between *GaAs* and *AlAs* is only 0.14%. This is based on the values of the room-temperature *lattice constant* of *GaAs* (i.e., 5.6532 Å) and *AlAs* (i.e., 5.6612 Å).

[10] Named after *Jan Czochralski*.

[11] That is of very high purity in the order of 99.9999999%.

[12] Which is about 2500°C.

[13] M. Boćkowski, I. Grzegory, B. Łucznik, T. Sochacki, M. Kryśko, P. Strąk, I. Dzięcielewski, E. Litwin-Staszewska, and S. Porowski, "High nitrogen pressure solution growth of bulk GaN in "feed-seed" configuration," *Phys Status Solidi A*, vol. **208**, no. 7, pp. 1507–1510, 2011.

For the *quaternary* systems, the following two-step procedure (applied in here to the special case of *GaInAsP quaternary* grown on the *InP* substrate) is used in identifying the *lattice-match* composition:

1. *Lattice-match* the ternary $Ga_{1-z}In_zAs$ to *InP*. This can be done by approximately assuming a linear relationship between the *lattice constants* while using the *In mole-fraction* z as a parameter:

$$a_1 + (a_2 - a_1)z = a_3. \tag{5.1}$$

For this example, a_1 is the *lattice constant* of *GaAs*, a_2 is the *lattice constant* of *InAs* (i.e., 6.0584 Å), and a_3 is the lattice constant of *InP* (i.e., $a_3 = 5.86875$ Å). As a result, the *In mole fraction* (i.e., z) for *lattice matching* to *InP* needs to be equal to 0.532 (i.e., at room temperature). Experimentally, it is also observed that the ternary *compound* of $Ga_{0.47}In_{0.53}As$ is *lattice-matched* to *InP*.

The linear (or *Vegard's law*) equation of (5.1), while often offering a very good first-order approximation, is not always valid.

2. Now the *quaternary compound* $Ga_xIn_{1-x}As_{1-y}P_y$ can be taken into account as an *alloy* of y parts *InP* and 1 – y parts $Ga_{1-z}In_zAs$. As a result, $x = (1 - y)(1 - z)$. Since for $z = 0.532$, $Ga_{1-z}In_zAs$ is already *lattice-matched* to *InP*, this *quaternary compound* also remains *lattice-matched* to *InP* for this value of z.

It is of course also of interest to evaluate the size of the *bandgap* of this *quaternary compound*. If we use *Vegard's* law between the sizes of the *bandgaps* of the parent semiconductors, the *bandgap* of the *quaternary* will be given by

$$E_g = E_{g1} \times x + E_{g2} \times (1 - x - y) + E_{g3} \times y \tag{5.2}$$

where E_{g1}, E_{g2}, and E_{g3} stand for the *bandgaps* of *GaAs*, *InAs*, and *InP*, respectively.

Obviously, before applying such a procedure to other material systems, we have to make sure that *Vegard's law* is applicable. Otherwise, the appropriate nonlinear trend should replace *Vegard's law*.

5.1.2 Properties of a Few Famous Nonpolar Heterostructures: A Brief Visit

In order to gain a closer insight into the behavior of *heterostructures*, in this section, an overview of the properties of two famous examples (i.e., *AlGaAs/GaAs* and *GaInAsP/InP*) is presented.

As indicated in Section 1.6.2.2.2, considering the deviations from the linear assumption of *Vegard's law*, generally speaking, a given property *P* of a *ternary compound* $A_xB_{1-x}C$ can be expressed in terms of the corresponding *P* properties of the *binary compounds*:

$$P_{ABC} = xP_{AC} + (1-x)P_{BC} - x(1-x)c = a + bx + cx^2 \tag{5.3}$$

where $a = P_{BC}$ and $b = P_{AC} - P_{BC} - c$.

The parameter c is generally referred to as the *bowing* or the *nonlinearity parameter*. Contributions to P from the *lattice* disorders generated by intermixing of atoms A and B on the *lattice sites* (which are supposed to be normally occupied by one of these atoms in a *binary*) are the causes of this nonlinearity. In the case of physical quantities related to *lattice vibrations* such as *phonon energy*, the quadratic term in (5.3) needs to be included.

5.1.2.1 AlGaAs/GaAs Figure 5.2 depicts the approximate variations of the *lattice constants* of *GaAs* and *AlAs* as functions of temperature. As shown in this figure, the two semiconductors become exactly *lattice-matched* at around 900°C. However, due to the very limited difference among the *thermal expansion coefficients* of these two *binaries*, even at room temperature, only a small *lattice mismatch* is observed between the $Al_xGa_{1-x}As$ and *GaAs* (i.e., $(\Delta a/a) = 13x \times 10^{-4}$).

This property is extremely precious to designing *AlGaAs/GaAs heterojunctions*. As an example, for $Al_{0.3}Ga_{0.7}As$ grown on *GaAs*, a very small *lattice mismatch* of about 0.04% is expected. This amount of *lattice mismatch* produces an average stress of only about $6 \times 10^3\ \mathrm{N/cm^2}$.

Aside from this exceptional case, in the presence of large differences among the *thermal-expansion coefficients* of the layers forming a *heterojunction*, even for the most suitable of *heterostructures*, *lattice matching* can be only achieved at one temperature. As a result, one should choose between *lattice matching* near the growth temperature and *lattice matching* at room temperature.

In forming the *ternary compounds* of *AlGaAs*, one should be reminded that while *GaAs* is a *direct* semiconductor, *AlAs* is an *indirect* semiconductor. As a result, in the *E–k* diagrams of $Al_xGa_{1-x}As$, there is a transition from *direct* to *indirect bandgap* as x

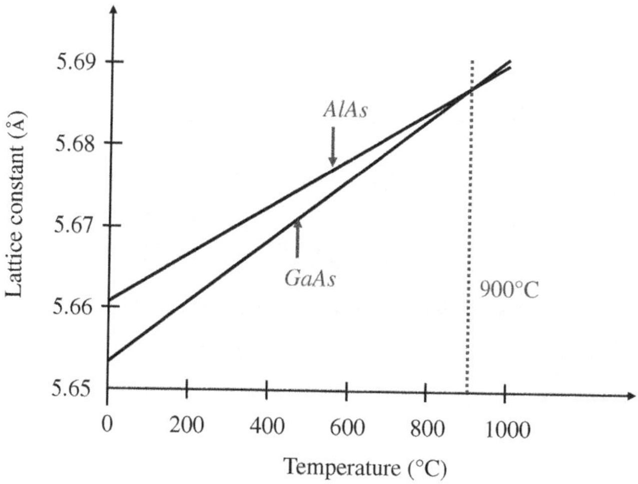

FIGURE 5.2 Approximate temperature dependence of the lattice constants of *GaAs* and *AlAs*. Reproduced with permission from: Ettenberg and Paff (1970). Copyright 1970, AIP Publishing LLC.

increases. As already mentioned in Section 5.1, this transition occurs at around $x = 0.45$. At room temperature, the *bandgap* of the *direct ternary* (i.e., $E_g = E(\Gamma_6) - E(\Gamma_8)$) varies with the *Al mole fraction* as

$$E_g = 1.424 + 1.247x. \tag{5.4}$$

This is the result of the application of *Vegard's law* between the *direct bandgaps* of *AlAs* and *GaAs*.

However, for the *indirect ternary*, the room-temperature value of the *bandgap* (i.e., $E_g = E(X_6) - E(\Gamma_8)$) varies as

$$E_g = 1.9 + 0.125x + 0.143x^2. \tag{5.5}$$

In the evaluation of the *bandgap* of a *ternary*, *Vegard's law* is often simply applied between the *bandgaps* of the *binaries*. In case of a *ternary* formed by *alloying* two *binaries* that do not both have *direct bandgaps*, *Vegard's law* should in principle be applied for all *k*-values to identify the bottom of *conduction band* (and a result the *bandgap*). Performing this task will be sufficient for *k*-values around the *X*-, Γ-, and *L-valleys*.

The *bandgap* difference between *GaAs* and *AlGaAs*, which is identifiable through Equations (5.4) and (5.5), is pivotal to the realization of *quantum wells* at *AlGaAs/GaAs heterointerfaces*. As noted already in Section 2.3.1, *electron affinity* is the other important parameter in evaluating the *band bending* at a semiconductor *heterointerface* (although such a parameter should be handled with due diligence).

For $Al_xGa_{1-x}As$, *electron affinity* in the *direct bandgap* region (i.e., $x < 0.45$) varies with the *Al mole fraction* as

$$q\chi = 4.07 - 1.06x. \tag{5.6}$$

However, *electron affinity* in the *indirect bandgap* region of *AlGaAs* is given by

$$q\chi = 3.64 - 0.14x. \tag{5.7}$$

As a result, a *conduction-band* discontinuity (i.e., $\Delta E_c = q\Delta\chi$) of $1.06x$ is expected between a *direct bandgap AlGaAs* and *GaAs*.

Corresponding to the definition presented in Section 1.6.2.2.2, *AlGaAs/GaAs heterostructures* present a *Type-I band lineup*. Accordingly, the *bandgap* discontinuity at the *heterointerface* in terms of the discontinuities of the *conduction* and *valence band* is presented as

$$\Delta E_g = \Delta E_c + \Delta E_v. \tag{5.8}$$

According to (5.4), (5.6), and (5.8), the *valence-band* discontinuity (i.e., ΔE_v) between the *direct bandgap AlGaAs* and *GaAs* is expected to be $0.187x$ (equivalently

$\Delta E_v \cong 0.15 \times \Delta E_g$[14]). However, in contradiction of this prediction, *C–V* and *I–V* measurements carried out on *AlGaAs/GaAs heterojunctions* provide the following empirical relationship between the *bandgap* and *conduction-band discontinuities*:

$$\Delta E_c \cong 0.60 \times \Delta E_g. \tag{5.9}$$

The observed discrepancy between the experimentally measured *conduction-band* discontinuity and the value predicted based on the difference of *electron affinities* across the *heterointerface* points to the inaccuracy in determining *electron affinity*. This goes beyond the expected error from the assumption of *Vegard's law*. As mentioned in Section 1.1.6, *electron affinity* is a quantity that can be only used in relative terms with a certain attributed approximation.

Another important property that can be tuned through the realization of a *ternary compound* from the *binaries* is the *dielectric constant* (i.e., ϵ_s). For $Al_xGa_{1-x}As$, using *Vegard's law*, the *static dielectric constant* is given by

$$\epsilon_s = (12.9 - 2.84x)\epsilon_0. \tag{5.10}$$

This presents the linear interpolation between $12.9\epsilon_0$ of *GaAs* and $10.06\epsilon_0$ of *AlAs*. In this case, the *high-frequency dielectric constant* is given by

$$\epsilon_\infty = (10.89 - 2.73x)\epsilon_0. \tag{5.11}$$

For the *direct bandgap* $Al_xGa_{1-x}As$, the values of *effective masses* of electrons and *holes* are respectively provided as

$$m_n^* = (0.063 + 0.083x)m_0 \tag{5.12}$$

$$m_p^* = (0.51 + 0.25x)m_0. \tag{5.13}$$

While not harboring the ambition to provide a list of all important properties of *AlGaAs*, it is worth mentioning that among the physical parameters important to our discussion, *phonon energies* and *thermal conductivity* are known to undergo severe nonlinearity in an *alloy*. *Adachi* has proposed the following nonlinear relationships for the *effective* energy values of the *longitudinal* and *transverse optical phonons* in $Al_xGa_{1-x}As$:[15]

$$h\nu_{LO} = 36.25 + 1.83x + 17.12x^2 - 5.11x^3 \text{ meV} \tag{5.14}$$

[14] Or equivalently $\Delta E_c \cong 0.85 \times \Delta E_g$.

[15] For a comprehensive source of information, see S. Adachi, *Properties of Semiconductor Alloys: Group IV, III–V, and II–VI Semiconductors*, John Wiley & Sons, Inc., Hoboken, NJ, 2009. As another reliable source of information, check out the electronic archive of New Semiconductor Materials on the website of *Ioffe Physico-Technical Institute*: http://www.ioffe.rssi.ru/SVA/NSM/

and

$$h\nu_{TO} = 33.29 + 10.70x + 0.03x^2 + 0.86x^3 \text{ meV}. \quad (5.15)$$

Linear interpolations, however, present an acceptable degree of approximation for the inverse of the *effective masses*, the *thermal-expansion coefficient*, and the *elastic constants*.

5.1.2.2 ***GaInAsP/InP*** As indicated earlier, the *GaInAsP/InP* system is also a system of technological importance. In the case of applicability of linear interpolations, the *bandgap* of the *GaInAsP quaternary* can be evaluated in terms of Equation (5.2).

According to *Adachi*, the following equations apply to the room-temperature *direct bandgaps* in this system:

$$E_g^\Gamma(InAs_yP_{1-y}) = 1.35 - 1.083y + 0.091y^2 \text{ eV} \quad (5.16)$$

$$E_g^\Gamma(In_{1-x}Ga_xP) = 1.35 + 0.643x + 0.786x^2 \text{ eV}. \quad (5.17)$$

Equation (5.17) assuming $x = 1$ yields $E_g^\Gamma = 2.78$ eV as compared to $E_g = 2.26$ eV for the *bandgap* of *GaP*. This is due to the fact that *GaP* is an *indirect bandgap* semiconductor.

For the $Ga_xIn_{1-x}As_yP_{1-y}$ *lattice-matched* to *InP*,[16] the following equation is used in the evaluation of the size of the *direct bandgap*:

$$E_g^\Gamma = 1.334 - 0.738y + 0.138y^2. \quad (5.18)$$

For this specific case, the *static-* and *high-frequency dielectric constants* are respectively given as

$$\epsilon_s = (12.50 + 1.44y)\epsilon_0 \quad (5.19)$$

and

$$\epsilon_\infty = (9.61 + 2.0y)\epsilon_0. \quad (5.20)$$

In this case, the values of the *longitudinal effective mass* of electrons and the *effective-masses* of *heavy* and *light holes* as well as *holes* of the *split-off band* are respectively provided as

$$m_l^* = (0.080 - 0.039y)m_0 \quad (5.21)$$

[16] That is, according to the discussion of Section 5.1.1, with *In mole fraction* of 0.532 in the *InGaAs* constituent.

$$m^*_{\mathrm{hh}} = (0.85 - 0.23y)m_0 \quad (5.22)$$

$$m^*_{\mathrm{lh}} = \left(0.12 - 0.099y + 0.03y^2\right)m_0 \quad (5.23)$$

$$m^*_{\mathrm{so}} = \left(0.21 - 0.01y - 0.05y^2\right)m_0. \quad (5.24)$$

5.2 QUANTUM PHENOMENA IN SEMICONDUCTOR HETEROSTRUCTURES

As previously mentioned in Section 1.6.2.2.2, the *bandgap* discontinuity materialized at semiconductor *heterojunctions* can effectively serve to confine the *carrier wave function* behind the interface of a thick-enough *barrier layer*. Before presenting an overview of quantum phenomena in thin-film *heterostructures*, it is important to identify the demarcation line between the notions of a *thick* and a *thin barrier layer* in this discussion. We refer to a layer as *thick* when its physical thickness is much greater than the penetration depth of the *carrier wave function*. A simple structure to present quantum phenomena consists of a single layer of a narrow-*bandgap* semiconductor (material A) sandwiched between two thick layers of a wider-*bandgap* semiconductor (material B), where the *bandgap* of material A in terms of energy is nested within the *bandgap* of material B. This is depicted in Figure 5.3. Such *heterostructures*, among other material combinations, can be formed between the following pairs: *GaAs/AlGaAs*, *GaInAsP/InP*, *GaInAs/AlInAs*, *GaSb/AlGaAs*, *GaInAsP/InP*, *GaInAs/AlInAs*, *GaSb/AlSb*, and *GaInN/AlGaN*. As observed in Chapter 1, this form of *band lineup*, which is known as *Type I*, results in the confinement of both electrons and *holes* in material A.

In such *Type-I heterostructures*, confinement of electrons and *holes* in material A results in the quantization of energy through the development of energy *subbands* in the *conduction* and the *valence band* of material A. In the discussion of this chapter, however, we will maintain a focus on the confinement of electrons only. According to

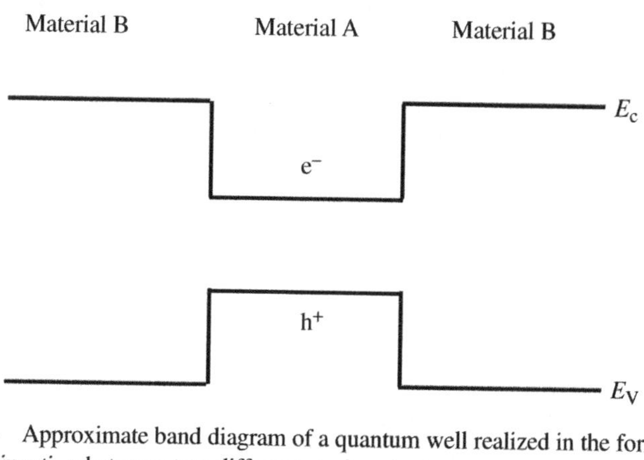

FIGURE 5.3 Approximate band diagram of a quantum well realized in the form of a *Type-I* double heterojunction between two different semiconductors.

Bastard,[17] the *energy levels* of the *conduction-band subbands* can be calculated quite easily via using the *envelope wave-function* approximation.[18] This approximation relies on employing a *Kane* model description of electron and *hole states* of the materials A and B.[19] This is valid based on two important assumptions:

1. On the scale of the variation of the *envelope wave function*, the interface potential is assumed to be well localized at the A/B *heterointerface*.
2. The interface potential is assumed to be only able of shifting the *band-edge wave functions* and not mixing them. This is plausible due to the very different symmetries of the *conduction* and the *valence band*.

According to these assumptions, *Bastard* has shown that the electron *wave function* is approximately representable by

$$\psi = \sum\nolimits_{A,B} \exp\left(j\vec{k}_\perp \cdot \vec{r}\right) \cdot U^{A,B}\left(\vec{r}\right) \cdot \chi_n(z) \tag{5.25}$$

where z indicates the growth direction of the *heterostructure*, $\vec{k}_\perp$ represents the *transverse* electron *wave vector*, $U^{A,B}\left(\vec{r}\right)$ are the *Bloch wave functions* in materials A and B, and $\chi_n(z)$ is the *envelope wave function*. This *envelope wave function* can be acceptably determined through solving the following *Schrödinger*-like equation:

$$\left(-\frac{\hbar^2}{2m_n^*(z)}\frac{\partial^2}{\partial z^2} + U_c(z)\right)\chi_n(z) = E_n\chi_n(z). \tag{5.26}$$

In calculation of the *envelope wave functions* in materials A and B, $m_n^*(z)$ is the electron *effective mass* of material A or B. $U_c(z)$ is representative of the *energy level* at the bottom of the *conduction band*, and E_n is referred to as the *confinement energy* of the *carriers*.

The continuity conditions over $\chi_n(z)$ and $\left[1/\left(m_n^*(z)\right)\right]\left[\left(\partial\chi_n(z)\right)/\partial z\right]$ should be enforced at the *heterointerfaces*. In quantum mechanics textbooks, instead of enforcing the continuity of $\left[1/\left(m_n^*(z)\right)\right]\left[\left(\partial\chi_n(z)\right)/\partial z\right]$, the continuity condition is applied over the derivative of the *wave function*. Considering the notion of *effective mass*, here this different definition of continuity is needed to enforce the conservation of particle current. This framework has well satisfied the predictions of *Dingle*, one of the early pioneers of *heterostructure*-based devices.

[17] See G. Bastard, "Superlattice band structure in the envelope-function approximation," *Phys Rev B*, vol. **24**, p. 5693, 1981; G. Bastard and J. Brum, "Electronic states in semiconductor heterostructures," *IEEE J Quantum Electron*, vol. **22**, no. 9, pp. 1625–1644, 1986.

[18] In case of the *valence band*, due to the presence of *heavy-* and *light-hole* and the *split-off bands*, the situation is more complicated.

[19] E. Kane, "Band structure of indium antimonide," *J Phys Chem Solids*, vol. **1**, no. 4, pp. 249–261, 1957; S. White and L. Sham, "Electronic-properties of flat-band semiconductor heterostructures," *Phys Rev Lett*, vol. **47**, no. 12, pp. 879–882, 1981.

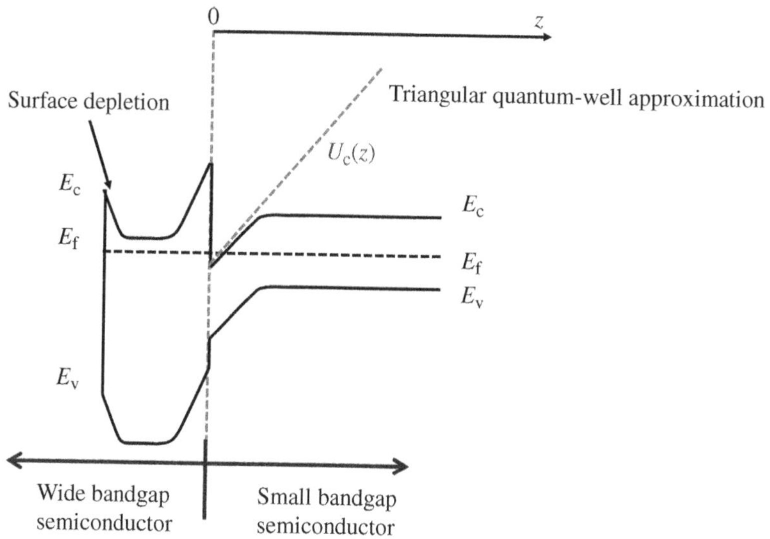

FIGURE 5.4 Approximate depiction of the typically observed thermal-equilibrium band diagram at the *Type-I* heterointerface of an n-doped wide bandgap semiconductor and an undoped small-bandgap semiconductor (such as *n-AlGaAs/i-GaAs*) and the evolution of the triangular quantum well.

As a practical example of electron confinement in a *quantum well*, and according to the discussion previously presented in Section 2.3, Figure 5.4 presents the approximate *thermal-equilibrium band lineup* toward the *heterointerface* of an *n-type doped AlGaAs* layer with an *undoped GaAs* layer. The *band lineup* of this *single heterojunction* presents one of the most studied structures for the realization of *HFETs*. Although in this structure the electron *wave function* is facing a single *AlGaAs barrier*, the evolution of the almost linearly varying potential at the *GaAs* side of the *heterointerface* poses the needed second *barrier* on the *wave function* to form a *quantum well*.[20] The presence of these two *barriers* results in the confinement of the electrons approximately to a plane formed parallel with the *heterointerface*. Since much of the *single heterostructures* used in the realization of *HFETs* present a *band bending* of this form,

[20] To be more thorough in our description, it is worth indicating that in a *MOSFET* (or a *single-heterojunction heterostructure*), while one of these *barriers* is formed at the Si/SiO_2 (or, generally speaking, the *barrier/channel*) interface, the formation of the second *barrier* is contingent on *bias* among a number of other parameters. For example, in a *MOSFET biased* in the *accumulation* regime, the second *barrier* does not materialize. Whereas in a *MOSFET* structure, the formation of the second *barrier* of the *potential well* (i.e., upon *inversion*) is also contingent on the *work function* of the *gate* metal and the *unblocked* presence of *carrier* reservoirs for populating the *inversion channel*, in the case of an *HFET*, the *doping* type and concentration in the *barrier layer* in addition to the *gate* metal's *work function* play important parts in determining the *band bending* between the *channel* at the *heterointerface* and the *channel layer* deeper within (which defines the size of the second *barrier*). In a *double-heterojunction HFET*, both *barrier* heights are determined by the discontinuity of the *conduction band* (for the case of n-type *channel*) or *valence-band* (for the case of p-type *channel*).

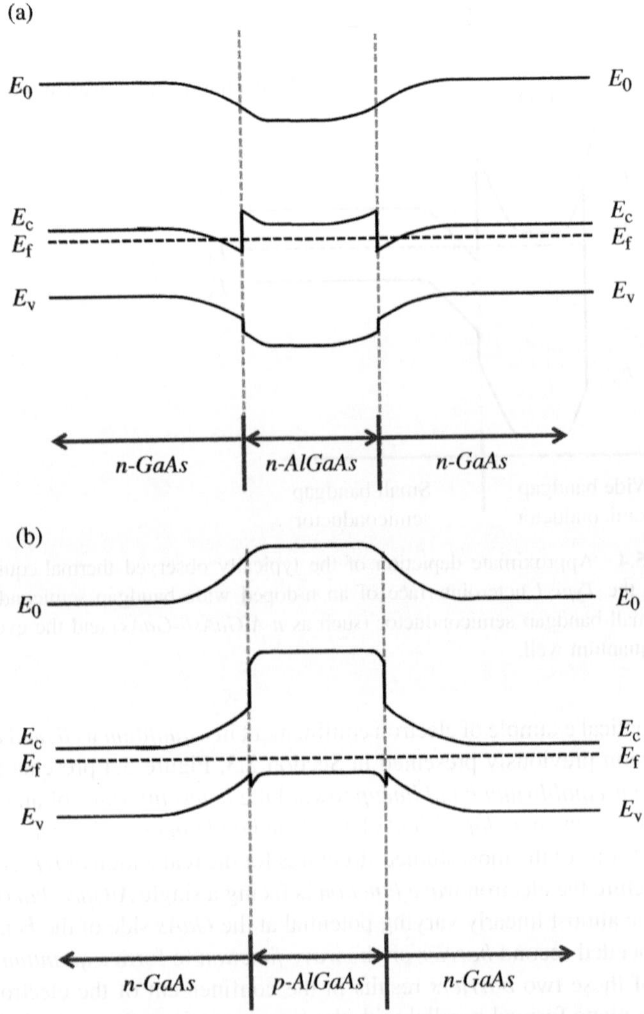

FIGURE 5.5 Approximate band diagram in the vicinity of the heterointerfaces of (a) a *n-GaAs/n-AlGaAs/n-GaAs* double heterostructure and (b) a *n-GaAs/p-AlGaAs/n-GaAs* double heterostructure.

the description of quantum phenomena in these so-called triangular *quantum wells* is quite important toward understanding the behavior of *HEFTs* realized on these *heterostructures*.[21]

To highlight the effect of *dopant* type on the *band lineup* of *heterostructures*, Figure 5.5 presents the approximate *band diagrams* of two *double-heterojunction*

[21] The following article of *Ando*, *Fowler*, and *Stern* presents one of the most detailed studies on the behavior of *heterostructure* systems. Readers are strongly encouraged to read this article: T. Ando, A. Fowler, and F. Stern, "Electronic properties of two-dimensional systems," *Rev Mod Phys*, vol. **54**, pp. 437–672, 1982.

heterostructures of different *doping* profiles: an *n-GaAs/n-AlGaAs/n-GaAs heterostructure* and an *n-GaAs/p-AlGaAs/n-GaAs heterostructure.*

In these *band diagrams*, according to the positions of the *acceptor* and *donor energy levels* in the *AlGaAs layer* with respect to the *band* edges of *GaAs*, *carrier* transfer across the *heterointerfaces* is prompted. This *carrier* transfer causes *band bending* (and potential drop) across different layers of the *heterostructure* to establish *thermal equilibrium*. As a result of these observations, we notice that in a *heterostructure*, both the *dopant* type and dopant distribution are important to the evaluation of the possibility of forming *quantum wells*.

5.2.1 Electron Behavior in a Triangular Quantum Well

The potential function in the case of a variety of semiconductor *heterojunctions* can be well approximated by a triangular profile, in which the potential $U_c(z)$ is linearly varying for $z>0$ and is defined by an infinite *barrier* at $z=0$. Such an approximate profile is superimposed on the *band diagram* of Figure 5.4. In terms of this potential function, the *Schrödinger*-like equation of (5.26), which is needed for calculating the form of the *envelope wave function*, is reformulated as

$$-\frac{\hbar^2}{2m_n^*}\frac{d^2\chi_n(z)}{dz^2}+qEz\chi_n(z)=E_n\chi_n(z) \tag{5.27}$$

in which E is the electric field for $z>0$, with the boundary condition of $\chi_n(z=0)=0$.

The solution to (5.27) is presented in terms of *Airy functions*.[22] While a differential equation of the form $u''-zu=0$ has two independent solutions, the solution that is *nonsingular* for $z\to\infty$ is referred to as the *Airy function* (often represented by Ai). As approximately shown in Figure 5.6, this function oscillates for $z<0$. For $z\to\infty$, it tends toward 0 following an exponentially decaying profile:

Considering the *Schrödinger*-like equation of (5.27), it can be deducted that the following *Airy function*

$$\mathrm{Ai}(\alpha+\beta z)=\mathrm{Ai}\left(\left(\frac{2m_n^*}{\hbar^2q^2E^2}\right)^{1/3}(qEz-E_n)\right) \tag{5.28}$$

yields an acceptable solution for the *envelope wave function*. This function behaves appropriately when $z\to\infty$. The application of the boundary conditions at $z=0$ allows the calculation of eigenvalues given by

$$E_n=-\left(\frac{\hbar^2q^2E^2}{2m_n^*}\right)^{1/3}a_n \tag{5.29}$$

where a_n is the nth 0 of Ai(z).

[22] Named after *George Airy*.

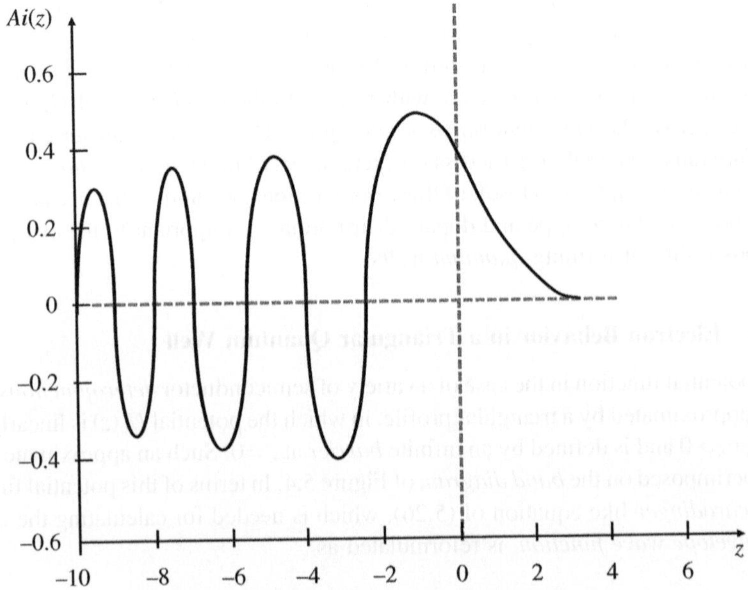

FIGURE 5.6 Behavioral plot of the *Airy function*.

These *eigenstates* stand for the *subband energy levels* developed within the triangular *quantum well*. Since often only the first (or the first two) *subbands* are occupied, an asymptotical approximation for small values of n is usually used for calculation of a_n, according to which

$$a_n \cong -\left[\frac{3\pi}{2}\left(n+\frac{3}{4}\right)\right]^{2/3}, \quad n=0,1,2,\ldots \tag{5.30}$$

so that

$$E_n = \left(\frac{\hbar^2}{2m_n^*}\right)^{1/3}\left[\frac{3\pi qE}{2}\left(n+\frac{3}{4}\right)\right]^{2/3}. \tag{5.31}$$

Figure 5.7 presents a behavioral depiction of the resulting *wave functions* in the *subbands* of such a triangular *quantum well*.

The degree of validity of these approximations is primarily dependent on the applicability of the triangular *potential-well* approximation. Whereas for some *heterojunctions* this framework is approximately valid, a more rigorous evaluation is needed for many other cases. We will revisit this framework in Section 5.2.3. Equation (5.31) is often presented in the empirical form of $E_n \propto (n_s)^{2/3}$, where n_s is the *sheet-carrier* concentration.

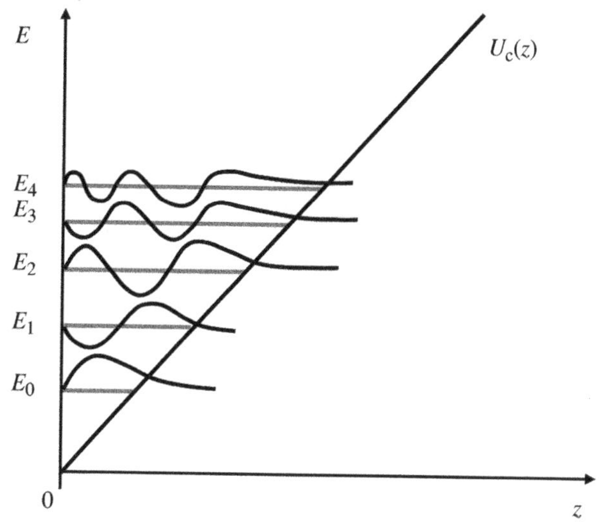

FIGURE 5.7 Behavioral depiction of the first five energy subbands and their electron wave functions in a triangular potential well.

5.2.2 Subbands and Two-Dimensional Electron Gas

The confinement of electrons to a triangular *quantum well* formed in the vicinity of a semiconductor *heterointerface* results in the formation of a sheet of *charge carriers* close to this interface. Many of the characteristics of this sheet of *charge carriers* are not unlike those of the *inversion-layer* charge formed in the *channel* of an *inversion-biased MOSFET*. The main difference between the two cases is that in the former, no *bias* is needed to induce this sheet of *charge carriers*. While this *sheet charge* is often encountered in terms of a 2-*D electron gas* (2*DEG*), the formation of a 2-*D hole gas* (i.e., 2*DHG*) is also plausible.

One of the important metrics in the characterization of semiconductor *heterointerfaces* is the *sheet-carrier concentration* (i.e., n_s). This metric, presented in units of per square centimeter, denotes the density of free electrons in the sheet of charge often formed toward the *bandgap* discontinuity of the *heterointerface*. The *sheet-carrier concentration* of the *n*th *subband* $n_{s\text{-}n}$ can be calculated using the *two-dimensional density of states*[23] and the *Fermi–Dirac* distribution function, $f_D(E)$:

$$n_{s-n} = \frac{m_n^*}{\pi\hbar^2}\int f_D(E)dE_\perp = \frac{m_n^* kT}{\pi\hbar^2} Ln\left[1+\exp\left(\frac{E_f - E_n}{kT}\right)\right]. \tag{5.32}$$

Here, $E = E_n + E_\perp$ for the *n*th *subband*. E_n is the energy of the *n*th *subband* and $E_\perp$ is the transverse kinetic energy in the *x–y* plane.

[23] See Equation (1.15).

According to Equation (5.32), the total *sheet-carrier concentration* among the *subbands* is calculated by summing up the values of n_{s-n}s. Assuming occupation of only the first two *subbands* and assuming the same *DOS effective electron mass* among the two *subbands*, we have

$$n_s = \frac{m_n^* kT}{\pi\hbar^2} Ln\left[\left(1 + \exp\left(\frac{E_f - E_0}{kT}\right)\right)\left(1 + \exp\left(\frac{E_f - E_1}{kT}\right)\right)\right]. \quad (5.33)$$

In terms of this total *sheet-carrier concentration*, the fraction of *carriers* occupying the lowest *subband* is given by

$$\frac{n_{s-0}}{n_s} = \frac{Ln(1 + \exp((E_f - E_0)/kT))}{Ln[(1 + \exp((E_f - E_0)/kT))(1 + \exp((E_f - E_1)/kT))]}. \quad (5.34)$$

There exist important differences between this 2-D electronic system (which is generally speaking developed in a triangular or a rectangular *quantum well*) and a 3-D free-electron population. Whereas the *density of states functions* of a 3-D system tends to 0 toward the bottom of *conduction band* and the top of the *valence band*, in such a 2-D system, the *DOS* remains finite even at its lowest *subbands*. The very important consequence of this property is that, in contrast to a 3-D system, in a 2-D system, all dynamic phenomena remain finite at low kinetic energies (e.g., low electric fields) and low temperatures. As examples, these dynamic phenomena include *scattering* and optical absorption.

In the case of a *double-heterojunction heterostructure* where the narrower-*bandgap* material is thick and the *bandgap discontinuity* (i.e., ΔE_g) between the narrow- and the wide-*bandgap* semiconductors is small, the differences between the energy *subbands* (i.e., *eigen energy states*) is quite small. In those cases, due to the improved probability of electron transition between the energy *subbands* (if at all formed), the 2-D systems tend to take on a more 3-D character. Under these situations, the 2-D system cannot be differentiated from a 3-D system.

5.2.3 Semiconductor Heterojunctions and Self-Consistent Evaluation

The evaluation of the electron *wave function* in a *quantum well*, as suggested by Equation (5.26), requires determining the potential profile of the *well* (i.e., U_c). However, in Equation (5.27) without invoking any connection between the presence of mobile and immobile charge profiles across the *heterointerface*, this potential function was merely defined in terms of an average electric-field E. Based on *Poisson's equation*, even the strength of such an electric field is linked to the aforementioned charge profiles. A more accurate assessment of the electron *wave functions* in a *quantum well* requires *self-consistent* evaluation of U_c in terms of the charge profiles present in the vicinity of the *heterojunction*.

In *self-consistent* evaluation of the *subband energy levels* of a *quantum well*, the most widely used scheme is based on the *Hartree* approximation.[24] This is an approximation that is based on using the *variational Fang–Howard*-like *wave functions*.

Evaluating the potential-function U_c in part requires appreciation of electron–electron interactions. At the lowest order, the potential function in electron–electron interactions (i.e., $U_{ee}(z)$) is described by the *Hartree* approximation as a solution to

$$\frac{d^2 U_{ee}(z)}{dz^2} = \frac{q^2 n_S}{\epsilon_S} |\chi_n(z)|^2 \tag{5.35}$$

where n_s is the *sheet-charge* density of the *2DEG*.

The differential Equation (5.35) presents the imposition of the average electric field created by all other electrons on a single electron (represented by the *envelope wave function*, $\chi_n(z)$).

Now considering $U_{ee}(z)$ and the potential functions imposed by the *heterojunction* and the narrow-*bandgap* semiconductor's *ionized impurities* on the *2DEG* electrons (i.e., U_0 and U_{imp}, respectively), (5.26) evolves to

$$\left[-\frac{\hbar^2}{2m_n^*}\frac{d^2}{dz^2} + U_0(z) + U_{imp}(z) + U_{ee}(z)\right]\chi_n(z) = E_n\chi_n(z). \tag{5.36}$$

While *Fang–Howard functions* were successfully employed in the studies of *Si/SiO₂* triangular *quantum wells* (i.e., due to the large *conduction-band* discontinuity: the case of no penetration of *wave function* into the *SiO₂ barrier*),[25] they required modifications to account for the *wave function* penetration in case of *heteroepitaxial* layer structures such as *AlGaAs/GaAs*. For these cases, the applicable forms of these modified *envelope wave functions*, representing the first *subband*, are given by[26]

$$\begin{cases} \chi_n(z) = Bb^{1/2}(bz+\beta)\exp\left(-\dfrac{bz}{2}\right) & \text{if } z>0 \\ \chi_n(z) = B'b'^{1/2}\exp\left(\dfrac{b'z}{2}\right) & \text{if } z<0 \end{cases} \tag{5.37}$$

where B, B', b, b', and β are the *variational* parameters. The application of the boundary and the normalization conditions leave only two of these *variational* parameters (i.e., b and b') in need of optimization. A very suitable approximation for the value

[24] Named after *Douglas Hartree*.

[25] F. Fang and W. Howard, "Negative field-effect mobility on (100) Si surfaces," *Phys Rev Lett*, vol. **16**, p. 797, 1966.

[26] G. Bastard, "Self-consistent variational calculations and alloy scattering in semiconductor heterojunctions," *Surf Sci*, vol. **142**, no. 1–3, pp. 284–289, 1984. Due to the difference in *band-bending*, different forms of the *envelope wave functions* are considered in our discussion of *polar AlGaN/GaN HFETs* in Section 5.4.2.

of b' is $2\hbar^{-1}\left(2m_n^*\Delta E_c\right)^{1/2}$, which is determined based on the standard *wave function* penetration in the *AlGaAs barrier*.

Other more elaborate schemes such as employing *local exchange correlation potential* have also been employed in the evaluation of electron–electron interactions. According to this technique, studies of *Ando* provided numerical solutions to the *Schrödinger equation*.[27] Also for *AlGaAs/GaAs heterojunctions*, *Stern* and *Das Sarma* have published a very complete set of calculations reflecting on a large number of important considerations and situations pertaining to such a system (including finite-*barrier* effects, *effective-mass* and dielectric-constant discontinuities, *interface grading*, and *nonzero* temperature).[28]

5.2.4 Modulation Doping

In the late 1970s, studying the incorporation of *n-type doping* only within the wide-*bandgap* layers of a *superlattice heterostructure*[29] by Dingle et al. resulted in a major leap toward the realization of high-performance devices.[30] Immediately after these studies, and according to this *doping* scheme, a group of high-performance electronic devices were developed based on implementing a single *Type-I heterojunction*. These developments were quite concurrently reported by Mimura et al. of *Fujitsu Laboratories* of Japan,[31] Delagebeaudeuf et al. from *Thomson-CSF* of France,[32] and *Dingle* and colleagues from *Bell Laboratories* of the United States. In these developments, the underlying idea was the requirement for *carrier* transfer across the *heterointerface* to equalize the *Fermi levels* on both sides. This requirement is highlighted in the *band diagram* presented in Figure 5.4. As illustrated in this *band diagram*, in order to establish *thermal equilibrium* between the two sides of the *heterointerface*, electrons migrate across the interface (i.e., from where *donors* were introduced, to the opposite side of the interface). As it is evident on this *band diagram*, this migration results in *depletion* of electrons from the wider-*bandgap* side of the *heterointerface*.[33] This transfer of *charge carriers* enforces a separation between the centroid of *carrier* concentration profile and that of the *dopant* concentration. This phenomenon of spatial separation between the *ionized impurities* and *charge carriers* is known as *modulation doping*.

Modulation doping has been proven to result in the improvement of *low-field carrier mobility* within the *quantum well*, especially under low-temperature conditions

[27] T. Ando, "Self-consistent results for GaAs/Al_xGa_{1-x}As heterojunctions 2. Low-temperature mobility," *J Phys Soc Jpn*, vol. **51**, no. 12, pp. 3900–3907, 1982.

[28] F. Stern and S. Das Sarma, "Electron-energy levels in GaAs-$Ga_{1-x}Al_x$As heterojunctions," *Phys Rev B*, vol. **30**, no. 2, pp. 840–848, 1984.

[29] For a discussion on *super-lattices* see Section 1.6.2.1.

[30] See: R. Dingle, H. Störmer, A. Gossard, and W. Wiegmann, "Electron mobilities in modulation-doped semiconductor heterojunction superlattices," *Appl Phys Lett*, vol. **33**, pp. 665–667, 1978.

[31] T. Mimura, S. Hiyamizu, T. Fujii, and K. Nanbu, "A new field-effect transistor with selectively doped GaAs/n–Al_xGa_{1-x}As heterojunctions," *Jpn J Appl Phys*, vol. **19**, no. 5, pp. L225–L227, 1980.

[32] D. Delagebeaudeuf, P. Delescluse, P. Etienne, M. Laviron, J. Chaplart, and N. T. Linh, "Two-dimensional electron gas MESFET structure," *Electron Lett*, vol. **16**, no. 17, pp. 667–668, 1980.

[33] As illustrated in Figure 5.4, also at the exposed side of the *AlGaAs barrier carrier-depletion* occurs under *thermal-equilibrium*. This is due to the presence of *surface states* at this exposed surface.

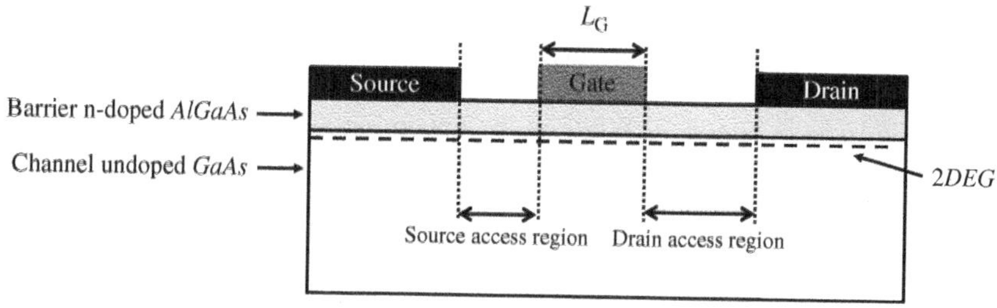

FIGURE 5.8 Cross-sectional view of a simple *AlGaAs/GaAs HFET.*

where *ionized-impurity scattering* is dominant. In this text, we refer to the sheet of charge formed at a *quantum well* as the *channel*. At *cryogenic* temperatures, *modulation-doped channels* have been observed to ideally manifest electron *mobilities* independent of temperature. The first experimental observation of the resulting dramatic increase of electron *mobility* at low temperatures was reported by Dingle et al. (i.e., *AlGaAs/GaAs 2DEG* electron *mobilities* in excess of 10^5 cm^2/V s at 77 K). In *bulk* samples, such a high value of electron *mobility* only corresponds to an extremely low *dopant* concentration of about 10^4 cm^{-3}. *Field effect transistors* that employ the *channel* formed at these *heterointerfaces* are as a result oftentimes referred to as *high-electron-mobility transistors* (i.e., *HEMTs*). Such a *field effect* transistor is schematically depicted in Figure 5.8.

The realization of the possibility of gaining high electron concentration in a high-purity/high-*mobility* semiconductor *channel* was an old dream of semiconductor technologists until it was materialized through the pioneering works of the aforementioned teams from *Fujitsu*, *Thomson-CSF*, and *Bell Laboratories*. Over the past three decades, *HEMTs* realized with *gate length* in the order of tens of *nanometers* have been capable of offering astonishing performance records such as *maximum oscillation frequency* (i.e., f_{max})[34] approaching 1 THz and *unity-gain cutoff frequency* (i.e., f_T) exceeding 600 GHz.[35] Also, impressive records have been set in terms of the high-frequency noise performance of these devices.

5.2.4.1 A Few Alternative HEMT Structures: Rewards and Early Challenges

Both *n-* and *p-type modulation dopings* have been explored on a large number of semiconductor *heterostructures*. *Field effect transistors* developed on the basis of *modulation doping* have received a variety of more or less equivalent names from different groups, including *HEMT*, *two-dimensional electron gas FET (TEGFET)*,

[34] That is the maximum frequency at which the transistor offers a power gain.

[35] As examples, see A. Leuther, S. Koch, A. Tessmann, I. Kallfass, T. Merkle, H. Massler, R. Loesch, M. Schlechtweg, S. Saito, and O. Ambacher, "20 nm metamorphic HEMT with 660 GHz f_T," Compound Semiconductor Week (CSW/IPRM) 2011 and 23rd International Conference on Indium Phosphide and Related Materials, May 22–26, 2011, Berlin, Germany. VDE Verlag, pp. 1–4 and T. Kim, D. Kim, and J. del Alamo, "60 nm self-aligned-gate InGaAs HEMTs with record high-frequency characteristics," IEEE International Electron Devices Meeting (IEDM), San Francisco, CA, December 6–8, 2010. Piscataway, NJ, IEEE, pp. 30.7.1–30.7.4.

selectively doped heterostructure transistor (SDHT), and *modulation-doped FET (MODFET)*. However, the proper umbrella name for these devices is *HFET*.

Figure 5.8 presents the cross-sectional view of an *HFET* structure realized on an *AlGaAs/GaAs heterojunction*. While in the upcoming sections of this chapter more will be presented on the intricacies of the realization of *ohmic* and *Schottky* contacts in such a device, it serves the discussion well to point out that such a transistor structurally mimics a *bulk planar MOSFET* (where part of the functions of the oxide is taken by the wide-*bandgap barrier layer*). As hinted already, from the point of view of the principles of operation, unlike *MOSFETs*, the formation of an *HFET channel* at the *heterointerface* of the *channel* and the *barrier layer* is not due to the *gate bias*. Since the origin of the *carrier* concentration of an *HFET channel* is the *dopant* profile incorporated in the *barrier layer*, these transistors do not demand the presence of *charge-carrier* reservoirs in the *unblocked* vicinity of the *channel* (which is the very argument for the necessity of *self-aligned* implementation of *MOSFETs*[36]). As expected, the *carrier* concentration within the *modulation-doped HFET channel* is tunable via the electrostatics of the *gate*.[37]

An important advantage of *FETs* that rely on *carrier transport* parallel to *coherently* grown semiconductor *heterointerfaces* (i.e., compared to *MOSFETs*) is that in these devices electron *transport* takes place in the vicinity of an orderly crystalline *interface*. As duly noted in our discussion of Chapter 3, *long-channel MOSFETs* using a thermally grown SiO_2 *amorphous barrier* were not severely penalized by the lack of crystallinity in the *barrier*. The necessities of *deeply scaled MOSFETs* for replacing this *barrier* with a high-*K* dielectric, however, have certainly magnified the negative consequences of the presence of an *amorphous barrier* next to the *channel*.

The structure presented for an *AlGaAs/GaAs HFET* in Figure 5.8 is the simplest possible structure for such a transistor. In this *non-self-aligned HFET*, the so-called drain and source access regions, identified by the parts of the *channel length* located between the edges of the *gate* and the *drain* and *source* electrode, respectively, are highlighted. In the later discussions of this chapter, we will express the ways enhancing the performance of these transistors has dictated the implementation of more sophisticated versions of this transistor (and its *self-aligned* alternative).

While the isolation technologies routinely employed on *MOSFET chips* are also used in the realization of *HFET chips*, *mesa isolation* has been a successful technology used among many *HFET*-based *MMICs*.[38] This mode of isolation, which was encountered earlier in Section 4.2.2.2, relies on *etching* away the *barrier layer* and a certain thickness of the *channel layer* (hence the 2*DEG*) from passive areas of the *chip*. The elimination of the 2*DEG* from these areas prevents cross talk between neighboring transistors realized on separate *mesas* that are only connected through the *undoped channel layer* (or, more properly said, *buffer layer*[39]).

[36] This has been pointed out in Section 3.1.

[37] Details of this process are presented in Section 5.3.2.

[38] *Monolithic Microwave Integrated Circuit.*

[39] Better acquaintance with these names will be achieved as we go further through this chapter.

Although so far in our presentation we have assumed that while the wide-*bandgap barrier layer* is *doped* the narrow-*bandgap channel layer* is kept devoid of *dopants*, in reality, challenges of crystal growth often result in a finite level of *doping* within this layer. This unintentional level of *doping* demands the careful attention of the device designer in proper implementation of an *HFET*. The presence of residual *unintentional doping* in the narrow-*bandgap channel layer*, in addition to the expected roles in degrading the *carrier mobility* and induction of a more pronounced 3-D character of *carrier transport*, has other important implications on the electrical performance of *HFETs*. This *unintentional doping* concentration not only creates an electric field in the resulting *depletion region* formed in the *channel layer* but also contributes to the definition of the potential within the *barrier layer*. These implications are already incorporated in Equation (5.36). Depending on the level of the *unintentional doping*, devices can express large *substrate-leakage* currents that are manifested in the form of *soft* (or even a no-) *pinched-off* behavior and cross talk between the transistors formed on neighboring *mesas*.

Since the level of this *unintentional doping* is usually quite small (especially in more mature *heterostructure* technologies such as *AlGaAs/GaAs*), the *depletion-region width* in the narrow-*bandgap channel layer* is much larger than all other dimensions in the system. As a result, the potential due to this *depletion* charge profile, on the basis of the assumption of a constant electric field, can be considered triangular in the region of interest. This is the logic behind the triangular *quantum-well* approximation presented earlier in Section 5.2.1. In contrast to this framework, in the case of small *carrier* transfer, the detailed knowledge of the residual impurities' potential becomes quite important. This detailed knowledge is also very relevant to the evaluation of *excited* electron *states* (i.e., higher energy *subbands*).This is because in these *excited states*, the *wave function* extends further away from the *heterointerface*.

In addition to the device structure presented in Figure 5.8, *GaAs-channel HFETs* have been also realized on the so-called inverted GaAs/AlGaAs heterostructures. In these cases, while still the *channel* is formed on the *undoped GaAs* side of the *heterointerface*, the proximity of the *gate* electrode to the entire thickness of the thin *GaAs channel layer* helps to prevent issues attributed to the *unintentional doping* of the *channel layer*. Whereas the thin *channel layer* of an *inverted HFET* is grown on top of a thick *relaxed* wider-*bandgap barrier*, the small *conductivity* of this *barrier* prevents this layer from strongly leaking at *gate-source* voltage below the *threshold voltage* (hence from imposing a *soft pinch-off* on the *channel*). Another advantage of the *inverted* implementation of an *HFET* is the relative ease of making *ohmic* contacts to the *2DEG channel* via the narrower-*bandgap* material. In the hindsight, however, establishing a blocking *gate* contact on *inverted HFETs* has proven to be more difficult.

While so far we have identified only two layers of the *heteroepitaxial* layer structure of an *HFET* (i.e., the intentionally highly *doped* wider-*bandgap barrier layer* and the *unintentionally doped* narrower-*bandgap channel layer*), a number of other layers are also incorporated in the structure of a high-performance *HFET*. One of these layers is referred to as the *spacer*. The *spacer* is often only a very thin *undoped* layer of the

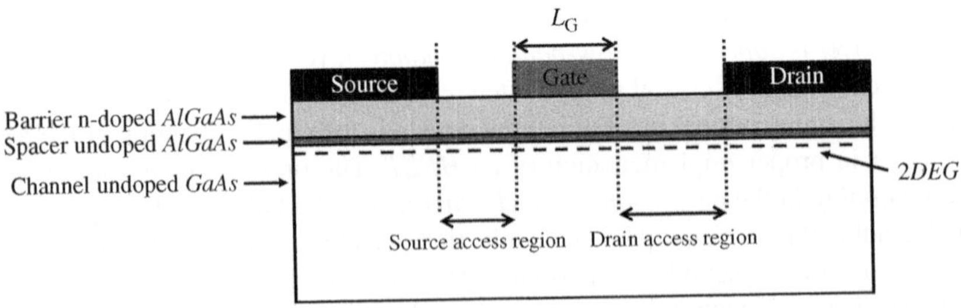

FIGURE 5.9 Cross-sectional view of an *AlGaAs/GaAs HFET* employing a spacer layer.

same material as the *barrier*, which is sandwiched between the two aforementioned layers. Figure 5.9 presents a modified layer structure of an *HFET* with the incorporation of the *spacer layer*.

Increasing the thickness of this *undoped spacer layer* diminishes the *Coulombic* interaction between the ionized *donors* and the *2DEG* electrons, resulting in improved low-field *mobility*. However, there is an upper limit to the thickness of this layer. This limit is set in terms of the potential drop across the *spacer*. In the absence of *space charge* in this *undoped* layer, a constant electric field will form throughout the thickness of the *spacer*. As a result of this constant electric field, an electrostatic potential builds up throughout this layer, which does not correspond to transferred charges. Consequently, increasing the *spacer layer's* thickness tends to decrease the *2DEG* electron density. Hence, it is only in the presence of a very thin *undoped spacer layer* that through further separation of *charge carriers* and *ionized-impurities* (and essentially diminishing the chance of *remote ionized-impurity scattering*) the *2DEG's* low-field *carrier mobility* is improved, while the efficiency of *modulation doping* is not gravely sacrificed.

5.2.4.2 *HEMT and Challenges of the Realization of High-Mobility Channel*

A large number of technological advances in the area of crystal growth have paved the way for the success of *AlGaAs/GaAs heterostructure* technologies. While *silicon* is a popular *donor* impurity among these technologies, one of the early hassles in these developments was the issue of *segregation* of the *in situ* introduced *silicon dopants* toward the $Al_xGa_{1-x}As$ growth surface. This was observed to be especially more important for larger values of x. As a result of this *dopant segregation*, smaller than expected *charge transfer* across the *AlGaAs/GaAs heterointerface* was recorded. This was due to the essential widening of space between the *donors* and the *heterointerface*. In an *inverted GaAs/AlGaAs* structure, the same issue was found out to be behind an observation of *mobility* degradation. In this case, *ionized impurities* were *segregated* more toward the *heterointerface* imposing a stronger *remote scattering* on the electrons populating the *energy subbands* of the *heterointerface*.

While the origin of the *mobility* improvement in the *channel* of *HEMTs* is the spatial separation of the *ionized-impurity* and the *charge-carrier* profiles, secondary issues raised by the presence of side effects such as *alloy scattering* and *interface-roughness scattering* necessitate a more careful analysis of the links between these limitations and the choice of semiconductor pairs forming the *heterostructure*.

Electron *scattering* mechanisms in a large number of semiconductors have been quite rigorously studied both on theoretical and on experimental bases. Oftentimes, electron *transport* in semiconductors is treated according to a perturbative description.[40] Generally speaking, the involved perturbation mechanisms in this description are classified into 11 groups:

1. *Optical-phonon scattering* (which is dominant at high temperatures)
2. *Acoustic-phonon scattering* caused by the *deformation* potential.
3. *Acoustic-phonon scattering* caused by the *piezoelectric field* among *III–V* and *II–V* semiconductors.[41]
4. *Scattering* by *ionized impurities.*
5. *Scattering* by *neutral impurities.*
6. Remote *scattering* by *phonons* originating in the *barrier layer.*
7. Remote *scattering* by *ionized* or *neutral impurities* located in the *barrier layer.*
8. *Alloy scattering* either originating in the *barrier layer* (when only the *barrier* and not the *channel* is an *alloy*, e.g., in the case of *AlGaAs/GaAs*) or in the *channel layer* (when only the *channel* is an *alloy*, e.g., in the case of *InP/InGaAs*) or in both the *barrier* and the *channel layers* (when both *layers* are *alloys*, e.g., in the case of *InGaAs/AlInAs*).
9. *Surface phonon scattering* due to the presence of *propagating surface modes* existing at the *heterointerfaces.*

   Proper choice of materials in a *heterostructure* (i.e., materials of similar density and *dielectric* functions) renders these *phonon modes* to impose only a negligible *scattering* probability. As a result of a proper choice, this mechanism of *scattering* in the evaluation of *transport* through *quantum wells* of *heterostructures* need not be considered.
10. *Interface-roughness scattering.*
11. *Intersubband scattering* between the *subbands* formed in the *quantum-well channel* of an *HFET.*

Only the first five of these mechanisms are responsible for electron *mobility* degradation in *bulk nonalloyed* semiconductors. However, the evaluation of the additional six perturbation mechanisms of this list is of utmost importance in assessing the *carrier transport* in the *channel* of *modulation-doped heterostructures.*

Through the proper selection of a *heterostructure*, some of the aforementioned *remote scattering* mechanisms can be alleviated. For example, in the case of an *AlGaAs/GaAs heterostructure*, *remote scattering* by *AlGaAs phonons* has been

[40] See the discussion in Section 1.7.

[41] The lack of *inversion symmetry* in the crystal structure of *III–V* and *II–VI* semiconductors is responsible for establishing a *piezoelectric field* among these semiconductors.

observed not to play a significant role. Whereas at low temperatures the *AlGaAs phonons* are altogether suppressed, since in the *AlGaAs/GaAs* case the penetration of *carrier wave function* in the *barrier* is negligible, at high temperatures also their impact on the *2DEG* is deemed insignificant.

As indicated earlier in Section 1.7.3.2, *alloy-disorder scattering* is an unavoidable fundamental property of *alloyed* semiconductors. This mode of *scattering* is caused by statistical compositional fluctuations in the *disordered alloys*. This gives rise to random fluctuating potentials. Although in a *heterostructure* such as *AlGaAs/GaAs* only the *barrier* is an *alloy*, this fluctuating potential can still impact the electron *transport* in the 2-D *channel* formed in the *GaAs channel layer*. The relevance of this potential to the evaluation of *carrier mobility* in a *2DEG channel* is contingent on the *wave function* penetrating into the *barrier layer*. While for low *sheet-carrier* concentrations this penetration is typically a few percent, at higher concentrations, due to the increased penetration, *alloy scattering* plays a more prominent role in *mobility* degradation.

The chance for *intersubband scattering* also increases with the increased *sheet-carrier* concentration (i.e., when the *energy subbands* higher than the *ground subband*[42] are populated). As a result of populating the higher *energy subbands*, the *phase space* for the final *states* in *scattering* events becomes larger. This factor effectively contributes to an increased *scattering* probability.[43] Störmer et al. were able to show a decrease in electron *mobility* caused by the increased *sheet-charge* concentration and to correlate this effect to populating the second *energy subband*.[44] This was substantiated through observation of double-period *Shubnikov–de Haas oscillations*.[45]

5.3 HFET: BRIEF EXPOSÉ OF DESIGN INTRICACIES

Understanding the process of *modulation doping* is crucial to the successful design of *HFETs*. Whereas *modulation doping* provides the high-*mobility 2DEG channel* with a high *carrier* concentration, due to charge sharing with a low-*mobility* parallel *channel* running through the *barrier layer*, partial transfer of *carriers* across the *heterointerface* limits the improvement in the *conductivity* of the *channel*. Due to its close proximity to the *gate* electrode, the effect of this parallel *channel* on the *gate transconductance* is often quite large. Resulting from the leakage of the electron *wave function* back into the *barrier layer*, *carrier* concentration of the parallel *channel*

[42] That is of $n = 0$.

[43] See the discussion of Section 1.7.3.

[44] H. Störmer, A. Gossard, and W. Wiegmann, "Observation of intersubband scattering in a two-dimensional electron-system," *Solid State Commun*, vol. **41**, no. 10, pp. 707–709, 1982.

[45] Named after *Lev Shubnikov* and *Wander de Haas*. For insight into these oscillations and the *quantum Hall effect*, see B. Halperin, "The quantized Hall effect," *Sci Am*, vol. **254**, pp. 52–60, 1986; H. Störmer and T. Tsui, "The quantized Hall effect," *Science*, vol. **220**, pp. 1241–1246, 1983; T. Ando, A. Fowler, and F. Stern, "Electronic properties of two-dimensional systems," *Rev Mod Phys*, vol. **54**, pp. 437–672, 1982.

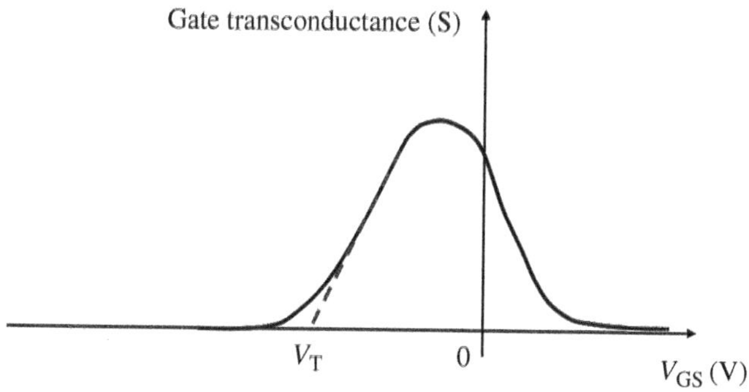

FIGURE 5.10 Behavioral depiction of the gate transconductance of a depletion-mode n-channel *HFET*.

running through this layer can be also increased through the application of large positive *gate-source* voltages.[46]

The *gate transconductance* of an *HFET* is approximately representable by a bell-shaped profile (see Fig. 5.10). In this characteristic, increasing the *gate* voltage beyond the *threshold voltage* results in successfully populating the 2*DEG channel* and improving the *effective carrier* concentration and the *effective mobility* of the *channel* established between the *source* and *drain* terminals. These factors provide an improvement in the *gate transconductance*. However, with further increasing the *gate bias*, the *gate transconductance* is deteriorated since the *channel*'s *carrier* population starts to leak back into the low-*mobility barrier layer*. Also effective in this degradation is the resulting enhanced proximity of the 2*DEG* to the *heterointerface*. For a number of reasons including *interface-roughness scattering*, *Coulomb scattering* from *barrier*'s *ionized impurities*, and *alloy scattering* both within the *alloyed barrier* and remotely, this causes the degradation in the *channel's effective mobility*.

As observed in Figure 5.10, the aforementioned low-*mobility* parallel *channel* in addition to effectively degrading the *carrier mobility* and hence the *current drive* of the transistor imposes a nonlinearity on the operation of the device. Depending on the interplay

[46] Studying the *C–V* characteristics of the *reverse-biased gate diode* of an *HFET* is capable of revealing the formation of the parallel *channel* within the *barrier layer* at higher *gate-source* voltages. While for values of V_{GS} smaller than the *threshold voltage* no 2*DEG* is formed at the *heterointerface*, the *small-signal capacitance* of this regime indicates a small value (i.e., owing to the low *unintentional doping* level of the *channel* and the *buffer layer*). Increasing V_{GS} beyond this *subthreshold* regime, however, is signaled by an increase in the value of the *small-signal capacitance*. This capacitance, as suggested by our previous discussion in Chapters 2 and 3, is owing to the series combination of the 2*DEG* and the *quantum capacitance*. Further increasing the value of V_{GS}, although short of the *gate-diode* turn-on, also indicates further increase in the value of the *small-signal capacitance*. This later increase is a sign of the formation of the dreaded parallel *channel* within the *barrier layer* which, due to its closer proximity to the *gate* electrode, is expressed by a larger *small-signal capacitance*. For values of V_{GS} in excess of the *threshold voltage*, there is often a region where the *small-signal capacitance* has reached a plateau. This is the intermediate *bias* range before the parallel *channel* is constituted. Of course, this order of events is provided with the assumption that the parallel *channel* is formed before the *gate-diode* turn-on, an assumption which is not always valid.

between the *carrier* populations of the *2DEG* and the *barrier channel*, sometimes even a double-hump *gate-transconductance* characteristic is observed. As a result of the aforementioned adverse effects, one of the major requirements of a properly designed *HFET* is the elimination of the *barrier layer channel* and the aforementioned *mobility*-degrading processes over the intended *bias* range of operation of the transistor.

Considering these limitations, design rules for the growth sequence, thickness, and *doping* concentration among various layers of the *heterostructure* are developed on the basis of a proper understanding of *modulation doping* process. In order to evaluate *modulation doping* in *heterostructures*, three main facts should be considered that yield a *self-consistent* position of *Fermi level* (and, as a result, the *carrier* concentration in different positions within the *heterostructure*):

1. *Band bending* in the *barrier* and in the *channel layer*. These are evaluated using the *heterointerface* profiles of charge concentration and electric field.
2. The development of *energy subbands* as a result of confinement of *carriers*. As mentioned earlier in Section 5.2.1, the *energy subbands* are defined on the basis of quantum mechanics.
3. The establishment of *thermal equilibrium* (i.e., constant *Fermi level*) across the *heterostructure*.

On these bases, the concentration of transferred electrons forming the *2DEG channel* (i.e., n_s) is *self-consistently* calculated.

Various frameworks for these sorts of calculations have been developed with approximations to different degrees of accuracy. Oftentimes, these frameworks rely on a very simple, yet well-justified, assumption of negligible penetration of *carrier wave function* into the *barrier*.[47] As mentioned earlier, whereas this assumption is more or less justified in the case of a large number of *heterostructures* (such as *AlGaAs/GaAs*), in other cases (such as *polar AlGaN/GaN heterojunctions*), it might not be as valid.

According to this simplifying assumption, the electron *energy levels* remain unaffected by the possible changes in the *barrier's* electrostatic potential defined in the *depletion region* of the *barrier* (which is induced by the charge transfer). Supporting this approximation, in the case of an *AlGaAs/GaAs heterointerface*, it is observed that the electron *wave function* penetrates at most about 20 Å into the *barrier layer*. As this penetration is much smaller than the width of the *depletion region* formed within this layer, it can be safely neglected. In cases such as this, the *wave function* can be determined purely on the basis of the *barrier height*, while the impact of the electric field formed at the *heterointerface* is only secondary. This grants the possibility of uncoupling the process of establishing the *thermal equilibrium* from the electrostatics and the process of *energy-level* calculation (which are listed above). Whereas in general a *self-consistent* evaluation of these three phenomena is required, the aforementioned approximation considerably lifts the burden of the *self-consistent* solution.

[47] See the assumption of infinite barrier at $z = 0$ in Section 5.2.1.

Corresponding to the earlier presented discussion on the role of *unintentional doping* in the evaluation of the *2DEG*,[48] for the special case of *polar AlGaN/GaN* heterostructures, both the high polar *2DEG* concentration and the large *unintentional doping* level of the *channel layer* aggravate the situation with regard to the requirements of a valid triangular *potential-well* approximation. In those cases, a more thorough *self-consistent* analysis of the three aforementioned phenomena is needed. We will deal with this problem in further detail in Section 5.4.2.

As a result of the finite thickness of the *barrier layer*, in the process of establishing *thermal equilibrium* throughout an *HFET* structure, not only the impact of the semiconductor *heterojunction* but also that of the *Schottky* metal–semiconductor contact on the *2DEG carrier* concentration should be considered. Depending on the sufficient proximity of this *Schottky* contact to the semiconductor *heterointerface*, the *depletion* of *2DEG carrier* concentration, which is needed for the realization of *enhancement-mode HFETs*, can be accomplished. Most *HFETs*, however, are realized in the *depletion mode* of operation (i.e., with relatively *thick barriers*). The pioneering works of *Vinter* on quantum-mechanical calculation of the *gated heterojunctions* revealed the intricacies of the abovementioned interactions.[49] According to these studies, as schematically illustrated in Figure 5.11, due to the interaction of the two junctions, *energy subbands* are formed not only within the *channel* but also in the *barrier layer*. In the case of an *AlGaAs/GaAs heterostructure*, due to the small thickness of the potential barrier formed at the *heterointerface*, the *2DEG* and the *barrier*'s *quantum states* are coupled together.[50] As a result, via the aforementioned leakage of electron *wave function* into the *barrier*, as V_{GS} increases, the *2DEG* concentration in the *channel* saturates in the case of *AlGaAs/GaAs* and other similar *HFET* structures. This is due to populating the *barrier's energy subbands*. Among the *HFETs* employing a *spacer layer*, a thicker *spacer layer* causes earlier onset of this saturation and, as a result, smaller maximum *2DEG* concentration. However, due to potential drop across the *spacer*, the value of the *threshold voltage* is more negative.

In Figure 5.11, *conduction-band lineups* of two *HFET* structures of *enhancement-* and *depletion-mode* characteristics under their *gate* electrode are presented. As mentioned above, in the *enhancement-mode* case, the *depletion* of the *2DEG* occurs due to the presence of a rather high *Schottky barrier* in the proximity of the 2-D *channel*. This *Schottky barrier* can completely *deplete* not only the *barrier* but also the *2DEG channel* of electron concentration. Through selective *thinning* of the *barrier layer*, one can theoretically realize both *depletion-* and *enhancement-mode* transistors on the same *chip*. This is needed for the implementation of the *direct-connected transistor logic*

[48] See Section 5.2.4.1.

[49] B. Vinter, "Subbands and charge control in a two-dimensional electron-gas field-effect transistor," *Appl Phys Lett*, vol. **44**, no. 3, pp. 307–309, 1984; B. Vinter, "Subbands in back-gated heterojunctions," *Surf Sci*, vol. **142**, no. 1–3, pp. 452–455, 1984.

[50] As will be indicated later in Section 5.4, among *polar III-Nitrides heterostructures*, the predominantly opposite slope of *band bending* across the *barrier layer* reduces the chance of the formation of *quantum states* within the *barrier* and as a result their coupling to the *2DEG states*.

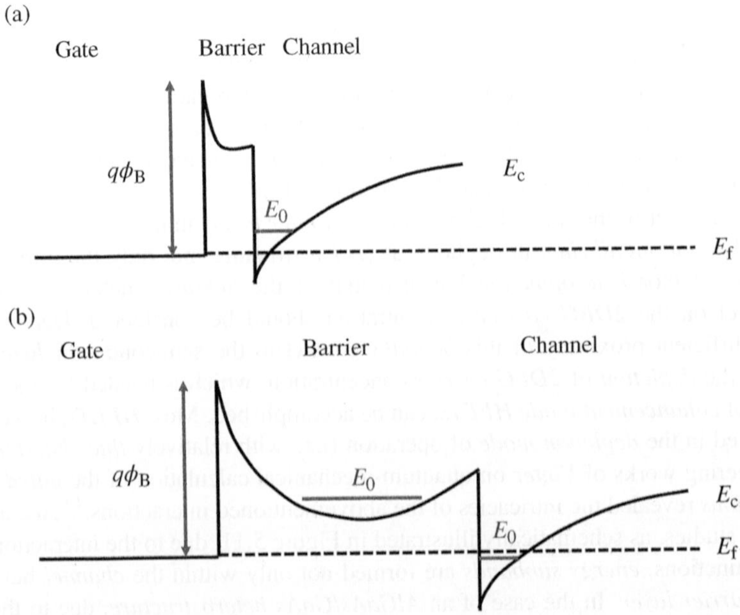

FIGURE 5.11 Behavioral depiction of the thermal-equilibrium conduction-band bending across the heterojunction in the gated area of (a) an enhancement-mode n-channel *HFET* and (b) a depletion-mode n-channel *HFET*. $q\phi_B$ is the *Schottky* barrier height. In (a), as indicated by the position of the first energy subband versus the *Fermi level*, there is no *2DEG* channel formed at the heterointerface when $V_{GS} = 0$ V. Also in this case, in light of the position of E_c versus *Fermi level* in the barrier, the barrier layer is depleted. In (b), however, there are subbands formed both in the channel layer and the barrier layer. Depending on dimensions and the potential barrier height, not only the channel's lower subband but also that of the barrier can be populated.

(i.e., *DCTL*).[51] This *barrier-thinning* process is limited to the *gated* area of the *channel*. Whereas this process of bringing the *Schottky* contact closer to the *channel* serves to positive-shift the originally negative *threshold voltage* of an *HFET*, maintaining a thicker *barrier* (especially in the high electric-field *drain-access region*) can assist with confining the *hot carriers* formed in this part of the *channel* to the *2DEG channel*.[52] This process of local *thinning* the *barrier* is often referred to as the *gate-recess* process. In the *enhancement-mode* devices realized according to this technique, the presence of a thicker *barrier layer* in the *drain* and *source access regions*, given a

[51] This *logic* family, through employing the *normally off* transistor of the pairs as a *switch*, while the *normally on* transistor with tied *gate/source* electrodes offers a *current source*, has advantages in terms of the *standby*-power consumption over a *depletion-mode logic*.

[52] While for a *MOSFET* we have already observed the evolution of the high electric-field region in the vicinity of the *drain* edge of the *gate* electrode, logically the same phenomenon can be forecasted for an *HFET*.

diminishing chance of *2DEG depletion* due to the effect of the *surface states*, is needed in maintaining the continuity of the *channel* between *drain* and *source* terminals.

The *threshold voltage* of a*n HFET* is as a result controlled not only via the *doping level* of the *barrier layer* but also by tailoring the thickness of this layer. In spite of its merits, since *dry etching* is usually required in the *gate-recess* process, the characteristics of *enhancement-mode* devices obtained in this fashion often suffer from hysteresis. This is due to the adverse role *dry etching* has on adding to the number of *surface states*.[53]

In terms of parameters such as thickness and *dopant* concentration of various layers of an *HFET*, several investigators have studied design-rules for the realization of optimum-performance *HFETs*. In the case of a simple *HFET* structure with uniformly *in situ doped barrier*[54] such as the one depicted in Figure 5.8, *Schottky's depletion approximation* suggests that increasing the *doping*-level N_d of the *barrier layer* results in improved charge-transfer in proportion to $N_d^{1/2}$. This is because the *depletion region-width* (i.e., t_d) formed on the *barrier* side of the *heterointerface* varies in proportionality with $N_d^{-1/2}$, leading to a *carrier* transfer defined by $t_d N_d \propto N_d^{1/2}$. A sufficiently high *carrier* transfer to the *2DEG channel* leads to efficient *screening* of the *unintentional impurities* of the *channel*. At the same time, improving the *carrier* transfer through uniformly increasing the *barrier's dopant* concentration leads to more *remote ionized-impurity scattering* of *2DEG electrons*. As pointed out in Section 5.2.4.1, while incorporation of a thick *spacer layer* might seem as a good remedy, the increased thickness of the *spacer* essentially results in degradation of *carrier* transfer. Considering these effects, *Stern* has worked out an optimal *doping* strategy for a*n HFET* structure.[55]

According to the experimental studies on the profiles of transferred-electrons as a function of the *barrier layer's dopant* concentration N_d and the *spacer-layer* thickness t_{sp}, it is observed that in a *single-heterojunction AlGaAs/GaAs* system there is not much room for obtaining *2DEG* concentrations larger than 10^{12} cm^{-2}. For different *heterojunctions*, this maximum-value is of course different. Sandwiching the *channel layer* between two wider-*bandgap* semiconductors (in this case *AlGaAs*), however, has been shown to be able to increase the electron concentration of the *channel* by combining two *2DEGs*. The transistors, such as the one depicted in Figure 5.12, which

[53] The incorporation of a *p–n* junction under the *gate* electrode (where its *n-layer* is the *doped barrier* and its *p-layer* is to be *etched* away elsewhere) and employing the traditional *doping* processes to compensate the *in situ* introduced *donors* within the *gated* portion of the *barrier layer* have been explored as two alternative techniques to deplete the *2DEG* under the *gate* electrode and to yield an *enhancement-mode* character. Each of these techniques deals with its own challenges and rewards.

[54] In realization of *HFETs*, often the traditional *doping* processes (i.e., *diffusion doping* and *ion implantation*) are altogether avoided. In such structures, layers of the *heterostructure* are *in situ doped* during growth. Although such a technique normally allows precise tailoring of the *dopant* concentration along the growth axis, in-plane variation of the *dopant* concentration normal to this axis is not routinely offered.

[55] F. Stern, "Doping considerations for heterojunctions," *Appl Phys Lett*, vol. **43**, no. 10, pp. 974–976, 1983.

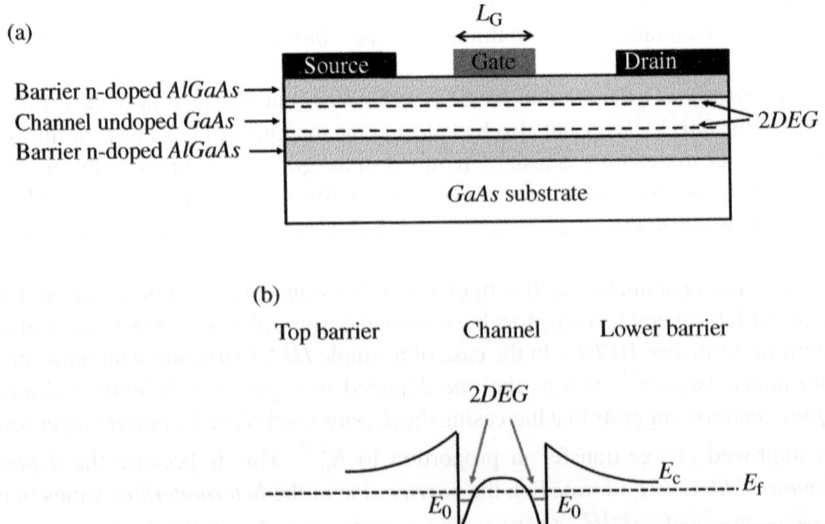

FIGURE 5.12 (a) Cross-sectional view of an *AlGaAs/GaAs/AlGaAs DHFET*. In a properly designed *DHFET*, the two indicated *2DEG* channels can combine together and form a channel whose centroid is in the middle of the channel layer. (b) Approximate depiction of the conduction-band bending in the vicinity of the heterojunctions of the layer structure of (a).

are realized on these structures are referred to as *double-heterojunction field effect transistor* (i.e., *DHFET*).[56] In terms of the realization of a *VMT*, in Section 4.2, we have already encountered the idea of implementation of two closely packed parallel *heterojunction channels*. While in a *VMT* the goal is to maintain two almost separate *channels* (i.e., one with *high mobility* and one with *low mobility*), in the case of a *DHFET* (otherwise called *multiple-channel HFET*[57]), one of the important goals is to achieve a higher *2DEG* concentration via merger of the *2DEGs* formed in the vicinity of the two *heterointerfaces*. Upon merger, the centroid of the combined charge concentration profile also shifts further away from the two *heterointerfaces*. This helps in improving the *low-field carrier mobility* via suppressing the remote *scattering* due to *ionized impurities* and *alloy scattering* of the *barrier layers*. It also further eliminates the *interface-roughness scattering* and interaction with *interface states*.

[56] K. Inoue and H. Sakaki, "A new highly-conductive (AlGa)As/GaAs/(AlGa)As selectively-doped double-heterojunction field-effect transistor (SD-DH-FET)," *Jpn J Appl Phys: Lett*, vol. **23**, no. 2, pp. L61–L63, 1984; K. Miyatsuji, H. Hihara, and C. Hamaguchi, "Single quantum well transistor with modulation doped AlGaAs/GaAs/AlGaAs structures," *Superlattice Microst*, vol. **1**, no. 1, pp. 43–47, 1985.

[57] N. Sheng, C. Lee, R. Chen, D. Miller, and S. Lee, "Multiple-channel GaAs AlGaAs high electron-mobility transistors," *IEEE Electron Device Lett*, vol. **6**, no. 6, pp. 307–310, 1985.

5.3.1 Deep Donors and Modulation Doping

While a seemingly important feature in the realization of *HFETs* is the possibility of tuning the *bandgap* discontinuity between the *barrier* and the *channel layer* over a wide range of energy (i.e., through varying the *mole fraction* of the *compound alloyed* semiconductors), issues such as the manifestation of *deep-donor* behavior have posed certain limitations on this approach.[58] These limitations are in addition to the previously indicated ones defined on the basis of maximum affordable strain at a *pseudomorphically grown heterointerface* (i.e., before *relaxation* or crack-formation processes set in).[59] In the presentation of this section, as famous examples, we will maintain a focus on *AlGaAs/GaAs HFETs*. An important issue that was faced early in the development of *AlGaAs/GaAs HFETs* was that the number of electrically active *donors* was not unambiguously related to the number of incorporated *donors*. Studies of the origins of this phenomenon have indicated that some of the *silicon dopants* become associated with *deep defects*. The tangible consequence of this association is the low-temperature collapse of device characteristics. Observation of *persistent photoconductivity* among these samples is also attributed to *deep defects*.

An important early observation in the *AlGaAs/GaAs* system was that the position of the *donor energy level* in $Al_xGa_{1-x}As$ (i.e., E_d) demonstrated considerable dependence on the *Al mole-fraction x*. The value of E_d was observed to change from only 6 meV for $x<0.1$ to more than 160 meV for the *indirect bandgap* $Al_xGa_{1-x}As$ (i.e., $x>0.45$).

According to Schubert et al., specifically it is observed that in *silicon doping* of $Al_xGa_{1-x}As$, the thermal *ionization energy* for $x<0.2$ is even below the value calculated in terms of the *hydrogen model*.[60,61] This is qualitatively illustrated in Figure 5.13. However, for $x>0.2$, E_d is observed to peak suddenly toward 100 meV and to demonstrate *deep-donor* energy (referred to as E_{dd}) of about 150 meV. The transition from *shallow* to *deep donor level* has been reported to occur at $x=0.235$. This behavior of E_d has negative consequences on the efficient transfer of *carriers* into the *2DEG channel*. We can express this observation through mathematical terms.

Considering the situation depicted in Figure 5.14, and assuming $T=0$ while only the first *energy subband* (indicated by E_0) is being occupied, in terms of the 2-D *density of states* D_{2D}[62], we can say that

$$n_s = D_{2D}(E-E_0) = \frac{m_n^*}{\pi\hbar^2}(E-E_0) \Rightarrow E = E_0 + \frac{\pi\hbar^2}{m_n^*} n_s \tag{5.38}$$

[58] *Deep donors* are often referred to as *DX centers*.

[59] See the discussion in Section 1.6.2.2.1

[60] See the discussion in Section 1.2.3.

[61] E. Schubert and K. Ploog, "Shallow and deep donors in direct-gap n-type $Al_xGa_{1-x}As$-Si grown by molecular-beam epitaxy," *Phys Rev B*, vol. **30**, no. 12, pp. 7021–7029, 1984.

[62] See Equation (1.15).

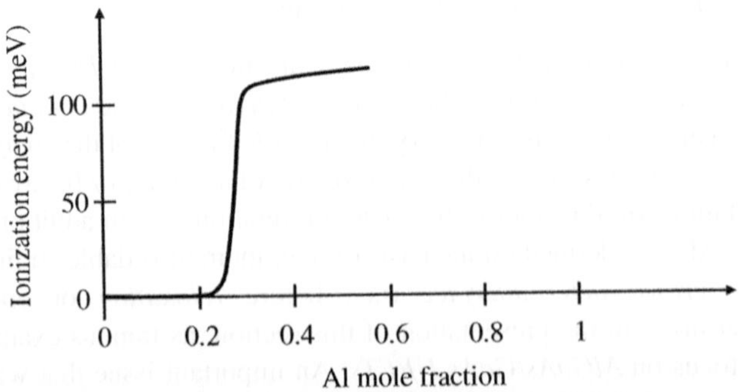

FIGURE 5.13 Approximate indication of the variation of the ionization energy of silicon as donor in *$Al_xGa_{1-x}As$* with the *Al mole fraction*. This figure is adapted with permission from Schubert and Ploog (1984). Copyright 1984, American Physical Society.

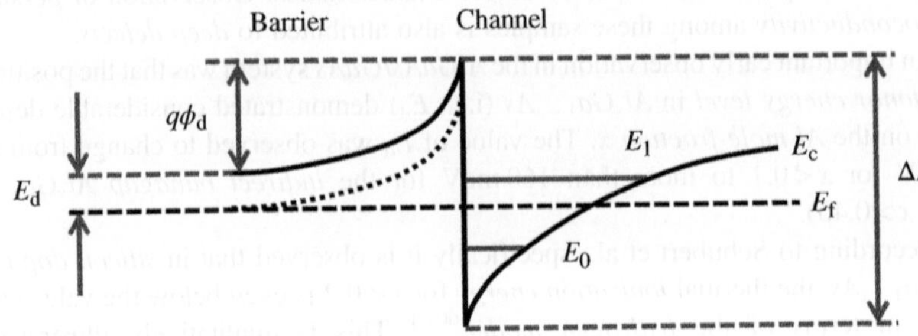

FIGURE 5.14 Schematic depiction of the foundation of expression (5.40) on the thermal-equilibrium conduction-band diagram of an *HFET*.

where under the *thermal-equilibrium* condition, $E = E_f$ and n_s is the *sheet-carrier* concentration.

Based on *Schottky's depletion approximation*, when the uniformly *doped AlGaAs barrier layer* is *depleted* over a certain portion of its thickness (i.e., $t_d < t_{AlGaAs}$) on its *heterojunction* side, the electrostatic potential ϕ_d induced across the *depletion-region width* can be approximated by

$$\phi_d = \frac{qN_d}{2\epsilon_s} t_d^2. \tag{5.39}$$

Here, N_d is the concentration of the *donors* in the *barrier layer*.

Accordingly, invoking the concept of a parallel-plate capacitor across the *hetero-interface* and applying *Schottky's depletion approximation*, the *sheet-carrier* concentration is given by $n_s = N_d \cdot t_d$.

Based on the *thermal-equilibrium band diagram* of Figure 5.14 and assuming that the *donor energy level* in the *AlGaAs barrier layer* is *deep* enough to cause *Fermi-level pinning*, it can be deduced that

$$\Delta E_c = E_0 + \frac{\pi\hbar^2}{m_n^*} n_s + E_d + q\phi_d \tag{5.40}$$

which yields

$$\Delta E_c - \frac{q^2 n_s^2}{2\epsilon_s N_d} - E_d = E_0(n_s) + \frac{\pi\hbar^2}{m_n^*} n_s. \tag{5.41}$$

Without the assumption of *Fermi-level pinning*, a *self-consistent* solution for finding the position of *Fermi level* is demanded.

According to (5.41), increasing E_d yields a diminishing *sheet-carrier* concentration, n_s. In addition to this, due to *carrier freeze-out*[63], *deep donors* limit the low-temperature operation of the *HFET*.

Even though our earlier arguments indicated a need for a larger value of *conduction-band* discontinuity (hence larger *Al mole fraction* in the *AlGaAs barrier*), the discussion presented in this section calls for a compromise. The truly exceptional ternary $Al_xGa_{1-x}As$ material system offers *lattice matching to GaAs* over a wide range of x and promises very little challenge with the adverse effects of growth under mechanical strain. Yet issues such as the aforementioned *deep* character of *donors* originally posed a considerable limit over increasing the barrier height.

As a remedy for the problem of *doping* of the high *Al-composition AlGaAs* layers, a very interesting solution has been developed in terms of *AlGaAs/GaAs superlattices*.[64] In these structures, only the *GaAs* layers of the *superlattice* are strongly *n-type doped*. Extreme spatial confinement of electrons in very thin (i.e., thinner than 20 Å) *GaAs* layers causes the formation of *energy subbands* in these layers, which are well above the *conduction-band* edge of *GaAs*. The positions of these *energy subbands* can be as high as the *conduction-band* edge of *AlGaAs*. As a result, *carrier* transfer can be easily achieved between these *GaAs*-confined *donor levels* and the *GaAs-channel layer*. This *doping* scheme is presented in Figure 5.15.

5.3.2 Threshold-Voltage Calculation in HFET

Key to the understanding of the operation of an *HFET* is the study of the controllability of *2DEG* concentration via a *Schottky gate* junction. As we have already pointed out, in this structure, the electrons forming the *2DEG* originate from the *n-doped barrier layer*. In an *HFET* structure, while under the *gate* electrode two *depletion regions* are formed across the *barrier layer* (i.e., one on the *MS* side and the other on the *heterojunction* side of the *barrier layer*), assuming a small thickness for this layer, these two *depletion regions* can completely deplete the *barrier layer* of *charge carriers*. As a result, these

[63] See Section 1.2.2.

[64] T. Baba, T. Mizutani, and M. Ogawa, "Elimination of persistent photoconductivity and improvement in Si activation coefficient by Al spatial separation from Ga and Si in AlGaAs-Si solid system: A novel short-period AlAs/-n-GaAs super-lattice," *Jpn J Appl Phys: Lett*, vol. **22**, no. 10, pp. L627–L629, 1983.

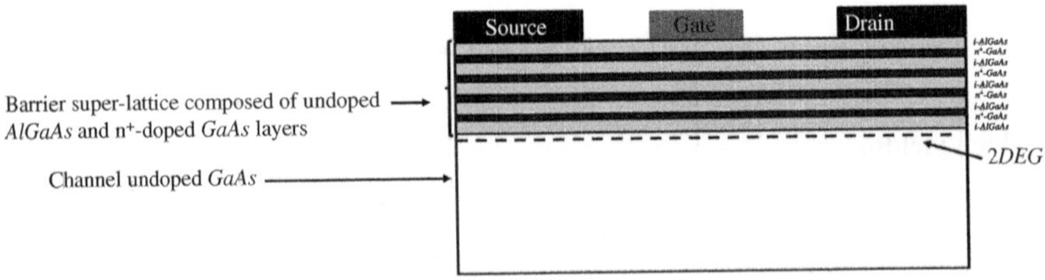

FIGURE 5.15 Superlattice-based modulation doping scheme used among *AlGaAs/GaAs HFETs* of high *Al mole fraction.*

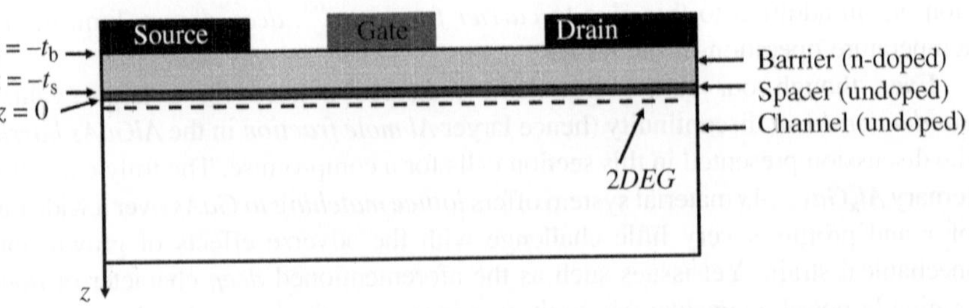

FIGURE 5.16 Cross-sectional view of the *HFET* used in formulating the discussions of Section 5.3.2.

charge carriers are divided among the *2DEG* and the electron concentration profile induced at the surface of the *gate* metal. Such a *carrier* depletion scheme is welcomed due to the fact that it eliminates the chance of forming the dreaded parallel *channel* within the low-*mobility barrier layer*. In this charge sharing, studying the proportional share of the electrons supplied to the *2DEG channel* is extremely important to the evaluation of the *threshold voltage* and the *current drive* of an *HFET* (i.e., among other device performance metrics). As a result of the *bias* dependence of the *depletion-region width* of the *gate's Schottky MS* contact, such a charge sharing is expected to be *bias*-dependent. In order to assess this *bias* dependence, we start with calculating the *2DEG* concentration as a function of *Fermi energy level* in a simple *HFET heterostructure* such as the one depicted in Figure 5.16. As shown in this figure, the layer structure is composed of an *n-doped barrier layer* of thickness $t_b - t_s$, an *undoped spacer layer* of thickness t_s, and a thick *undoped channel layer*.

In this discussion, in dealing with the evaluation of controllability of the *2DEG* by the *gate* voltage, we assume that the two semiconductors from which this *heterostructure* is composed are *lattice-matched*. According to this assumption, there is no need to consider the role of mechanical strain in forming the *2DEG*. Such an approximation in the case of a *pseudomorphically* grown *lattice-mismatch heterostructure* enjoying large *piezoelectric* coefficients can result in an erroneous evaluation.[65]

[65] As pointed out earlier, we will encounter these situations in Section 5.4.

For the *HFET* structure of Figure 5.16, assuming total depletion of *charge carriers* from the *barrier layer*, according to the *Gauss's law*, we have

$$\frac{dE}{dz}=\frac{qN_{\mathrm{d}}}{\epsilon_{\mathrm{s}}}\quad \text{for}-t_{\mathrm{b}}\le z\le -t_{\mathrm{s}} \tag{5.42}$$

where $z=0$ is defined at the *heterointerface* and ϵ_{s} is the *dielectric constant* of the *barrier layer*.

Applying the boundary condition of $\phi(-t_{\mathrm{b}})=0$ on the value of potential, the electric-field profile, which is calculated as

$$E(z)=\frac{qN_{\mathrm{d}}}{\epsilon_{\mathrm{s}}}(z+t_{\mathrm{b}})+E\left(-t_{\mathrm{b}}^{+}\right)\quad \text{for}-t_{\mathrm{b}}\le z\le -t_{\mathrm{s}}, \tag{5.43}$$

yields the potential variation across the *barrier layer*:

$$\phi(z)=-\frac{qN_{\mathrm{d}}}{2\epsilon_{\mathrm{s}}}(z+t_{\mathrm{b}})^{2}-E\left(-t_{\mathrm{b}}^{+}\right)(z+t_{\mathrm{b}})\quad \text{for}-t_{\mathrm{b}}\le z\le -t_{\mathrm{s}}. \tag{5.44}$$

Since the *spacer layer* is *undoped*, the *zero*-gradient electric field across this layer suggests that the electric field at the *heterointerface* is given by

$$E(0^{-})=E(-t_{\mathrm{s}})=\frac{qN_{\mathrm{d}}}{\epsilon_{\mathrm{s}}}(-t_{\mathrm{s}}+t_{\mathrm{b}})+E\left(-t_{\mathrm{b}}^{+}\right). \tag{5.45}$$

Hence,

$$E\left(-t_{\mathrm{b}}^{+}\right)=E(0^{-})-\frac{qN_{\mathrm{d}}}{\epsilon_{\mathrm{s}}}(-t_{\mathrm{s}}+t_{\mathrm{b}}). \tag{5.46}$$

According to Equation (5.46), (5.44) can be reformulated as

$$\phi(z)=-\frac{qN_{\mathrm{d}}}{2\epsilon_{\mathrm{s}}}(z+t_{\mathrm{b}})^{2}-\left[E(0^{-})-\frac{qN_{\mathrm{d}}}{\epsilon_{\mathrm{s}}}(-t_{\mathrm{s}}+t_{\mathrm{b}})\right](z+t_{\mathrm{b}})\quad \text{for}-t_{\mathrm{b}}\le z\le -t_{\mathrm{s}} \tag{5.47}$$

based on which

$$\phi(-t_{\mathrm{s}})=\frac{qN_{\mathrm{d}}}{2\epsilon_{\mathrm{s}}}(-t_{\mathrm{s}}+t_{\mathrm{b}})^{2}-E(0^{-})(-t_{\mathrm{s}}+t_{\mathrm{b}}). \tag{5.48}$$

Considering Equation (5.48), the constant electric field of the *spacer layer* yields the potential at the *heterointerface* as

$$\phi(0)=\frac{qN_{\mathrm{d}}}{2\epsilon_{\mathrm{s}}}(-t_{\mathrm{s}}+t_{\mathrm{b}})^{2}-E(0^{-})(-t_{\mathrm{s}}+t_{\mathrm{b}})-E(0^{-})t_{\mathrm{s}}=\frac{qN_{\mathrm{d}}}{2\epsilon_{\mathrm{s}}}(-t_{\mathrm{s}}+t_{\mathrm{b}})^{2}-E(0^{-})t_{\mathrm{b}} \tag{5.49}$$

where the term $(qN_{\mathrm{d}}/2\epsilon_{\mathrm{s}})(-t_{\mathrm{s}}+t_{\mathrm{b}})^{2}$ is referred to as the *pinch-off voltage*, V_{P}.

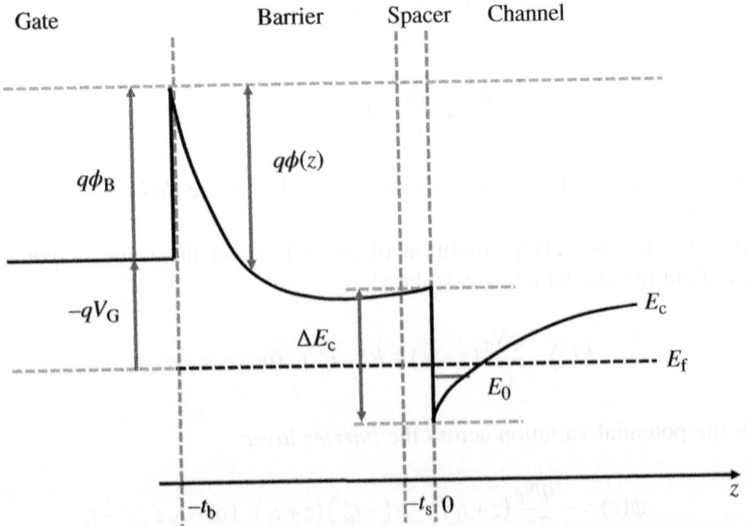

FIGURE 5.17 Conduction-band edge diagram of the *HFET* of Figure 5.16 under negative gate bias.

Based on the definition of the *pinch-off voltage* and Equation (5.49), the electric field at the *heterointerface* is given by

$$E(0^-) = \frac{V_P - \phi(0)}{t_b}. \tag{5.50}$$

Corresponding to the *band diagram* shown in Figure 5.17, for the structure depicted in Figure 5.16 *biased* with a negative *gate*-to-*channel* potential of V_G, we have

$$q\phi(0) + \Delta E_c = q\phi_B - qV_G + E_f(n_s) \tag{5.51}$$

where $q\phi_B$ is the *Schottky* barrier height.

According to (5.50) and (5.51), and the continuity of the electric field at the *heterointerface*, the *sheet-carrier* concentration of the *2DEG* can be transcendentally calculated according to

$$qn_s = \epsilon_s E(0^-) = \frac{\epsilon_s}{t_b}\left(V_P - \phi_B + V_G + \frac{\Delta E_c}{q} - \frac{E_f(n_s)}{q}\right). \tag{5.52}$$

As a first-order approximation in calculating the *threshold voltage* of an *HFET*, if we neglect the variation of the position of the *Fermi level* with the *sheet-carrier* concentration, a fully electrostatic linear relationship is obtained. As pointed out earlier in Chapter 2, this is also used in the case of *MOSFETs*:

$$qn_s = \frac{\epsilon_s}{t_b}(V_G - V_T) \tag{5.53}$$

where for *HFETs* the per unit area *gate capacitance* is defined as ϵ_s/t_b and

$$V_T = \phi_B - \frac{\Delta E_c}{q} + \frac{E_f(n_s)}{q} - V_P. \tag{5.54}$$

In terms of these definitions, as the thickness of the *spacer layer* increases, the *pinch-off voltage* and hence the *sheet-carrier* concentration reduces. As suggested earlier in Section 5.2.4.1, this is a negative side effect of increasing the *spacer-layer* thickness.

For *gate* voltages smaller than the *threshold voltage*, in (5.53), the exact relationship between the E_f and n_s must be used.[66] This is not unlike the *subthreshold regime* of operation of a *MOSFET*. However, unlike *MOSFETs*, in *HFETs*, we face an additional limitation at higher *gate* voltages. At this higher end, increasing the *gate* voltage to values greater than $V_{G,\max} \cong qn_{s0}(t_b/\epsilon_s)$ (calculated in terms of the so-called *equilibrium sheet-carrier* concentration n_{s0}) results in the separation of the *depletion regions* of the *blocking MS* contact and the *2DEG*. For these values of *gate* voltage, the dreaded parallel *channel* will be formed within the semiconducting *barrier layer*. Such a situation cannot transpire within the insulating *barrier* of a *MOSFET*. For these large *gate* voltages, the *depletion* approximation is no longer valid, and the *2DEG* concentration saturates to its equilibrium value of qn_{s0}.

While in this section we assumed a uniform *doping* across the *barrier layer*, in implementation of *HFETs*, other schemes of *doping* are also often employed. One of these schemes is referred to as *delta doping*.[67] Such a *doping* scheme, among other factors, has proven effective in eliminating the parallel *channel* from the *barrier layer*. In *delta-doped HFETs*, instead of *doping* the entire thickness of the *barrier layer*, only a thin portion of this layer next to the *spacer layer* is heavily *doped*. Figure 5.18 presents a schematic illustration of the layer structure of a *delta-doped HFET*. The closer the *delta-doped* layer is to the *heterointerface* (i.e., the thinner is the *spacer*), the larger the proportional share of the *2DEG* from the depletion of the *doped* layer. *Delta-doped HFETs*, due to the possibility of *doping* this layer *degenerately*, offer a larger *2DEG* concentration and as a result a larger *drain current drive*. In *delta-doped HFETs*, since the *gate* metal is in contact with an *undoped* semiconductor, the quality of the *gate Schottky* contact is improved. Using *delta doping*, a thinner *barrier layer* (hence a larger *gate transconductance*) can be achieved without degrading the *gate* properties. *Delta-doped HFETs* also offer a larger *breakdown voltage* for the *gate diode* due to

[66] See the discussion in Section 5.2.3.

[67] E. Schubert, "Delta doping of III–V compound semiconductors: Fundamentals and device applications," *J Vac Sci Technol A*, vol. **8**, no. 3, pp. 2980–2996, 1990.

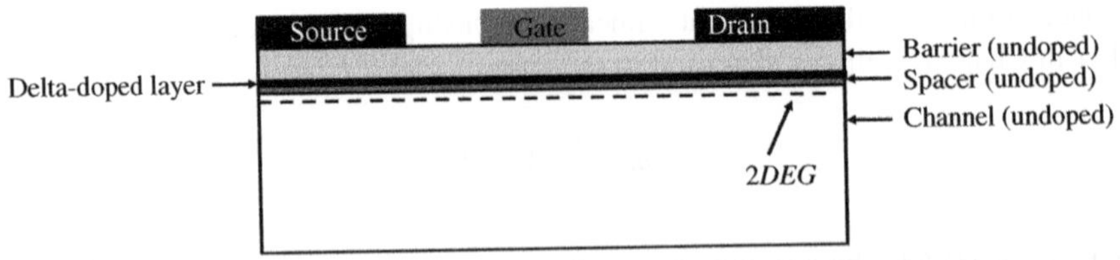

FIGURE 5.18 Cross-sectional view of an *HFET* employing delta doping technique. *Barrier*, *spacer*, and the *delta-doped* layer can be made in the same semiconductor.

their effectively *undoped barrier*. The opportunity to improve the proximity of the *gate* to the *channel* also heralds improvement in *short-channel effects*[68] in the operation of these *HFET*.

While the discussion of this section focused only on the calculation of the *threshold voltage* among *HFETs*, through employing this concept, the current–voltage characteristics of *HFETs* can be described on the basis of the frameworks presented in Chapter 3 for *long-* and *short-channel MOSFETs*.

5.3.3 *HFET*: A Brief Visit to Microfabrication Challenges

Figure 5.19, compared to the *HFET* structures presented so far in this chapter, presents the schematic depiction of a slightly more complete version of an *HFET*. On the basis of this layer structure, it is worthwhile pointing out that due to difficulties in achieving *ohmic* contacts across a wide-*bandgap barrier layer*, often a *highly doped* thin layer of a *narrow-bandgap* semiconductor (which could be of the same material as the *channel layer*) is grown over the *barrier* of a *normal HFET*. This highly *doped* layer, which is known as the *cap layer*, assists with the realization of *tunneling ohmic* contacts to the 2DEG. Such a necessity is absent in the implementation of *inverted HFETs*. The *cap layer*, while only required under the *drain* and *source* electrode, is patterned prior to the *ohmic* contact *deposition* step.

Electron-beam evaporation of a stack of different metals is the preferred mode of *metallization* among the *compound* semiconductor technologies. Only for the *ohmic* contacts this *metallization* step is succeeded by a *rapid thermal-annealing* process, intended for forming a combination of *tunneling/alloyed*-contact across the highly *doped cap* and the *barrier* layer. In this *metallization* step, the combination of the metals and the thicknesses of each layer are in need of a proper selection. In the case of *ohmic* contacts, this selection is performed via considering requirements such as forming a conductive metal–semiconductor *alloy* at the *MS* interface, producing a smooth surface for minimizing the *contact resistance*, avoiding intermixing of metals at high *annealing* temperatures (which for some metals such as *Al* and *Au* could form a dark low-*conductivity alloy*), and producing a surface that is not prone to oxidation.

[68] We have already spoken of these effects in case of a *MOSFET* in Chapter 3. That discussion can be easily extended to *HFETs*.

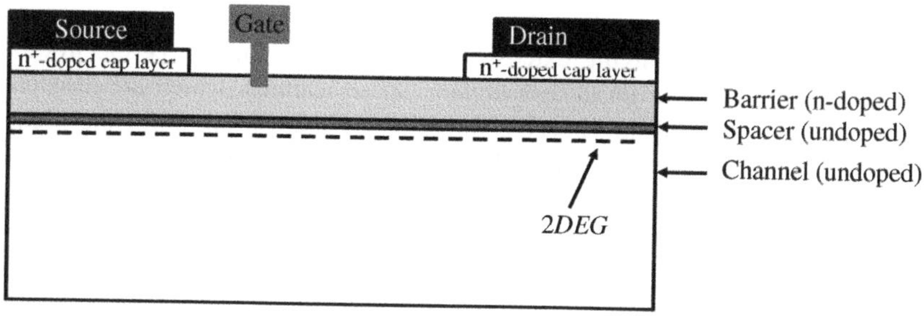

FIGURE 5.19 Cross-sectional view of an *HFET* employing offset/recessed mushroom gate, *spacer layer*, and n^+-doped *cap layer*. *Cap* and *channel* layer can be made of the same narrow bandgap semiconductor, while *spacer* and the *barrier* layer are made of the wider bandgap semiconductor.

In addition to using the *cap layer*, selective thinning of the *barrier layer* (known as *ohmic recess* process) has been explored to reduce the contact resistance between the *drain/source* electrodes and the *2DEG*. In our discussion in Section 5.3, we have already encountered another process of *recess* (i.e., *gate recess*). While the *ohmic recess* is intended to provide an easier conductive path to the *2DEG*, such a side effect is certainly not welcome in the *gate-recess* process. To avoid this side effect, in the *gate-recessed* implementation of the *enhancement-mode HFETs*, a thin layer of an insulator can be sandwiched between the *gate* metal and the thinned *barrier*. The name often used for *FETs* realized on these *MIS heterostructures* is *MISHFET*.[69]

While *HFETs* are often employed as high-speed/high-frequency transistors, as mentioned earlier, boosting their frequency of operation to *hundreds of gigahertz* has dictated *scaling* of their *gate length* down to tens of *nanometers*. However, at these small values of *gate length*, the series access resistance to the *gate* electrode poses degradation on the microwave properties of the extrinsic transistor. As illustrated in Figure 5.19, in order to limit this degradation, *HFETs* are implemented with the so-called mushroom *gate* structures (in which the *gate* is named according to its shape).

Considering the interest in the operation of *HFETs* as high-power transistors, it is worthwhile emphasizing that the aforementioned removal of the requirement of *self-alignment* provides these transistors with an improved *breakdown voltage*. Such a gain in *breakdown voltage*, which is achieved through establishing a gap between the *drain* and *gate* electrodes, as indicated in Figure 5.19 can be further boosted by placing the *drain* at a larger distance from the *gate* (while the total *channel length* is not increased). This is often referred to as the *offset gate* structure. Another factor that can be employed toward boosting the *breakdown voltage* of an *HFET* is to employ *wider-bandgap* semiconductors for the *barrier* and even the *channel layer*.

[69] *Metal Insulator Semiconductor HFET*.

The *breakdown-voltage* characteristics of *HFETs* have been also improved by implementing an overhanging extension of the *gate* or the *source* electrode over the high electric field *drain access region*. These features, which are respectively known as the *gate-* and the *source-connected field plates*, allow for the reduction of the maximum electric field across this region, hence boosting the *breakdown voltage* of the transistor.

Regarding the high-power performance of *HFETs*, one of the important factors in selecting the proper semiconductor technology is the *thermal conductivity* of the *epilayers* and especially that of the substrate. While not all semiconductor technologies offer a viable high *thermal-conductivity* substrate, in many high-power *HFET chips* substrate removal through the *lapping* process and sometimes mounting this thinned *chip* on a *thermally conductive carrier substrate* is explored.

5.3.4 Hot Electron Applications Among *HFETs*

Heterostructures have been used in a number of different ways in the realization of *negative differential resistance* (i.e., *NDR*). An example of that is the realization of *Esaki's tunnel diode*[70] on a *heterojunction*.[71] While the *NDR* effect proposed by Esaki was based on the application of a large electric field perpendicular to the *heterointerface*, in 1979 Hess et al. proposed an *NDR* effect on the basis of the application of a large electric field parallel to the *heterointerface* of an *AlGaAs/GaAs HFET* structure.[72] According to this scheme, the *hot electrons* that are accelerated by the large electric field move up in the two- or three-dimensional continuum of *states* in *GaAs* until they acquire enough momentum (e.g., through *scattering*) to propagate perpendicular to the *heterointerface*. Upon this perpendicular migration, electrons are reunited with their parent *donors*. This reunification, due to the strong *ionized-impurity scattering*, results in *mobility* degradation and therefore an *NDR* manifestation.

This *NDR* manifestation can be seen as a *real-space* analogy to the *Gunn effect*.[73] In fact, as suggested earlier in Chapter 1, many of the peculiarities of the *E–k* relationships have a *Fourier-transformed* analogy in *real space*. In many instances, these *real-space* analogies have high potentials for device applications.

Regarding the evaluation of the required amount of energy to instigate this *real-space transfer*, for example, in the case of an *AlGaAs/GaAs heterostructure*, it is educational to pay attention to the parallelism with *Gunn effect* research. On the basis of this research, it is well established that electrons can acquire up to 300 meV under moderate electric fields of the order of 4 kV/cm. It has been shown that in this case the average electron energy increases steeply beyond the 3.4 kV/cm electric field. These amounts

[70] L. Esaki, "New phenomenon in narrow germanium p–n junctions," *Phys Rev Lett*, vol. **109**, p. 603, 1958.

[71] See also the discussion in Section 4.2.1.

[72] K. Hess, H. Morkoc, H. Shichijo, and B. Streetman, "Negative differential resistance through real-space electron-transfer," *Appl Phys Lett*, vol. **35**, no. 6, pp. 469–471, 1979.

[73] Named after *John Gunn*. In *Gunn diodes*, *intervalley* transfer of electrons results in *NDR* manifestations. The *NDR* in the v_d–E characteristic of *III–V compounds* is not just because of the heavier *effective mass* of the *higher valleys* but also due to the fact that the electrons in *L-* and the *X-valleys* experience a *deformation-potential scattering*.

of energy are close to the *conduction-band* discontinuity and, as a result, are close to the required amount of energy for *carrier* migration from the *channel* to the *barrier layer* of an *AlGaAs/GaAs heterostructure*. For a *two-dimensional electron gas*, because of the slightly higher *phonon scattering rate*, the initial increase in electron energy is slower. However, upon populating many *subbands*, this limit is known to tend toward the three-dimensional case. A true advantage of *real-space transfer NDR* over the *Gunn effect* is the fact that in this process due to the fundamentally different approach to *NDR* formation, *NDR* can be achieved under considerably lower electric fields.

5.4 POLAR III-NITRIDE *HFET*

During the past two decades, *III-Nitride HFETs* have managed to attract the attention of the device research community as alluring candidates for high-power, high-voltage microwave applications. The term *III-Nitride* often refers to *GaN*, *InN*, *AlN*, and their *alloys*. Intensive theoretical and experimental investigations on the material properties and growth techniques of *III-Nitride heterostructures* have opened new frontiers to electronic devices. Extraordinary high *sheet-carrier* concentration and strong confinement of the *carriers* at specific *heterointerfaces* of the *unintentionally doped Wurtzite AlGaN/GaN heterostructures*[74] are known to be consequences of the strong *spontaneous* and *piezoelectric polarization fields*.[75] Despite the inferiority of the *low-field* electron *mobility* of *GaN* compared to the main contenders in the *III–V* niche of electronic device market, *Wurtzite III-Nitride HFETs* employing *GaN* as the *channel layer* offer unprecedented values of *drain* current density in the order of 1 A/mm *width* of the *channel* due to the much larger peak electron velocity, larger saturation velocity, better thermal stability, higher *2DEG* concentration, and better *carrier* confinement as a consequence of the larger *bandgap* discontinuity. Additionally, the wide *bandgap* of *GaN* and *AlN* (i.e., 3.4 and 6.2 eV, respectively) extend very high *breakdown voltages* (i.e., in the order of 100 V) to the transistors fabricated on these *heterojunctions*. While as already pointed out in Section 5.1.1, problems in finding a proper *lattice-match substrate* still plague the success of this technology, the use of a superbly thermally conductive substrate such as *SiC* has been proven to be extremely rewarding in maximizing power density of *III-Nitride HFETs*. As a result of intensive research, *AlGaN/GaN HFET*s offering power densities exceeding 40 W/mm *width* of the

[74] *III-Nitride* crystals can be also grown in the *cubic* form. However, since in this form they do not demonstrate the *polar* properties to the extent demonstrated by the *hexagonal* crystals grown along the *c-axis*, in this section, in our assessment of the *polar HFETs*, only the *hexagonal III-Nitrides* (and only when grown along the *c-axis*) are treated.

[75] F. Bernardini, V. Fiorentini, and D. Vanderbilt, "Spontaneous polarization and piezoelectric constants of III–V nitrides," *Phys Rev B*, vol. **56**, no. 16, pp. R10-024–R10-027, 1997; O. Ambacher, B. Foutz, J. Smart, J. Shealy, N. Weinmann, K. Chu, M. Murphy, A. Sierakowski, W. Schaff, L. Eastman, R. Dimitrov, A. Mitchell, and M. Stutzmann, "Two dimensional electron gases induced by spontaneous and piezoelectric polarization in undoped and doped AlGaN/GaN heterostructures," *J Appl Phys*, vol. **87**, no. 1, pp. 334–344, 2000.

channel at *X-band*,[76] with values of f_T and f_{max} in the *hundreds* of *gigahertz*, capable of operating at temperatures as high as 400°C are now a possibility.

As will be observed in this section, although the existence of a large *lattice mismatch* between the materials forming the *III-Nitride heterojunctions* is naturally expected to be limiting, the material properties of these semiconductors have proven this exploration extremely rewarding. Due to *polarization* dependence of most *III-Nitride 2DEGs*, instead of the usual *modulation doping*, the operation of the *HFETs* realized on these *heterostructures* is considerably independent of *doping*.

In the *Wurtzite III-Nitride* material system, *spontaneous* and *piezoelectric polarizations* exceeding those of conventional *III–V* semiconducting *compounds* (i.e., by as high as an order of magnitude) are realizable. The strength of the *spontaneous polarization* and the *strain-induced piezoelectric polarization* of *pseudomorphic AlGaN/GaN* and *InGaN/GaN heterostructures* can be as high as 3 and 2 MV/cm, respectively. Such high *polarization* fields are capable of inducing extraordinary high *sheet-carrier* concentrations on the orders of 10^{13} cm^{-2} at the *heterointerface*. Due to the importance of the role of *polarization* fields in the operation of *III-Nitride* electronic devices, in Section 5.4.1, a summary of *polar* properties of these materials is provided. In Section 5.4.2, it will be shown how the *polar* properties change the *band diagram* of such *heterostructures* for which case a modified version of the calculation framework previously presented in Section 5.2 will be provided.

5.4.1 Polarization Among III-Nitride Heterostructures

Noncentrosymmetric compound crystals, which include the *Wurtzite III-Nitrides*, exhibit two different atomic layering orders in opposing directions parallel to certain crystallographic axes. For *Wurtzite GaN*, the sequence of atomic bilayers of *Ga* and *N* are reversed along the directions [0001] and $[000\bar{1}]$. *Ga face* and *N face* correspond to (0001) and $(000\bar{1})$ crystalline faces. For *AlGaN/GaN epitaxial* layers, the most common growth direction is normal to the {0001} basal plane (i.e., *c-plane*). In the *III-Nitride epilayers*, the existing bilayers consist of two tightly packed *hexagonal* layers of *cations* and *anions*. The existence of these bilayers of *cations* and *anions* is the origin for the *face polarity* of these crystals. For *AlGaN*, if the *cation* is located on the top of the {0001} basal-plane bilayer, the layer is called *metallic face* or for simplicity *Ga face*. If it is the *anion* (i.e., *N*) that is on top, it is called *N face*. *Ga face* corresponds to [0001] direction and *N face* corresponds to $[000\bar{1}]$ direction. Conventionally, the [0001] direction is shown by a vector from a *Ga* atom to the nearest neighboring *N* atom.

Due to the existence of significant *spontaneous* and *piezoelectric polarizations* in *III-Nitride heterostructures*, the *2DEG* concentration (and consequently the performance of the electronic devices realized on these structures) depends considerably on the *polarity*, *alloy composition*, *strain*, *thickness*, and unlike traditional *HFETs* to a lesser extent *barrier doping*.

Theoretical investigations on the *piezoelectric* properties of *III-Nitride* materials have revealed that the *piezoelectric coefficients* of this material system, resembling

[76] That is, the 8–12 GHz frequency range.

those of *ZnO*, are about an order of magnitude higher than those of the conventional *III–V* and *II–VI compounds*. The *piezoelectric coefficients* of this material system, unlike those of the traditional *III–V compounds*, are of positive values. This is in resemblance to those of the *II–VI compounds*.

In addition to the *piezoelectric polarization*, which is induced at the *strained pseudomorphically grown III-Nitride heterointerfaces*, because of the dependence of *spontaneous polarization* on the structural parameters, considerable quantitative differences among the *spontaneous polarizations* of different *Wurtzite III-Nitrides* forming the *heterojunction* are observed. The discontinuity in the *spontaneous polarization* vectors across the *III-Nitride heterointerface* is the cause for the induction of a *spontaneous polarization* at the interface. Among *Wurtzite III-Nitrides*, it is observed that the increase in the nonideality from *GaN*, to *InN*, and to *AlN* (i.e., because of the increasing elongation of the *anion/cation bond* length along the (0001) *axis*, often referred to by u, and reduction in the value of c/a) along this material order reflects itself through an increase in the *spontaneous polarization*.[77] The ideal values of u and c/a for a *Wurtzite* structure, corresponding to the *maximum sphere packing*, are equal to $0.375c$ and $\sqrt{8/3}$, respectively.

Among *III-Nitride alloys*, it has been shown that the linear prediction of *polarization* based on the *polarizations* of the parent *binary compounds* is not applicable. The *spontaneous polarization* among these *alloys* is characterized by large bowing effects, which are strongly dependent on the *alloy's* microscopic structure. These *bowing effects* are due to the different responses of the *bulk binaries* to hydrostatic pressure and to internal strain effects (i.e., *bond* alternation). Assuming that the *ternary III-Nitride alloys* have random microscopic structures, the values of *spontaneous polarization* (in C/m^2) among the *III-Nitride alloys* as quadratic functions of the *alloy* composition parameters can be modeled as

$$\begin{cases} P^{sp}_{Al_xGa_{1-x}N} = -0.090x - 0.034(1-x) + 0.019x(1-x) \\ P^{sp}_{In_xGa_{1-x}N} = -0.042x - 0.034(1-x) + 0.038x(1-x). \\ P^{sp}_{Al_xIn_{1-x}N} = -0.090x - 0.042(1-x) + 0.071x(1-x) \end{cases} \quad (5.55)$$

For the *piezoelectric polarization*, in the case where appreciable nonlinearity of the *bulk piezoelectric polarization* of the parent *binaries* versus the mechanical strain is taken into account, *Vegard's law* succeeds in predicting the *piezoelectric polarization* of the *alloys*.[78] Unlike the *spontaneous polarization*, the nonlinearity of the

[77] See the discussion in Section 1.6.1.1. This *Wurtzite* crystalline structure is defined through the edge length a of the basal hexagon, the height c of the hexagonal prism, and the internal parameter u defined as the *anion/cation bond* length along the (*0001*) axis, in units of c.

[78] F. Bernardini and V. Fiorentini, "Macroscopic polarization and band offsets at nitride heterojunctions," *Phys Rev B*, vol. **57**, no. 16, p. R9427, 1998; F. Bernardini, V. Fiorentini, and D. Vanderbilt, "Accurate calculation of polarization-related quantities in semiconductors," *Phys Rev B*, vol. **63**, pp. 193.1–193.4, 2001; F. Bernardini and V. Fiorentini, "Nonlinear macroscopic polarization in III–V nitride alloys," *Phys Rev B*, vol. **64**, pp. 085207.1–085207.7, 2001; F. Bernardini, V. Fiorentini, and O. Ambacher, "Evidence for nonlinear macroscopic polarization in III–V nitride alloy heterostructures," *Appl Phys Lett*, vol. **80**, no. 7, pp. 1204–1207, 2002.

piezoelectric polarization is independent of the microscopic *alloy* structure or disorder effects and is entirely due to the nonlinear strain dependence of the *bulk piezoelectric* response. In this case, for a generic *III-Nitride alloy* of $X_xY_{1-x}N$ that is grown along the *c-axis* where the strain is imposed over the grown *epilayers* in the basal plane, according to *Vegard's law*, the *piezoelectric polarization* is given by

$$P^{\mathrm{pz}}_{X_xY_{1-x}N} = xP^{\mathrm{pz}}_{\mathrm{XN}}[\epsilon(x)] + (1-x)P^{\mathrm{pz}}_{\mathrm{YN}}[\epsilon(x)]. \tag{5.56}$$

The applicability of *Vegard's law* by no means implies the linear dependence of the *piezoelectric polarization* on the *alloy* composition. In Equation (5.56), the nonlinearity is included through the nonlinearities of the *piezoelectric polarization* versus the *in-plane stress* exerted on the parent *binaries*. These nonlinearities among *III-Nitride binaries* are modeled as

$$\begin{cases} P^{\mathrm{pz}}_{\mathrm{AlN}} = -1.808\epsilon + 5.624\epsilon^2 & \text{for } \epsilon < 0 \\ P^{\mathrm{pz}}_{\mathrm{AlN}} = -1.808\epsilon - 7.888\epsilon^2 & \text{for } \epsilon > 0 \\ P^{\mathrm{pz}}_{\mathrm{GaN}} = -0.918\epsilon + 9.541\epsilon^2 \\ P^{\mathrm{pz}}_{\mathrm{InN}} = -1.373\epsilon + 7.559\epsilon^2 \end{cases} \tag{5.57}$$

where the basal strain ϵ, as previously presented in (1.105), is a function of x in the *heterostructure*:

$$\epsilon(x) = \frac{[a_{\mathrm{sub}} - a(x)]}{a(x)}. \tag{5.58}$$

In (5.58), $a(x)$ and a_{sub} represent the *lattice constants* of the *free-standing alloy* of composition x and the substrate, respectively. In the case of *pseudomorphic* growth, *lattice constants* (in nm) approximately follow *Vegard's law*:

$$\begin{cases} a_{Al_xGa_{1-x}N}(x) = 0.31986 - 0.00891x \\ a_{In_xGa_{1-x}N}(x) = 0.31986 + 0.03862x. \\ a_{Al_xIn_{1-x}N}(x) = 0.35848 - 0.04753x \end{cases} \tag{5.59}$$

As will be further elaborated in Section 5.4.2, it is in association with the *divergence* of the *polarization* in space that the *2DEG* concentration at the *III-Nitride heterointerface* is calculated. The concentration of *polarization*-induced *charge carriers* in terms of the *spontaneous* and the *piezoelectric-polarization* vectors (i.e., $\vec{P}_{\mathrm{sp}}$ and $\vec{P}_{\mathrm{pz}}$, respectively) is given as

$$\rho(P) = \nabla \cdot \left(\vec{P}_{\mathrm{sp}} + \vec{P}_{\mathrm{pz}}\right). \tag{5.60}$$

On this basis and using the developed notations for the *spontaneous* and *piezoelectric polarizations* in *III-Nitride epilayers*, *polarization*-induced *sheet-carrier* concentration at the *heterointerface* is equal to

$$\sigma_{\text{s-polar}} = P(\text{barrier}) - P(\text{channel})$$
$$= P_{\text{sp}}(\text{barrier}) - P_{\text{sp}}(\text{channel}) + P_{\text{pz}}(\text{barrier}) - P_{\text{pz}}(\text{channel}). \quad (5.61)$$

As for the *spontaneous polarization*, among all *III-Nitride alloys*, the direction of the *polarization* is always positive. As illustrated in Figure 5.20, for the *c-plane Wurtzite AlGaN/GaN heterostructures* of *Ga face*, the direction of the P_{sp} discontinuity is normal to the *heterointerface* opposite to the growth direction. This direction

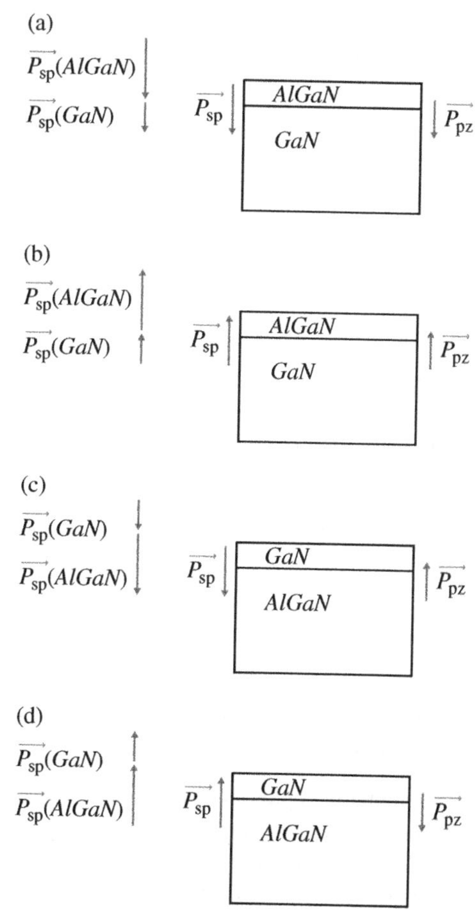

FIGURE 5.20 Indication of the directions of *piezoelectric* and *spontaneous polarization* vectors (i.e., $\vec{P}_{pz}$ and $\vec{P}_{sp}$, respectively) in (a) a pseudomorphically grown *AlGaN/GaN* epilayer in *Ga face*, (b) a pseudomorphically grown *AlGaN/GaN* epilayer in *N face*, (c) a pseudomorphically grown *GaN/AlGaN* epilayer in *Ga face*, and (d) a pseudomorphically grown *GaN/AlGaN* epilayer in *N face*. As indicated in the text, *spontaneous polarization* vector is calculated as the difference between the spontaneous polarizations of the overlayer and the underlayer. The bottom layer in each case is assumed to be relaxed to its freestanding lattice constant.

reverses in the case of *N-face* layers. Regarding the direction of the *piezoelectric polarization*, growth of a *tensile-strained* overlayer induces a positive *piezoelectric polarization* at the *heterointerface* (i.e., in *Ga-face* layers from top to bottom and the other way around for *N face*). Considering the *free-standing lattice constant* of *AlGaN* and *GaN*, this is the case among the *pseudomorphically* grown *normal AlGaN/GaN heterostructures*. However, for other *heterostructures*, such as *inverted GaN/AlGaN*, developing a *compressive* strain across the overlayer grown on the *relaxed* template reverses the direction of the *piezoelectric polarization* vector.

Consequent to these polarity dependences, as shown in Figure 5.20, growing a *normal AlGaN/GaN pseudomorphic heterostructure* in *Ga face* induces a significant *2DEG* at the *heterointerface*. This is since both the *spontaneous* and the *piezoelectric polarizations* in Equation (5.61) are pursuing this charge induction. As expected, changing the *face* or the order of layers can put the two components of *polarization* in opposing directions resulting in a smaller *polarization*-induced *2DEG* concentration (or possibly a *2DHG*). Examples of these situations are provided in Figure 5.20.

For *AlGaN/GaN heterostructures*, it has been observed that *sheet-carrier* concentrations even higher than 10^{13} cm^{-2} are realizable for higher *Al mole fractions*. Certainly, to achieve these high *polarization*-induced *sheet-carrier* concentrations, the *AlGaN barrier layer* should be thick enough to avoid depletion due to the effect of *surface layer* (or where present the *Schottky gate*). Values of electron *mobility* beyond 20,000 cm^2/V s for the *2DEG* of $Al_{0.27}Ga_{0.73}N/GaN$ *heterostructures* measured at the temperature of 13 K have been reported by Smorchkova et al.[79] This electron *mobility* drops to values below 1000 cm^2/V s by significantly increasing the *sheet-carrier* concentration for *barrier* thicknesses exceeding 20 nm and *Al mole fractions* exceeding 0.3. Studies of the *scattering* mechanisms in *AlGaN/GaN HFETs* suggest that only for *sheet-carrier* concentrations higher than 7×10^{12} cm^{-2}, the *interface roughness* becomes the dominant *scattering* mechanism. For lower concentrations, *2DEG mobility* is dominated by the *interface-charge scattering* even at temperatures as high as 300 K.[80]

5.4.2 Subband Energy Levels and *2DEG* Characteristics of Polar *AlGaN/GaN* Heterojunctions

Figure 5.21 presents a schematic depiction of the *conduction-band* edge diagram of a *Ga-face Wurtzite AlGaN/GaN heterostructure* grown on the *c-plane*. As observed in this figure, the presence of *polarization* at the interface of the *heterostructure* yields a drastically different version of *band bending*. As a result of the change in the slope of *band bending* across the *AlGaN* layer (i.e., compared to that of the *AlGaAs* layer, such

[79] I. Smorchkova, C. Elsass, J. Ibbetson, R. Vetury, B. Heying, P. Fini, E. Hauss, S. Denbaars, J. Speck, and U. Mishra, "Polarization-induced charge and electron mobility in AlGaN/GaN heterostructures grown by plasma-assisted molecular beam epitaxy," *J Appl Phys*, vol. **86**, no. 8, pp. 4520–4526, 1999.

[80] J. Antoszewski, M. Gracey, J. Dell, L. Faraone, T. Fisher, G. Parish, Y. Wu, and U. Mishra, "Scattering mechanisms limiting two-dimensional electron gas mobility in $Al_{0.25}Ga_{0.75}N/GaN$ modulation doped-field effect transistors," *J Appl Phys*, vol. **87**, no. 8, pp. 3900–3904, 2000.

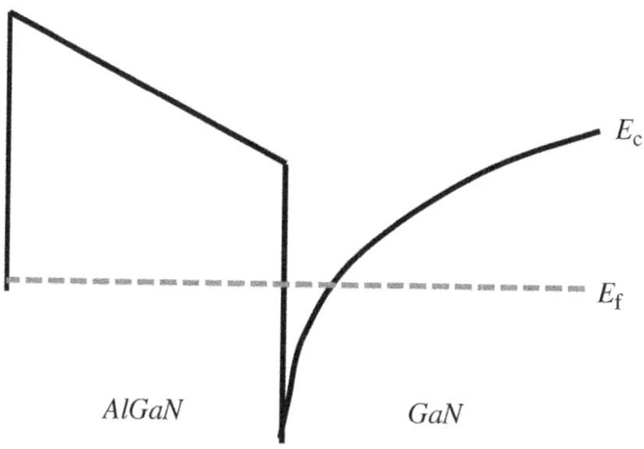

FIGURE 5.21 Typically observed conduction-band edge bending at the *Ga-face* polar *AlGaN/GaN* heterointerface.

as the one depicted in a *normal AlGaAs/GaAs HEMT* structure in Fig. 5.4), these *heterostructures* demonstrate considerable departure from the notion of a triangular *quantum well*.[81] Accordingly, in this section through invoking the foundations of the *variational method*, outcomes of a study on evaluating the *subband energy levels* in the case of *polar AlGaN/GaN HFETs* are presented.[82] In this study, considering the high *2DEG* concentration of *AlGaN/GaN heterojunctions*, both the *ground subband* and the first *excited subband* are considered.

In here, the following *envelope wave functions* are used for electrons in the first and the second *energy subbands*, respectively:[83]

$$\chi_{n,0}(z) = \frac{1}{\sqrt{A}}\left[u(z)(z+\alpha)e^{-(bz/2)} + u(-z)\alpha e^{k_b z}\right] \tag{5.62}$$

and

$$\chi_{n,1}(z) = \frac{1}{\sqrt{B}}\left[u(z)(z+\alpha')(1-Cz)e^{-(bz/2)} + u(-z)\alpha' e^{k_b' z}\right] \tag{5.63}$$

where z is the direction normal to the *heterointerface*, b is the *variational* parameter, $u(z)$ is the *Heaviside unitary step function*, A and B are normalization factors, and C is an arbitrary constant calculated by applying the orthogonality condition between $\chi_{n,0}(z)$ and $\chi_{n,1}(z)$. The constants α and α' are calculated by applying the continuity condition on $\chi_{n,0}(z)$ and $\chi_{n,1}(z)$ and their normalized first spatial derivatives. The

[81] Which was earlier presented in Section 5.2.1.

[82] F. Manouchehri, P. Valizadeh, and M. Kabir, "Determination of subband energies and *2DEG* characteristics of $Al_xGa_{1-x}N$/GaN heterojunctions using variational method," *J Vac Sci Technol A*, vol. **32**, no. 2, pp. 021104-1–021104-8, 2014.

[83] See B. K. Ridley, *Electrons and Phonons in Semiconductor Multilayers*, Cambridge University Press, Cambridge, UK, 1997.

parameters k_b and k'_b are the *wave numbers* associated with the first and the second *energy subbands* in the *barrier layer*, respectively, and are calculated as

$$k_b = \frac{\sqrt{2m_n^*(\Delta E_c - E_0)}}{\hbar} \tag{5.64}$$

and

$$k'_b = \frac{\sqrt{2m_n^*(\Delta E_c - E_1)}}{\hbar}. \tag{5.65}$$

The *Hamiltonian* of the *AlGaN/GaN* system is defined as

$$\langle H_n \rangle = \langle T_n \rangle + \langle U_n \rangle \tag{5.66}$$

where $\langle T_n \rangle$ and $\langle U_n \rangle$ are the *expected values* of the kinetic and potential energies associated with the *n*th *energy subband*, respectively. These are calculated as[84]

$$\langle T_n \rangle = \int_{-\infty}^{+\infty} \frac{1}{2m_n^*} \left| \frac{\hbar}{j} \frac{\partial \chi_{n,n}(z)}{\partial z} \right|^2 dz \tag{5.67}$$

and

$$\langle U_n \rangle = \int_{-\infty}^{+\infty} U(z) \left| \chi_{n,n}(z) \right|^2 dz \tag{5.68}$$

where $U(z)$ is the total potential energy profile developed by the presence of *2DEG*, *polarization* fields, and the *depletion-region* charge. In the *barrier* and the *channel/buffer layer*, $U(z)$ is appropriately represented by U_{AlGaN} and U_{GaN}, respectively.

Based on the *band diagram* presented in Figure 5.21, the potential profile of the *barrier layer* (i.e., $U_{AlGaN}(z)$) is calculated by assuming a constant electric field inside the *barrier*. The potential profile of the *channel/buffer* layer (i.e., $U_{GaN}(z)$), which is the potential developed due to the presence of *2DEG* (i.e., $U_{GaN-s}(z)$) and the *depletion-region* charge (i.e., $U_{GaN-d}(z)$), can be calculated by solving *Poisson's equation*.

According to these definitions, in order to calculate the proper value of the *variational parameter* (i.e., $b = b_m$), the total energy per electron defined as

$$\frac{E}{n_S} = \langle T_0 \rangle + 0.5\left(\left\langle U_{GaN-s(0)} \right\rangle + \left\langle U_{AlGaN(0)} \right\rangle\right) + \left\langle U_{GaN-d(0)} \right\rangle \tag{5.69}$$

is minimized. In this definition, the factor 0.5 prevents the double counting of the electron–electron interactions.

[84] T. Ando, A. Fowler, and F. Stern, "Electronic properties of two-dimensional systems," *Rev Mod Phys*, vol. **54**, no. 2, pp. 437–672, 1982.

Following this step, E_0 is calculated by setting $n=0$ and $b=b_m$ in Equation (5.66). The second *energy subband* can be also calculated by setting $n=1$ and using the same value of b_m in (5.66).

Calculation of the *sheet-carrier* concentration should be performed self-consistently. This calculation procedure starts by assuming an initial position for the *Fermi energy level* (e.g., at the lower edge of the *conduction band* of *GaN* at the *heterointerface*) and calculating the total *2DEG* concentration using the following equation:

$$n_s = \frac{\sigma_{s\text{-polar}}}{q} - \left(\frac{\epsilon_s}{t_{AlGaN} q^2}\right)[q(\phi_B - V_{GS}) + E_f - \Delta E_c] \tag{5.70}$$

where $\sigma_{s\text{-polar}}$ is the *polarization*-induced *2DEG* concentration at the *polar AlGaN/GaN heterointerface* defined in terms of Equation (5.61), t_{AlGaN} is the thickness of the *AlGaN barrier*, and $q\phi_B$ is the *effective Schottky barrier height*.

In addition to (5.70), Equation (5.33) provides another definition for n_s. Both of these equations should be satisfied concurrently as part of the aforementioned recursive calculations.

Figure 5.22 presents the results of the evaluation of the *energy subbands* and *band bending* at the *conduction-band* edge in the *gated* area of a *normal polar*

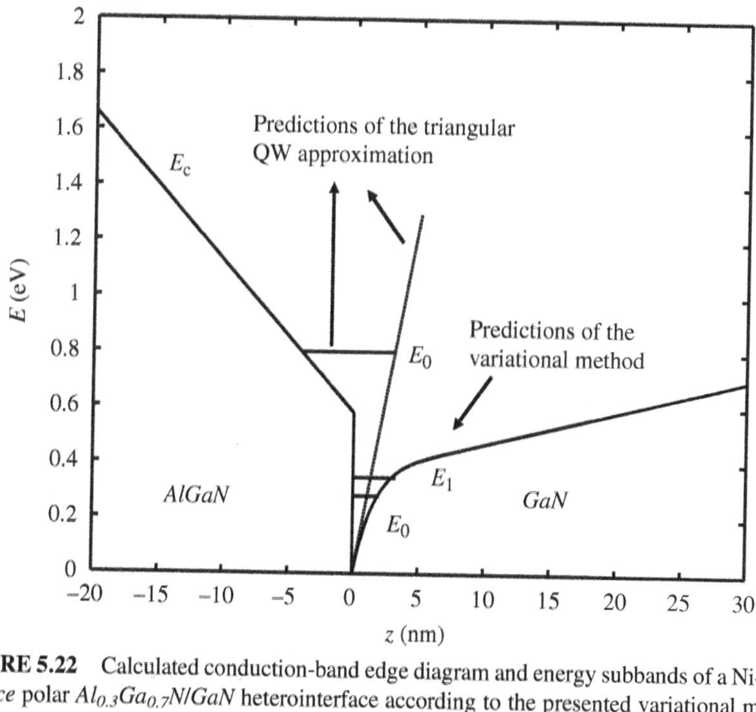

FIGURE 5.22 Calculated conduction-band edge diagram and energy subbands of a Ni-gated *Ga-face* polar $Al_{0.3}Ga_{0.7}N/GaN$ heterointerface according to the presented variational method and the triangular quantum well approximation. Other conditions taken into account in these calculations are provided in the text. Reprinted with permission from Manouchehri et al. (2014). Copyright 2014, American Vacuum Society.

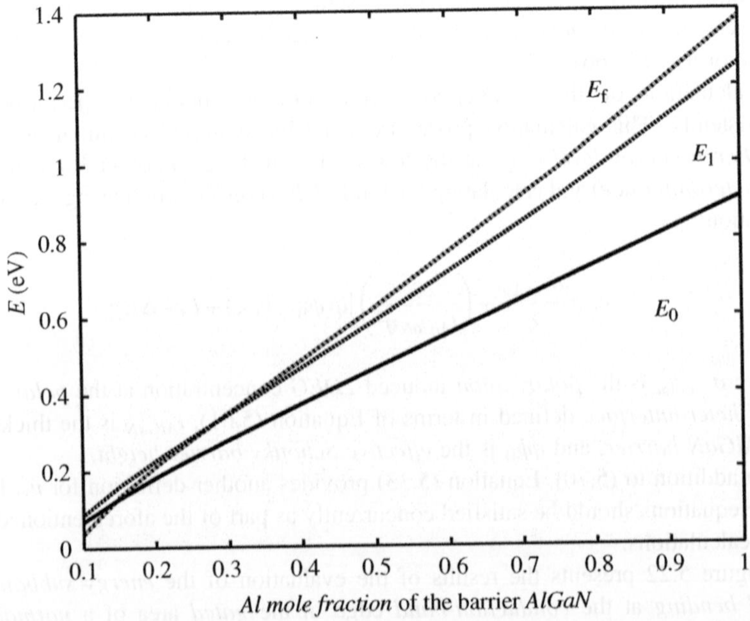

FIGURE 5.23 Calculated positions of the *Fermi level*, first and second energy subbands in *AlGaN/GaN HFETs* with variation of the *Al mole fraction*. Other device parameters are as expressed for the device considered in Figure 5.22. Zero gate bias is assumed. In these calculations, crystal relaxation is overlooked. Reprinted with permission from Manouchehri et al. (2014). Copyright 2014, American Vacuum Society.

AlGaN/GaN HFET. The *Al mole fraction* of the *barrier*, the thickness of this layer, and V_{GS} are 0.3, 20 nm, and 0 V, respectively. On this diagram, the results of the evaluation of these parameters based on the triangular *quantum-well approximation* are superimposed. In these evaluations, the concentration of the *unintentional donors* of the *channel/buffer layer* were assumed as 10^{14} cm^{-3}, while *Ni Schottky gate* contact, for which $q\phi_B$ as a function of *Al mole fraction* is taken as $q\phi_B = (1.3x + 0.84)$ eV, was considered.

The superposition of the results of the presented *variational* model and the assumption of the triangular *quantum well* clearly highlight how erroneous the assumption of the triangular *potential well* can be in the presence of strong *polarization*. Such a difference in nature is even more important at higher values of *Al mole fraction* where *polarization* is more dominant. Especially for these higher values of *Al mole fraction*, *AlGaN/GaN HFETs* can achieve a high value of *2DEG* concentration even in the absence of intentional *barrier doping*. Supporting this argument and according to the presented *variational* method, Figure 5.23 provides the position of the *Fermi level* and the first two *subbands* for a wide range of *Al mole fractions*. *Bias* and *barrier* thickness information are the same as in Figure 5.22.

REFERENCES

M. Ettenberg and R. J. Paff, "Thermal expansion of AlAs," *J Appl Phys*, vol. **41**, no. 10, pp. 3926–3927, 1970.

F. Manouchehri, P. Valizadeh, and M. Z. Kabir, "Determination of subband energies and 2DEG characteristics of $Al_xGa_{1-x}N/GaN$ heterojunctions using variational method," *J Vac Sci Technol A*, vol. **32**, no. 2, pp. 021104-1–8, 2014.

E. F. Schubert and K. Ploog, "Shallow and deep donors in direct-gap n-type $Al_xGa_{1-x}As$:Si grown by molecular-beam epitaxy," *Phys Rev B*, vol. **30**, pp. 7021–7029, 1984.

FURTHER READING

The following references have been used in preparation of the materials presented in this chapter. They are also suggested as sources for further reading.

PHYSICS OF HETEROSTRUCTURES AND HIGH-SPEED TRANSISTORS

G. Bastard, *Wave Mechanics Applied to Semiconductor Heterostructures*, Wiley Interscience, Les Ulis Cedex, 1991.

K. Lee, M. Shur, T. Fjeldly, and T. Ytterdal, *Semiconductor Device Modeling for VLSI*, Prentice Hall, Englewood Cliffs, NJ, 1993.

U. Mishra and J. Singh, *Semiconductor Device Physics and Design*, Springer, Dordrecht, 2008.

P. Roblin and H. Rohdin, *High-Speed Heterostructure Devices, from Device Concepts to Circuit Modeling*, Cambridge University Press, Cambridge, UK, 2002.

F. Schwierz and J. Liou, *Modern Microwave Transistors: Theory, Design, and Performance: Theory, Design and Performance*, Wiley Interscience, Hoboken, NJ, 2002.

S. Sze (Ed), *High Speed Semiconductor Devices*, Wiley Interscience, Hoboken, NJ, 1990.

S. Sze (Ed), *Modern Semiconductor Device Physics*, Wiley Interscience, Hoboken, NJ, 1997.

S. Wang, *Fundamentals of Semiconductor Theory and Device Physics*, Prentice Hall, Englewood Cliffs, NJ, 1989.

C. Weisbuch and B. Vinter, *Quantum Semiconductor Structures, Fundamentals and Applications*, Academic Press, San Diego, CA, 1991.

MATERIAL PROPERTIES AND PROCESSING OF SEMICONDUCTOR MATERIALS AND HETEROSTRUCTURES

S. Adachi, *Properties of Semiconductor Alloys: Group IV, III–V, and II–VI Semiconductors*, John Wiley & Sons, Inc., Hoboken, NJ, 2009.

Ioffe Physico-Technical Institute. Electronic Archive of New Semiconductor Materials, Characteristics and Properties. Available at: http://www.ioffe.rssi.ru/SVA/NSM/ (accessed on September 29, 2015).

M. Levinshtein, S. Rumyantsev, and M. Shur, *Properties of Advanced Semiconductor Materials: GaN, AlN, InN, BN, SiC, SiGe*, John Wiley & Sons, Inc., Hoboken, NJ, 2001.

E. Schubert, *Doping in III–V Semiconductors*, Cambridge University Press, Cambridge, UK, 2005.

PROBLEMS

5.1 Describe the basis of Equation (5.2).

5.2 Prove Equation (5.33).

5.3 Through drawing the band diagram of an *AlGaAs/GaAs* system, demonstrate how superlattice-based modulation doping depicted in Figure 5.15 works.

5.4 Perform a literature survey on the concept of modulation efficiency, and for a delta-doped *AlGaAs/GaAs HFET*, demonstrate how it changes with repositioning the delta-doped layer within the barrier (anywhere from the exposed surface to the heterointerface).

5.5 Perform a literature survey on the techniques of improving the breakdown voltage through employing field plates and address the basics of these techniques.

5.6 Produce the band diagram of a simple *DHFET* composed of two $Al_{0.2}Ga_{0.8}As$ barriers doped to the level of 10^{17} cm^{-3} and an undoped *GaAs* channel layer of thickness 10 nm. The top barrier is 20 nm, and the bottom barrier is 100 nm thick grown on the *GaAs* substrate.

5.7 If we were to modify the *DHFET* structure of Problem 5.6 by changing the doped $Al_{0.2}Ga_{0.8}As$ for undoped $Al_{0.2}Ga_{0.8}N$ and *GaAs* for *GaN* and assume the substrate to be *GaN*, while the layer structure is grown pseudomorphically in hexagonal *Ga face*, what form(s) of band bending do you expect at the heterointerfaces? What implementation challenges would you expect in the growth of this layer structure?

5.8 Using the foundations presented in Section 5.4.1, calculate the polarization-induced charge density at the following metallic-face pseudomorphically grown heterointerfaces:

a. Thin $In_{0.1}Ga_{0.9}N$/relaxed *GaN*

b. Thin AlN/relaxed *GaN*.

6

FETs AT MOLECULAR SCALES

Over the recent years, *scaling* of the *minimum feature size* of *FETs* has rendered the device models developed on the basis of the *drift–diffusion carrier-transport* formalism less and less applicable. Accordingly, *quantum transport* foundations are increasingly needed to appreciate the wave nature of electrons in *ballistic transport* through the short *channels* of these transistors. The *transmission* formalism and the use of the *nonequilibrium Green's function*[1] (*NEGF*) seem capable of providing a suitable foundation for filling in the gap (i.e., upon inevitable farewell to the *drift–diffusion transport* formalism). Due to the obligation to address this inevitability, in this closing chapter it seems appropriate to take a departure from the foundations used in explaining the behavior of electronic devices in the previous five chapters and to present a few of the outcomes of this change of paradigm. This chapter is merely aimed at presenting an introduction to the new era of molecular-scale *FETs*. Such an introduction is expected to open the eyes of students and professionals within the discipline of traditional electron devices on the aforementioned eminent change and to make a bridge between the world of *microelectronics*, which with a sharp focus on *FETs* has been dealt with in the first five chapters of this volume, and the new reality of *nanoelectronics*. A marvelously presented detailed investigation of the *NEGF* formalism can be found in *Datta's* seminal work, which is listed among the references of this chapter.

[1] Named after *George Green*.

Field Effect Transistors, A Comprehensive Overview: From Basic Concepts to Novel Technologies, First Edition. Pouya Valizadeh.

6.1 *FET*: A CHANGE OF PARADIGM

As we took note in Chapters 3–5, *gate-length scaling* has been the prevalent mode of transformation of *FET* technologies. Over the past five decades, the *gate length* of these transistors has gone through a tremendous transformation in scale from *tens* of *microns* to about 10 nm. For many generations of *MOSFETs*, *scaling* contributed to both increasing the maximum switching frequency of the transistors and the *chip's* packing density. However, caused by a number of factors previously enumerated in Chapter 3, over the past decade or so *gate-length scaling* has started to yield a diminished gain in terms of the frequency of operation of these transistors. Nevertheless, since *scaling* has kept contributing to improving the *chip's* packing density and as a result to its functionality, *chip* manufacturers are still aggressively pursuing this trend. Although for many decades *scaling* had largely remained a mere technological challenge, due to the shrinking *length* of the *channel*, approaching the molecular scales heralds a clash between the laws of physics and the operation of *FETs* as we know them. Quite critically sitting at this juncture is the *Heisenberg's uncertainty principle*: you can either have certainty about momentum or position, but not both. Shrinking device dimensions are imposing new limits on certainty of position. However, irrelevant of the *gate length* at which this hard limit will put a stop to the *scaling* trend of *FETs*,[2] as long as they are offering the signature characteristic of a tunable resistor, it is still critical to be able to provide a model for understanding the behavior of these *deeply scaled* transistors.

The tremendous transformation in the *minimum feature size* has recently resulted in a shift of paradigm in the operation of modern *FETs*. Whereas until not too far ago the *drift–diffusion* formalism (or an ad hoc modified version of it) was approximately sufficient in explaining the behavior of *FETs*, in *deeply scaled FETs* electron *transport* is taking on an increasingly *ballistic* character. In this era, to explain the behavior of transistors, a number of *carrier-transport* frameworks such as the *Landauer*[3] and *transmission* formalism, according to the use of the *nonequilibrium Green's function* (*NEGF*), have been proposed to replace the *drift–diffusion formalism*.

Among the frameworks proposed for *charge-carrier transport* through *nanoscale channels*, formulating *transmission* according to *NEGF* is deemed the most capable in bridging the gap between the fully *coherent quantum transport* (which is of interest to *mesoscopic*[4] physicists) and the *fully incoherent Boltzmann transport* models[5] (which are preferred by the device physicists).

Since the publication of this volume is coinciding with this wave of transformations, as a text dedicated to *FETs* it bears the obligation to provide a picture of what will come out of *scaling* while this trend goes on to molecular dimensions. As will be noted in this chapter, the aforementioned transformations in the nature

[2] That is, as devices that we have got to know them over the years and not in the new roles to be played by them in futuristic capacities.
[3] Named after *Rolf Landauer*.
[4] See Section 1.4.2.
[5] See Section 1.7.2.1.

of *charge-carrier transport* at these length scales have immense impact on our very understanding of current conduction in *FETs*. According to such transformations, things as simple as the definition of *Ohmic* behavior, which are normally taken for granted, demand major attention. The broad objective of this chapter is to review a few of the important observations and concepts relevant to understanding current conduction through a molecular-scale *channel* consisting of *one* (or a limited number of) *energy level(s)* in the *energy range of interest*. This *energy range of interest* in a *FET* is identified by the difference among *Fermi levels* of the so-assumed *thermal-equilibrium* infinite contacts of *source* and *drain*. The so-called *Coulomb blockade* phenomenon and the observation of a lower limit for contact resistance are the important phenomena that will be visited in the upcoming sections of this chapter.[6]

6.2 RESISTANCE REDEFINED

So far in this volume, we have observed the *gate-length scaling* of *MOSFETs* from a top-down perspective (i.e., according to the chronological order of events, from the circumstances of longer *gates* to those of the shorter). The opposite approach, known as the *bottom-up* approach, provides an educational experience on the true meaning of fundamental concepts such as electrical resistance. Thus far, biased by our *macroscopic* experiences, we have been compelled to take the conductance of a uniform *channel* of *conductivity* σ to be proportional to its cross-sectional area A and inversely proportional to its length L:

$$G = \sigma \frac{A}{L}. \tag{6.1}$$

In this definition, *conductivity* is treated as a material property defined in terms of the *mobility* and the concentration of the *channel's* free *charge carriers*. As indicated in Section 1.7, the definition of *carrier mobility* relies on the presence of a large number of *scattering* events that take place when *charge carriers* existing in the partially occupied *energy bands* travel along the length of the *channel*. In this context, both problems of evaluation of the *charge-carrier* concentration and the frequency of *scattering* events are dealt with on statistical bases. However, the number of *carriers* in a *nanoscale channel* and the number of *scattering* events in this semi-*ballistic* (or even *ballistic*) medium are small. Hence, in this case, the aforementioned fashion of using these statistically evaluated parameters in predicting the measurable value of conductance is insupportable. In addition, it is very important to appreciate that with reduction in the dimensions of the *channel*, *energy bands* (or *subbands*) will give their place to *energy levels*. Consequently, this length scale does not allow us to define the electrical *conductivity* in the same old fashion. The *bottom-up* approach of this chapter, by means of dealing with conduction through *energy levels* instead of *energy bands*,

[6] Based on the groundbreaking *bottom-up* approach of *Datta* to presenting this problem, going beyond the presentation of this chapter requires very little background in quantum mechanics.

and via considering the implications of this change, paves the way for the presentation of the simple physics of very small conductors.

An early major obstacle to understanding the conductance of molecular-scale *channels* had its roots at the original lack of ability to measure the electrical conductance by applying a potential difference across the two ends of these *channels*. Such a measurement obviously requires improvising large-size connecting pads between the molecular-scale *channel* and the electrodes of a *power supply*. Whereas such a possibility is readily extended in a *macroscale FET*, for *nanoscale channels* until the late twentieth century, such an option was not available. As will be shown through the *bottom-up* presentation of the conduction problem, existence of the large scale difference between the *macroscale* pads and the molecular-scale *channel* is a major contributor to the change of paradigm in molecular-scale devices.

In order to present this *bottom-up* approach, as presented in Figure 6.1, a generic structure of a *nanoscale FET* is adopted. This transistor is composed of a semiconducting *channel*, a metallic *gate*, an insulator sandwiched between the *gate* and the *channel*, and two high-*conductivity macroscale* contact pads identified as the *source* and *drain* electrodes. Like a *MOSFET*, application of a voltage to the *gate* electrode is expected to control the concentration of the *charge carriers* within the *channel* where a *drain–source* potential difference possibly induces a net flow of *carriers* along the *channel*. In the analysis of *carrier transport* through this molecular-scale *FET*, we will assume that under any given *bias* condition the *macroscopic source* and *drain* electrodes maintain their individual states of *thermal equilibrium*. Upon application of a *drain–source* voltage, without disturbing the state of *thermal equilibrium* in each of these electrodes, only the *Fermi levels* of the two contacts (i.e., E_{f1} and E_{f2}) separate from one another. However, under this condition, *channel* ceases to stay under *thermal equilibrium*. The energy range between E_{f1} and E_{f2} is referred to as the *energy range of interest*. Considering an extremely small *channel*, in this treatment it will be assumed that the *channel* can have only one allowed *energy level* within this range. Before disturbing the *thermal-equilibrium* state of the *channel*, the *Fermi level* of the *channel* is determined by the equal *Fermi levels* of the *source* and *drain* electrodes.

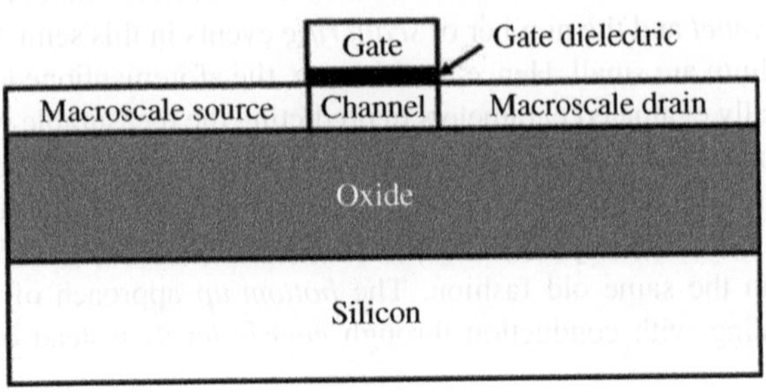

FIGURE 6.1 Cross-sectional view of the adopted generic structure of a *nanoscale FET*. While the *channel* is at *nanoscale*, *source* and *drain* electrodes are *macroscale* entities.

In explaining the current conduction within the context of the single *energy-level channel* of the *FET* presented in Figure 6.1, a number of key concepts of quantum mechanics and nonequilibrium statistical mechanics should be dealt with.

Based on our understanding of the operation of *macroscale FETs*, improving the quality of the *drain* and *source* contacts to the *channel* is expected to yield a higher *current drive*. One of the major breakthroughs in the area of *mesoscopic FETs*, however, was to show the existence of an upper limit for the conductance of a *single-level channel*. In this section, through conceptual reasoning it will be proven that the value of this maximum, which is determined based on the fundamentals of physics, is given by

$$G \equiv \frac{q^2}{h} = 38.7\ \mu\text{S} = (25.8\ \text{k}\Omega)^{-1}. \qquad (6.2)$$

Considering the twofold *spin degeneracy* of electrons at the sole *energy level* of the *channel*, the actual maximum conductance is $2G$.

Observing such an upper limit is in stark contrast to (6.1) in which case, assuming no boundary for improving the contact resistance, it is expected that by reducing the *channel-length* conductance will asymptotically tend toward infinity.

Toward the goal of understanding the I_D–V_D characteristics of the *MOSFET* depicted in Figure 6.1, Figure 6.2 summarizes the essence of the role of the *gate* electrode in establishing a conduction *channel* between *drain* and *source* electrodes. In this figure, in spite of the application of the *gate* voltage, the *channel's* state of *thermal equilibrium* is assumed to be maintained by the equipotential *macroscale drain* and

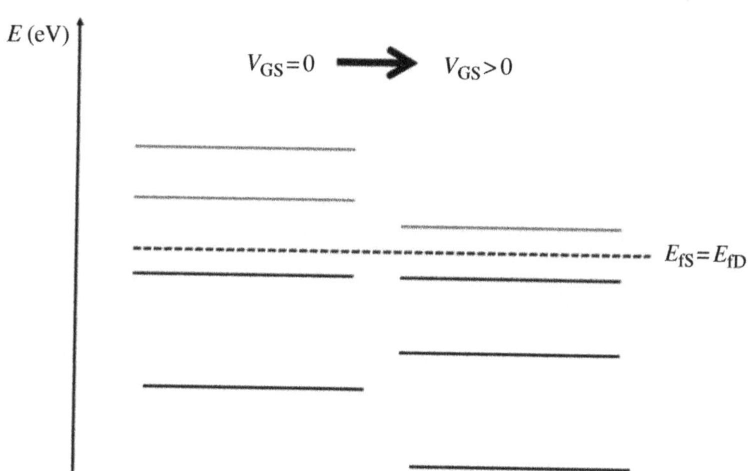

FIGURE 6.2 Effect of the application of a positive *gate* voltage to the gate of the *nanoscale FET* of Figure 6.1 on the status of occupation of the energy levels of the *channel*, when the macroscale *source* and *drain* electrodes are in *thermal equilibrium*. Temperature is 0 K. Occupied levels are indicated in black, while unoccupied levels are shown in gray.

source electrodes. In these electrodes unlike the *channel*, which exhibits a semiconducting character highlighted by the presence of allowed *energy levels* and forbidden *bands* of energy, a metallike behavior defined by a continuum of *energy states* is assumed.

In view of this perspective on the role of the *gate* in operation of a *MOSFET*, an interesting insight into the differences between p- and n-channel *MOSFETs* can be acquired. In the *n-channel* operation, the application of a positive *gate-source bias* lowers the *energy levels* of the *channel* resulting in their realignment with the mutual *Fermi level* of the tied *drain–source* contacts. This can drive the *channel's Fermi level* from the *forbidden gap* into an allowed but originally unoccupied *band*. Whereas before this realignment there was no possibility for *transport* of electrons between the *drain* and *source* electrodes via the *gated channel* (i.e., upon application of a finite V_{DS}), coincidence of the *Fermi level* with the originally empty *band* increases the *channel's conductivity* and turns the transistor *on*. At this point, it can be expected that the amount of current passing through a *biased channel* (i.e., for a finite V_{DS}) depends on the number of *channel energy levels* available around the *Fermi levels* of the two *macroscopic* contacts (i.e., the *energy range of interest*). What is worth correcting in our mindset is that, although hidden throughout the discussion of the previous three chapters, it is really the increase in the number of these *energy levels* and not the concentration of the *charge carriers* in the *channel* that is driving the improvement of the current passing through this *channel*. What defines this *channel* as an n-type *channel* is the fact that the *drain* and *source Fermi levels* have overlapped an unoccupied *energy band*. As a result, upon application of a positive *drain–source* voltage and lowering of the *Fermi level* of the *drain* compared to that of the *source*, the contribution of electrons to the *channel* from the *source* and transfer of those electrons to the *drain* are the causes of current conduction. In this situation, the *threshold voltage* V_T is determined by the energy difference between the equilibrium *Fermi level* and the lowest available empty *energy level* in the *channel*.

In contrast to this order of events, under p-type operation a negative *gate-source* voltage drives the *thermal-equilibrium Fermi level* of the *channel* into an occupied *band*. As a result, upon application of a finite *drain–source* voltage, first electrons are taken out of the *channel* on the *drain* side and then replaced on the *source* side. As shown in Figure 6.3, this is due to availability of *energy states* around $E = E_f$.

The lesson to take from this discussion is that it is the availability of *energy states* and not the state of their occupancy that turns a *MOSFET on*. This language suggests that excluding all other factors (such as the shape of the *depletion regions* formed at the contacts between metal and the semiconducting *channel* and the possibility of turning the insulating *gate* contact *on*), a given *FET* can be operated both as n- and p-channel. Such an interpretation can be only true in short devices when contacts are composed of continuous distribution of *states*.

For *FETs* where this language is adoptable, upon the application of a potential difference V_{DS} between the *drain* and the *source*, we will have

$$E_{f1} - E_{f2} = qV_{DS}. \tag{6.3}$$

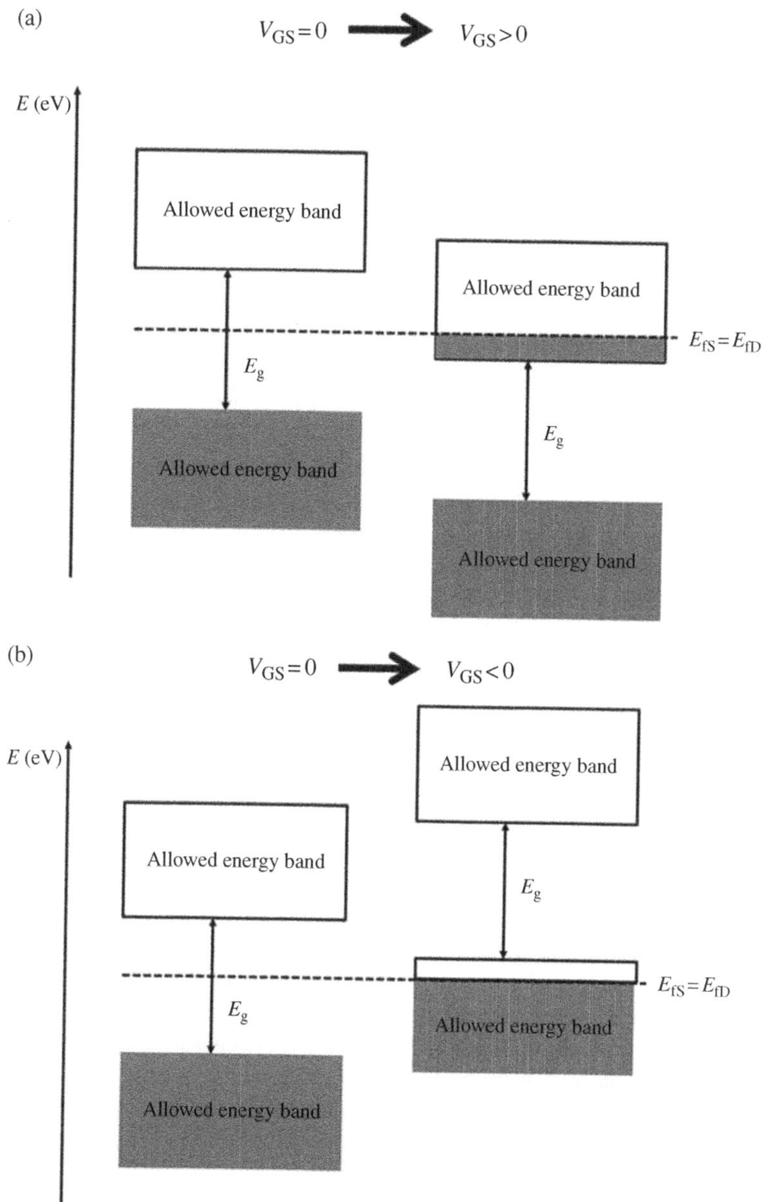

FIGURE 6.3 Effect of the *gate* voltage in inducing current conduction across a channel in the form of (a) an n-type conduction and (b) a p-type conduction. In both cases drain and source are kept under *thermal equilibrium*. Occupied portion of an allowed energy band of the channel is indicated by the gray fill color. In case (a), application of $V_{GS}>0$ pushes an originally empty band below the *mutual Fermi level* of the drain and source. In (b), however, application of $V_{GS}<0$ pulls an originally full band above the *mutual Fermi level* of the drain and source. While in case (a) channel offers an n-type conduction characteristic, the same channel in (b) is offering a p-type conduction character. In recording the occupation status, temperature is assumed to be very low.

Considering this difference among the *Fermi levels* of the two contacts for which we are assuming that the state of *thermal equilibrium* has remained prevalent, we can write two *Fermi–Dirac distribution functions* (i.e., one for each contact):

$$f_{D1}(E) = \frac{1}{1+\exp((E-E_{f1})/kT)} = f_0(E-E_{f1}) \tag{6.4}$$

and

$$f_{D2}(E) = \frac{1}{1+\exp((E-E_{f2})/kT)} = f_0(E-E_{f2}). \tag{6.5}$$

Each of these two individual *thermal-equilibrium* systems has the desire to bring the *channel* into *thermal equilibrium* with itself. This is the driving force behind current conduction. As an example, in n-channel operation *source* electrode keeps feeding the *channel* with electrons to establish an equilibrium state between itself and the *channel* while following the same agenda the *drain* electrode insists on removing electrons from the *channel*. According to the concurrent existence of both of these desires, neither of the two electrodes can succeed in establishing *thermal equilibrium* with the *channel*. The balancing act that the *channel* has to maintain between the two electrodes, as a result, will establish a state of nonequilibrium somewhere between the two electrodes within the *channel*.

It is via this balancing act that we will now try to explain the current conduction formalism applicable to *FETs* of very short *channels*. In this explanation, the concept of the so-called rate equations will be employed. These equations are first provided for the case of the aforementioned *single-level channel*.

Consider a *biased single-level channel* whose *energy level* is positioned between the *Fermi levels* of the *drain* and the *source* electrode. As explained earlier, whereas the *source* electrode seeks to see $f_{D1}(E_1)$ electrons in the *channel's* sole *energy level*, the *drain* electrode is attempting to retain $f_{D2}(E_1)$ electrons at this *level* (E_1 being the sole *energy level* of the *channel*). As a result, at steady state the average number of electrons in this *energy level* (i.e., N) is expected to take a value intermediate between these two. Corresponding to this explanation, the net flux of electrons across the *channel* can be defined according to its two components I_1 and I_2, which are defined as

$$I_1 = \frac{qr_1}{\hbar}(f_{D1} - N) \tag{6.6}$$

and

$$I_2 = \frac{qr_2}{\hbar}(f_{D2} - N). \tag{6.7}$$

In these equations, the parameters $r_1/\hbar$ and $r_2/\hbar$ are referred to as *rate constants*. These *rate constants* identify the rates at which an electron placed initially in the single *energy-level channel* escapes into the *source* or the *drain* contact, respectively. According to Equations (6.6) and (6.7), these quantities are obviously presented in units of *per second*. As a result, r_1 and r_2 have the dimension of energy. It will be observed shortly that the definition of these two parameters is linked to the concept of *energy-level broadening* (which is a concept explaining the quality of contact to the *channel's* sole *energy level*). In the absence of *energy-relaxation* and *inelastic* interactions during the inflow/outflow processes at the *drain–source* electrodes, it can be shown that these factors explain the amount of *broadening* in the originally *delta function*-looking *density of states* of the single *energy-level channel* (i.e., *broadening* when in contact with the *drain–source* electrodes). Under these conditions, this *energy-level broadening* can be proven to be independent of $f_{D1}(E)$ and $f_{D2}(E)$.

Following the definition of current continuity along the *channel*, under the steady-state conditions the algebraic sum of the two current terms (i.e., I_1 and I_2) would be equal to 0. Accordingly, Equations (6.6) and (6.7) yield

$$N = \frac{r_1 f_{D1} + r_2 f_{D2}}{r_1 + r_2}. \tag{6.8}$$

As expected, this equation provides evidence that under steady-state conditions N should be a weighted average between $f_{D1}(E_1)$ and $f_{D2}(E_1)$. Obviously, this average will be skewed toward the wish of the contact with larger *rate constant*. Based on this value of N, Equations (6.6) and (6.7) allow us to write

$$I_D = I_1 = -I_2 = \frac{q}{\hbar}\left(\frac{r_1 \cdot r_2}{r_1 + r_2}\right)[f_{D1}(E_1) - f_{D2}(E_1)]. \tag{6.9}$$

So far, in this explanation, we have not paid attention to *spin degeneracy* of the single *energy level* of the *channel*. Considering *spin degeneracy*, current is twice the value indicated in Equation (6.9). Made evident by this equation, since for *energy levels* well below or well above the *Fermi levels* of both *source* and *drain* electrodes functions $f_{D1}(E)$ and $f_{D2}(E)$ take identical values (i.e., 1 and 0, respectively), a *nonzero* current can be only expected when the *channel's energy level* is positioned between E_{f1} and E_{f2}. According to the definition of the *Fermi–Dirac* distribution function, at temperatures higher than 0 K, this window expands to energies slightly above and below these limits. As indicated earlier, the process of conduction is indeed locked to the existence of *channel energy states* within this window of energy. Figure 6.4 illustrates current conduction in terms of the difference in agendas of *source* and *drain* electrodes.

On the basis of Equations (6.4), (6.5), and (6.9), the so-called quantum conductance can be defined. To avoid complexities of current saturation, we should consider a small *drain–source* voltage for this purpose. Assuming that $E_{f2} < E_1 < E_{f1}$ and the temperature to be low enough so that $f_{D1}(E_1)$ and $f_{D2}(E_1)$ can be approximated by 1 and 0, respectively, we will have

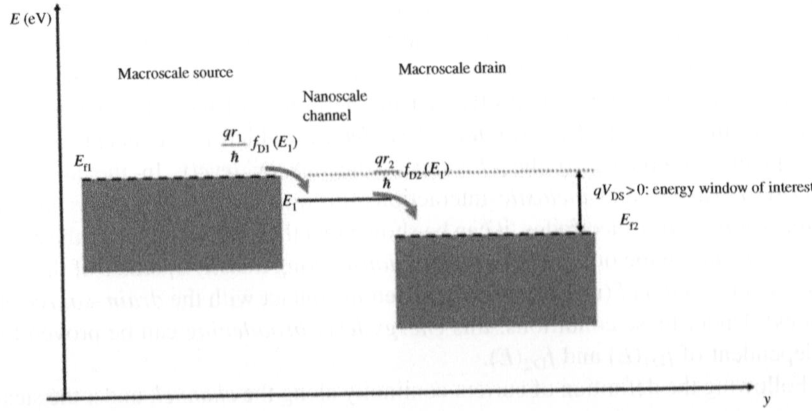

FIGURE 6.4 Role of difference of agenda among the macroscale *drain* and *source* electrodes in dictating the occupation status of the single energy level of the *channel* (i.e., within the window of interest).

$$I_D = \frac{q}{\hbar}\frac{r_1 \cdot r_2}{r_1 + r_2} \tag{6.10}$$

where for $r_1 = r_2$, this value reduces to $(qr_1)/2\hbar$.

At first glance, Equation (6.10) seemingly suggests that by increasing the *coupling* quality of the contacts to the *channel* (i.e., increasing $r_1(=r_2)$), an unlimited current can be supplied through the *channel*. However, *mesoscopic* physicists have already proven this impression to be wrong. In this expectation, what is taken for granted is that when you improve the quality of *coupling* to the *channel*, the *channel* retains all its properties (e.g., the *density of states*). This impression could have been acceptable when we were using an *annealing* technique to establish an *alloyed Ohmic* contact to a *macroscale* semiconducting *channel*. Through this process, whereas the two ends of the *channel* were turning into metallic *alloys*, properties of the *channel* were retained over most of the length of the *channel*. We can easily speculate that this cannot be acceptable in a molecular-scale single *energy-level channel*.

What *mesoscopic* physicists have proven is that, through improving the *coupling* to these small *channels*, the *delta function* of the *density of states* of the *channel* broadens. The higher the value of *r*, the larger is the *broadening*. Whereas this *broadened energy level* can still accommodate a maximum number of two electrons, part of the *broadened level* can eventually fall outside the aforementioned energy window of interest to current conduction. As a result, the larger the *broadening*, the smaller the overlap of the *broadened energy level* with this window, and the less is the current. Accordingly, the actual current level delivered through this *energy level* is scaled by a factor of $(E_{f1}-E_{f2})/(c \cdot r_1)$. Defining $c \cdot r_1$ as the *effective width* of the *energy level* (with *c* being a constant), this factor represents the portion of the *broadened energy level* that still lies within the window between E_{f1} and E_{f2}. Based on this definition, under the above-assumed very low temperature condition, having known that $E_{f1}-E_{f2}=qV_{DS}$,

$$I_D=\frac{qr_1}{2\hbar}\frac{qV_{DS}}{c\cdot r_1}\Rightarrow G=\frac{I_D}{V_{DS}}=\frac{q^2}{2c\hbar}. \tag{6.11}$$

So far, this equation proves that due to *level broadening* there will be a hard limit to the maximum level of conductance through the *single-level channel.*

The spillover of the *energy states* from the *channel* to the *drain* and *source* electrodes is the cause of the aforementioned *energy-level broadening*. In the exchange between the *channel* and these electrodes, whereas the *channel* loses part of its *state*, on the same token it gains part of the contacts' *energy states*. However, since the *channel* loses *energy state* at a singular value of energy and gains *states* from the contact in a broader range of energy, in this exchange the originally *delta*-formed *DOS* of the *channel* broadens upon contact formation. As a result, although a *zero-sum* exchange, contact formation results in *broadening* the *DOS* of the *channel*. The *broadened DOS* can in principle take any number of shapes. The simplest of these shapes follows a *Lorentzian* distribution[7] around $E=E_1$. Satisfying the aforementioned *zero-sum* exchange, the integral over all energies of the following *Lorentzian* is equal to 1:

$$D_{E_1}(E)=\frac{r/2\pi}{(E-E_1)^2+(r/2)^2}. \tag{6.12}$$

It is evident that at the limit of $r\rightarrow 0$ the function expressed in (6.12) reduces to the original *Dirac's delta-function DOS* of the *channel*. In adopting this distribution, $r=r_1+r_2$.

According to the *channel's DOS* expressed in (6.12), the definition of the *drain–source* current previously presented in (6.9) can be reformulated by the following integral:

$$I_D=\frac{q}{\hbar}\int_{-\infty}^{\infty}D_{E_1}(E)\cdot\frac{r_1\cdot r_2}{r_1+r_2}\cdot[f_{D1}(E)-f_{D2}(E)]dE. \tag{6.13}$$

Based on the definition of *Dirac's delta function*, Equation (6.13) reduces to the form presented in (6.9) as $r\rightarrow 0$.

Under the low-temperature conditions assumed earlier, Equation (6.13) yields

$$I_D=\frac{q}{\hbar}\frac{r_1\cdot r_2}{r_1+r_2}\int_{E_{f2}}^{E_{f1}}D_{E_1}(E)dE. \tag{6.14}$$

In Equation (6.14), for small values of V_{DS}, the integrand can be taken as a constant, resulting in

[7] Named after *Hendrik Lorentz*.

$$I_{\mathrm{D}} = \frac{q}{\hbar} \frac{r_1 \cdot r_2}{r_1 + r_2} (E_{\mathrm{f1}} - E_{\mathrm{f2}}) \frac{(r_1 + r_2)/2\pi}{(E_{\mathrm{f}} - E_1)^2 + ((r_1 + r_2)/2)^2}, \tag{6.15}$$

where E_{f} is the average of E_{f1} and E_{f2}.

Based on (6.15), it can now be observed that the *drain–source* current is maximized when E_1 is equal to E_{f}. Using the identity of $E_{\mathrm{f1}} - E_{\mathrm{f2}} = qV_{\mathrm{DS}}$, this relationship provides a definition for the maximum value of conduction through a single *energy-level channel*,

$$G \equiv \frac{I_{\mathrm{D}}}{V_{\mathrm{DS}}} = 4 \frac{q^2}{h} \frac{r_1 \cdot r_2}{(r_1 + r_2)^2}, \tag{6.16}$$

which produces the expected value of q^2/h if $r_1 = r_2$.

To be more accurate, this value should be multiplied by 2 to consider the *spin degeneracy*. This value is the maximum value of conductance for the single *energy-level channel FET* depicted in Figure 6.1 (which we have spoken of previously in Equation (6.2)).

Using this language and according to (6.8), for the average number of electrons in the *channel*, we have

$$N = \int_{-\infty}^{\infty} D_{E_1}(E) \frac{r_1 \cdot f_{\mathrm{D1}}(E) + r_2 \cdot f_{\mathrm{D2}}(E)}{r_1 + r_2} dE. \tag{6.17}$$

6.3 EVALUATION OF CURRENT–VOLTAGE CHARACTERISTICS OF A SINGLE ENERGY-LEVEL CHANNEL *FET*

So far in the present behavioral treatment of the *FET* depicted in Figure 6.1, we have paid attention to a situation in which only a very small V_{DS} is applied between the *drain* and *source* electrodes. To go beyond this *linear regime* of operation, we need to deal with calculation of *channel* potential when *source*, *drain*, and *gate* are *biased*. As we have seen in Chapters 3–5, maintaining the electrostatic dominance of the *gate* over the *channel* has turned into the most important topic in studying the *scaling* of *MOSFETs*. Due to the importance of the *gate's* electrostatic control, in order to calculate the *channel* potential, prior to presenting a generic model, we will first deal with this calculation on the basis of two simple scenarios:

1. Pure electrostatic control of the *channel* potential by the *source* and *drain* electrodes
2. Pure electrostatic control of the *channel* potential by the *gate* electrode

Under the first scenario, in order to assess the *channel* potential, we consider the aforementioned *single-level channel* in which the *thermal-equilibrium Fermi level* E_{f}

is located slightly above the energy level E_1. Upon application of a *drain–source* voltage V_{DS}, we already know that the two contacts will develop their own states of *thermal equilibrium* with $E_{f1} - E_{f2} = qV_{DS}$. For this situation and considering the case of very low temperatures, current will be only established if E_1 lies between E_{f1} and E_{f2}. If we ignore the impact of the *gate*, one possibility for the *channel* potential under this condition will be to lie halfway between the *source* and the *drain* potential. As a result, the energy level E_1 will change to $E_1 - (qV_{DS}/2)$. In Figure 6.5, this evolution of *channel's energy level* for both positive and negative values of V_{DS} is depicted. Under this assumption, independent of the polarity of V_{DS}, the *channel's*

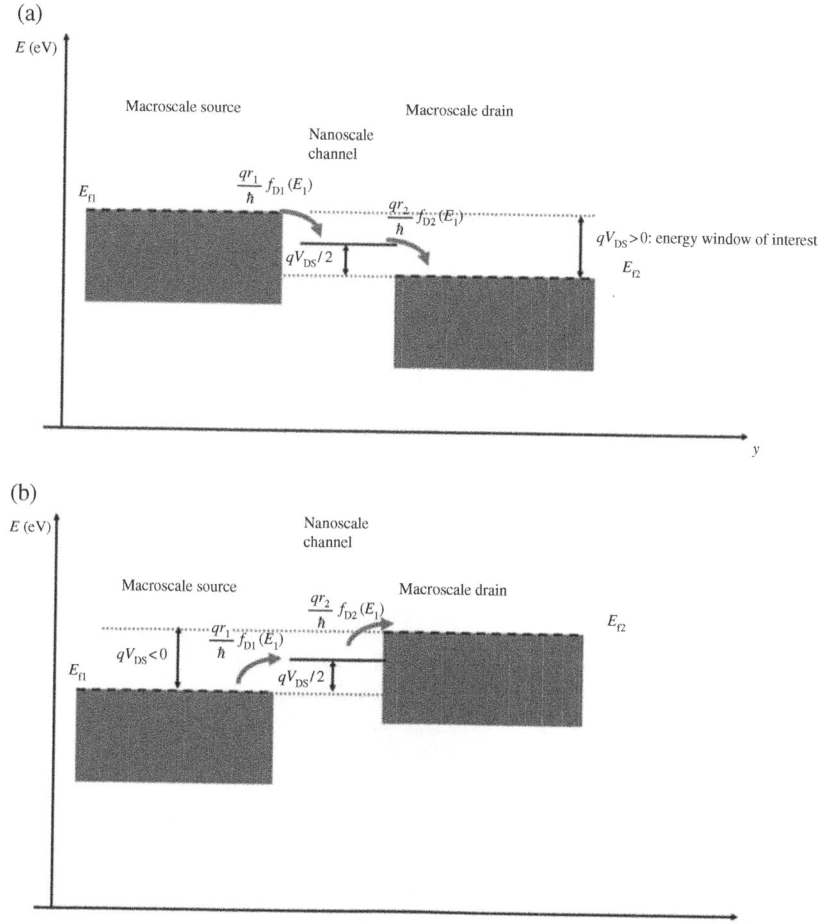

FIGURE 6.5 Graphic representation of the mechanisms involved in current conduction through the single energy level of a nanoscale *channel* with the application of (a) a positive V_{DS} and (b) a negative V_{DS}, assuming the energy level of the *channel* to be repositioned to halfway within the window of interest after the application of V_{DS}.

energy level will lie halfway between E_{f1} and E_{f2}, resulting in the same magnitude of current.

Under the second scenario, however, when the *gate* is closely coupled to the *channel*, a different situation emerges. Under this condition, the *channel's energy level* is defined by the *gate* and not the *drain* and *source* electrodes. Consequently, under this condition, current conduction is not symmetric with regard to change of polarity of the *drain–source* voltage.

In order to provide a generic framework for determining the potential of the *channel*, the capacitive circuit-level model presented in Figure 6.6 is often employed. According to this figure, the three capacitors established between the three electrodes of the *FET* and the *channel* are all competing to exert electrostatic dominance over the *channel*. In treating this model, we can first take the *channel* as an insulator and solve the *Laplace equation* for determining the *channel* potential *V*:

$$\nabla\cdot(\epsilon_s \nabla V)=0. \tag{6.18}$$

The solution to (6.18) can be developed on the basis of the boundary conditions imposed individually in terms of the potentials of the *source*, *drain*, and *gate* electrodes. The *channel* of the small *FET* depicted in Figure 6.1 can be observed as a single point (i.e., with no need to look for spatial variation in the potential). Clearly, the potential that is evaluated on the basis of linear superposition of the impacts of the aforementioned stimuli and (6.18) still does not pay attention to the presence of charge concentration in the *channel*. Regarding the use of the *Laplace equation*, we name this part of the potential V_L. In terms of the aforementioned superposition and considering *zero source* potential, V_L is presented as

$$V_L=\frac{C_G}{C_E}(V_{GS})+\frac{C_D}{C_E}(V_{DS}), \tag{6.19}$$

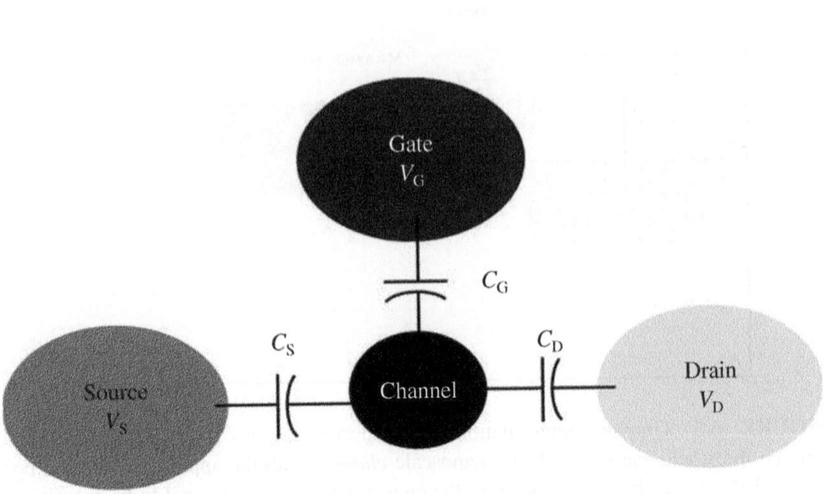

FIGURE 6.6 Capacitive circuit model of the *nanoscale FET*.

where C_E is defined by the sum of the capacitances established between the *point channel* presented in Figure 6.6 and the three electrodes of the *FET*.

In a single *energy-level channel*, the absence of charge and as a result the calculation of *channel* potential using the *Laplace equation* of (6.18) are justified. This *channel* potential can be modified though assuming an electron concentration ρ in the *channel*. Considering this charge concentration, the following *Poisson equation* should be solved:

$$\nabla \cdot (\epsilon_s \nabla V) = -\rho. \tag{6.20}$$

According to the model presented in Figure 6.6, for the change in the number of electrons in the *channel* with respect to the reference number of electrons (when the *channel's* single *energy level* was located at E_1), ΔN is calculated as

$$-q\Delta N = C_s V + C_G(V - V_{GS}) + C_D(V - V_{DS}). \tag{6.21}$$

As a result of this linear relationship, the *channel* potential is given by summing the potential V_L and the potential calculated on the basis of (6.20):

$$V = V_L - \frac{q}{C_E}\Delta N. \tag{6.22}$$

This equation is pivotal to understanding an important property of a group of very *deeply scaled MOSFETs*, known as *single-electron transistors* (*SET*). In this equation, the contribution of the second term in terms of energy (i.e., $(q^2/C_E)\Delta N$), which for $\Delta N = 1$ is usually referred to as the *single-electron charging energy* U_0, identifies the potential energy difference induced by the subtraction or addition of a single electron to the *channel*. This value is identified by the extent of the spatial confinement of the electron *wave function*. According to (6.22), for a small C_E encountered in a *SET*, the change of potential induced by the addition of a single electron to the *channel* can induce the walking out of the *channel's energy level* from the *energy range of interest* (i.e., effectively in ceasing of conduction). This phenomenon is known as *Coulomb blockade*. *Coulomb blockade* can be readily observed when the *single-electron charging energy* is larger than the *broadening* r and also larger than kT.

Interdependent calculations of the number of electrons and the energy level of the *channel* among (6.17) and (6.22) require a *self-consistent* solution among these equations. In terms of the *channel* potential expressed in (6.22), (6.17) is modified to

$$N = \int_{-\infty}^{\infty} 2 \cdot D_{E_1}(E + qV) \cdot \frac{r_1 f_{D1}(E) + r_2 f_{D2}(E)}{r_1 + r_2} dE = \int_{-\infty}^{\infty} n(E) dE, \tag{6.23}$$

while (6.13) evolves to

$$I_D = 2\frac{q}{\hbar}\int_{-\infty}^{\infty} D_{E_1}(E + qV) \cdot \frac{r_1 \cdot r_2}{r_1 + r_2}[f_{D1}(E) - f_{D2}(E)] dE. \tag{6.24}$$

It is in terms of (6.24) that the departure from the linear dependence of I_D on V_{DS} (i.e., Eq. 6.16) is identified.

Considering the role of *spin degeneracy*, in Equations (6.23) and (6.24), multiplication by a factor of 2 has been performed. The aforementioned phenomenon of *Coulomb blockade* can be also observed through the lifting of *spin degeneracy* and splitting of the *DOS* of the *spin-degenerated energy state* of the *channel*. This is due to the fact that the two *spin-degenerate states* are feeling different *self-consistent* potentials.

Drain current saturation in this single *energy-level channel FET* can be easily expected since an increase in V_{DS} can only provide a maximum overlap of the full *broadened channel's energy level* and the now-widened *energy range of interest* and nothing beyond that.

6.4 FROM CURRENT CONDUCTION IN SINGLE ENERGY-LEVEL CHANNELS TO DEFINITION OF CONDUCTANCE IN MACROSCALE CONDUCTORS

In Section 6.2, it was indicated why *Ohm's* law cannot be extended to the single *energy-level channels*. Now the important question is: what happens in longer *channels* that creates the necessary conditions for the application of the *Ohm's* law? The *bottom-up* approach that is being followed in this chapter provides an easy path to understanding these conditions.

One way to address this question is to imagine a large conductor as an array of small unit cells (as schematically depicted in Figure 6.7). Quite similarly to the fashion that *coupling* between the *drain* and *source* electrodes and the single *energy level* of the *channel* of the *FET* depicted in Figure 6.1 was proposed, *coupling* between these unit cells can be expressed in terms of an interunit *coupling energy* (i.e., t).[8] For values of $t \geq U_0$, without worrying about issues similar to the *single-electron charging effect*,

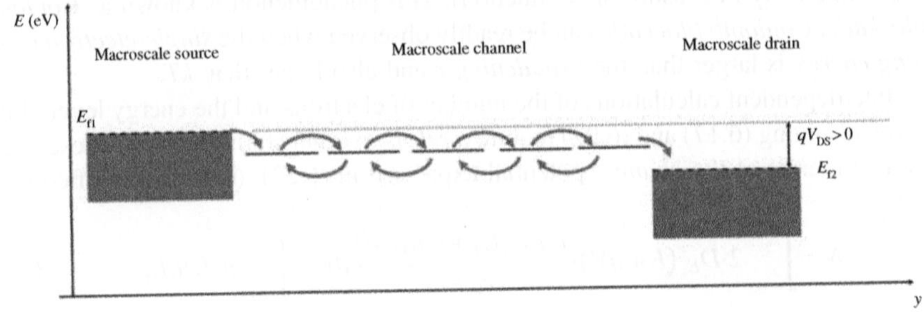

FIGURE 6.7 Illustrative description of conduction in a *long channel* in terms of transmission among a number of small unit cells.

[8] Reader should be reminded that the quantities r_1 and r_2 also have a dimension of energy.

following the *self-consistent* evaluation of potential, which for the example of Figure 6.1 was presented in Section 6.3, current conduction can be formulated. The language of this approach is known as the *transmission* formalism. In terms of the *transmission* formalism, Equation (6.24), which has been developed in Section 6.3 for describing the steady-state current conduction in a single *energy-level channel*, can be rewritten as

$$I_D = 2\frac{q}{h}\int_{-\infty}^{\infty} \bar{T}(E)[f_{D1}(E) - f_{D2}(E)]dE, \tag{6.25}$$

where $\bar{T}(E)$ is referred to as the *transmission* and is defined as

$$\bar{T}(E) = 2\pi \cdot D(E+qV) \cdot \frac{r_1 \cdot r_2}{r_1 + r_2}. \tag{6.26}$$

In Equation (6.26), instead of $D_{E_1}(E+qV)$, $D(E+qV)$ has been deliberately used to indicate that under some extreme conditions this formalism can be applied to the *multilevel channels* of a *nanotransistor*. These extreme conditions refer to cases in which all *energy levels* of the *channel* are conducting independently. The model applicable to these circumstances is referred to as the *independent-level model*.

However, as we have already indicated, this *independent-level model* is very limiting. For this model to be extended to larger conductors and to produce *Ohm's* law, the major missing component is *scattering* (in both *coherent* and especially *incoherent phase-breaking* forms). So far in this chapter's presentation, *carrier transport* along the *channel* has been envisioned *ballistically* with the only opposition to conduction arising at the *MS* contacts. In order to get the issue of *scattering* right, a *carrier-transport* model applicable beyond the tight boundaries of the *independent-level model* is required. For a proper inclusion of these effects, interference between multiple paths of conduction should be considered. As a result of this need, *rate constants r*, instead of being scalar quantities, need to evolve into matrices. However, inclusion of these coherent *scatterers* will not yet result in yielding *Ohm's* law. In order to produce *Ohm's* law, the other missing component is the incorporation of the *incoherent phase-breaking scattering* processes. As proposed by *Datta*, this can be done in the form of presenting these *scattering* processes in terms of a new sort of contact, referred to as a *scattering contact*.[9] Excluding this *scattering contact*, the generic model of an *n-level* conductor can be presented by the following set of transformations from the scalar values of the quantities used in the description of *carrier transport* through a single *energy-level channel* (presented in Sections 6.2 and 6.3) to a set of $n \times n$ matrices:

- From energy level E to *Hamiltonian matrix* $[H]$
- From *broadening parameters* r_1 and r_2 to the *broadening matrices* $[\Gamma_{1,2}(E)]$
- From the *DOS*-defined function $2\pi D(E)$ to the *spectral function matrix* $[A(E)]$

[9] S. Datta, "Steady state quantum kinetic equation," *Phys Rev B*, vol. **40**, no. 8, pp. 5830–5833, 1989.

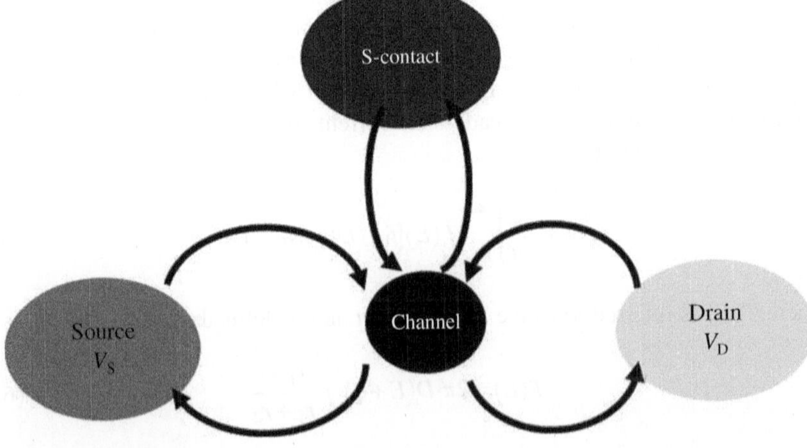

FIGURE 6.8 Illustrative depiction of *Datta's* model on the interaction between the *channel* and *source*, *drain*, and *S contact*.

- From $2\pi n(E)$ to the *correlation function matrix* $[G^n(E)]$
- From *self-consistent potential* $U = -q \cdot V$ to the *self-consistent potential matrix* $[U]$
- From N to the *density matrix* $[\rho] = \int [G^n(E)] \left(\frac{dE}{2\pi}\right)$

In this formalism, under the aforementioned conditions, all of the parameters of interest in a device are calculated using the above matrices.

Figure 6.8 provides an illustrative depiction of this model with inclusion of the aforementioned fictitious *scattering contact* (i.e., *S contact*). *Datta* has proposed that the *scattering* mechanisms can be also seen as mechanisms of injection and extraction of electrons in/out of the *channel*. Accordingly, to this fictitious contact also a *broadening matrix* Γ_S is assigned. The major difference between the *S contact* and the real contacts to the transistor is that to this contact a *Fermi–Dirac* distribution function cannot be assigned.

As a proper technique for solving the matrix equations of this formalism, *nonequilibrium Green's function* method (i.e., *NEGF*) is often used. *NEGF* offers a sound conceptual basis for developing a proper simulator for *nanoscale FETs*. The degree of complexity of this model easily allows writing codes to be run on a personal computer. Considering the *scaling* trajectory taken by modern *CMOS* technology, being educated on these concepts will soon enough be mandatory for device designers. Although in this section we only attempt to name the fundamental building blocks of this formalism, the interested reader is strongly urged to go beyond this introduction.[10]

[10] In addition to the comprehensive coverage provided in *Datta's* book (i.e., *Quantum Transport, Atom to Transistor*), the reader can use numerous articles on this topic including: S. Datta, "Nanoscale device modeling: The Green's function method," *Superlattice Microst*, vol. **28**, no. 4, pp. 253–278, 2000.

The first step in devising an *NEGF* simulator is to determine the *Hamiltonian* of the system. This is a step that the device physicists are very familiar with while formulating the *Schrödinger–Poisson* solver under equilibrium. The *Hamiltonian* needs to provide an adequate description of the isolated device. As an example, the simplest situation would be for a device whose operation involves only the electrons in a parabolic *conduction band*. In this case, the *Hamiltonian* is simply given by the *effective-mass Hamiltonian* $H \equiv -\left(\hbar^2/2m^*\right)\nabla^2$.[11]

As indicated earlier for the case of a single *energy-level channel*, connection and *bias* of a device require *self-consistent* calculation of the *channel* potential, $U(\vec{r})$, in terms of the transferred charge concentration. For an electron concentration $n(\vec{r})$ and an ionized *donor* concentration $N_d(\vec{r})$ in an n-doped semiconductor, *Poisson's equation*[12] needs to be *self-consistently* solved with the *Schrödinger equation* for the potential function $U(\vec{r})$.

Linking these two equations under *thermal equilibrium* is the role of the law of equilibrium statistical mechanics, according to which

$$n(\vec{r}) = \sum_n |\psi_n(\vec{r})|^2 f_0(E_n - E_f), \tag{6.27}$$

where $\psi_n(\vec{r})$ is the *eigenstate* of the *Schrödinger equation* $[H+U]\psi_n(\vec{r}) = E_\alpha \psi_n(\vec{r})$ and $f_0(E_n - E_f) = (1 + \exp((E_n - E_f)/kT))^{-1}$.

However, Equation (6.27) cannot be used when there is a *drain–source bias*. This is because the two electrodes are maintaining two different states of *thermal equilibrium*. Under these circumstances, in Section 6.2 for the simple case of a single *energy-level channel*, it was observed that the *energy level* of the *channel* will have a probability of occupation intermediate between the *source* and *drain Fermi-functions* (i.e., $f_0(E_n - E_{f1})$ and $f_0(E_n - E_{f2})$, respectively).[13] In terms of this probability, $n(\vec{r})$ can be written as

$$n(\vec{r}) = \sum_n \psi_n(\vec{r}) \cdot \psi_n^*(\vec{r}) \cdot f_n, \tag{6.28}$$

where f_n refers to the probability of occupation for the *energy level* E_n.

Equation (6.28), however, is not general enough since it ignores the correlation of occupation of different *energy states*. This correlated occupation is described through the *density matrix*, $[\rho]$:

$$n(\vec{r}) = \sum_{n,m} \psi_n(\vec{r}) \cdot \psi_m^*(\vec{r}) \cdot \rho_{nm}. \tag{6.29}$$

[11] See Equation (1.13).
[12] That is, $\nabla \cdot (\epsilon \nabla U) = q^2[N_d - n]$.
[13] See Equation (6.8).

As a result, under a non-*thermal-equilibrium* situation, the central problem of statistical mechanics is determining this matrix. All other parameters such as current and charge density can then be determined on its basis.

In simulating the behavior of a *nanoscale FET*, for establishing the *density matrix* (besides the *Hamiltonian* of the isolated system and the *self-consistent potential*), additional information on the *coupling* to both the physical (i.e., *source* and *drain*) and the fictitious contact (i.e., *scattering contact*) is required. These pieces of information are provided in terms of establishing the *self-energy* functions $\sum_1, \sum_2,$ and $\sum_s$, respectively. These tasks are easily performed through implementing a numerical simulator via adoption of the *NEGF* formalism. The importance of the application of *Green's function* is that it allows the user to focus on the device of interest and to embody the effects of all external contacts (including *scattering*) in *self-energy* functions (i.e., $\sum_{1,2,s}$). The key point in adopting this formalism is that in this representation the size of the matrices representing these functions in the discrete domain is the same as the size of the device *Hamiltonian*. So, using this formalism, although contacts are much larger entities than the *channel* of a *nanoscale FET*, there is no need to bother with considering their full complexity (and large-size *Hamiltonian* matrices). Whereas in *macroscale FETs* the complexities of contacts and the *channel* were avoided through adopting the semiclassical approach of the *drift–diffusion* formalism and definition of the *effective mass*, in small-size devices in absence of such a luxury the adoption of *Green's function* formalism circumvents the formidable challenge of handling the large *Hamiltonians* of the contacts.

FURTHER READING

Following reference has been used in preparation of the materials presented in this chapter. This marvelous text is suggested for a complete coverage beyond the introduction presented in here:

S. Datta, *Quantum Transport, Atom to Transistor,* Cambridge University Press, Cambridge, UK, 2005.

For further reading on nanoelectronics consult the following:

M. Dragoman and D. Dragoman, *Nanoelectronics Principles and Devices,* Artech House, Norwood, MA, 2006.

D. Esseni, P. Palestri, and L. Selmi, *Nanoscale MOS Transistors, Semi-Classical Transport and Applications,* Cambridge University Press, Cambridge, UK, 2011.

T. Heinzel, *Mesoscopic Electronics in Solid State Nanostructures,* Wiley-VCH, Weinheim, 2003.

M. Lundstrom and J. Guo, *Nanoscale Transistors: Device Physics, Modeling and Simulation,* Springer, New York, 2006.

V. Mitin, V. Kochelap, and M. Stroscio, *Introduction to Nanoelectronics, Science, Nanotechnology, Engineering, and Applications,* Cambridge University Press, Cambridge, UK, 2008.

INDEX

Field Effect Transistors, A Comprehensive Overview: From Basic Concepts to Novel Technologies, First Edition. Pouya Valizadeh.

Printed and bound by CPI Group (UK) Ltd, Croydon, CR0 4YY
16/09/2026
14963424-0002